CENTIGRADE AND FAHRENHEIT EQUIVALENTS

$$F = \frac{9}{5}C + 32 \qquad C = \frac{5}{9}(F - 32)$$

CENTIGRADE	FAHRENHEIT
100°	212° ← water boils at sea level
95°	203°
90°	194°
85°	185°
80°	176°
75°	167°
70°	158°
65°	149°
60°	140°
55°	131°
50°	122°
45°	113°
40°	104°
35°	95°
comfort range { 30°	comfort range { 86°
25°	77°
20° }	68° }
15°	59°
10°	50°
5°	41°
0°	32° ← water freezes at sea level

Arithmetic for College Students

Arithmetic for College Students

Fifth Edition

D. Franklin Wright
CERRITOS COLLEGE

D. C. HEATH AND COMPANY
Lexington, Massachusetts Toronto

This book is dedicated to the memory of
Sara M. Wright and C. Donald Wright

Acquisitions Editor: Mary Lu Walsh
Development Editor: Anne Marie Jones
Production Editor: Cathy Cantin
Designer: Henry Rachlin
Production Coordinator: Mike O'Dea
Photo Researcher: Martha Shethar

Cover photograph © Michael DeCamp/The Image Bank

Published simultaneously in Canada.

Printed in the United States of America.

International Standard Book Number: 0-669-12189-4

Library of Congress Catalog Card Number: 86-82101

Preface

New Material

This fifth edition has several new topics and improved, simplified explanations. Additional examples have been included with more detailed explanations and more exercises. Whenever possible, procedures and rules are given in an easy-to-read list format. These lists are highlighted in boxes for easy reference. The students can quickly locate references for particular procedures and skills. The basic theme from previous editions of *why* procedures work as well as *how* to do them has been maintained throughout this fifth edition.

Many of the explanations from the fourth edition are now incorporated within the examples for better understanding. The students will be more likely to read these discussions and associate them with the related problems. Additional diagrams have been included in the text and in the exercises.

New material includes:

1. Rounding-off whole numbers (Section 1.2) for better understanding of rounding-off decimals in Section 5.2
2. An improved explanation of the relationship between factors and division with whole numbers in Section 1.8
3. Fractions and mixed numbers covered in separate chapters (Chapter 3 and Chapter 4)
4. Expanded coverage of order of operations (Sections 2.2, 3.6, 4.5, and 11.5)
5. A revised basic approach to solving percent problems in Chapter 7 using the formula $R \times B = A$ where R is in decimal form
6. Check-balancing and simple statistics covered in Chapter 8 (Applications with Calculators) as topics more relevant to student experiences
7. A reorganized Chapter 9 (Measurement: The Metric System) using conversion charts that have been expanded to include area and volume
8. A new Chapter 10 (Measurement: The U.S. Customary System) that introduces special measurement topics
9. An expanded and improved introduction to algebraic topics in Chapters 11–13, including signed decimals and fractions, more graphing, the Pythagorean Theorem, and the midpoint formula

Content

With the exceptions of Chapter 1 (Whole Numbers), Chapter 9 (Measurement: The Metric System), and Chapter 10 (Measurement: The U.S. Customary System), the chapters are best covered in the order presented because many topics depend on previously discussed material. The text includes enough material for a three or four semester-hour course.

Chapter 1 (Whole Numbers) is a review of basic operations with whole numbers, including a new section (Section 1.2) on rounding off whole numbers. The instructor may choose to start the course with Chapter 2 and have students review Chapter 1 on their own time.

Chapter 2 (Prime Numbers) forms the foundation for the text. It contains the basic ideas of exponents, order of operations, and prime numbers used throughout the text. To avoid confusion with the more important topic of least common multiple (LCM), the related topic of greatest common divisor (GCD) is discussed in Appendix III. This material can be taught with Chapter 2 if desired.

The approach to fractions in Chapter 3 (Fractions) uses prime factorizations. Many students have told me that fractions "make more sense" with this approach, and that working with fractions is no longer a series of mysterious steps that they cannot remember. These techniques carry over nicely into algebra and give the students a procedure that guarantees that they are reducing correctly and completely.

Chapter 4 (Mixed Numbers) covers mixed numbers, which was previously included with fractions in Chapter 3. This split allows division with fractions to be discussed twice (Sections 3.3 and 4.5). It also clarifies the distinction between working with mixed numbers and working with simple fractions. Addition and subtraction with mixed numbers requires careful explanation. Emphasis is placed on the use of the word *of* to indicate multiplication by a fraction. This is discussed again in Chapter 7 (Percent).

Two chapters, Chapter 5 (Decimal Numbers) and Chapter 6 (Ratio and Proportion), take the place of Chapter 4 from the fourth edition. In Chapter 5, reading and writing decimal numbers is emphasized, as well as the basic operations and rounding off. Chapter 6 has been expanded and now includes the topic of unit pricing, a particularly practical application of ratios. Proportions directly related to nursing are discussed and included in the exercises. The instructor might choose to introduce the use of hand-held calculators in Chapter 6.

Percent is developed in Chapter 7 (Percent) using the tools of ratio and proportion and solving simple equations that are developed in Chapter 6. Because most percent problems are of Type 1 where multiplication by a decimal is used, the basic approach has been adjusted to the use of the formula $R \times B = A$ with R in decimal form. With two of these three quantities known, the third value can be found by solving the related equation. The students should be encouraged to use calculators on the word problems in Sections 7.4 and 7.5 before they try to solve the more complicated applied problems in Chapter 8.

Chapter 8 (Applications with Calculators) now covers the topics of balancing a checkbook and simple statistics. Also included are simple and compound interest and reading graphs. These five topics provide practical applications seen by almost everyone at some time. Generally, several calculations are involved in a single problem. This makes the use of calculators particularly appropriate for increased speed and accuracy.

The topics related to measurement in Chapter 9 (Measurement: The Metric System) and Chapter 10 (Measurement: The U.S. Customary System) are particularly useful both for initial learning and for reference. Included are elementary geometric formulas for area and volume and many tables of equivalent measures. Chapter 9 includes an expanded use of charts for changing from one unit to another in the metric system. Chapter 10 has a new section on applications in medicine.

Chapter 11 (Signed Numbers), Chapter 12 (Solving Equations), and Chapter 13 (Real Numbers and Graphing) have been completely rewritten and reorganized. They form the pre-algebra section of the text. The term *signed number* is used to indicate positive and negative decimals and fractions now as well as integers. Number lines are used to introduce elementary operations in Chapter 11. Section 12.4 (Working with Formulas) is new and provides reinforcement of the procedures for solving equations. Chapter 13 provides an introduction to topics in algebra useful to continuing students. Included are graphing lines and parabolas, the Pythagorean Theorem, and the midpoint formula.

There are four appendices:

I. Ancient Numeration Systems
II. Base Two and Base Five
III. Greatest Common Divisor (GCD)
IV. Topics from Plane Geometry

Exercises

There are more than 4462 exercises carefully chosen and graded, proceeding from easy exercises to more difficult ones, plus practice quizzes. Answers to all the exercises except those numbered as multiples of 4 are in the Answer Key at the back of the text. Answers to all Chapter Review questions and all Chapter Test questions are also in the Answer Key. I have made a special effort to provide interesting and meaningful word problems. Word problems appear as early as Section 1.3 and are spread throughout the text as frequently as possible.

Special Features

1. Each chapter has
    - A Chapter Summary of definitions and key ideas
    - A Chapter Review of questions similar to the exercises with all the answers in the back of the text
    - A Chapter Test similar to a test the students can expect to take

2. All the important procedures and rules are in list form and are highlighted.
3. Explanations have been simplified. Many are included with the examples for better understanding.
4. Calculators are used in special sections on practical applications.
5. Most sections include Practice Quizzes of 3–5 exercises to provide immediate classroom feedback to the instructor.
6. There are three new comprehensive exams provided for the students to help them prepare for mid-terms and the final exam. These cover Chapters 1–4, Chapters 1–8, and Chapters 1–13.

Additional Aids

An *Instructor's Guide* written by Carol L. Johnson contains brief discussions of each chapter, the answers to the exercises numbered as multiples of four, and sample tests for each chapter. These tests are in camera-ready form for easy reproduction. There are three tests for each chapter including one multiple-choice test. Also, there are three forms of each of three comprehensive tests corresponding to those in the text. I am confident that any instructor will find this *Guide* a valuable tool and am grateful to Carol for an outstanding job.

Acknowledgments

I would like to thank the editorial staff at D. C. Heath for all their efforts on this fifth edition. In particular, Mary Lu Walsh, Ann Marie Jones, and Cathy Cantin deserve special mention for their personal attention and obvious concern for me and the project. Of course, Pat Wright did another terrific job of typing and proofreading the manuscript. All reviewers were helpful with their constructive and critical comments: Jerry L. Collins, Odessa College; John T. Gordon, Georgia State University; Charles E. Sweatt, Odessa College; and Roberta T. Yellot, McNeese State University.

Thank you all.

D. Franklin Wright

Contents

Arithmetic for College Students

1 Whole Numbers

1.1 READING AND WRITING WHOLE NUMBERS

The Hindu-Arabic system that we use was invented about A.D. 800. It is called the **decimal system** (**deci** means ten in Latin). This system allows us to add, subtract, multiply, and divide more easily and faster than any of the ancient number systems. For example, multiplying 22 by 3 in the Roman system would appear as

$$\begin{array}{r} \text{XXII} \\ \underline{\text{III}} \\ \text{LXVI} \end{array} \quad \text{while we might write} \quad \begin{array}{r} 22 \\ \underline{\times\ 3} \\ 66 \end{array}$$

A discussion of several ancient number systems is included in Appendix I.

> The decimal system (or base ten system) is a **place value system** that depends on three things:
>
> 1. The **ten digits:** 0, 1, 2, 3, 4, 5, 6, 7, 8, 9;
> 2. The **placement** of each digit; and
> 3. The **value** of each place.

The **decimal point** is the beginning point for writing digits. The value of each place to the left of the decimal point has a **power of ten** (1, 10, 100, 1000, and so on) as its place value. (See Figure 1.1.)

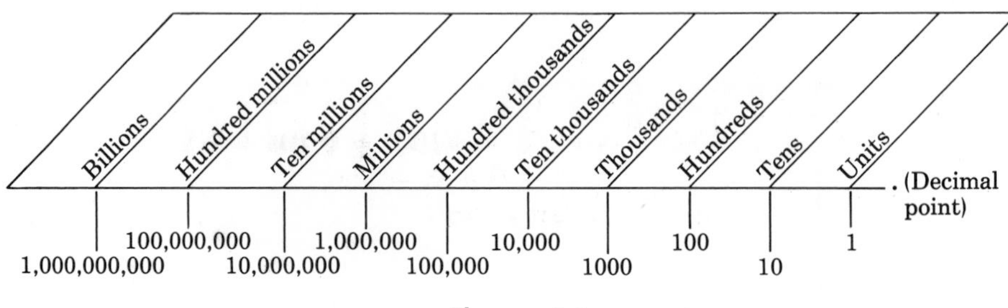

Figure 1.1

When a digit is written in a place, the value of the digit is to be multiplied by the value of the place. The value of any number is found by adding the results of multiplying the digits and place values. For example,

$$\frac{6}{100}\ \frac{9}{10}\ \frac{3}{1}. \quad \begin{array}{l}\leftarrow \text{digits (standard notation)} \\ \leftarrow \text{place values}\end{array}$$

$$693 = 6(100) + 9(10) + 3(1) = 600 + 90 + 3 \quad \text{(expanded notation)}$$

NOTE: Writing a number next to a number in parentheses means to multiply.

In **expanded notation** the values represented by each digit in standard notation are added. The English word equivalents can then be easily read from these sums. If a number has more than four digits, commas are placed to separate every three digits. Commas are used in the same manner in the word equivalents.

EXAMPLES

Each number is written in standard notation and in expanded notation and in its English word equivalent. If no decimal point is written, it is understood to be to the right of the rightmost digit.

1. 573 (standard notation)

$$\begin{aligned} 573 &= 5(100) + 7(10) + 3(1) \\ &= 500 + 70 + 3 \quad \text{(expanded notation)} \end{aligned}$$

five hundred seventy-three

2. 4862

$$
\begin{aligned}
4862 &= 4(1000) + 8(100) + 6(10) + 2(1) \\
&= 4000 + 800 + 60 + 2 \qquad \text{(expanded notation)}
\end{aligned}
$$

four thousand eight hundred sixty-two

3. 8007

$$
\begin{aligned}
8007 &= 8(1000) + 0(100) + 0(10) + 7(1) \\
&= 8000 + 0 + 0 + 7
\end{aligned}
$$

eight thousand seven

4. 1,590,768

$$
\begin{aligned}
1{,}590{,}768 &= 1(1{,}000{,}000) + 5(100{,}000) + 9(10{,}000) + 0(1000) \\
&\quad + 7(100) + 6(10) + 8(1) \\
&= 1{,}000{,}000 + 500{,}000 + 90{,}000 + 0 + 700 + 60 + 8
\end{aligned}
$$

one million, five hundred ninety thousand, seven hundred sixty-eight

Whole numbers are those numbers used for counting and the number 0. They are the decimal numbers with digits written to the left of the decimal points. Decimal numbers with digits to the right of the decimal point will be discussed in Chapter 5.

We use the letter **W** to represent all whole numbers:

0, 1, 2, 3, 4, 5, 6, 7, 8, 9, 10, 11, 12, . . .

The three dots indicate that the pattern is to continue without end.

Note four things when reading or writing whole numbers:

1. The word **and** does not appear in English word equivalents. **And** is said only when reading a decimal point. (See Chapter 5.)
2. Digits are read in groups of three.
3. Commas are used to separate groups of three digits **if a number has more than four digits.**
4. Hyphens (-) are used to write English words for two-digit numbers.

PRACTICE QUIZ	Write the following numbers in expanded notation and in their English word equivalents.	ANSWERS
	1. 512	**1.** 500 + 10 + 2 Five hundred twelve
	2. 6394	**2.** 6000 + 300 + 90 + 4 Six thousand three hundred ninety-four
	3. Write one hundred eighty thousand, five hundred forty-three as a decimal numeral.	**3.** 180,543

Exercises 1.1

Did you read the explanation and work through the examples before beginning these exercises?

Write the following decimal numbers in expanded notation and in their English word equivalents.

1. 37 **2.** 84 **3.** 98 **4.** 56

5. 122 **6.** 493 **7.** 821 **8.** 1976

9. 1892 **10.** 5496 **11.** 12,517 **12.** 42,100

13. 243,400 **14.** 891,540 **15.** 43,655 **16.** 99,999

17. 8,400,810 **18.** 5,663,701 **19.** 16,302,590

20. 71,500,000 **21.** 83,000,605 **22.** 152,403,672

23. 679,078,100 **24.** 4,830,231,010 **25.** 8,572,003,425

Write the following numbers as decimal numbers.

26. seventy-six

27. one hundred thirty-two

28. five hundred eighty

29. three thousand eight hundred forty-two

30. two thousand five

31. one hundred ninety-two thousand, one hundred fifty-one
32. seventy-eight thousand, nine hundred two
33. twenty-one thousand, four hundred
34. thirty-three thousand, three hundred thirty-three
35. five million, forty-five thousand
36. five million, forty-five
37. ten million, six hundred thirty-nine thousand, five hundred eighty-two
38. two hundred eighty-one million, three hundred thousand, five hundred one
39. five hundred thirty million, seven hundred
40. seven hundred fifty-eight million, three hundred fifty thousand, sixty
41. ninety million, ninety thousand, ninety
42. eighty-two million, seven hundred thousand
43. one hundred seventy-five million, two
44. thirty-six
45. seven hundred fifty-seven
46. Name the position of each nonzero digit in the following number: 109,750.
47. Name the position of each nonzero digit in the following number: 7,840,062.

Write the English word equivalent for the number in each sentence.

48. The population of Los Angeles is about 2,816,000.
49. The circumference of the earth is approximately 25,120 miles.
50. The distance between the earth and the sun is about 93,000,000 miles or 149,730,000 kilometers.

1.2 ROUNDING OFF WHOLE NUMBERS

Rounding off a given number means to find another number close to the given number. The desired place of accuracy must be stated. For example, if you were asked to round off 872, you would not know what to do unless you were told the position or place of accuracy desired. The number lines in Figure 1.2 help to illustrate the problem.

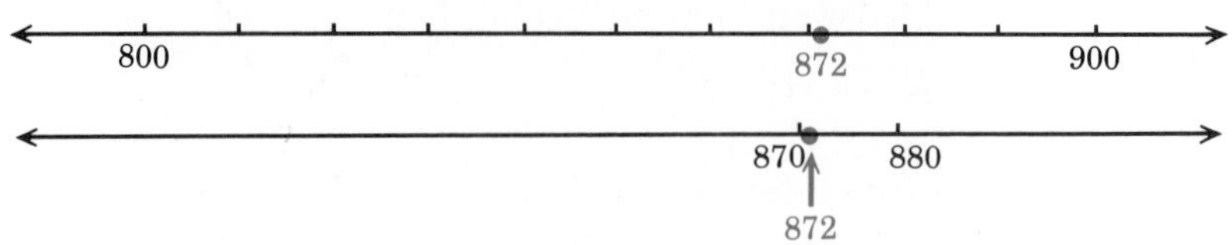

Figure 1.2

We can see that 872 is closer to 900 than to 800. So, **to the nearest hundred,** 872 rounds off to 900. Also, 872 is closer to 870 than to 800. So, **to the nearest ten,** 872 rounds off to 870.

Inaccuracy in numbers is common, and rounded-off answers can be quite acceptable. For example, what do you think is the distance across the United States from the east coast to the west coast? Most people will use the rounded-off answer of 3000 miles. This answer is appropriate and acceptable.

EXAMPLES

Number lines are used as visual aids in rounding off the following numbers.

1. Round 47 to the nearest ten.

   Solution

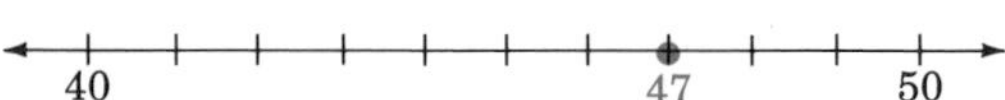

   47 is closer to 50 than to 40. So, 47 rounds off to 50 (to the nearest ten).

2. Round 238 to the nearest hundred.

   Solution

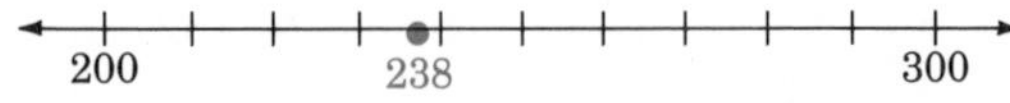

   238 is closer to 200 than to 300. So, 238 rounds off to 200 (to the nearest hundred).

Using figures as an aid to understanding is fine but, for practical purposes, the following rule is more useful.

ROUNDING-OFF RULE FOR WHOLE NUMBERS

1. Look at the single digit just to the right of the digit that is in the place of desired accuracy.
2. If this digit is 5 or greater, make the digit in the desired place of accuracy one larger and replace all digits to the right with zeros.
3. If this digit is less than 5, leave the digit that is in the place of desired accuracy as it is and replace all digits to the right with zeros.

EXAMPLES

The Rounding-Off Rule is used in the following examples.

3. Round 5749 to the nearest hundred.

Solution

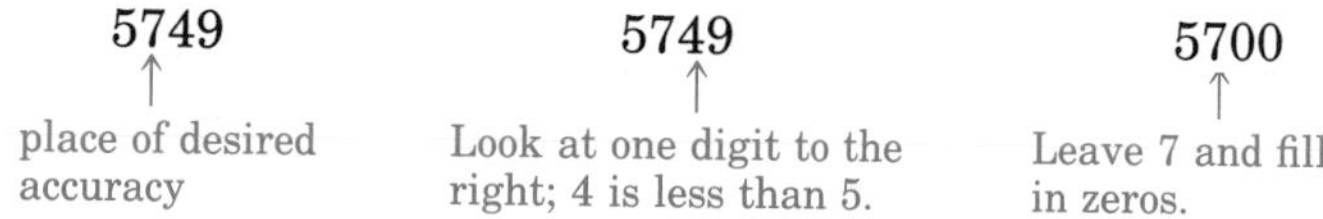

So, 5749 rounds off to 5700 (to the nearest hundred).

4. Round 6500 to the nearest thousand.

Solution

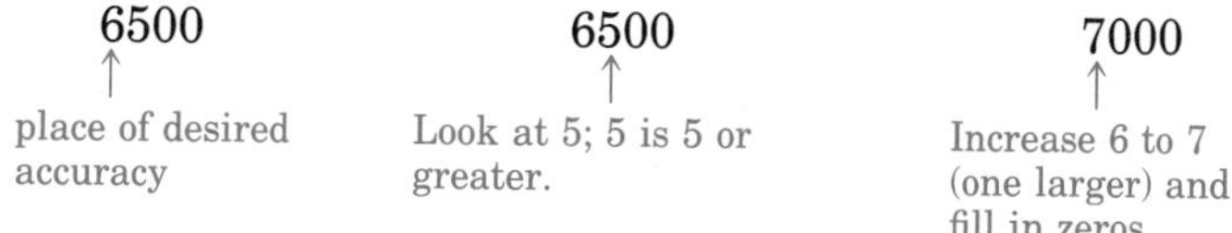

So, 6500 rounds off to 7000 (to the nearest thousand).

5. Round 397 to the nearest ten.

Solution

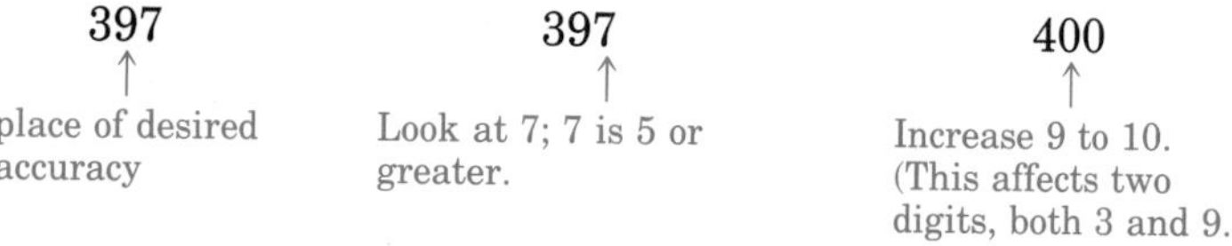

So, 397 rounds off to 400 (to the nearest ten).

Exercises 1.2

Round off as indicated.

To the nearest ten:

1. 763 **2.** 31 **3.** 82 **4.** 503

5. 296 **6.** 722 **7.** 987 **8.** 347

To the nearest hundred:

9. 4163 **10.** 4475 **11.** 495 **12.** 572

13. 637 **14.** 3789 **15.** 76,523 **16.** 7007

To the nearest thousand:

17. 6912 **18.** 5500 **19.** 7500 **20.** 7499

21. 13,499 **22.** 13,501 **23.** 62,265 **24.** 47,800

To the nearest ten thousand:

25. 78,419 **26.** 125,000 **27.** 256,000 **28.** 62,200

29. 118,200 **30.** 312,500 **31.** 184,900 **32.** 615,000

33. 87 to the nearest hundred **34.** 46 to the nearest hundred

35. 532 to the nearest thousand

1.3 ADDITION

Addition with whole numbers is indicated either by writing the numbers horizontally with a plus sign (+) between them or by writing the numbers vertically in columns with instructions to add.

$$6 + 23 + 17 \quad \text{or} \quad \text{Add} \quad \begin{array}{r} 6 \\ 23 \\ \underline{17} \end{array} \quad \text{or} \quad \begin{array}{r} 6 \\ 23 \\ \underline{+17} \end{array}$$

The numbers being added are called **addends,** and the result of the addition is called the **sum.**

$$\text{Add} \quad \begin{array}{rl} 6 & \text{addend} \\ 23 & \text{addend} \\ \underline{17} & \text{addend} \\ 46 & \text{sum} \end{array}$$

Be sure to keep the digits aligned (in column form) so you will be adding units to units, tens to tens, and so on. Neatness is a necessity in mathematics.

To be able to add with speed and accuracy, you must **memorize** the basic addition facts, which are given in Table 1.1. Practice all the combinations so you can give the answers immediately. Concentrate.

Table 1.1 Basic Addition Facts

+	0	1	2	3	4	5	6	7	8	9
0	0	1	2	3	4	5	6	7	8	9
1	1	2	3	4	5	6	7	8	9	10
2	2	3	4	5	6	7	8	9	10	11
3	3	4	5	6	7	8	9	10	11	12
4	4	5	6	7	8	9	10	11	12	13
5	5	6	7	8	9	10	11	12	13	14
6	6	7	8	9	10	11	12	13	14	15
7	7	8	9	10	11	12	13	14	15	16
8	8	9	10	11	12	13	14	15	16	17
9	9	10	11	12	13	14	15	16	17	18

As an exercise to help find out which combinations you need special practice with, write, on a piece of paper in mixed order, all one hundred possible combinations to be added. Perform the operations as quickly as possible. Then, using the table, check to find the ones you missed. Study these frequently until you are confident that you know them as well as all the others.

Your adding speed can be increased if you learn to look for combinations of digits that total ten.

EXAMPLE

1. To add the following numbers, we note the combinations that total ten and find the sums quickly.

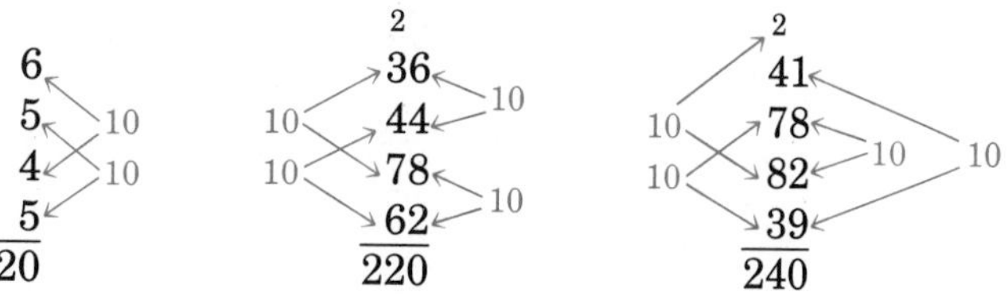

When two numbers are added, the order of the numbers does not matter. That is, $4 + 9 = 13$ and $9 + 4 = 13$. Also,

$$\begin{array}{r} 5 \\ 7 \\ \hline 12 \end{array} \quad \text{and} \quad \begin{array}{r} 7 \\ 5 \\ \hline 12 \end{array}$$

By looking at Table 1.1 again, we can see that reversing the order of any two addends will not change their sum. This fact is called the **commutative property of addition.** We can state this property using letters to represent whole numbers.

COMMUTATIVE PROPERTY OF ADDITION

If a and b are two whole numbers, then $a + b = b + a$.

To add three numbers, such as $6 + 3 + 5$, we can add only two at a time, then add the third. Which two should we add first? The answer is that the sum is the same either way:

$$6 + 3 + 5 = (6 + 3) + 5 = 9 + 5 = 14$$

and

$$6 + 3 + 5 = 6 + (3 + 5) = 6 + 8 = 14$$

We can write

$$8 + 4 + 7 = (8 + 4) + 7 = 8 + (4 + 7)$$

This illustrates the **associative property of addition.**

ASSOCIATIVE PROPERTY OF ADDITION

If a, b, and c are whole numbers, then

$$a + b + c = (a + b) + c = a + (b + c)$$

Another property of addition with whole numbers is addition with 0. Whenever we add 0 to a number, the result is the original number:

$$8 + 0 = 8, \quad 0 + 13 = 13, \quad \text{and} \quad 38 + 0 = 38$$

Zero (0) is called the **additive identity** or the **identity element for addition.**

ADDITIVE IDENTITY

If a is a whole number, then there is a unique whole number 0 with the property that $a + 0 = a$.

EXAMPLES

2. $(15 + 20) + 6 = 15 + (20 + 6)$ — associative property

3. $(15 + 20) + 6 = (20 + 15) + 6$ — commutative property

4. $17 + 0 = 17$ — additive identity

Note that Example 2 shows a change in the **grouping** (association) of the numbers, while Example 3 shows a change in the **order** of two numbers.

5. Another illustration of the associative property:

$$\begin{array}{r} 8 \\ 2 \\ +3 \\ \hline \end{array} \quad \begin{array}{r} 10 \\ 3 \\ \hline 13 \end{array} \qquad\qquad \begin{array}{r} 8 \\ 2 \\ 3 \\ \hline \end{array} \quad \begin{array}{r} 8 \\ 5 \\ \hline 13 \end{array}$$

Adding large numbers or several numbers can be done by writing the numbers vertically so that the place values are **lined up** in columns. Then only the digits with the same place value are added.

EXAMPLES

6. If no sum of digits is more than 9:

Add 5623 + 3172.

$$\begin{array}{r} 5\,6\,2\,3 \\ +\,3\,1\,7\,2 \\ \hline 5 \end{array}$$ ← addend, ← addend; Add ones.

$$\begin{array}{r} 5\,6\,2\,3 \\ +\,3\,1\,7\,2 \\ \hline 9\,5 \end{array}$$ Add tens.

$$\begin{array}{r} 5\,6\,2\,3 \\ +\,3\,1\,7\,2 \\ \hline 7\,9\,5 \end{array}$$ Add hundreds.

$$\begin{array}{r} 5\,6\,2\,3 \\ +\,3\,1\,7\,2 \\ \hline 8\,7\,9\,5 \end{array}$$ Add thousands.

$$\begin{array}{r} 5\,6\,2\,3 \\ +\,3\,1\,7\,2 \\ \hline 8\,7\,9\,5 \end{array}$$ ← You do not write all the steps. You write only the addends and the sum. ← sum

7. If the sum of the digits in one column is more than 9:

   (a) Write the ones digits in that column.
   (b) **Carry** the other digit to the column to the left.

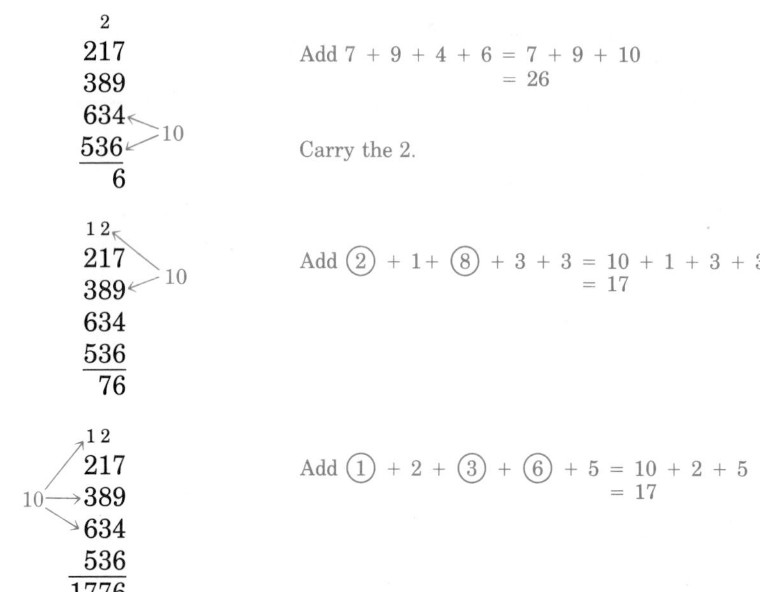

2
217
389
634
536
6
10
Add 7 + 9 + 4 + 6 = 7 + 9 + 10
= 26
Carry the 2.
12
217
389
634
536
76
10
Add (2) + 1 + (8) + 3 + 3 = 10 + 1 + 3 + 3
= 17
12
217
389
634
536
1776
10
Add (1) + 2 + (3) + (6) + 5 = 10 + 2 + 5
= 17

PRACTICE QUIZ

Which property of addition is illustrated?

1. 12 + 0 = 12
2. 15 + 3 = 3 + 15

Find the following sums.

3. 57 + 98
4. 36 + 78 + 89

ANSWERS

1. Additive identity
2. Commutative property
3. 155
4. 203

Exercises 1.3

Did you read the explanation and work through the examples before beginning these exercises?

Do the following exercises mentally and write only the answers.

1. 6 + (3 + 7) **2.** (4 + 5) + 6 **3.** (2 + 3) + 8

4. (2 + 6) + (4 + 5) **5.** 8 + (3 + 4) + 6 **6.** 9 + (8 + 3)

7. 9 + 2 + 8
8. 7 + (6 + 3)
9. (2 + 3) + 7
10. 8 + 7 + 2
11. 9 + 6 + 3
12. 4 + 3 + 6
13. 4 + 4 + 4
14. 9 + 1 + 5 + 6
15. 5 + 3 + 7 + 2
16. 3 + 6 + 2 + 8
17. 5 + 4 + 6 + 4 + 8
18. 6 + 2 + 9 + 1
19. 4 + 3 + 6 + 6 + 2
20. 8 + 8 + 7 + 6 + 3

Show that the following statements are true by performing the addition. State which property of addition is being illustrated.

21. 9 + 3 = 3 + 9
22. 8 + 7 = 7 + 8
23. 4 + (5 + 3) = (4 + 5) + 3
24. 4 + 8 = 8 + 4
25. 2 + (1 + 6) = (2 + 1) + 6
26. (8 + 7) + 3 = 8 + (7 + 3)
27. 9 + 0 = 9
28. (2 + 3) + 4 = (3 + 2) + 4
29. 7 + (6 + 0) = 7 + 6
30. 8 + 20 + 1 = 8 + 21

Copy the following problems and add.

31.	32.	33.	34.	35.
65	24	73	165	876
43	78	68	276	279
54	95	98	394	143

36.	37.	38.	39.	40.
268	981	2112	114	1403
93	146	147	5402	7010
74	92	904	710	622
192	17	1005	643	29

41.	42.	43.	44.
213,116	21,442	438,966	123,456
116,018	32,462	1,572,486	456,123
722,988	564,792	327,462	879,282
24,336	801,801	181,753	617,500
526,968	43,433	90,000	740,765

45. Mr. Jones kept the mileage records indicated in the table shown here. How many miles did he drive during the six months?

MONTH	MILEAGE
Jan.	546
Feb.	378
Mar.	496
Apr.	357
May	503
June	482

46. The Modern Products Corp. showed profits as indicated in the table for the years 1974–1977. What were the company's total profits for the years 1974–1977?

YEAR	PROFITS
1974	$1,078,416
1975	1,270,842
1976	2,000,593
1977	1,963,472

47. During six years of college, including two years of graduate school, Fred estimated his expenses each year as $2035; $2842; $2786; $3300; $4000; $3500. What were his total expenses for six years of schooling? [NOTE: He had some financial aid.]

48. Apple County has the following items budgeted: highways, $270,455; salaries, $95,479; maintenance, $127,220. What is the county's total budget for these three items?

49. The following numbers of students at South Junior College are enrolled in mathematics courses: 303 in arithmetic; 476 in algebra; 293 in trigonometry; 257 in college algebra; 189 in calculus. Find the total number of students taking math.

50. In one year, the High Price Manufacturing Co. made 2476 refrigerators; 4217 gas stoves; 3947 electric stoves; 9576 toasters; 11,872 electric fans; 1742 air conditioners. What was the total number of appliances High Price produced that year?

Complete each of the following addition tables. You may want to make up your own tables for practicing difficult combinations.

51.

+	5	8	7	9
3				
6				
5				
2				

52.

+	6	7	1	4	9
3					
5					
4					
8					
2					

1.4 SUBTRACTION

A long-distance runner who is 25 years old had run 17 miles of his usual 19-mile workout when a storm forced him to quit for the day. How many miles short was he of his usual daily training?

What *thinking* did you do to answer the question? You may have thought something like this: "Well, I don't need to know how old the runner is to answer the question, so his age is just extra information. Since he had already run 17 miles, I need to know what to add to 17 miles to get 19 miles. Since 17 + 2 = 19, he was 2 miles short of his usual workout."

In this problem, the sum of two addends was given, and only one of the addends was given. The other addend was the unknown quantity.

17	+	□	=	19
addend		missing addend		sum

As you may know, this kind of addition problem is called **subtraction** and can be written:

19	−	17	=	□	(Read: "19 minus 17 equals blank.")
sum		addend		missing addend (or difference)	

Or

19	sum
− 17	addend
□	difference or missing addend

In other words, subtraction is a reverse addition, and the missing addend is called the **difference** between the sum and one addend.

Perform the following subtraction mentally: 17 − 8 = □. You should think, "8 plus what number gives 17? Since 8 + 9 = 17, we have 17 − 8 = 9."

Suppose you want to find the difference 496 − 342. Thinking of a number to add to 342 to get 496 is difficult. A better technique, using the place values, is developed and discussed in the following examples.

EXAMPLES

1. Subtract 496 − 342.

Solution

STEP 1:
$$\begin{array}{r} 4\ 9\ 6 \\ -\ 3\ 4\ 2 \\ \hline 4 \end{array}$$
Subtract ones. 6 − 2 = 4

STEP 2:
$$\begin{array}{r} 4\ 9\ 6 \\ -\ 3\ 4\ 2 \\ \hline 5\ 4 \end{array}$$
Subtract tens. 9 − 4 = 5

STEP 3:
$$\begin{array}{r} 4\ 9\ 6 \\ -\ 3\ 4\ 2 \\ \hline 1\ 5\ 4 \end{array}$$
Subtract hundreds. 4 − 3 = 1

← difference

Or, using expanded notation, we get the same result:

$$\begin{array}{rcrrr} 4\ 9\ 6 & = & 400 & +\ 90 & +\ 6 \\ -\ 3\ 4\ 2 & = & -[300 & +\ 40 & +\ 2] \\ \hline & & 100 & 50 & 4 \end{array} = 154$$

← difference

2. Subtract 65 − 28 using expanded notation.

Solution

$$\begin{array}{rcl} 65 &=& 60 + 5 \\ \underline{-28} &=& \underline{-[20 + 8]} \end{array}$$

Starting from the right, 5 is smaller than 8. We cannot subtract 8 from 5, so **borrow** 10 from 60.

Borrow 10 from 60. ↓   10 borrowed from 60, plus 5. ↓

$$\begin{array}{rcccccl} 65 &=& 60 + 5 &=& 50 + 10 + 5 &=& 50 + 15 \\ \underline{-28} &=& \underline{-[20 + 8]} &=& \underline{-[20 + \quad\quad 8]} &=& \underline{-[20 + \ \ 8]} \\ &&&&&& 30 + \ \ 7 = 37 \end{array}$$

Now subtract.

3. Use expanded notation and borrowing to find the difference 536 − 258.

Solution

$$\begin{array}{rcl} 536 &=& 500 + 30 + 6 \\ \underline{-258} &=& \underline{-[200 + 50 + 8]} \end{array}$$

Since 6 is smaller than 8, borrow 10 from 30.   Since 20 is smaller than 50, borrow 100 from 500.

$$\begin{array}{rcccccl} 536 &=& 500 + 30 + 6 &=& 500 + 20 + 16 &=& 400 + 120 + 16 \\ \underline{-258} &=& \underline{-[200 + 50 + 8]} &=& \underline{-[200 + 50 + \ \ 8]} &=& \underline{-[200 + \ \ 50 + \ \ 8]} \\ &&&&&& 200 + \ \ 70 + \ \ 8 \\ &&&&&& = 278 \end{array}$$

A common practice, with which you are probably familiar, is to indicate the borrowing by crossing out digits and writing new digits instead of expanded notation. The expanded notation technique has been shown so that you can have a "picture" in your mind to help you understand the procedure of subtraction with borrowing.

50 in place of 60.   Borrowed 10

EXAMPLES

4. $$\begin{array}{rcccl} 65 &=& 60 + 5 &=& 50 + 15 \\ \underline{-28} &=& \underline{-[20 + 8]} &=& \underline{-[20 + \ \ 8]} \end{array}$$ can be written $$\begin{array}{rr} {}^{5} & {}^{1} \\ \not{6} & 5 \\ -2 & 8 \\ \hline 3 & 7 \end{array}$$

5. $\begin{array}{r} 736 \\ -258 \\ \hline \end{array}$

STEP 1: Since 6 is smaller than 8, borrow 10 from 30. This leaves 20, so cross out 3 and write 2.

$$\begin{array}{rrrr} & & {}^{2} & {}^{1} \\ & 7 & \not{3} & 6 \\ - & 2 & 5 & 8 \\ \hline \end{array}$$

STEP 2: Since 2 is smaller than 5, borrow 100 from 700. This leaves 600, so cross out 7 and write 6.

$$\begin{array}{rrrr} & {}^{6} & {}^{12} & {}^{1} \\ & \not{7} & \not{3} & 6 \\ - & 2 & 5 & 8 \\ \hline \end{array}$$

STEP 3: Now subtract.

$$\begin{array}{rrrr} & {}^{6} & {}^{12} & {}^{1} \\ & \not{7} & \not{3} & 6 \\ - & 2 & 5 & 8 \\ \hline & 4 & 7 & 8 \end{array}$$

6. $\begin{array}{r} 8000 \\ -\ \ 657 \\ \hline \end{array}$

STEP 1: Trying to borrow from 0 each time, we end up borrowing 1000 from 8000. Cross out 8 and write 7.

$$\begin{array}{rrrrr} & {}^{7} & {}^{1} & & \\ & \not{8} & 0 & 0 & 0 \\ - & & 6 & 5 & 7 \\ \hline \end{array}$$

STEP 2: Now borrow 100 from 1000. Cross out 10 and write 9.

$$\begin{array}{rrrrr} & & {}^{9} & & \\ & {}^{7} & \not{1} & {}^{1} & \\ & \not{8} & \not{0} & 0 & 0 \\ - & & 6 & 5 & 7 \\ \hline \end{array}$$

STEP 3: Borrow 10 from 100. Cross out 10 and write 9.

$$\begin{array}{rrrrr} & & {}^{9} & {}^{9} & \\ & {}^{7} & \not{1} & \not{1} & {}^{1} \\ & \not{8} & \not{0} & \not{0} & 0 \\ - & & 6 & 5 & 7 \\ \hline \end{array}$$

STEP 4: Now subtract.

$$\begin{array}{rrrrr} & & {}^{9} & {}^{9} & \\ & {}^{7} & \not{1} & \not{1} & {}^{1} \\ & \not{8} & \not{0} & \not{0} & 0 \\ - & & 6 & 5 & 7 \\ \hline & 7 & 3 & 4 & 3 \end{array}$$

7. What number should be added to 546 to get a sum of 732?

Solution

We know the sum and one addend. To find the missing addend, subtract.

$$\begin{array}{r} \scriptstyle 1 \\ \scriptstyle 6\ 2\ 1 \\ \not{7}\ \not{3}\ 2 \\ -\ 5\ 4\ 6 \\ \hline 1\ 8\ 6 \end{array} \quad \text{difference}$$

The number to be added is 186.

8. The cost of repairing Ed's used TV set was going to be \$230 for parts (including tax) plus \$50 for labor. To buy a new set, he was going to pay \$400 plus \$24 in tax and the dealer was going to pay him \$75 for his old set. How much more would Ed have to pay for a new set than to have his old set repaired?

Solution

USED SET		*NEW SET*		*DIFFERENCE*
\$230	parts	\$400	cost	\$349
+50	labor	+24	tax	−280
\$280	total	\$424		\$ 69
		−75	trade-in	
		\$349	total	

Ed would pay \$69 more for the new set than for having his old set repaired. What would you do?

PRACTICE QUIZ

Find the following differences.

1. $\begin{array}{r} 83 \\ -54 \\ \hline \end{array}$

2. $\begin{array}{r} 600 \\ -368 \\ \hline \end{array}$

3. $\begin{array}{r} 7856 \\ -6397 \\ \hline \end{array}$

ANSWERS

1. 29
2. 232
3. 1459

Exercises 1.4

Subtract. Do as many problems as you can mentally.

1. 8 − 5 **2.** 19 − 6 **3.** 14 − 14 **4.** 17 − 9

5. 20 − 11 **6.** 17 − 0 **7.** 17 − 8 **8.** 16 − 16

9. 11 − 6 **10.** 13 − 7

11. $\begin{array}{r}17\\-17\\\hline\end{array}$ **12.** $\begin{array}{r}42\\-31\\\hline\end{array}$ **13.** $\begin{array}{r}89\\-76\\\hline\end{array}$ **14.** $\begin{array}{r}53\\-33\\\hline\end{array}$ **15.** $\begin{array}{r}47\\-27\\\hline\end{array}$

16. $\begin{array}{r}96\\-27\\\hline\end{array}$ **17.** $\begin{array}{r}23\\-18\\\hline\end{array}$ **18.** $\begin{array}{r}74\\-29\\\hline\end{array}$ **19.** $\begin{array}{r}61\\-48\\\hline\end{array}$ **20.** $\begin{array}{r}52\\-27\\\hline\end{array}$

21. $\begin{array}{r}126\\-\ \ 32\\\hline\end{array}$ **22.** $\begin{array}{r}174\\-\ \ 48\\\hline\end{array}$ **23.** $\begin{array}{r}347\\-129\\\hline\end{array}$ **24.** $\begin{array}{r}256\\-118\\\hline\end{array}$

25. $\begin{array}{r}692\\-217\\\hline\end{array}$ **26.** $\begin{array}{r}543\\-167\\\hline\end{array}$ **27.** $\begin{array}{r}900\\-307\\\hline\end{array}$ **28.** $\begin{array}{r}603\\-208\\\hline\end{array}$

29. $\begin{array}{r}474\\-286\\\hline\end{array}$ **30.** $\begin{array}{r}657\\-179\\\hline\end{array}$ **31.** $\begin{array}{r}7843\\-6274\\\hline\end{array}$ **32.** $\begin{array}{r}6793\\-5827\\\hline\end{array}$

33. $\begin{array}{r}4376\\-2808\\\hline\end{array}$ **34.** $\begin{array}{r}3275\\-1744\\\hline\end{array}$ **35.** $\begin{array}{r}3546\\-3546\\\hline\end{array}$ **36.** $\begin{array}{r}4900\\-3476\\\hline\end{array}$

37. $\begin{array}{r}5070\\-4376\\\hline\end{array}$ **38.** $\begin{array}{r}8007\\-2136\\\hline\end{array}$ **39.** $\begin{array}{r}4065\\-1548\\\hline\end{array}$ **40.** $\begin{array}{r}7602\\-2985\\\hline\end{array}$

41. $\begin{array}{r}7{,}085{,}076\\-4{,}278{,}432\\\hline\end{array}$ **42.** $\begin{array}{r}6{,}543{,}222\\-2{,}742{,}663\\\hline\end{array}$ **43.** $\begin{array}{r}4{,}000{,}000\\-2{,}993{,}042\\\hline\end{array}$

44. $\begin{array}{r}8{,}000{,}000\\-\ \ 647{,}561\\\hline\end{array}$ **45.** $\begin{array}{r}6{,}000{,}000\\-\ \ 328{,}989\\\hline\end{array}$

46. What number should be added to 978 to get a sum of 1200?

47. What number should be added to 860 to get a sum of 1000?

48. If the sum of two numbers is 537 and one of the numbers is 139, what is the other number?

49. A man is 36 years old, and his wife is 34 years old. Including high school, they have each attended school for 16 years. How many total years of schooling have they had?

50. Basketball Team A has twelve players and won its first three games by the following scores: 84 to 73, 97 to 78, and 101 to 63. Team B has ten players and won its first three games by 76 to 75, 83 to 70, and 94 to 84. What is the difference between the total of the differences of Team A's scores and those of its opponents and the total of the differences of Team B's scores and those of its opponents?

51. The Kingston Construction Co. made a bid of $7,043,272 to build a stretch of freeway, but the Beach City Construction Co. made a lower bid of $6,792,868. How much lower was the Beach City bid?

52. Two landscaping companies made bids on the landscaping of a new apartment complex. Company A bid \$550,000 for materials and plants and \$225,000 for labor. Company B bid \$600,000 for materials and plants and \$182,000 for labor. Which company had the lower total bid? How much lower?

53. In June, Ms. White opened a checking account and deposited \$1342, \$238, \$57, and \$486. She also wrote checks for \$132, \$76, \$25, \$42, \$480, \$90, \$17, and \$327. What was her balance at the end of June?

54. A manufacturing company had assets of \$5,027,479, which included \$1,500,000 in real estate. The liabilities were \$4,792,023. By how much was the company "in the black"?

55. A man and woman sold their house for \$135,000. They paid the realtor \$8100, and other expenses of the sale came to \$800. If they owed the bank \$87,000 for the mortgage, what were their net proceeds from the sale?

56. A woman bought a condominium for a price of \$150,000. She also had to pay other expenses of \$750. If the local Savings and Loan agreed to give her a mortgage loan of \$105,500, how much cash did she need to make the purchase?

57. Junior bought a red sports car with a sticker price of \$10,000. But the salesman added \$1200 for taxes, license, and extras. The bank agreed to give Junior a loan of \$7500 on the car. How much cash did Junior need to buy the car?

58. In pricing a four-door car, Pat found she would have to pay a base price of \$9500 plus \$570 in taxes and \$250 for license fees. For a two-door of the same make, she would pay a base price of \$8700 plus \$522 in taxes and \$230 for license fees. Including all expenses, how much cheaper was the two-door model?

1.5 BASIC FACTS OF MULTIPLICATION

Repeated addition of the same addend can be shortened considerably by learning to multiply. Here a raised dot between two numbers indicates multiplication. Thus,

$$7 + 7 + 7 + 7 = 4 \cdot 7 = 28$$
$$279 + 279 + 279 + 279 = 4 \cdot 279 = 1116$$

The repeated addend (279) and the number of times it is used (4) are both called **factors,** and the sum is called the **product** of the two factors.

$$279 + 279 + 279 + 279 = \underbrace{4}_{\text{factor}} \cdot \underbrace{279}_{\text{factor}} = \underbrace{1116}_{\text{product}}$$

Several notations can be used to indicate multiplication. In this text, we will use the raised dot and parentheses much of the time.

(a) $4 \cdot 279$ (b) $4(279)$ (c) $(4)279$

(d) $(4)(279)$ (e) 4×279 (f) $\begin{array}{r} 279 \\ \times \ \ 4 \\ \hline \end{array}$

(g) Directions are given:

Multiply $\begin{array}{r} 279 \\ 4 \\ \hline \end{array}$

To change a multiplication problem to a repeated addition problem every time we are to multiply two numbers would be ridiculous. For example, $48 \cdot 137$ would mean using 137 as an addend 48 times. The first step in learning the multiplication process is to **memorize** the basic multiplication facts in Table 1.2. The factors in the table are only the digits 0 through 9. Using other factors involves the place value concept, as we shall see.

Table 1.2 Basic Multiplication Facts

·	**0**	**1**	**2**	**3**	**4**	**5**	**6**	**7**	**8**	**9**
0	**0**	0	0	0	0	0	0	0	0	0
1	0	**1**	2	3	4	5	6	7	8	9
2	0	2	**4**	6	8	10	12	14	16	18
3	0	3	6	**9**	12	15	18	21	24	27
4	0	4	8	12	**16**	20	24	28	32	36
5	0	5	10	15	20	**25**	30	35	40	45
6	0	6	12	18	24	30	**36**	42	48	54
7	0	7	14	21	28	35	42	**49**	56	63
8	0	8	16	24	32	40	48	56	**64**	72
9	0	9	18	27	36	45	54	63	72	**81**

If you have difficulty with **any** of the basic facts in the table, write all the possible combinations in a mixed-up order on a sheet of paper. Write the products down as quickly as you can and then find the ones you missed. Practice these in your spare time until you are sure you know them.

The table is a mirror image of itself on either side of the main diagonal (the numbers 0, 1, 4, 9, 16, 25, 36, 49, 64, 81). This indicates that multiplication is **commutative.** For example, we note that

$$5 \cdot 3 = 3 \cdot 5, \qquad 6 \cdot 2 = 2 \cdot 6, \qquad 7 \cdot 1 = 1 \cdot 7, \qquad 8 \cdot 0 = 0 \cdot 8$$

COMMUTATIVE PROPERTY OF MULTIPLICATION

If a and b are whole numbers, then $a \cdot b = b \cdot a$.

A close look at Table 1.2 also shows that multiplication by 0 gives 0 and that multiplication by 1 gives the number being multiplied. These two properties are called the **zero factor law** and **multiplicative identity,** respectively.

EXAMPLES

ZERO FACTOR LAW	*MULTIPLICATIVE IDENTITY*
$0 \cdot 7 = 0$	$1 \cdot 7 = 7$
$9 \cdot 0 = 0$	$9 \cdot 1 = 9$
$83 \cdot 0 = 0$	$83 \cdot 1 = 83$
$0 \cdot 654 = 0$	$1 \cdot 654 = 654$

We state these two properties formally.

ZERO FACTOR LAW

If a is a whole number, then $a \cdot 0 = 0$.

MULTIPLICATIVE IDENTITY

If a is a whole number, then there is a unique whole number 1 with the property that $a \cdot 1 = a$.

One (1) is called the **multiplicative identity** or the **identity element for multiplication.**

Only two numbers can be multiplied at a time. Therefore, if three or more factors are to be multiplied, we have to decide which two to **group** or **associate** together first. Does it matter? The answer is no. For example,

$$2 \cdot 3 \cdot 7 = (2 \cdot 3) \cdot 7 = 6 \cdot 7 = 42$$

and $$2 \cdot 3 \cdot 7 = 2 \cdot (3 \cdot 7) = 2 \cdot 21 = 42$$

Also,

$$9 \cdot 2 \cdot 5 = (9 \cdot 2) \cdot 5 = 18 \cdot 5 = 90$$

and

$$9 \cdot 2 \cdot 5 = 9 \cdot (2 \cdot 5) = 9 \cdot 10 = 90$$

These examples illustrate the **associative property of multiplication.**

ASSOCIATIVE PROPERTY OF MULTIPLICATION

If a, b, and c are whole numbers, then $a \cdot b \cdot c = a(b \cdot c) = (a \cdot b)c$.

PRACTICE QUIZ

Which property of multiplication is illustrated?

1. $5 \cdot (3 \cdot 7) = (5 \cdot 3) \cdot 7$
2. $32 \cdot 0 = 0$
3. $6 \cdot 8 = 8 \cdot 6$

ANSWERS

1. Associative property
2. Zero factor law
3. Commutative property

Exercises 1.5

Did you read the material in the text and study the examples before starting the exercises?

Do the following problems mentally and write only the answers.

1. $8 \cdot 9$ **2.** $7 \cdot 6$ **3.** $8(6)$ **4.** $9(7)$ **5.** $6 \cdot 5$

6. $5 \cdot 9$ **7.** $3(9)$ **8.** $8(5)$ **9.** $6(4)$ **10.** $7(4)$

11. $8(7)$ **12.** $3(6)$ **13.** $0 \cdot 3$ **14.** $5 \cdot 0$ **15.** $1 \cdot 9$

16. $4 \cdot 1$ **17.** $9 \cdot 0$ **18.** $7 \cdot 0$ **19.** $6 \cdot 1$ **20.** $8 \cdot 1$

21. $2 \cdot 7 \cdot 3$ **22.** $4 \cdot 3 \cdot 5$ **23.** $2 \cdot 5 \cdot 6$ **24.** $5 \cdot 7 \cdot 1$

25. $3 \cdot 2 \cdot 5$ **26.** $6 \cdot 1 \cdot 4$ **27.** $5 \cdot 1 \cdot 8$ **28.** $3 \cdot 2 \cdot 9$

29. $8(6)(2)$ **30.** $3(7)(5)$ **31.** $4(3)(6)$ **32.** $6(1)(9)$

33. $0 \cdot 3 \cdot 96$ **34.** $5(0)(42)$ **35.** $16(0)(82)(193)$

For each property listed, give two examples that illustrate the property with whole numbers.

36. Associative property of multiplication

37. Associative property of addition

38. Commutative property of addition

39. Commutative property of multiplication

40. Multiplicative identity **41.** Zero factor law **42.** Additive identity

43. In your own words, describe the meaning of the term **factor.**

44. Fill in the missing numbers in the chart. Five is added to the given number. The sum is then doubled, and ten is subtracted from the product. The answer is written in the last column. [HINT: To fill in the last two rows, you must think backward.]

GIVEN NUMBER	ADD 5	DOUBLE	SUBTRACT 10
2	7	14	4
0	5	?	?
1	6	12	?
7	?	?	?
?	?	?	16
?	?	?	10

45. Fill in the missing numbers in the chart. Five is added to the given number. The sum is then tripled (multiplied by 3), and fifteen is subtracted from the product. The answer is written in the last column. [HINT: To fill in the last two rows, you must think backward.]

GIVEN NUMBER	ADD 5	TRIPLE	SUBTRACT 15
2	7	21	6
1	6	18	?
7	12	?	?
8	?	?	?
?	?	?	15
?	?	?	18

46. Complete the table.

•	5	8	7	9
3				
6				
5				
2				

47. Complete the table.

•	4	1	9	6
5				
9				
6				
0				

48. Complete the table.

•	2	8	6	3
1				
8				
0				
7				

1.6 MULTIPLICATION BY POWERS OF 10

A **power** of any number is either 1 or that number multiplied by itself one or more times. We will discuss this idea in more detail in Chapter 2. Some of the powers of ten are: 1, 10, 100, 1000, 10,000, 100,000, and so on.

$$\begin{aligned} &1 \\ &10 \\ 10 \cdot 10 &= 100 \\ 10 \cdot 10 \cdot 10 &= 1000 \\ 10 \cdot 10 \cdot 10 \cdot 10 &= 10{,}000 \\ 10 \cdot 10 \cdot 10 \cdot 10 \cdot 10 &= 100{,}000 \\ \text{and so on} & \end{aligned}$$

Multiplication by powers of ten is useful in explaining multiplication with whole numbers in general, as we will do in Section 1.7. Such multiplication should be done mentally and quickly. The following examples illustrate an important pattern:

$$\begin{aligned} 6 \cdot 1 &= 6 \\ 6 \cdot 10 &= 60 \\ 6 \cdot 100 &= 600 \\ 6 \cdot 1000 &= 6000 \end{aligned}$$

If one of two whole number factors is 1000, the product will be the other factor with three zeros (000) written to the right of it. Two zeros (00) are written to the right of the other factor when multiplying by 100, and one zero (0) is written when multiplying by 10. Will multiplication by one million (1,000,000) result in writing six zeros to the right of the other factor? The answer is yes.

To multiply a number

by 10, write 0 to the right;
by 100, write 00 to the right;
by 1000, write 000 to the right;
by 10,000, write 0000 to the right;
and so on.

Many products can be found mentally by using the properties of multiplication and the techniques of multiplying by powers of ten. The processes are written out in the following examples, but they can easily be done mentally with practice.

EXAMPLES

1. $6 \cdot 90 = 6(9 \cdot 10) = (6 \cdot 9)10 = 54 \cdot 10 = 540$

2. $3 \cdot 400 = 3(4 \cdot 100) = (3 \cdot 4)100 = 12 \cdot 100 = 1200$

3. $2 \cdot 300 = 2(3 \cdot 100) = (2 \cdot 3)100 = 6 \cdot 100 = 600$

4. $6 \cdot 700 = 6(7 \cdot 100) = (6 \cdot 7)100 = 42 \cdot 100 = 4200$

5. $40 \cdot 30 = (4 \cdot 10)(3 \cdot 10) = (4 \cdot 3)(10 \cdot 10) = 12 \cdot 100 = 1200$

6. $50 \cdot 700 = (5 \cdot 10)(7 \cdot 100) = (5 \cdot 7)(10 \cdot 100) = 35 \cdot 1000$
$= 35{,}000$

7. $200 \cdot 800 = (2 \cdot 100)(8 \cdot 100) = (2 \cdot 8)(100 \cdot 100) = 16 \cdot 10{,}000$
$= 160{,}000$

8. $7000 \cdot 9000 = (7 \cdot 1000)(9 \cdot 1000) = (7 \cdot 9)(1000 \cdot 1000)$
$= 63 \cdot 1{,}000{,}000 = 63{,}000{,}000$

NOTE: To find the product in each example, the nonzero digits are multiplied, and the appropriate number of zeros is written.

A whole number that ends with 0's has a power of 10 as a factor. Thus, we can easily find at least two factors of a whole number that ends with one or more 0's.

EXAMPLES

9. 70 = 7 · 10
(7 ↑ factor, 10 ↑ factor)

10. 500 = 5 · 100
(5 ↑ factor, 100 ↑ factor)

11. 46,000 = 46 · 1000
(46 ↑ factor, 1000 ↑ factor)

Exercises 1.6

Use the techniques of multiplying by the powers of ten to find the following products mentally.

1. $25 \cdot 10$	**2.** $76 \cdot 100$	**3.** $47 \cdot 1000$	**4.** $18 \cdot 10$
5. $72 \cdot 10$	**6.** $13 \cdot 1$	**7.** $50 \cdot 60$	**8.** $90 \cdot 80$
9. $20 \cdot 20$	**10.** $60 \cdot 60$	**11.** $30 \cdot 40$	**12.** $70 \cdot 80$

13. $90 \cdot 70$ **14.** $300 \cdot 30$ **15.** $200 \cdot 20$ **16.** $500 \cdot 70$

17. $500 \cdot 30$ **18.** $120 \cdot 30$ **19.** $130 \cdot 40$ **20.** $200 \cdot 60$

21. $200 \cdot 80$ **22.** $300 \cdot 600$ **23.** $100 \cdot 100$ **24.** $100 \cdot 50$

25. $3000 \cdot 20$ **26.** $500 \cdot 50$ **27.** $400 \cdot 30$ **28.** $50 \cdot 200$

29. $40 \cdot 6000$ **30.** $2000 \cdot 400$ **31.** $80 \cdot 600$

32. $3000 \cdot 5000$ **33.** $20{,}000 \cdot 30$ **34.** $4000 \cdot 4000$

35. $70 \cdot 9000$ **36.** $800 \cdot 4000$

Write each of the following numbers as a product of two factors with a power of ten as one factor.

37. 5000 **38.** 300 **39.** 170

40. 190,000 **41.** 630,000 **42.** 80

1.7 MULTIPLICATION

We can use expanded notation and our skills with multiplication by powers of ten to help in understanding the technique for multiplying two whole numbers. We also need the **distributive property of multiplication over addition** illustrated in the following discussion.

To multiply $3(70 + 2)$, we can add first, then multiply:

$$3(70 + 2) = 3(72) = 216$$

But we can also multiply first, then add, in the following manner:

$$\begin{aligned} 3(70 + 2) &= 3 \cdot 70 + 3 \cdot 2 \\ &= 210 + 6 \quad \text{(210 and 6 are called \textbf{partial products.})} \\ &= 216 \end{aligned}$$

Or, vertically,

$$\begin{array}{r} 70 + 2 \\ 3 \\ \hline 210 + 6 \end{array} = 216$$

partial products (210, 6) product of 72 and 3 (216)

DISTRIBUTIVE PROPERTY OF MULTIPLICATION OVER ADDITION	If a, b, and c are whole numbers, then $a(b + c) = a \cdot b + a \cdot c$.

EXAMPLES

1. Multiply $6 \cdot 39$.

$$\begin{array}{r} 39 \\ \times\ 6 \\ \hline \end{array} \qquad \begin{array}{r} 30 + 9 \\ 6 \\ \hline \underbrace{180 + 54}_{\text{partial products}} = \underbrace{234}_{\text{product}} \end{array} \qquad \text{or} \qquad \begin{array}{rll} 39 & & \\ 6 & & \\ \hline 54 & \leftarrow 6\cdot 9 & \text{partial} \\ 180 & \leftarrow 6\cdot 30 & \text{products} \\ \hline 234 & \leftarrow \text{product} & \end{array}$$

The expanded notation is shown so that you can see that 6 is multiplied by 30 and not just by 3. However, the vertical notation is generally used because it is easier to add the partial products in this form.

2. Multiply $37 \cdot 42$.

$$\begin{array}{r} 42 \\ \times 37 \\ \hline \end{array} \qquad \begin{array}{rcrcrcl} & & 40 & + & 2 & & \\ & & 30 & + & 7 & & \\ \cline{3-5} & & 280 & + & 14 & & \\ 1200 & + & 60 & & & & \\ \hline 1200 & + & 340 & + & 14 & = & 1554 \end{array} \qquad \text{or} \qquad \begin{array}{rl} 42 & \text{factor} \\ 37 & \text{factor} \\ \hline 14 & (7\cdot 2 = 14) \\ 280 & (7\cdot 40 = 280) \\ 60 & (30\cdot 2 = 60) \\ 1200 & (30\cdot 40 = 1200) \\ \hline 1554 & \text{product} \end{array}$$

3. Multiply $26 \cdot 276$.

$$\begin{array}{r} 276 \\ \times\ 26 \\ \hline \end{array} \qquad \begin{array}{rcrcrcrcl} & & 200 & + & 70 & + & 6 & & \\ & & & & 20 & + & 6 & & \\ \cline{3-7} & & 1200 & + & 420 & + & 36 & & \\ 4000 & + & 1400 & + & 120 & & & & \\ \hline 4000 & + & 2600 & + & 540 & + & 36 & = & 7176 \end{array} \qquad \text{or} \qquad \begin{array}{rl} 276 & \text{factor} \\ 26 & \text{factor} \\ \hline 36 & (6\cdot 6 = 36) \\ 420 & (6\cdot 70 = 420) \\ 1200 & (6\cdot 200 = 1200) \\ 120 & (20\cdot 6 = 120) \\ 1400 & (20\cdot 70 = 1400) \\ 4000 & (20\cdot 200 = 4000) \\ \hline 7176 & \text{product} \end{array}$$

There is a shorter and faster method of multiplication used most of the time. In this method:

(a) digits are carried to be added to the next partial product; and

(b) sums of some partial products are found mentally.

EXAMPLES

4. *WRITING PARTIAL PRODUCTS* | *SHORT METHOD*

```
  56              1   ←1 carried from 18
×  3             56
----            × 3
  18  (3·6 = 18) ----
 150  (3·50 = 150) 168
----
 168  product
```

From $6 \cdot 3 = 18$, write 8 below the 3 and carry 10 by writing 1 above the 5. Then multiply $3 \cdot 5$ and add 1: $3 \cdot 5 = 15$
$15 + 1 = 16$

5. *WRITING PARTIAL PRODUCTS* | *SHORT METHOD*

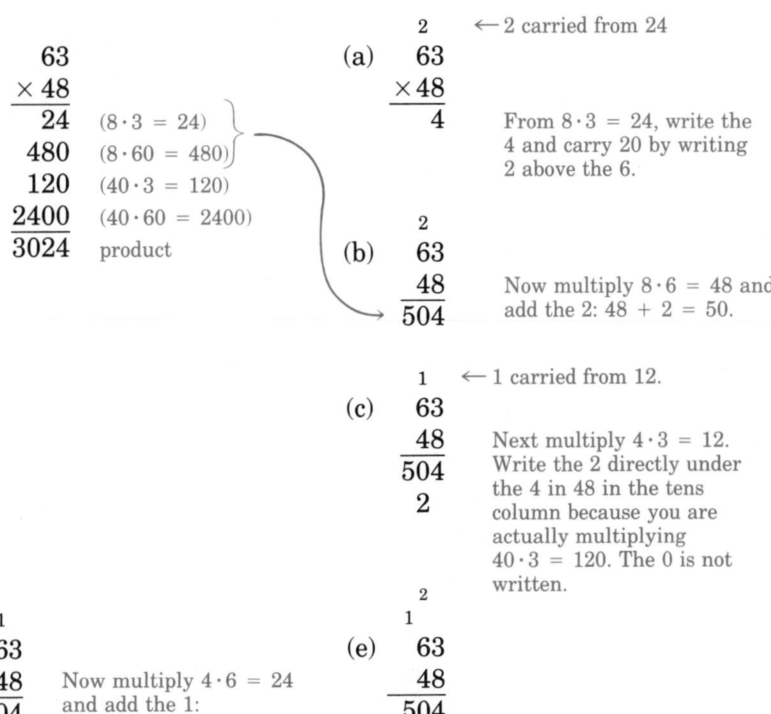

63
× 48
24 (8 · 3 = 24)
480 (8 · 60 = 480)
120 (40 · 3 = 120)
2400 (40 · 60 = 2400)
3024 product
←2 carried from 24
(a) 63 ×48 4
From 8 · 3 = 24, write the 4 and carry 20 by writing 2 above the 6.
(b) 63 48 504
Now multiply 8 · 6 = 48 and add the 2: 48 + 2 = 50.
←1 carried from 12.
(c) 63 48 504 2
Next multiply 4 · 3 = 12. Write the 2 directly under the 4 in 48 in the tens column because you are actually multiplying 40 · 3 = 120. The 0 is not written.


```
      1                                      2
(d)  63                                      1
     48   Now multiply 4·6 = 24        (e)  63
    ---   and add the 1:                    48
    504   24 + 1 = 25.                     ----
    252                                    504
                                           252   Add to find the final
                                          ----   product.
                                          3024
```

Steps (a), (b), (c), and (d) are shown for clarification of the Short Method. You will write only Step (e) as shown in Example 6.

```
       4
      1
6.   87
     26
    ---
    522   ←6·7 = 42. Write 2, carry 4.   6·8 = 48. Add 4: 48 + 4 = 52.
    174   ←2·7 = 14. Write 4, carry 1.   2·8 = 16. Add 1: 16 + 1 = 17.
   ----
   2262   ←product
```

Suppose you want to multiply 2000 by 423. Would you write

```
  2000      or      423
   423             2000
  6000              000
 4000              000
8000              000
846000           846
                 846000
```

Both answers are correct; neither technique is wrong. However, writing all the 0's is a waste of time, so knowledge about powers of ten is helpful. We can write

```
423
  2 000
846,000
```

We know 2000 = 2 · 1000, so we are simply multiplying 423 · 2, then the result by 1000.

EXAMPLES

7. Multiply 596 · 3000.

```
    596
      3000
1,788,000
```

8. Multiply 265 · 15,000.

```
  265
    15,000
1 325 000
2 65
3,975,000
```

PRACTICE QUIZ	Find the following products.	ANSWERS
	1. 18 × 24	1. 432
	2. 300 × 500	2. 150,000
	3. 129 × 39	3. 5031

Exercises 1.7

Please read the explanations and work through the examples before starting these exercises.

Find the following products by writing in all the partial products.

1. 56 × 4 **2.** 27 × 6 **3.** 48 × 9 **4.** 65 × 5 **5.** 43 × 8 **6.** 72 × 6

7. 91 × 5 **8.** 39 × 2 **9.** 84 × 3 **10.** 95 × 8 **11.** 42 × 56 **12.** 25 × 33

13. 15 × 22 **14.** 29 × 41 **15.** 67 × 36 **16.** 54 × 27 **17.** 48 × 20 **18.** 93 × 30

19. 83 × 85 **20.** 96 × 62

Multiply each of the following.

21. 17 × 32 **22.** 28 × 91 **23.** 20 × 44 **24.** 16 × 26 **25.** 25 × 15

26. 93 × 47 **27.** 24 × 86 **28.** 72 × 65 **29.** 12 × 13 **30.** 81 × 36

31. 126 × 41 **32.** 232 × 76 **33.** 114 × 25 **34.** 72 × 106 **35.** 207 × 143

36. 420 × 104 **37.** 200 × 49 **38.** 849 × 205 **39.** 673 × 186 **40.** 192 × 467

Multiply, using your knowledge of powers of ten.

41. 52 × 600 **42.** 72 × 930 **43.** 76 × 5000 **44.** 500 × 8000

45. 68 × 7300 **46.** 320 × 4700 **47.** 41 × 5300 **48.** 157 × 6000

49. 48 × 5200 **50.** 39 × 23,000

51. Find the sum of eighty-three and two hundred seventy-six. Find the difference between ninety-four and seventy-five. Find the product of the sum and the difference.

52. Find the sum of forty-three and sixty-six. Find the sum of one hundred and two hundred. Find the product of the sums.

53. If your salary is \$1300 per month and you are to get a raise once a year of \$130 per month and you love your work, what will you earn over a two-year period? Over a five-year period?

54. If you rent an apartment with 3 bedrooms for \$650 per month and you know the rent will increase once a year by \$30 per month, what will you pay in rent over a three-year period? Over a five-year period?

55. If you drive at 55 miles per hour for 5 hours, how far will you drive? If you drive at 50 miles per hour in a new four-door car for 4 hours, how far will you drive?

56. Your company bought 18 new cars at a price of \$9750 per car. What did the company pay for the new cars? Each one had an air conditioner.

1.8 DIVISION

We know that $6 \cdot 10 = 60$ and that 6 and 10 are **factors** of 60. They are also called **divisors.** The process of division can be thought of as the reverse of multiplication. In division, we want to find how many times one number is contained in another. How many 6's are in 60? There are 10 sixes in 60, and we say that 60 **divided by** 6 is 10 (or $60 \div 6 = 10$).

The relationship between multiplication and division can be seen from the following table format:

DIVISION									
						MULTIPLICATION			
DIVIDEND		*DIVISOR*		*QUOTIENT*		*FACTORS*		*PRODUCT*
21	÷	7	=	3	since	$7 \cdot 3$	=	21
24	÷	6	=	4	since	$6 \cdot 4$	=	24
36	÷	4	=	9	since	$4 \cdot 9$	=	36
12	÷	2	=	6	since	$2 \cdot 6$	=	12

This table indicates that the number being divided is called the **dividend;** the number doing the dividing is called the **divisor;** and the result of division is called the **quotient.**

Division does not always involve factors (or exact divisors). Suppose we want to divide 23 by 4. By using repeated subtraction, we can find how many 4's are in 23. We continuously subtract 4 until the number left (called the **remainder**) is less than 4.

$$\begin{array}{r} 23 \\ -\ 4 \\ \hline 19 \end{array} \quad \begin{array}{r} 19 \\ -\ 4 \\ \hline 15 \end{array} \quad \begin{array}{r} 15 \\ -\ 4 \\ \hline 11 \end{array} \quad \begin{array}{r} 11 \\ -\ 4 \\ \hline 7 \end{array} \quad \begin{array}{r} 7 \\ -4 \\ \hline 3 \end{array} \leftarrow \text{remainder}$$

(subtraction 5 times)

Another form used to indicate division as repeated subtraction is shown in Examples 1 and 2.

EXAMPLES

1. How many 7's are in 185?

$$\begin{array}{rlrll} 7\overline{)185} & & & & \\ -140 & \leftarrow \text{subtract} & 20 & \text{sevens} & (20\cdot 7 = 140) \\ \hline 45 & & & & (30\cdot 7 = 210\text{; too much since } 210 \text{ is greater than } 185) \\ -42 & \leftarrow \text{subtract} & \underline{6} & \text{sevens} & (6\cdot 7 = 42) \\ \hline 3 & & 26 & \text{sevens total} & \\ \uparrow & & \uparrow & & \\ \text{remainder} & & \text{quotient} & & \end{array}$$

Check

Division is checked by multiplying the quotient and the divisor and then adding the remainder. The result should be the dividend.

$$\begin{array}{rl} 26 & \text{quotient} \\ \times\ 7 & \text{divisor} \\ \hline 182 & \end{array} \qquad \begin{array}{rl} 182 & \\ +\ \ 3 & \text{remainder} \\ \hline 185 & \text{dividend} \end{array}$$

2. Find 275 ÷ 6 using repeated subtraction, and check your work.

$$\begin{array}{rll} 6\overline{)275} & & \\ -180 & \text{Subtract 30 sixes.} & (30\cdot 6 = 180) \\ \hline 95 & & \\ -60 & \text{Subtract 10 sixes.} & (10\cdot 6 = 60) \\ \hline 35 & & \\ -18 & \text{Subtract 3 sixes.} & (3\cdot 6 = 18) \\ \hline 17 & & \\ -12 & \text{Subtract 2 sixes.} & (2\cdot 6 = 12) \\ \hline 5 & \text{45 sixes total} & \\ \uparrow & \quad\uparrow & \\ \text{remainder} & \text{quotient} & \end{array}$$

NOTE: You can subtract any number of sixes less than the quotient. But this will not lead to a good explanation of the shorter division algorithm. You should subtract the largest number of thousands, hundreds, tens, or units that you can at each step.

Repeated subtraction provides a basis for understanding the **division algorithm,*** a much shorter method of division.

---

*An algorithm is a process or pattern of steps to be followed in working with numbers.

EXAMPLES

3. Find 2076 ÷ 8.

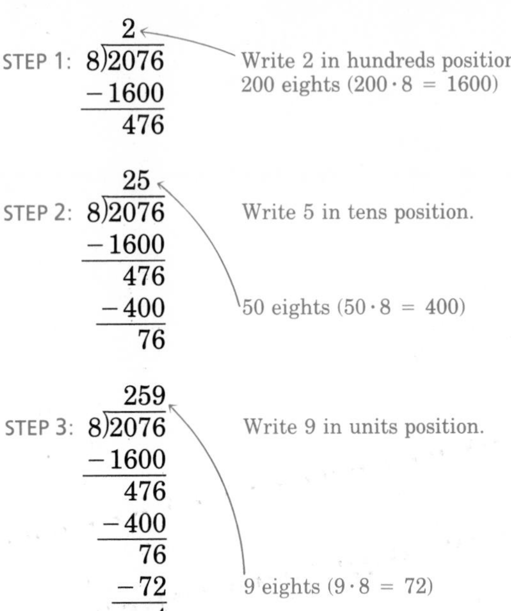

Summary

The process can be shortened by not writing all the 0's and "bringing down" only one digit at a time.

259 R4
8)2076
16
47 ← "Bring down" the 7 only; then divide 8 into 47.
40
76
72
4

4. 746 ÷ 32

STEP 1:

2
32)746
64
10

Trial divide 30 into 70 or 3 into 7, giving 2 in the tens position. Note that 10 is less than 32.

STEP 2:

23 R10
32)746
64
106
96
10

Trial divide 30 into 100 or 3 into 10, giving 3 in the units position.

Check

$$\begin{array}{r} 23 \\ \times\ 32 \\ \hline 46 \\ 69 \\ \hline 736 \end{array} \qquad \begin{array}{r} 736 \\ +\ 10 \\ \hline 746 \end{array}$$

5. 9325 ÷ 45

STEP 1: $45\overline{)9325}$ with quotient 2; 90, remainder 3

Trial divide 40 into 90 or 4 into 9, giving 2 in the hundreds position.

STEP 2: $45\overline{)9325}$ with quotient 20; 90, 32, 0

45 will not divide into 32, so write 0 in the tens column and multiply $0 \cdot 45 = 0$.

STEP 3: $45\overline{)9325}$ with quotient 208; 90, 32, 0, 325, 360

Trial divide 45 into 325 or 4 into 32. But the trial quotient 8 is too large since $8 \cdot 45 = 360$ and 360 is larger than 325.

$45\overline{)9325}$ with quotient 207 R10; 90, 32, 0, 325, 315, 10

Now the trial divisor is 7. Since $7 \cdot 45 = 315$ and 315 is smaller than 325, 7 is the desired number.

Check

$$\begin{array}{r} 207 \\ \times\ 45 \\ \hline 1035 \\ 828 \\ \hline 9315 \end{array} \qquad \begin{array}{r} 9315 \\ +\ \ 10 \\ \hline 9325 \end{array}$$

Special Note: In Step 2 of Example 5, we wrote 0 in the quotient because 45 did not divide into 32. Be sure to write 0 in the quotient whenever the divisor does not divide into any of the partial remainders.

If the remainder is 0, then both the divisor and quotient are factors of the dividend. We say that both factors divide **exactly** into the dividend.

EXAMPLES

6. Show that 17 and 36 are factors (or divisors) of 612.

$$\begin{array}{r} 36 \\ 17\overline{)612} \\ \underline{51} \\ 102 \\ \underline{102} \\ 0 \end{array}$$

Since 0 is the remainder, both 17 and 36 are factors (or divisors) of 612. Just to double-check, we find 612 ÷ 36.

$$\begin{array}{r} 17 \\ 36\overline{)612} \\ \underline{36} \\ 252 \\ \underline{252} \\ 0 \end{array}$$

Yes, both 17 and 36 divide exactly into 612.

7. A plumber purchased 17 special pipe fittings. What was the price of one fitting if the bill was $544 before taxes?

Solution

We need to know how many times 17 goes into 544.

$$\begin{array}{r} 32 \\ 17\overline{)544} \\ \underline{51} \\ 34 \\ \underline{34} \\ 0 \end{array}$$

The price for one fitting was $32.

PRACTICE QUIZ	Find the quotient and remainder for each of the following problems.	ANSWERS
	1. $325 \div 7$	**1.** 46 R3
	2. $16\overline{)324}$	**2.** 20 R4
	3. $41\overline{)24682}$	**3.** 602 R0

Exercises 1.8

Please read the text and study the examples before working these exercises.

Find the quotient and remainder for each of the following problems by using the method of repeated subtraction.

1. $210 \div 7$ **2.** $140 \div 14$ **3.** $168 \div 8$ **4.** $70 \div 5$
5. $132 \div 11$ **6.** $120 \div 4$ **7.** $75 \div 15$ **8.** $51 \div 3$
9. $52 \div 8$ **10.** $44 \div 6$ **11.** $600 \div 25$ **12.** $413 \div 20$
13. $161 \div 15$ **14.** $182 \div 13$ **15.** $150 \div 13$ **16.** $500 \div 14$
17. $205 \div 5$ **18.** $321 \div 7$
19. $1042 \div 22$ **20.** $1461 \div 12$

Divide and check using the division algorithm.

21. $6\overline{)32}$ **22.** $7\overline{)17}$ **23.** $4\overline{)25}$ **24.** $5\overline{)35}$
25. $8\overline{)48}$ **26.** $6\overline{)72}$ **27.** $9\overline{)81}$ **28.** $2\overline{)76}$
29. $3\overline{)98}$ **30.** $14\overline{)52}$ **31.** $12\overline{)108}$ **32.** $11\overline{)424}$
33. $16\overline{)128}$ **34.** $20\overline{)305}$ **35.** $18\overline{)206}$ **36.** $30\overline{)847}$
37. $10\overline{)423}$ **38.** $15\overline{)750}$ **39.** $13\overline{)260}$ **40.** $17\overline{)340}$
41. $12\overline{)360}$ **42.** $19\overline{)7603}$ **43.** $16\overline{)4813}$ **44.** $11\overline{)4406}$
45. $13\overline{)3917}$ **46.** $73\overline{)148}$ **47.** $68\overline{)207}$ **48.** $49\overline{)993}$
49. $50\overline{)3065}$ **50.** $40\overline{)2163}$ **51.** $105\overline{)210}$ **52.** $116\overline{)232}$
53. $213\overline{)4760}$ **54.** $716\overline{)3056}$ **55.** $630\overline{)4768}$ **56.** $414\overline{)83276}$
57. $502\overline{)98762}$ **58.** $317\overline{)70365}$ **59.** $471\overline{)50612}$ **60.** $215\overline{)64930}$

61. Find the quotient if eight hundred twenty-eight is divided by thirty-six. Is the remainder zero? Does this mean that thirty-six is a factor of eight hundred twenty-eight? If so, what is its corresponding factor?

62. Find the quotient if two thousand five hundred forty-two is divided by forty-one. Is the remainder zero? Does this mean that forty-one is a factor of two thousand five hundred forty-two? If so, what is its corresponding factor?

63. If one factor of 810 is 27, what is the corresponding factor?

64. If one factor of 1610 is 35, what is the corresponding factor?

65. What number multiplied by 73 gives a product of 1606? Is 73 a factor of 1606?

66. What number multiplied by 18 gives a product of 3654? Is 18 a factor of 3654?

67. If you bought four textbooks for a price of $16 each, including tax, what did you pay for the four books? The price of a pad of paper was 53 cents.

68. If you bought five plants for your yard and paid $13 for each plant, including tax, what did you pay for the plants? Fertilizer cost $10 per bag.

69. A purchasing agent bought 19 new cars for his company for a total price of $190,665. What price did he pay per car?

70. A restaurant owner bought new chairs for a total cost of $9792. If each chair cost $96, how many chairs did she buy?

1.9 AVERAGE (OR MEAN)

A topic closely related to addition and division is **average.** Your grade in this course may be based on the average of your exam scores. Newspapers and magazines have information about the Dow Jones stock averages, the average income of American families, the average life expectancy of laboratory rats, and so on. The average of a set of numbers is a kind of "middle number" of the set.* **The average of a set of numbers can be defined as the number found by adding the numbers in the set, then dividing this sum by the number of numbers in the set.** This average is also called the **arithmetic average,** or **mean.**

*Such terms as the **average citizen** or **average voter** are not related to numbers and are not so easily defined as the average of a set of numbers.

EXAMPLES

1. Find the average of the three numbers 32, 47, 23.

```
                34   (average)
     32      3)102
     47        9
   + 23        12
    102        12
                0
```

The sum, 102, is divided by 3 because there are three numbers being added.

The average of a set of whole numbers need not be a whole number. However, in this section, the problems will be set up so that the averages will be whole numbers. Other cases will be discussed later in the chapters on fractions and decimals (Chapters 3 and 5).

The average of a set of numbers can be very useful, but it can also be misleading. Judging the importance of an average is up to you, the reader of the information.

EXAMPLES

2. Suppose five people had the following incomes for one year: \$8,000; \$9,000; \$10,000; \$11,000; \$12,000. Find their average income.

Solution

```
   $ 8,000       $10,000
     9,000     5)50,000
    10,000       50,000
    11,000            0
  + 12,000
   $50,000
```

The average income is \$10,000.

3. Suppose five people had the following incomes for one year: \$1000; \$1000; \$1000; \$1000; \$46,000. Find their average income.

Solution

```
   $ 1,000       $10,000
     1,000     5)50,000
     1,000       50,000
     1,000            0
  + 46,000
   $50,000
```

The average income is \$10,000.

In Example 2, the average of $10,000 serves well as a "middle score" or "representative" of all the incomes. However, in Example 3, none of the incomes is even close to $10,000. The one large income completely destroys the "representativeness" of the average. Thus, it is useful to see the numbers or at least know something about them before attaching too much importance to an average.

We will discuss some related statistics in more detail in Chapter 6.

EXAMPLES

4. On an English exam, two students scored 95 points, five students scored 86 points, one student scored 82 points, one student scored 78 points, and six students scored 75 points. What was the mean score of the class?

Solution

$$\begin{array}{r} 95 \\ \times\ 2 \\ \hline 190 \end{array} \qquad \begin{array}{r} 86 \\ \times\ 5 \\ \hline 430 \end{array} \qquad \begin{array}{r} 82 \\ \times\ 1 \\ \hline 82 \end{array} \qquad \begin{array}{r} 78 \\ \times\ 1 \\ \hline 78 \end{array} \qquad \begin{array}{r} 75 \\ \times\ 6 \\ \hline 450 \end{array}$$

We have multiplied rather than write down all 15 scores. So, when the five products are added, we will divide the sum by 15 because the sum represents 15 scores.

$$\begin{array}{r} 190 \\ 430 \\ 82 \\ 78 \\ +\ 450 \\ \hline 1230 \end{array} \qquad \begin{array}{r} 82 \\ 15\overline{)1230} \\ 120 \\ \hline 30 \\ 30 \\ \hline \end{array} \quad \text{mean score}$$

The class mean is 82 points.

Exercises 1.9

Find the average (or mean) of each of the following sets of numbers.

1. 102, 113, 97, 100

2. 56, 64, 38, 58

3. 6, 8, 7, 4, 4, 5, 6, 8

4. 5, 4, 5, 6, 5, 8, 9, 6

5. 512, 618, 332, 478

6. 436, 520, 630, 422

7. 1000, 1000, 7000

8. 4000, 5000, 6000

9. 897, 182, 617, 534, 700

10. 648, 930, 556, 852, 544

11. On a math exam, five students' scores were 85, 90, 75, 64, and 96. What was the average of these five scores?

12. On an English exam, six students' scores were 76, 83, 68, 90, 92, and 95. What was their mean score?

13. On an exam in history, two students scored 95, six students scored 90, three students scored 80, and one student scored 50. What was the class average?

14. A woman bought ten shares of stock in a company at $35 per share. Later she bought ten shares in another company at $49 per share. What was the average price per share that she paid?

15. Nick bought shares in the stock market from two companies. He paid $450 for nine shares in one company and $690 for eleven shares in a second company. What did he pay as an average price per share for the twenty shares?

16. If you made deposits in your checking account of $640, $830, $1056, $890, and $734 in five months, what was your average deposit?

17. A salesman sold items from his sales list for $972, $834, $1005, $1050, and $799. What was the average price per item?

18. In flying twelve trips in one month (30 days), an airline pilot spent the following amount of hours in preparing and flying for each trip: 6, 8, 9, 6, 7, 7, 7, 5, 6, 6, 6, and 11 hours. What was the average amount of time he spent per flight?

19. Three families, each with two children, had incomes of $8942. Two families, each with four children, had incomes of $10,512. Four families, each with two children, had incomes of $11,111. One family had no children and an income of $12,026. What was the mean income per family?

20. During July, Mr. Rodriguez made deposits in his checking account of $400 and $750 and wrote checks totaling $625. During August, his deposits were $632, $322, and $798, and his checks totaled $978. In September, the deposits were $520, $436, $200, and $376; the checks totaled $836. What was the average monthly difference between his deposits and his withdrawals? What was his bank balance at the end of September if he had a balance of $500 on July 1?

SUMMARY: CHAPTER 1

The decimal system (or base ten system) is a **place value system** that depends on three things:

1. The **ten digits:** 0, 1, 2, 3, 4, 5, 6, 7, 8, 9;
2. The **placement** of each digit; and
3. The **value** of each place.

Whole numbers are those numbers used for counting and the number 0.

$$\mathbf{W} = \{0, 1, 2, 3, 4, 5, 6, 7, 8, 9, 10, 11, 12, \ldots\}$$

Rounding off a given number means to find another number close to the given number.

ROUNDING OFF RULE FOR WHOLE NUMBERS

1. Look at the single digit just to the right of the digit that is in the place of desired accuracy.
2. If this digit is 5 or greater, make the digit in the desired place of accuracy one larger and replace all digits to the right with zeros.
3. If this digit is less than 5, leave the digit that is in the place of desired accuracy as it is and replace all digits to the right with zeros.

Numbers being added are called **addends,** and the result of the addition is called the **sum.**

COMMUTATIVE PROPERTY OF ADDITION — If a and b are two whole numbers, then $a + b = b + a$.

ASSOCIATIVE PROPERTY OF ADDITION — If a, b, and c are whole numbers, then

$$a + b + c = (a + b) + c = a + (b + c)$$

ADDITIVE IDENTITY — If a is a whole number, then there is a unique whole number 0 with the property that $a + 0 = a$.

Subtraction is a reverse addition, and the missing addend is called the **difference** between the sum and one addend.

The result of multiplying two numbers is called the **product,** and the two numbers are called **factors** of the product.

COMMUTATIVE PROPERTY OF MULTIPLICATION — If a and b are whole numbers, then $a \cdot b = b \cdot a$.

ZERO FACTOR LAW	If a is a whole number, then $a \cdot 0 = 0$
MULTIPLICATIVE IDENTITY	If a is a whole number, then there is a unique whole number 1 with the property that $a \cdot 1 = a$.
ASSOCIATIVE PROPERTY OF MULTIPLICATION	If a, b, and c are whole numbers, then $a \cdot b \cdot c = a(b \cdot c) = (a \cdot b)c$
DISTRIBUTIVE PROPERTY OF MULTIPLICATION OVER ADDITION	If a, b, and c are whole numbers, then $a(b + c) = a \cdot b + a \cdot c$

In division:

The number dividing is called the **divisor.**

The number being divided is called the **dividend.**

The result is called the **quotient.**

The number left over is the **remainder,** and it must be less than the divisor.

An **average** (or **mean**) of a set of numbers is the number found by adding the numbers in the set, then dividing this sum by the number of numbers in the set.

REVIEW QUESTIONS: CHAPTER 1

Write the following decimal numerals in expanded notation and in their English word equivalents.

1. 495 **2.** 1975 **3.** 60,308

Write the following numbers as decimal numerals.

4. four thousand eight hundred fifty-six

5. fifteen million, thirty-two thousand, one hundred ninety-seven

6. six hundred seventy-two million, three hundred forty thousand, eighty-three

Round off as indicated.

7. 625 (to the nearest ten)

8. 14,620 (to the nearest thousand)

9. 749 (to the nearest hundred)

10. 2570 (to the nearest hundred)

State which property of addition or multiplication is illustrated.

11. $17 + 32 = 32 + 17$

12. $3(22 \cdot 5) = (3 \cdot 22)5$

13. $28 + (6 + 12) = (28 + 6) + 12$

14. $72 \cdot 89 = 89 \cdot 72$

15. Show that the following statement is **not true.**

$$32 \div (16 \div 2) = (32 \div 16) \div 2$$

What fact does this illustrate?

Add.

16. $\begin{array}{r} 8445 \\ 267 \\ 1351 \\ +\ 478 \\ \hline \end{array}$

17. $\begin{array}{r} 39 \\ 487 \\ 966 \\ +182 \\ \hline \end{array}$

Multiply.

18. $60 \cdot 40$

19. $47 \cdot 0$

20. $80 \cdot 6000$

Subtract.

21. $\begin{array}{r} 647 \\ -139 \\ \hline \end{array}$

22. $\begin{array}{r} 7036 \\ -4652 \\ \hline \end{array}$

23. $\begin{array}{r} 5000 \\ -2898 \\ \hline \end{array}$

Multiply.

24. $\begin{array}{r} 96 \\ \times 62 \\ \hline \end{array}$

25. $\begin{array}{r} 8973 \\ \times\ 426 \\ \hline \end{array}$

26. $\begin{array}{r} 4796 \\ \times\ \ 3000 \\ \hline \end{array}$

27. Divide using the method of repeated subtraction:

$$7\overline{)2046}$$

Divide.

28. $529\overline{)71496}$

29. $38\overline{)26721}$

30. If the product of 17 and 51 is added to the product of 16 and 12, what is the sum?

31. Find the average of 33, 42, 25, and 40.

32. If the quotient of 546 and 6 is subtracted from 100, what is the difference?

33. Two years ago, Ms. Miller bought five shares of stock at $353 per share. One year ago, she bought another ten shares at $290 per share. Yesterday, she sold all her shares at $410 per share. What was her total profit? What was her average profit per share?

34. On a history exam, two students scored 98 points, five students scored 87 points, one student scored 81 points, and six students scored 75 points. What was the average score in the class?

35. State two identity properties, one for addition of whole numbers and one for multiplication of whole numbers.

36. What number should be added to seven hundred forty-three to get a sum of eight hundred thirteen?

37. Fill in the missing numbers in the chart according to the directions.

GIVEN NUMBER	ADD 100	DOUBLE	SUBTRACT 200
3	103	206	?
20	120	?	?
15	?	?	?
?	?	?	16

38. How many times can 35 be repeatedly subtracted from 700?

CHAPTER TEST: CHAPTER 1

1. Write 8952 in expanded notation and in its English word equivalent.
2. The number 1 is called the multiplicative ______.
3. Give an example that illustrates the commutative property of multiplication.
4. Round off 997 to the nearest thousand.
5. Round off 135, 721 to the nearest ten-thousand.

Add.

6. $\begin{array}{r} 9586 \\ 345 \\ +2078 \\ \hline \end{array}$

7. $\begin{array}{r} 37 \\ 486 \\ 493 \\ 162 \\ +557 \\ \hline \end{array}$

8. $\begin{array}{r} 1{,}480{,}900 \\ 2{,}576{,}850 \\ 5{,}200{,}635 \\ +4{,}523{,}276 \\ \hline \end{array}$

Subtract.

9. $\begin{array}{r} 850 \\ -362 \\ \hline \end{array}$

10. $\begin{array}{r} 5097 \\ -3868 \\ \hline \end{array}$

11. $\begin{array}{r} 6000 \\ -\ \ 293 \\ \hline \end{array}$

Multiply.

12. $\begin{array}{r} 34 \\ \times 76 \\ \hline \end{array}$

13. $\begin{array}{r} 2593 \\ \times\ \ 85 \\ \hline \end{array}$

14. $\begin{array}{r} 793 \\ \times 266 \\ \hline \end{array}$

Divide.

15. $25\overline{)10{,}075}$

16. $462\overline{)79{,}852}$

17. $603\overline{)1{,}209{,}015}$

18. Find the average of 82, 96, 49, and 69.
19. If the quotient of 51 and 17 is subtracted from the product of 19 and 3, what is the difference?
20. Robert and his brother were saving money to buy a new TV set for their parents. If Robert saved $23 a week and his brother saved $28 a week, how much did they save in six weeks? What was their average weekly savings? If the set they wanted to buy cost $530 including tax, how much did they still need after the six weeks?

2 Prime Numbers

2.1 EXPONENTS

Repeated addition is shortened by using multiplication:

$$2 + 2 + 2 + 2 = 4 \cdot 2 = 8$$

Repeated multiplication can be shortened by using **exponents.** Thus, if 2 is used as a factor three times, we can write

$$2 \cdot 2 \cdot 2 = 2^3 = 8$$

In an expression such as $2^3 = 8$, 2 is called the **base,** 3 is called the **exponent,** and 8 is called the **power.** (Exponents are written slightly to the right and above the base.)

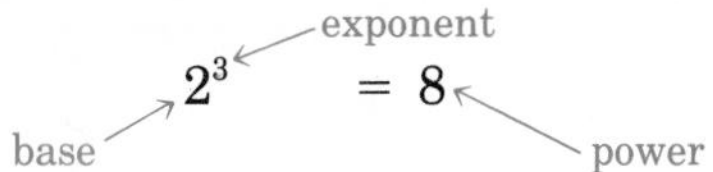

EXAMPLES

	REPEATED MULTIPLICATION	*USING EXPONENTS*
1.	$7 \cdot 7 = 49$	$7^2 = 49$
2.	$3 \cdot 3 = 9$	$3^2 = 9$
3.	$2 \cdot 2 \cdot 2 \cdot 2 = 16$	$2^4 = 16$
4.	$10 \cdot 10 \cdot 10 = 1000$	$10^3 = 1000$

DEFINITION An **exponent** is a number that tells how many times its base is to be used as a factor.

NOTE: Do **not** multiply by an exponent. Many beginning students make the mistake of thinking $6^3 = 6 \cdot 3 = 18$. This is **WRONG.** In fact, $6^3 = 6 \cdot 6 \cdot 6 = 216$.

Expressions with exponent 2 are read "squared," with exponent 3 are read "cubed," and with other exponents are read "to the ______ power." For example, we read

$5^2 = 25$ as "five squared is equal to twenty-five";
$4^3 = 64$ as "four cubed is equal to sixty-four"; and
$3^4 = 81$ as "three to the fourth power is equal to eighty-one."

If there is no exponent, the exponent is understood to be 1. Thus, $8 = 8^1$, $6 = 6^1$, and $942 = 942^1$.

When the exponent 0 is used for any base except 0, the value of the power is defined to be 1:

$$\begin{aligned} 2^0 &= 1 \\ 3^0 &= 1 \\ 5^0 &= 1 \\ 46^0 &= 1 \end{aligned}$$

One of the rules for using exponents (which will be studied in algebra) is related to division and involves subtracting exponents. This will help our understanding of the 0 exponent. To divide $\frac{2^6}{2^2}$ or $\frac{5^4}{5^3}$, we can write

$$\frac{2^6}{2^2} = \frac{2 \cdot 2 \cdot 2 \cdot 2 \cdot 2 \cdot 2}{2 \cdot 2} = 2 \cdot 2 \cdot 2 \cdot 2 = 2^4 \quad \text{or} \quad \frac{2^6}{2^2} = 2^{6-2} = 2^4$$

$$\frac{5^4}{5^3} = \frac{5 \cdot 5 \cdot 5 \cdot 5}{5 \cdot 5 \cdot 5} = 5 \quad \text{or} \quad \frac{5^4}{5^3} = 5^{4-3} = 5^1$$

The rule is to **subtract the exponents when dividing numbers with the same base.** So,

$$\frac{3^4}{3^4} = 3^{4-4} = 3^0 \qquad \text{and} \qquad \frac{5^2}{5^2} = 5^{2-2} = 5^0$$

But,

$$\frac{3^4}{3^4} = \frac{81}{81} = 1 \qquad \text{and} \qquad \frac{5^2}{5^2} = \frac{25}{25} = 1$$

For the rules of exponents to make sense, we must have $3^0 = 1$ and $5^0 = 1$.

DEFINITION For any nonzero whole number a, $a^0 = 1$.

EXAMPLES

5. $6^0 = 1$

6. $6^2 = 36$ $(6^2 = 6 \cdot 6)$

7. $5^3 = 125$ $(5^3 = 5 \cdot 5 \cdot 5)$

8. $3^3 = 27$ $(3^3 = 3 \cdot 3 \cdot 3)$

9. $2^5 = 32$ $(2^5 = 2 \cdot 2 \cdot 2 \cdot 2 \cdot 2)$

To improve your speed in factoring (Section 2.5) and working with fractions and algebraic expressions (next semester), you should try to memorize all the squares of the numbers from 1 to 20. The following table lists these squares.

NUMBER (N)	1	2	3	4	5	6	7	8	9	10
SQUARE (N^2)	1	4	9	16	25	36	49	64	81	100

NUMBER (N)	11	12	13	14	15	16	17	18	19	20
SQUARE (N^2)	121	144	169	196	225	256	289	324	361	400

These powers and others are listed in the table on the inside back cover of the text.

Exercises 2.1

Study the text and examples carefully before working these exercises.

In each of the following expressions, name (a) the exponent and (b) the base. Also, find each power.

1. 2^3 **2.** 2^5 **3.** 5^2 **4.** 6^2 **5.** 7^0

6. 11^2 **7.** 1^4 **8.** 4^3 **9.** 4^0 **10.** 3^6

11. 3^2 **12.** 2^4 **13.** 5^0 **14.** 1^{50} **15.** 62^1

16. 12^2 **17.** 10^2 **18.** 10^3 **19.** 4^2 **20.** 2^5

21. 10^4 **22.** 5^3 **23.** 6^3 **24.** 10^5 **25.** 19^0

Find a base and exponent form for each of the following powers without using the exponent 1. [HINT: The table inside the back cover may be helpful.]

26. 4 **27.** 25 **28.** 16 **29.** 27 **30.** 32

31. 121 **32.** 49 **33.** 8 **34.** 9 **35.** 36

36. 125 **37.** 81 **38.** 64 **39.** 100 **40.** 1000

41. 10,000 **42.** 216 **43.** 144 **44.** 169 **45.** 243

46. 625 **47.** 225 **48.** 196 **49.** 343 **50.** 100,000

Rewrite the following products using exponents.

51. $6 \cdot 6 \cdot 6 \cdot 6 \cdot 6$ **52.** $7 \cdot 7 \cdot 7 \cdot 7$ **53.** $2 \cdot 2 \cdot 7 \cdot 7$

54. $5 \cdot 5 \cdot 9 \cdot 9 \cdot 9$ **55.** $2 \cdot 2 \cdot 3 \cdot 3 \cdot 3$ **56.** $3 \cdot 3 \cdot 5 \cdot 5 \cdot 5$

57. $7 \cdot 7 \cdot 13$ **58.** $11 \cdot 11 \cdot 11$ **59.** $2 \cdot 3 \cdot 3 \cdot 11 \cdot 11$

60. $5 \cdot 5 \cdot 5 \cdot 11 \cdot 11$

Find the following squares. Write as many of them as you can from memory.

61. 8^2 **62.** 3^2 **63.** 7^2 **64.** 11^2 **65.** 15^2

66. 14^2 **67.** 18^2 **68.** 9^2 **69.** 12^2 **70.** 20^2

2.2 ORDER OF OPERATIONS

To evaluate the expression $5 \cdot 2^2 + 14 \div 2$, which of the following procedures would you use?

$$\begin{aligned} 5 \cdot 2^2 + 14 \div 2 &= 10^2 + 14 \div 2 \\ &= 100 + 14 \div 2 \\ &= 114 \div 2 \\ &= 57 \end{aligned}$$

OR

$$\begin{aligned} 5 \cdot 2^2 + 14 \div 2 &= 5 \cdot 4 + 14 \div 2 \\ &= 5 \cdot 4 + 7 \\ &= 5 \cdot 11 \\ &= 55 \end{aligned}$$

Both of these answers are **WRONG.** Mathematicians have agreed on a set of rules for simplifying (or evaluating) any numerical expression involving addition, subtraction, multiplication, division, and exponents. Under these rules,

$$\begin{aligned} 5 \cdot 2^2 + 14 \div 2 &= 5 \cdot 4 + 14 \div 2 \\ &= 20 + 7 \\ &= 27 \quad \textbf{RIGHT ANSWER} \end{aligned}$$

RULES FOR ORDER OF OPERATIONS

1. First, simplify within grouping symbols, such as parentheses ( ), brackets [ ], or braces { }. Start with the innermost grouping.
2. Second, find any powers indicated by exponents.
3. Third, moving from **left to right,** perform any multiplications or divisions in the order in which they appear.
4. Fourth, moving from **left to right,** perform any additions or subtractions in the order in which they appear.

The rules are very explicit. Read them carefully. Note that in Rule 3, neither multiplication nor division has priority over the other. Whichever of these operations occurs first, moving left to right, is done first. In Rule 4, addition and subtraction are handled in the same way.

The following examples show how to apply the rules for order of operations. Generally, more than one step is performed at the same time. Use the + and − signs to separate the expressions into various parts. The addition and subtraction occur last unless they are grouping symbols.

EXAMPLES

Use the rules for order of operations to find the value of each of the following expressions.

1. Evaluate $14 \div 7 + 3 \cdot 2 - 5$.

$$\underbrace{14 \div 7}_{\downarrow} + \underbrace{3 \cdot 2}_{\downarrow} - 5$$

$$= \quad 2 \quad + \quad 6 \quad - 5 = 8 - 5 = 3$$

2. Evaluate $3 \cdot 6 \div 9 - 1 + 4 \cdot 7$.

$$\underbrace{3 \cdot 6}_{\downarrow} \div 9 - 1 + \underbrace{4 \cdot 7}_{\downarrow}$$

$$= \quad 18 \div 9 - 1 + \quad 28 \quad = 2 - 1 + 28 = 1 + 28 = 29$$

3. Evaluate $(6 + 2) + (8 + 1) \div 9$.

$$\begin{aligned} &(6 + 2) + (8 + 1) \div 9 \\ &= 8 + 9 \div 9 = 8 + 1 = 9 \end{aligned}$$

4. Evaluate $30 \div 3 \cdot 2 + 3(6 - 21 \div 7)$.

$$\begin{aligned} &30 \div 3 \cdot 2 + 3(6 - 21 \div 7) \\ &= 10 \cdot 2 + 3(6 - 3) \\ &= 20 + 3(3) \\ &= 20 + 9 = 29 \end{aligned}$$

5. Evaluate $2 \cdot 3^2 + 18 \div 3^2$.

$$\begin{aligned} &2 \cdot 3^2 + 18 \div 3^2 \\ &= 2 \cdot 9 + 18 \div 9 \\ &= 18 + 2 = 20 \end{aligned}$$

6. Evaluate $3 \cdot 5^2 \div 15 + 30 - 2^3 \cdot 3$.

$$\begin{aligned} &3 \cdot 5^2 \div 15 + 30 - 2^3 \cdot 3 \\ &= 3 \cdot 25 \div 15 + 30 - 8 \cdot 3 \\ &= 75 \div 15 + 30 - 24 \\ &= 5 + 30 - 24 = 11 \end{aligned}$$

7. Evaluate $[(5 + 2^2) \div 3 + 8](14 - 10)$.

$$\begin{aligned} &[(5 + 2^2) \div 3 + 8](14 - 10) \\ &= [(5 + 4) \div 3 + 8](4) \\ &= [9 \div 3 + 8](4) \\ &= [3 + 8](4) \\ &= [11](4) = 44 \end{aligned}$$

PRACTICE QUIZ	Find the value for each of the following expressions using the rules for order of operations.	ANSWERS
	1. $15 \div 15 + 10 \cdot 2$	**1.** 21
	2. $3 \cdot 2^3 - 12 - 3 \cdot 2^2$	**2.** 0
	3. $4 \div 2^2 + 3 \cdot 2^2$	**3.** 13
	4. $(5 + 7) \div 3 + 1$	**4.** 5
	5. $19 - 5(3 - 1)$	**5.** 9

Exercises 2.2

Be sure to study the examples before you begin these exercises.

Find the value of each of the following expressions using the rules for order of operations.

1. $4 \div 2 + 7 - 3 \cdot 2$

2. $8 \cdot 3 \div 12 + 13$

3. $6 + 3 \cdot 2 - 10 \div 2$

4. $14 \cdot 3 \div 7 \div 2 + 6$

5. $6 \div 2 \cdot 3 - 1 + 2 \cdot 7$

6. $5 \cdot 1 \cdot 3 - 4 \div 2 + 6 \cdot 3$

7. $72 \div 4 \div 9 - 2 + 3$

8. $14 + 63 \div 3 - 35$

9. $(2 + 3 \cdot 4) \div 7 + 3$

10. $(2 + 3) \cdot 4 \div 5 + 3 \cdot 2$

11. $(7 - 3) + (2 + 5) \div 7$

12. $16(2 + 4) - 90 - 3 \cdot 2$

13. $35 \div (6 - 1) - 5 + 6 \div 2$

14. $22 - 11 \cdot 2 + 15 - 5 \cdot 3$

15. $(42 - 2 \div 2 \cdot 3) \div 13$

16. $18 + 18 \div 2 \div 3 - 3 \cdot 1$

17. $4(7 - 2) \div 10 + 5$

18. $(33 - 2 \cdot 6) \div 7 + 3 - 6$

19. $72 \div 8 + 3 \cdot 4 - 105 \div 5$

20. $6(14 - 6 \div 2 - 11)$

21. $48 \div 12 \div 4 - 1 + 6$

22. $5 - 1 \cdot 2 + 4(6 - 18 \div 3)$

23. $8 - 1 \cdot 5 + 6(13 - 39 \div 3)$

24. $(21 \div 7 - 3)42 + 6$

25. $16 - 16 \div 2 - 2 + 7 \cdot 3$

26. $(135 \div 3 + 21 \div 7) \div 12 - 4$

27. $(13 - 5) \div 4 + 12 \cdot 4 \div 3 - 72 \div 18 \cdot 2 + 16$

28. $15 \div 3 + 2 - 6 + (3)(2)(18)(0)(5)$

29. $100 \div 10 \div 10 + 1000 \div 10 \div 10 \div 10 - 2$

30. $[(85 + 5) \div 3 \cdot 2 + 15] \div 15$

31. $2 \cdot 5^2 - 4 \div 2 + 3 \cdot 7$

32. $16 \div 2^4 - 9 \div 3^2$

33. $(4^2 - 7) \cdot 2^3 - 8 \cdot 5 \div 10$

34. $4^2 - 2^4 + 5 \cdot 6^2 - 10^2$

35. $(2^5 + 1) \div 11 - 3 + 7(3^3 - 7)$

36. $(6 + 8^2 - 10 \div 2) \div 5 + 5 \cdot 3^2$

37. $(5 + 7) \div 4 + 2$

38. $(2^3 + 2) \div 5 + (7^2 \div 7)$

39. $(5^2 + 7) \div 8 - (14 \div 7 \cdot 2)$

40. $(3 \cdot 2^2 - 5 \cdot 2 + 2) - (1 \cdot 2^2 + 5 \cdot 2 - 10)$

41. $2^3 \cdot 3^2 \div 24 - 3 + 6^2 \div 4$

42. $2 \cdot 3^2 + 5 \cdot 3^2 + 15^2 - (21 \cdot 3^2 + 6)$

43. $3 \cdot 2^3 - 2^2 + 4 \cdot 2 - 2^4$

44. $2 \cdot 5^2 - 4(21 \div 3 - 7) + 10^3 - 1000$

45. $(4 + 3)^2 - (2 + 3)^2$

46. $40 \div 2 \cdot 5 + 1 \cdot 3^2 \cdot 2$

47. $20 - 2(3 - 1) + 6^2 \div 2 \cdot 3$

48. $8 \div 2 \cdot 4 + 16 \div 4 \cdot 2 + 3 \cdot 2^2$

49. $50 \cdot 10 \div 2 - 2^2 \cdot 5 + 14 - 2 \cdot 7$

50. $(20 \div 2^2 \cdot 5) + (51 \div 17)^2$

51. $[(2 + 3)(5 - 1) \div 2](10 + 1)$

52. $3[4 + (6 \div 3 \cdot 2)]$

53. $5[3^2 + (8 + 2^3)] - 15$

54. $(2^4 - 16)[13 - (5^2 - 20)]$

55. $100 + 2[(7^2 - 9)(5 + 1)^2]$

2.3 TESTS FOR DIVISIBILITY (2, 3, 4, 5, 9, 10)

In our work with factoring (Section 2.5) and fractions (Chapter 3), we will need to be able to divide quickly by small numbers. We will want to know if a number is **exactly divisible** (remainder 0) by some number **before** we divide. There are simple tests we can use to determine whether a number is divisible by 2, 3, 4, 5, 9, or 10 **without actually dividing.**

For example, can you tell (without dividing) if 6495 is divisible by 2? By 3? The answer is that 6495 is divisible by 3 but not by 2.

```
   2165                  3247
 3)6495                2)6495
   6                     6
   04                    04
    3                     4
    19                    09
    18                     8
     15                    15
     15                    14
      0  Remainder          1  Remainder (remainder not 0)
```

The following list of rules explains how to test for divisibility by 2, 3, 4, 5, 9, and 10. There are other tests for other numbers such as 6, 7, 8, and 15, but the rules given here are sufficient for our purposes.

TESTS FOR DIVISIBILITY BY 2, 3, 4, 5, 9, AND 10

For 2: If the last digit (units digit) of a whole number is 0, 2, 4, 6, or 8, then the whole number is divisible by 2. In other words, even whole numbers are divisible by 2; odd whole numbers are not divisible by 2.

For 3: If the sum of the digits of a whole number is divisible by 3, then the number is divisible by 3.

For 4: If the last two digits of a whole number form a number that is divisible by 4, then the number is divisible by 4. (00 is considered to be divisible by 4.)

For 5: If the last digit of a whole number is 0 or 5, then the number is divisible by 5.

For 9: If the sum of the digits of a whole number is divisible by 9, then the number is divisible by 9.

For 10: If the last digit of a whole number is 0, then the number is divisible by 10.

EXAMPLES

1. 356 is divisible by 2 since the last digit is 6.
2. 6801 is divisible by 3 since 6 + 8 + 0 + 1 = 15 and 15 is divisible by 3.
3. 9036 is divisible by 4 since 36 (last 2 digits) is divisible by 4.

4. 1365 is divisible by 5 since 5 is the last digit.

5. 9657 is divisible by 9 since $9 + 6 + 5 + 7 = 27$ and 27 is divisible by 9.

6. 3590 is divisible by 10 since 0 is the last digit.

If a number is divisible by 9, then it will be divisible by 3. In Example 5, $9 + 6 + 5 + 7 = 27$ and 27 is divisible by 9 and by 3, so 9657 is divisible by 9 and by 3.

But, a number that is divisible by 3 might not be divisible by 9. In Example 2, $6 + 8 + 0 + 1 = 15$ and 15 is *not* divisible by 9, so 6801 is divisible by 3 but not by 9.

Similarly, any number divisible by 10 is also divisible by 5, but a number that is divisible by 5 might not be divisible by 10. The number 2580 is divisible by 10 and also by 5, but 4365 is divisible by 5 and not by 10.

Many numbers are divisible by more than one of the numbers 2, 3, 4, 5, 9, and 10. These numbers will satisfy more than one of the six tests. For example, 4365 is divisible by 5 (last digit is 5) and by 3 $(4 + 3 + 6 + 5 = 18)$ and by 9 $(4 + 3 + 6 + 5 = 18)$.

EXAMPLES

Use all six tests to determine which of the numbers 2, 3, 4, 5, 9, and 10 will divide into the following numbers.

7. 5712

(a) divisible by 2 (last digit is 2, an even digit)

(b) divisible by 3 $(5 + 7 + 1 + 2 = 15$ and 15 is divisible by 3)

(c) divisible by 4 (12 is divisible by 4)

(d) not divisible by 5 (last digit is not 0 or 5)

(e) not divisible by 9 $(5 + 7 + 1 + 2 = 15$ and 15 is not divisible by 9)

(f) not divisible by 10 (last digit is not 0)

8. 2530

(a) divisible by 2 (last digit is 0, an even digit)

(b) not divisible by 3 $(2 + 5 + 3 + 0 = 10$ and 10 is not divisible by 3)

(c) not divisible by 4 (30 is not divisible by 4)

(d) divisible by 5 (last digit is 0)

(e) not divisible by 9 (2 + 5 + 3 + 0 = 10 and 10 is not divisible by 9)

(f) divisible by 10 (last digit is 0)

9. 3401

(a) not divisible by 2 (last digit is 1, an odd digit)

(b) not divisible by 3 (3 + 4 + 0 + 1 = 8 and 8 is not divisible by 3)

(c) not divisible by 4 (1 is not divisible by 4)

(d) not divisible by 5 (last digit is not 0 or 5)

(e) not divisible by 9 (3 + 4 + 0 + 1 = 8 and 8 is not divisible by 9)

(f) not divisible by 10 (last digit is not 0)

Therefore, 3401 is not divisible by any numbers in this list.

PRACTICE QUIZ	Using the techniques of this section, determine which of the numbers 2, 3, 4, 5, 9, and 10 (if any) will divide exactly into each of the following numbers.	ANSWERS
	1. 842	**1.** 2
	2. 9030	**2.** 2, 3, 5, 10
	3. 4031	**3.** None

Exercises 2.3

Be sure to study the text and the examples before working these exercises.

Using the techniques of this section, determine which of the numbers 2, 3, 4, 5, 9, and 10 (if any) will divide exactly into each of the following numbers.

1. 72	**2.** 81	**3.** 105	**4.** 333	**5.** 150
6. 471	**7.** 664	**8.** 154	**9.** 372	**10.** 375
11. 443	**12.** 173	**13.** 567	**14.** 480	**15.** 331
16. 370	**17.** 571	**18.** 466	**19.** 897	**20.** 695
21. 795	**22.** 777	**23.** 45,000	**24.** 885	**25.** 4422

26. 1234 **27.** 4321 **28.** 8765 **29.** 5678 **30.** 402

31. 705 **32.** 732 **33.** 441 **34.** 555 **35.** 666

36. 9000 **37.** 10,000 **38.** 576 **39.** 549 **40.** 792

41. 5700 **42.** 4391 **43.** 5476 **44.** 6930 **45.** 4380

46. 510 **47.** 8805 **48.** 7155 **49.** 8377 **50.** 2222

51. 35,622 **52.** 75,495 **53.** 12,324 **54.** 55,555

55. 632,448 **56.** 578,400 **57.** 9,737,001 **58.** 17,158,514

59. 36,762,252 **60.** 20,498,105

61. If a number is divisible by both 2 and 9, will it be divisible by 18? Give five examples to support your answer.

62. If a number is divisible by both 3 and 9, will it be divisible by 27? Give five examples to support your answer.

2.4 PRIME NUMBERS AND COMPOSITE NUMBERS

Every counting number, except 1, has at least two factors, as the following list shows.

COUNTING NUMBERS	*FACTORS*
18	1, 2, 3, 6, 9, 18
14	1, 2, 7, 14
23	1, 23
17	1, 17
21	1, 3, 7, 21
36	1, 2, 3, 4, 6, 9, 12, 18, 36
3	1, 3

We are particularly interested in those numbers that have exactly two different factors. In the list just given, 23, 17, and 3 fall into that category. They are called **prime numbers.**

DEFINITION A **prime number** is a counting number with exactly two different factors (or divisors).

DEFINITION A **composite number** is a counting number with more than two different factors (or divisors).

Thus, in the list discussed, 18, 14, 21, and 36 are **composite numbers.**

NOTE: 1 is **neither** a prime nor a composite number. $1 = 1 \cdot 1$, and 1 is the only factor of 1. 1 does not have **exactly** two **different** factors, and it does not have more than two factors.

EXAMPLES

1. Some prime numbers:

2	2 has exactly two different factors, 1 and 2.
7	7 has exactly two different factors, 1 and 7.
11	11 has exactly two different factors, 1 and 11.
29	29 has exactly two different factors, 1 and 29.

2. Some composite numbers:

12	$1 \cdot 12 = 12$, $2 \cdot 6 = 12$, and $3 \cdot 4 = 12$ So, 1, 2, 3, 4, 6, and 12 are all factors of 12; and 12 has more than two different factors.
33	$1 \cdot 33 = 33$ and $3 \cdot 11 = 33$ So, 1, 3, 11, and 33 are all factors of 33; and 33 has more than two different factors.

There is no formula to help us find all the prime numbers. However, there is a technique developed by a Greek mathematician named Eratosthenes. He used the concept of **multiples.** To find the multiples of any counting number, multiply each of the counting numbers by that number.

counting numbers	1, 2, 3, 4, 5, 6, 7, 8, . . .
multiples of 8	8, 16, 24, 32, 40, 48, 56, 64, . . .
multiples of 2	2, 4, 6, 8, 10, 12, 14, 16, . . .
multiples of 3	3, 6, 9, 12, 15, 18, 21, 24, . . .
multiples of 10	10, 20, 30, 40, 50, 60, 70, 80, . . .

None of the multiples of a number, except possibly the number itself, can be prime since they all have that number as a factor. To sift out the prime numbers according to the **Sieve of Eratosthenes,** we proceed by eliminating multiples as the following steps describe.

1. To find the prime numbers from 1 to 50, list all the counting numbers from 1 to 50 in rows of ten.

1	2	3	4	5	6	7	8	9	10
11	12	13	14	15	16	17	18	19	20
21	22	23	24	25	26	27	28	29	30
31	32	33	34	35	36	37	38	39	40
41	42	43	44	45	46	47	48	49	50

2. Start by crossing out 1 (since 1 is not a prime number). Next, circle 2 and cross out all the other multiples of 2; that is, cross out every second number.

~~1~~	(2)	3	~~4~~	5	~~6~~	7	~~8~~	9	~~10~~
11	~~12~~	13	~~14~~	15	~~16~~	17	~~18~~	19	~~20~~
21	~~22~~	23	~~24~~	25	~~26~~	27	~~28~~	29	~~30~~
31	~~32~~	33	~~34~~	35	~~36~~	37	~~38~~	39	~~40~~
41	~~42~~	43	~~44~~	45	~~46~~	47	~~48~~	49	~~50~~

3. The first number after 2 not crossed out is 3. Circle 3 and cross out all multiples of 3 that are not already crossed out; that is, after 3, every third number should be crossed out.

~~1~~	(2)	(3)	~~4~~	5	~~6~~	7	~~8~~	~~9~~	~~10~~
11	~~12~~	13	~~14~~	~~15~~	~~16~~	17	~~18~~	19	~~20~~
~~21~~	~~22~~	23	~~24~~	25	~~26~~	~~27~~	~~28~~	29	~~30~~
31	~~32~~	~~33~~	~~34~~	35	~~36~~	37	~~38~~	~~39~~	~~40~~
41	~~42~~	43	~~44~~	~~45~~	~~46~~	47	~~48~~	49	~~50~~

4. The next number not crossed out is 5. Circle 5 and cross out all multiples of 5 that are not already crossed out. If we proceed this way, we will have the prime numbers circled and the composite numbers crossed out.

The final table shows that the prime numbers less than 50 are:

2, 3, 5, 7, 11, 13, 17, 19, 23, 29, 31, 37, 41, 43, 47

You should also note that

(a) 2 is the only even prime number; and

(b) All other prime numbers are odd, but not all odd numbers are prime.

1	2	3	4	5	6	7	8	9	10
11	12	13	14	15	16	17	18	19	20
21	22	23	24	25	26	27	28	29	30
31	32	33	34	35	36	37	38	39	40
41	42	43	44	45	46	47	48	49	50

Computers can be used to tell whether very large numbers are prime or not. The following steps can be used to determine whether relatively small numbers are prime:

1. Use the tests for divisibility for 2, 3, and 5 (Section 2.3).
2. Continue to divide by progressively larger prime numbers until
   (a) You find a 0 remainder (meaning the number is composite); or
   (b) You find a quotient less than the divisor (meaning the number is prime).

EXAMPLES

3. Is 103 prime?

Tests for 2, 3, and 5 fail. (103 is not even; 1 + 0 + 3 = 4 and 4 is not divisible by 3; the last digit is not 0 or 5.)

Divide by 7:

$$\begin{array}{r} 14 \\ 7\overline{)103} \\ \underline{7} \\ 33 \\ \underline{28} \\ 5 \end{array}$$

14 quotient greater than divisor

5 remainder not 0

Divide by 11:

$$\begin{array}{rl} 9 & \text{quotient is less than divisor} \\ 11\overline{)103} & \\ \underline{99} & \\ 4 & \end{array}$$

So, 103 is prime.

4. Is 221 prime?

   Tests for 2, 3, and 5 fail.

   Divide by 7:

$$\begin{array}{rl} 31 & \text{quotient greater than divisor} \\ 7\overline{)221} & \\ \underline{21} & \\ 11 & \\ \underline{7} & \\ 4 & \text{remainder not 0} \end{array}$$

   Divide by 11:

$$\begin{array}{rl} 20 & \text{quotient greater than divisor} \\ 11\overline{)221} & \\ \underline{22} & \\ 01 & \\ \underline{0} & \\ 1 & \text{remainder not 0} \end{array}$$

   Divide by 13:

$$\begin{array}{rl} 17 & \\ 13\overline{)221} & \\ \underline{13} & \\ 91 & \\ \underline{91} & \\ 0 & \text{remainder is 0} \end{array}$$

   So, 221 is composite and not prime. (Note: $221 = 13 \cdot 17$.)

5. One interesting application of factors (very useful in beginning algebra) involves finding two factors that add up to some specific number. For example, find two factors of 72 whose product is 72 and whose sum is 27.

72 has several factors. Simply experiment until you find the right pair. The numbers are 3 and 24.

$$3 \cdot 24 = 72 \quad \text{and} \quad 3 + 24 = 27$$

Exercises 2.4

Be sure you know what prime numbers are and what multiples are before you begin these exercises.

List the multiples for each of the following numbers.

1. 5 **2.** 7 **3.** 11 **4.** 13 **5.** 12

6. 9 **7.** 20 **8.** 17 **9.** 16 **10.** 25

11. Construct a Sieve of Eratosthenes for the numbers from 1 to 100. List the prime numbers from 1 to 100.

Decide whether each of the following numbers is prime or composite. If the number is composite, find at least two pairs of factors for the number.

12. 17 **13.** 19 **14.** 28 **15.** 32 **16.** 47

17. 59 **18.** 16 **19.** 63 **20.** 14 **21.** 51

22. 67 **23.** 89 **24.** 73 **25.** 61 **26.** 52

27. 57 **28.** 98 **29.** 86 **30.** 53 **31.** 37

Two numbers are given. Find two factors of the first number whose sum is the second number.

EXAMPLE: 12, 8
Two factors of 12 whose sum is 8 are 6 and 2 since $6 \cdot 2 = 12$ and $6 + 2 = 8$.

32. 24, 10 **33.** 12, 7 **34.** 16, 10 **35.** 12, 13 **36.** 14, 9

37. 50, 27 **38.** 20, 9 **39.** 24, 11 **40.** 48, 19 **41.** 36, 15

42. 7, 8 **43.** 63, 24 **44.** 51, 20 **45.** 25, 10 **46.** 16, 8

47. 60, 17 **48.** 52, 17 **49.** 27, 12 **50.** 72, 22

51. List the set of prime numbers less than 100 that **are not** odd.

52. List the set of prime numbers less than 100 that **are** odd.

2.5 PRIME FACTORIZATIONS

To find common denominators for fractions, which will be discussed in Chapter 3, finding **all** the prime factors of a number will be very useful. For example, $28 = 4 \cdot 7$ and 7 is a prime factor, but 4 is not prime. We can write $28 = 4 \cdot 7 = 2 \cdot 2 \cdot 7$ and this product $(2 \cdot 2 \cdot 7)$ contains all prime factors. It is called the **prime factorization** of 28.

The **prime factorization** of a number is the product of all the prime factors of that number, including repeated factors.

Two methods, I and II, for finding prime factorizations are discussed here. Both methods give the same prime factorization for a composite number. Method I is easier to apply in working with fractions, and Method II is easier to follow for some students.

The tests for divisibility by 2, 3, 4, 5, 9, and 10 discussed in Section 2.3 are very useful here. Review them now if you do not remember them. Many times they provide a beginning factor.

METHOD I

To find the prime factorization of a composite number:

1. Factor the composite number into **any** two factors.
2. If either or both factors are not prime, factor each of these.
3. Continue this process until all factors are prime.

EXAMPLES

Using Method I, find the prime factorization for each of the following numbers.

1. 60

$$60 = 6 \cdot 10$$ Since the last digit is 0, we know 10 is a factor.

$$= 2 \cdot 3 \cdot 2 \cdot 5$$ 6 and 10 can both be factored so that each factor is a prime number. This is the prime factorization of 60.

or,

$$60 = 3 \cdot 20$$ 3 is prime, but 20 is not.

$$= 3 \cdot 4 \cdot 5$$ 4 is not prime.

$$= 3 \cdot 2 \cdot 2 \cdot 5$$ All factors are prime.

Since multiplication is commutative, the order of the factors is not important. What is important is that **all the factors must be prime numbers.**

Writing the factors in order, the prime factorization of 60 is $2 \cdot 2 \cdot 3 \cdot 5$ or, using exponents, $2^2 \cdot 3 \cdot 5$.

2. 85

$85 = 5 \cdot 17$ 5 is a factor since the last digit is 5. You can find 17 either by dividing mentally or by writing out the long division.

3. 72

$$\begin{aligned} 72 &= 8 \cdot 9 \\ &= 2 \cdot 4 \cdot 3 \cdot 3 \\ &= 2 \cdot 2 \cdot 2 \cdot 3 \cdot 3 = 2^3 \cdot 3^2 \end{aligned}$$

9 is a factor since $7 + 2 = 9$.

or

$$\begin{aligned} 72 &= 2 \cdot 36 \\ &= 2 \cdot 6 \cdot 6 \\ &= 2 \cdot 2 \cdot 3 \cdot 2 \cdot 3 = 2^3 \cdot 3^2 \end{aligned}$$

2 is a factor since the last digit is even.

4. 264

The lines leading to smaller factors form an arrangement called a **factor tree.** Two factor trees for 264 are shown here. Note that the prime factorization is the same regardless of the two factors chosen first.

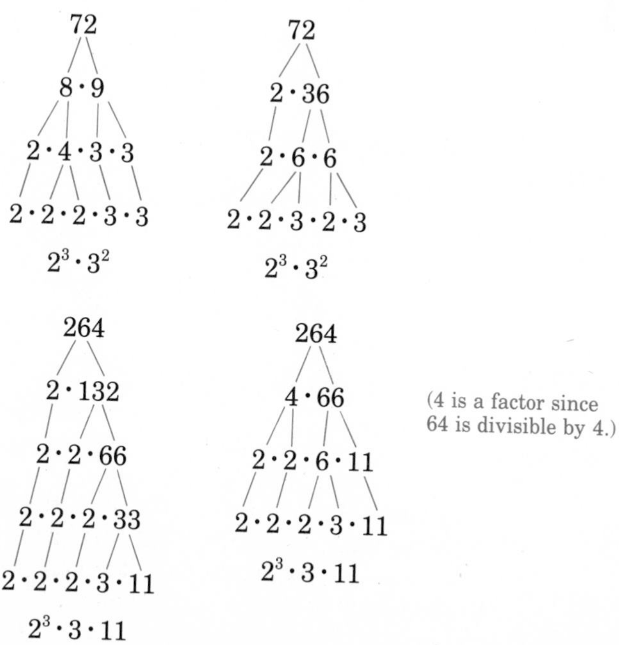

METHOD II

To find the prime factorization of a composite number:

1. Divide the composite number by **any** prime number that will divide into it.
2. Continue to divide the **quotient** by prime numbers until the quotient is prime.
3. The prime factorization is the product of all the prime divisors and the last prime quotient.

EXAMPLES

Using Method II, find the prime factorization for each of the following numbers.

5. 70

$$2\overline{)70}^{\,35} \qquad 5\overline{)35}^{\,7}$$

So, $70 = 2 \cdot 5 \cdot 7$.

6. 245

$$5\overline{)245}^{\,49} \qquad 7\overline{)49}^{\,7}$$

So, $245 = 5 \cdot 7 \cdot 7 = 5 \cdot 7^2$.

7. 168

$$2\overline{)168}^{\,84} \qquad 2\overline{)84}^{\,42} \qquad 2\overline{)42}^{\,21} \qquad 7\overline{)21}^{\,3}$$

So, $168 = 2 \cdot 2 \cdot 2 \cdot 7 \cdot 3 = 2^3 \cdot 3 \cdot 7$.

Regardless of what factors you start with or what method you use, **there is only one prime factorization for any composite number.** This fact is so important that it is called the **Fundamental Theorem of Arithmetic.**

FUNDAMENTAL THEOREM OF ARITHMETIC Every composite number has exactly one prime factorization.

We can use prime factorizations to find all the factors (or divisors) of a composite number.

> The only factors (or divisors) of a composite number are:
> (a) 1 and the number itself;
> (b) Each prime factor; and
> (c) Products of prime factors.

EXAMPLES

8. Find all the factors of 30.

Since $30 = 2 \cdot 3 \cdot 5$, the factors are

(a) 1 and 30;
(b) 2, 3, and 5; and
(c) $2 \cdot 3$, $2 \cdot 5$, and $3 \cdot 5$.

The factors are 1, 30, 2, 3, 5, 6, 10, and 15. These are the only factors of 30.

9. Find all the factors of 196.

$$\begin{aligned} 196 &= 2 \cdot 98 \\ &= 2 \cdot 2 \cdot 49 \\ &= 2 \cdot 2 \cdot 7 \cdot 7 \end{aligned}$$

The factors are

(a) 1 and 196;
(b) 2 and 7; and
(c) $2 \cdot 2$, $2 \cdot 7$, $7 \cdot 7$, $2 \cdot 2 \cdot 7$, and $2 \cdot 7 \cdot 7$.

The factors are 1, 196, 2, 7, 4, 14, 49, 28, and 98. These are the only factors (or divisors) of 196.

PRACTICE QUIZ	Find the prime factorization of each of the following numbers.	ANSWERS
	1. 42	**1.** $2 \cdot 3 \cdot 7$
	2. 56	**2.** $2^3 \cdot 7$
	3. 230	**3.** $2 \cdot 5 \cdot 23$
	4. Using the prime factorization of 63, find all the factors of 63.	**4.** 1, 63, 3, 7, 9, 21

Exercises 2.5

Study the text carefully before you start these exercises.

Find the prime factorization for each of the following numbers. Use the tests for divisibility given in Section 2.3 whenever you need them to help you get started.

1. 24	**2.** 28	**3.** 27	**4.** 16	**5.** 36
6. 60	**7.** 72	**8.** 90	**9.** 81	**10.** 105
11. 125	**12.** 160	**13.** 75	**14.** 150	**15.** 210
16. 40	**17.** 250	**18.** 93	**19.** 168	**20.** 360
21. 126	**22.** 48	**23.** 17	**24.** 47	**25.** 51
26. 144	**27.** 121	**28.** 169	**29.** 225	**30.** 52
31. 32	**32.** 98	**33.** 108	**34.** 103	**35.** 101
36. 202	**37.** 78	**38.** 500	**39.** 10,000	**40.** 100,000

Using the prime factorization of each number, find all the factors (or divisors) of each number.

41. 12	**42.** 18	**43.** 28	**44.** 98	**45.** 121
46. 45	**47.** 105	**48.** 54	**49.** 97	**50.** 144

2.6 LEAST COMMON MULTIPLE (LCM)

The techniques discussed in this section are used throughout Chapter 3 on fractions. Study these ideas thoroughly because they will make your work with fractions much easier.

Remember that the multiples of a number are the products of that number with the counting numbers. The first multiple of a number is that number, and all other multiples are larger than the number.

counting numbers	1, 2, 3, 4, 5, 6, 7, 8, 9, 10, . . .
multiples of 6	6, 12, 18, 24, (30,) 36, 42, 48, 54, (60,) . . .
multiples of 10	10, 20, (30,) 40, 50, (60,) 70, 80, (90,) 100, . . .

The common multiples for 6 and 10 are 30, 60, 90, 120, ... The **Least Common Multiple** is 30.

DEFINITION The **Least Common Multiple (LCM)** of a set of counting numbers is the smallest number common to all the sets of multiples of the given numbers.

Except for the number itself, **factors** (or **divisors**) of the number are smaller than the number, and **multiples** are larger than the number. For example, 6 and 10 are factors of 30, and both are smaller than 30. 30 is a multiple of both 6 and 10 and is larger than either 6 or 10.

Listing all the multiples as we did for 6 and 10 and then choosing the least common multiple (LCM) is not very efficient. Either of two other techniques, one involving prime factorizations and the other involving division by prime factors, is easier to apply.

METHOD I

To find the LCM of a set of counting numbers:

1. Find the prime factorization of each number.
2. Find the prime factors that appear in **any one** of the prime factorizations.
3. Form the product of these primes using each prime the most number of times it appears in **any one** of the prime factorizations.

EXAMPLES

1. Find the LCM for 18, 30, and 45.

   (a) prime factorizations:

   $$18 = 2 \cdot 9 = 2 \cdot 3 \cdot 3 \quad \text{One 2, two 3's.}$$
   $$30 = 6 \cdot 5 = 2 \cdot 3 \cdot 5 \quad \text{One 2, one 3, one 5.}$$
   $$45 = 9 \cdot 5 = 3 \cdot 3 \cdot 5 \quad \text{Two 3's, one 5.}$$

   (b) 2, 3, and 5 are the only prime factors.

   (c) most of each factor in any one factorization:

   one 2 (in 18 and in 30)
   two 3's (in 18 and in 45)
   one 5 (in 30 and in 45)

   $$\text{LCM} = 2 \cdot 3 \cdot 3 \cdot 5 = 90$$

   90 is the smallest number divisible by all the numbers 18, 30, and 45. (Note also that the LCM, 90, contains all the factors of each of the numbers 18, 30, and 45.)

2. Find the LCM for 36, 24, and 48.

   (a) prime factorizations:

   $$\left.\begin{aligned} 36 &= 4 \cdot 9 = 2^2 \cdot 3^2 \\ 24 &= 8 \cdot 3 = 2^3 \cdot 3 \\ 48 &= 16 \cdot 3 = 2^4 \cdot 3 \end{aligned}\right\} \quad \begin{aligned} \text{LCM} &= 2^4 \cdot 3^2 \\ &= 144 \end{aligned}$$

   (b) The only prime factors that appear are 2 and 3.

   (c) There are four 2's in 48 and two 3's in 36.

   Thus, the product $2^4 \cdot 3^2 = 144$ is the least common multiple.

3. Find the LCM for 27, 30, and 42.

   $$\left.\begin{aligned} 27 &= 3 \cdot 9 = 3^3 \\ 30 &= 6 \cdot 5 = 2 \cdot 3 \cdot 5 \\ 42 &= 2 \cdot 21 = 2 \cdot 3 \cdot 7 \end{aligned}\right\} \quad \begin{aligned} \text{LCM} &= 2 \cdot 3^3 \cdot 5 \cdot 7 \\ &= 1890 \end{aligned}$$

The second technique for finding the LCM involves division.

METHOD II

To find the LCM of a set of counting numbers:

1. Write the numbers horizontally and find a prime number that will divide into more than one number, if possible.
2. Divide by that prime and write the quotients beneath the dividends. Rewrite any numbers not divided beneath themselves.
3. Continue the process until no two numbers have a common prime divisor.
4. The LCM is the product of all the prime divisors and the last set of quotients.

EXAMPLES

4. Find the LCM for 20, 25, 18, and 6.

$$\begin{array}{r|rrrr} 5 & 20 & 25 & 18 & 6 \\ \hline 2 & 4 & 5 & 18 & 6 \\ \hline 3 & 2 & 5 & 9 & 3 \\ \hline & 2 & 5 & 3 & 1 \end{array}$$

$$\begin{aligned} \text{LCM} &= 5 \cdot 2 \cdot 3 \cdot 2 \cdot 5 \cdot 3 \cdot 1 \\ &= 2^2 \cdot 3^2 \cdot 5^2 \\ &= 900 \end{aligned}$$

In Example 3 by using prime factorizations, we found that for 27, 30, and 42 the LCM is 1890. We know that 1890 is a multiple of 27. That is, 27 will divide evenly into 1890. How many times will 27 divide into 1890? You could perform long division:

$$\begin{array}{r} 70 \\ 27\overline{)1890} \\ \underline{189} \\ 00 \\ \underline{0} \\ 0 \end{array}$$

Thus, 27 divides 70 times into 1890.

However, we do not need to divide. We can find the same result much more quickly by looking at the prime factorization of 1890 (which we have as a result of finding the LCM). First, find the prime factors for 27. Then the remaining prime factors will have 70 as their product.

$$\begin{aligned} 1890 &= 2 \cdot 3^3 \cdot 5 \cdot 7 = \underbrace{(3 \cdot 3 \cdot 3)}_{} \cdot \underbrace{(2 \cdot 5 \cdot 7)}_{} \\ &= \quad 27 \quad \cdot \quad 70 \end{aligned}$$

The following examples illustrate how to use prime factors to tell how many times each number in a set divides into the LCM for that set of numbers. **This technique is very useful in working with fractions.**

EXAMPLES

5. Find the LCM for 27, 30, and 42, and tell how many times each number divides into the LCM.

$$\left.\begin{aligned} 27 &= 3^3 \\ 30 &= 2\cdot3\cdot5 \\ 42 &= 2\cdot3\cdot7 \end{aligned}\right\} \text{LCM} = 2\cdot3^3\cdot5\cdot7 = 1890$$

$$1890 = 2\cdot3\cdot3\cdot3\cdot5\cdot7 = \underbrace{(3\cdot3\cdot3)}_{27}\cdot\underbrace{(2\cdot5\cdot7)}_{70} = 27\cdot70$$

$$1890 = 2\cdot3\cdot3\cdot3\cdot5\cdot7 = \underbrace{(2\cdot3\cdot5)}_{30}\cdot\underbrace{(3\cdot3\cdot7)}_{63} = 30\cdot63$$

$$1890 = 2\cdot3\cdot3\cdot3\cdot5\cdot7 = \underbrace{(2\cdot3\cdot7)}_{42}\cdot\underbrace{(3\cdot3\cdot5)}_{45} = 42\cdot45$$

So, 27 divides into 1890 70 times;
30 divides into 1890 63 times;
42 divides into 1890 45 times.

6. Find the LCM for 12, 18, and 66, and tell how many times each number divides into the LCM.

$$\left.\begin{aligned} 12 &= 2^2\cdot3 \\ 18 &= 2\cdot3^2 \\ 66 &= 2\cdot3\cdot11 \end{aligned}\right\} \text{LCM} = 2^2\cdot3^2\cdot11 = 396$$

$$396 = 2\cdot2\cdot3\cdot3\cdot11 = \underbrace{(2\cdot2\cdot3)}_{12}\cdot\underbrace{(3\cdot11)}_{33} = 12\cdot33$$

$$396 = 2\cdot2\cdot3\cdot3\cdot11 = \underbrace{(2\cdot3\cdot3)}_{18}\cdot\underbrace{(2\cdot11)}_{22} = 18\cdot22$$

$$396 = 2\cdot2\cdot3\cdot3\cdot11 = \underbrace{(2\cdot3\cdot11)}_{66}\cdot\underbrace{(2\cdot3)}_{6} = 66\cdot6$$

So, 12 divides into 396 33 times;
18 divides into 396 22 times;
66 divides into 396 6 times.

Whenever various events occur at regular intervals of time, one question that arises is, "How frequently will these events occur at the same time?" The answer can be stated in terms of the LCM, as illustrated in Example 7.

EXAMPLE

7. Suppose three weather satellites A, B, and C orbit the earth in different times: A takes 18 hours, B takes 14 hours, and C takes 10 hours. If they are directly above each other now (as shown in the figure), in how many hours will they again be directly above each other?

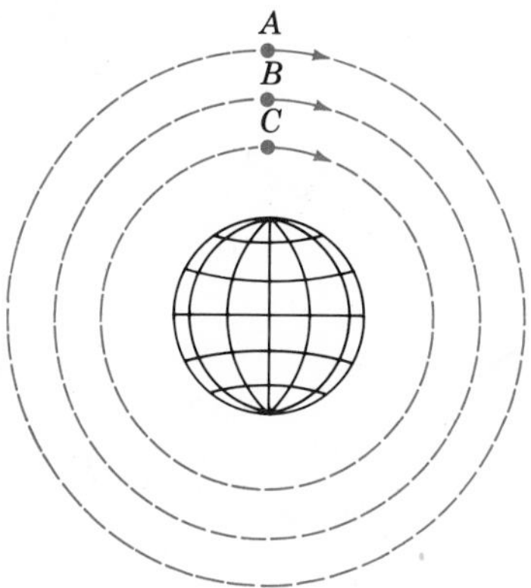

Solution

The answer is the LCM for 18, 14, and 10 since all three satellites must make complete orbits on a continual basis. One does not stop and wait for another.

$$\left.\begin{aligned} 18 &= 2\cdot 3^2 \\ 14 &= 2\cdot 7 \\ 10 &= 2\cdot 5 \end{aligned}\right\} \text{LCM} = 2\cdot 3^2\cdot 5\cdot 7 = 630$$

Thus, the satellites will not align again for 630 hours (or 26 days and 6 hours).

PRACTICE QUIZ	Find the LCM for each of the following sets of numbers.	ANSWERS
	1. 30, 40, 50	**1.** LCM = 600
	2. 28, 70	**2.** LCM = 140
	3. 168, 140	**3.** LCM = 840

Exercises 2.6

Study the procedures and examples in this section carefully before you begin these exercises.

Find the LCM for each of the following sets of numbers.

1. 8, 12
2. 3, 5, 7
3. 4, 6, 9
4. 3, 5, 9
5. 2, 5, 11
6. 4, 14, 18
7. 6, 15, 12
8. 6, 8, 27
9. 25, 40
10. 40, 75
11. 28, 98
12. 30, 75
13. 30, 80
14. 16, 28
15. 25, 100
16. 20, 50
17. 35, 100
18. 144, 216
19. 36, 42
20. 40, 100
21. 2, 4, 8
22. 10, 15, 35
23. 8, 13, 15
24. 25, 35, 49
25. 6, 12, 15
26. 8, 10, 120
27. 6, 15, 80
28. 13, 26, 169
29. 45, 125, 150
30. 34, 51, 54
31. 33, 66, 121
32. 36, 54, 72
33. 45, 145, 290
34. 54, 81, 108
35. 45, 75, 135
36. 35, 40, 72
37. 10, 20, 30, 40
38. 15, 25, 30, 40
39. 24, 40, 48, 56
40. 169, 637, 845

Find the LCM and tell how many times each number divides into the LCM.

41. 8, 10, 15
42. 6, 15, 30
43. 10, 15, 24
44. 8, 10, 120
45. 6, 18, 27, 45
46. 12, 95, 228
47. 45, 63, 98
48. 40, 56, 196
49. 99, 143, 363
50. 125, 135, 225

51. Two long-distance joggers are running in the same direction on the same course. The faster runner passes the slower one and they say "Hi." One jogger goes around the course in 10 minutes, and the other goes around in 14 minutes. They continue to jog until the faster runner passes the slower runner. In how many minutes will the faster runner pass the slower one? How many times will each have jogged around the course in that many minutes?

52. Three night watchmen walk around inspecting buildings at a shopping center. The watchmen take 9, 12, and 14 minutes, respectively, for the inspection trip. If they start at the same time, in how many minutes will they meet? How many inspection trips will each watchman have made?

53. Two astronauts miss connections at their first rendezvous in space. If one astronaut circles the earth every 12 hours and the other every 16 hours, in how many hours will they rendezvous again? How many orbits will each astronaut make between the first and second rendezvous?

54. Three truck drivers lunch together whenever all three are at the routing station at the same time. The route for the first driver takes 5 days, for the second driver 15 days, and for the third driver 6 days. How often do the three drivers lunch together? If the first driver's route was changed to 6 days, how often would they lunch together?

55. Four book salespersons leave the home office the same day. They take 10 days, 12 days, 15 days, and 18 days, respectively, to travel their own sales regions. In how many days will they all meet again at the home office? How many sales trips will each have made?

SUMMARY: CHAPTER 2

DEFINITION An **exponent** is a number that tells how many times its base is to be used as a factor.

DEFINITION For any nonzero whole number a, $a^0 = 1$.

RULES FOR ORDER OF OPERATIONS

1. First, simplify within grouping symbols, such as parentheses ( ), brackets [ ], or braces { }. Start with the innermost grouping.
2. Second, find any powers indicated by exponents.
3. Third, moving from **left to right,** perform any multiplications or divisions in the order in which they appear.
4. Fourth, moving from **left to right,** perform any additions or subtractions in the order in which they appear.

TESTS FOR DIVISIBILITY BY 2, 3, 4, 5, 9, AND 10

For 2: If the last digit (units digit) of a whole number is 0, 2, 4, 6, or 8, then the whole number is divisible by 2. In other words, even whole numbers are divisible by 2; odd whole numbers are not divisible by 2.

For 3: If the sum of the digits of a whole number is divisible by 3, then the number is divisible by 3.

For 4: If the last two digits of a whole number form a number that is divisible by 4, then the number is divisible by 4. (00 is considered to be divisible by 4.)

For 5: If the last digit of a whole number is 0 or 5, then the number is divisible by 5.

For 9: If the sum of the digits of a whole number is divisible by 9, then the number is divisible by 9.

For 10: If the last digit of a whole number is 0, then the number is divisible by 10.

DEFINITION A **prime number** is a counting number with exactly two different factors (or divisors).

DEFINITION A **composite number** is a counting number with more than two different factors (or divisors).

METHOD I

To find the prime factorization of a composite number:

1. Factor the composite number into **any** two factors.
2. If either or both factors are not prime, factor each of these.
3. Continue this process until all factors are prime.

METHOD II

To find the prime factorization of a composite number:

1. Divide the composite number by **any** prime number that will divide into it.
2. Continue to divide the **quotient** by prime numbers until the quotient is prime.
3. The prime factorization is the product of all the prime divisors and the last prime quotient.

FUNDAMENTAL THEOREM OF ARITHMETIC

Every composite number has a unique prime factorization.

The only factors (or divisors) of a composite number are:

(a) 1 and the number itself;

(b) Each prime factor; and

(c) Products of prime factors.

DEFINITION

The **Least Common Multiple (LCM)** of a set of natural numbers is the smallest number common to all the sets of multiples of the given numbers.

TO FIND THE LCM OF A SET OF NATURAL NUMBERS

Method I.

1. Find the prime factorization of each number.
2. Find the prime factors that appear in **any one** of the prime factorizations.
3. Form the product of these primes using each prime the most number of times it appears in **any one** of the prime factorizations.

Method II.

1. Write the numbers horizontally and find a prime number that will divide into more than one number, if possible.
2. Divide by that prime and write the quotients beneath the dividends. Rewrite any numbers not divided beneath themselves.
3. Continue the process until no two numbers have a common prime divisor.
4. The LCM is the product of all the prime divisors and the last set of quotients.

REVIEW QUESTIONS: CHAPTER 2

1. In the expression $3^4 = 81$, 3 is called the ______, 4 is called the ______, and 81 is called the ______.

2. A prime number is a counting number with exactly two ______ ______.

3. Every composite number has a unique ______ factorization.

4. Since 36 has more than two different factors, 36 is a ______ number.

Find each power.

5. 2^4 6. 3^3 7. 13^2 8. 21^2

Evaluate each of the following expressions.

9. $7 + 3 \cdot 2 - 1 + 9 \div 3$

10. $3 \cdot 2^5 - 2 \cdot 5^2$

11. $14 \div 2 + 2 \cdot 8 + 30 \div 5 \cdot 2$

12. $(16 \div 2^2 + 6) \div 2 + 8$

13. $(75 - 3 \cdot 5) \div 10 - 4$

14. $(7^2 \cdot 2 + 2) \div 10 \div (2 + 3)$

Determine which of the numbers 2, 3, 4, 5, 9, and 10 (if any) will divide exactly into each of the following numbers.

15. 45 16. 72 17. 479

18. 5040 19. 8836 20. 575,493

21. List the multiples of 3. Are any of these multiples prime numbers?

22. Is 223 a prime number? Show your work.

23. List the prime numbers less than 60.

24. Find two factors of 24 whose sum is 10.

25. Find two factors of 60 whose sum is 17.

Find the prime factorization for each of the following numbers.

26. 150 27. 65 28. 84 29. 92

Find the LCM for each of the following sets of numbers.

30. 8, 14, 24 31. 8, 12, 25, 36 32. 27, 54, 135

33. Find the LCM for the numbers 18, 39, and 63 and tell how many times each number divides into the LCM.

34. One racing car goes around the track every 30 seconds and the other every 35 seconds. If both cars start a race from the same point, in how many seconds will the first car be exactly one lap ahead of the second car? How many laps will each car have made when the first car is two laps ahead of the second?

TEST: CHAPTER 2

1. Name (a) the exponent and (b) the base, and (c) find the power of 7^3.

2. List the squares of the prime numbers between 5 and 20.

Evaluate each of the following expressions.

3. $15 \div 3 + 6 \cdot 2^3$

4. $7 \cdot 3^2 + 8 \div 6$

5. $4 + (75 \div 5^2 + 5) + 16 \div 2^3$

6. $12 \div 3 \cdot 2 - 5 + 2^3 \div 4$

Using the tests for divisibility, tell which of the numbers 2, 3, 4, 5, 9, and 10 (if any) will divide the following numbers.

7. 216

8. 221

9. 180

10. List the multiples of 19 that are less than 100.

11. List all of the composite numbers found in the following set:

1, 2, 6, 9, 13, 15, 17, 27, 39, 41, 51, 59

Find the prime factorization for each of the following numbers.

12. 70

13. 234

14. 750

15. List the factors of 45.

Find the LCM for each of the following sets of numbers.

16. 15 and 24

17. 12, 27, and 36

18. 15, 25, and 75

19. 7, 9, and 16

20. Two long-distance runners practice on the same track and start at the same point. One takes 54 seconds and the other takes 63 seconds to go around the track once. If each continues at that same pace, in how many seconds will they meet again? How many laps will each have run by the time they meet again?

3 Fractions

3.1 BASIC MULTIPLICATION

In this chapter we will deal only with fractions in which the numerator (top number) and denominator (bottom number) are whole numbers. Such fractions have the technical name of **rational numbers.**

Examples of rational numbers are $\frac{1}{2}$, $\frac{3}{4}$, $\frac{9}{10}$, and $\frac{17}{3}$.

Figure 3.1 shows how a whole may be separated into equal parts. We

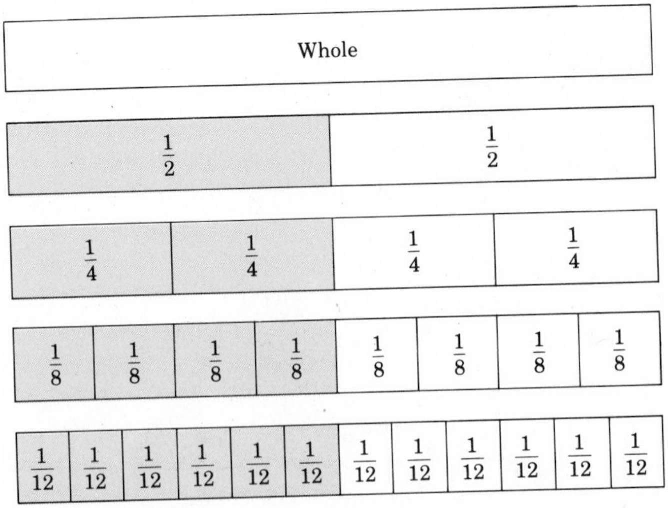

Whole

$\frac{1}{2}$ $\frac{1}{2}$

$\frac{1}{4}$ $\frac{1}{4}$ $\frac{1}{4}$ $\frac{1}{4}$

$\frac{1}{8}$ $\frac{1}{8}$ $\frac{1}{8}$ $\frac{1}{8}$ $\frac{1}{8}$ $\frac{1}{8}$ $\frac{1}{8}$ $\frac{1}{8}$

$\frac{1}{12}$ $\frac{1}{12}$ $\frac{1}{12}$ $\frac{1}{12}$ $\frac{1}{12}$ $\frac{1}{12}$ $\frac{1}{12}$ $\frac{1}{12}$ $\frac{1}{12}$ $\frac{1}{12}$ $\frac{1}{12}$ $\frac{1}{12}$

Figure 3.1

see that the rational numbers $\frac{1}{2}, \frac{2}{4}, \frac{4}{8}$, and $\frac{6}{12}$ all represent the same amount of the whole. These numbers are **equivalent,** or **equal.**

DEFINITION

A **rational number** is a number that can be written in the form $\frac{a}{b}$, where a is a whole number and b is a nonzero whole number.

$\frac{a}{b}$ numerator / denominator

NOTE: The numerator a can be 0, but the denominator b cannot be 0.

Until otherwise stated, we will use the terms **fraction** and **rational number** to mean the same thing.

Fractions can be used to indicate:

1. Equal parts of a whole.
2. Division.

EXAMPLES

1. $\frac{1}{2}$ 1 of 2 equal parts

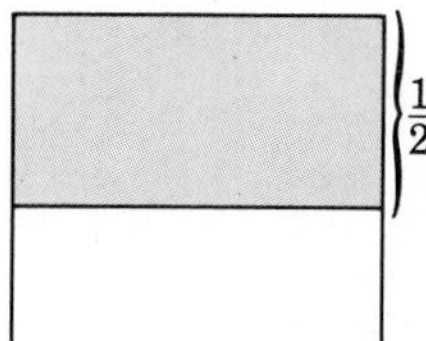

2. $\frac{3}{4}$ 3 of 4 equal parts

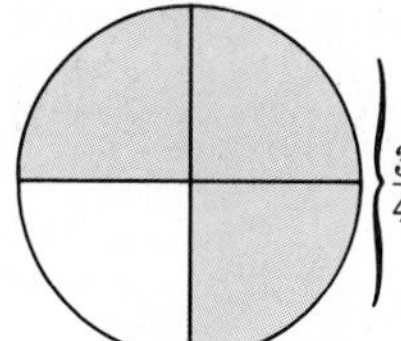

3. $\frac{2}{3}$ 2 of 3 equal parts

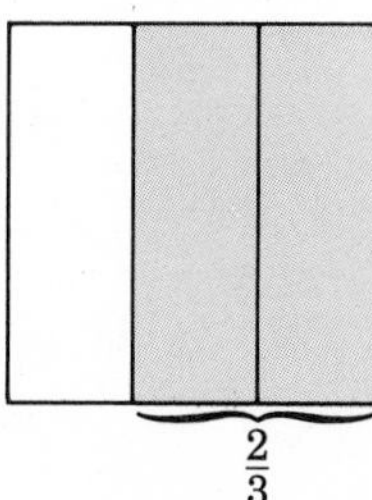

Improper fractions, such as $\frac{5}{3}$, $\frac{13}{6}$, and $\frac{18}{10}$, are fractions in which the numerator is larger than the denominator. The term "improper fraction" is misleading because there is nothing improper about such fractions. In fact, in algebra and other courses in mathematics, improper fractions are sometimes preferred over mixed numbers such as $2\frac{3}{4}$ and $5\frac{1}{8}$. Mixed numbers will be discussed in Chapter 4.

EXAMPLE

4. $\frac{5}{3}$ 5 of 3 equal parts (more than a whole)

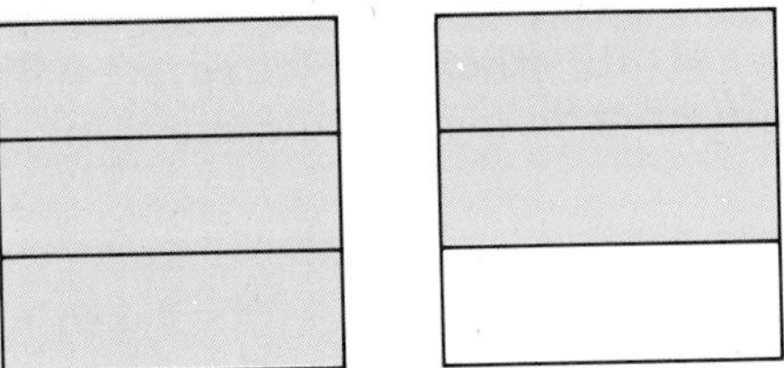

Under the division meaning of fractions, $\frac{15}{5}$ is the same as $15 \div 5$. By the definition of division, $\frac{15}{5} = 3$ because $15 = 5 \cdot 3$. Similarly, $\frac{0}{4} = 0$ because $0 = 4 \cdot 0$. In general, $\frac{0}{b} = 0$ if $b \neq 0$ ($\neq$ means **is not equal to**).

If the denominator is 0, then the fraction has no meaning. We say that **division by 0 is undefined.** The following discussion explains the reasoning.

DIVISION BY 0 IS UNDEFINED. NO DENOMINATOR CAN BE 0.

Consider $\frac{5}{0} = \square$. Then $5 = 0 \cdot \square = 0$. But this is impossible; $5 \neq 0$.

Next consider $\frac{0}{0} = \square$. Then $0 = 0 \cdot \square = 0$ for any value of $\square$. This means that $\frac{0}{0}$ could be any number. But in arithmetic an operation such as division cannot give more than one answer.

Thus, $\frac{a}{0}$ **is undefined for any value of** a.

A square can be used to illustrate the results when two fractions are multiplied. Figure 3.2 shows how to shade a square to find the product $\frac{2}{3} \cdot \frac{4}{5}$:

1. Separate the square into thirds in one direction and shade in 2 thirds $\left(\frac{2}{3}\right)$.

2. Separate it into fifths in the other direction and shade in 4 fifths $\left(\frac{4}{5}\right)$.

3. Note that there are 15 little rectangles and the common shaded portion of 8 rectangles represents the product 8 fifteenths $\left(\frac{2}{3} \cdot \frac{4}{5} = \frac{8}{15}\right)$.

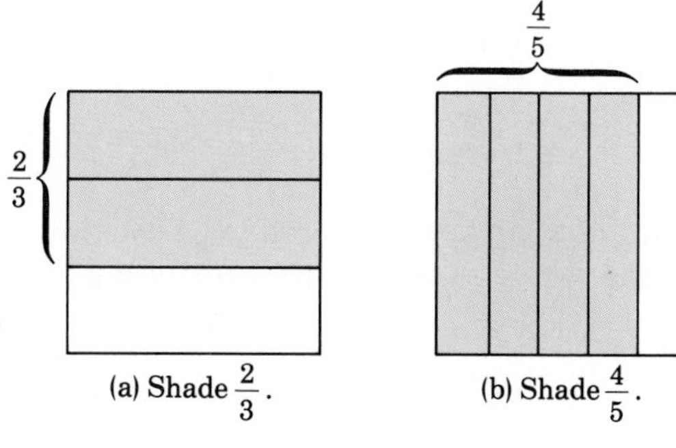

(a) Shade $\frac{2}{3}$. (b) Shade $\frac{4}{5}$.

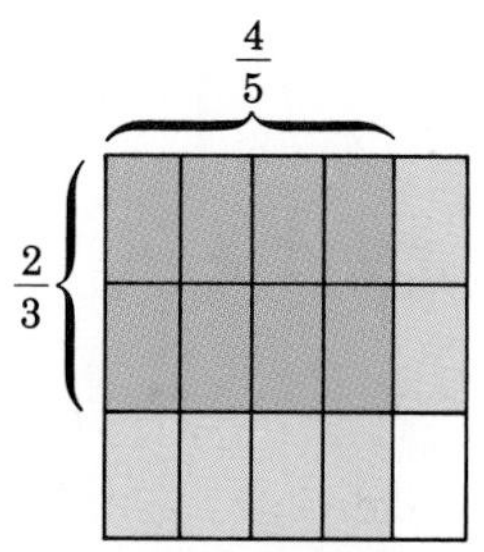

(c) The overlapped region is $\frac{2}{3} \cdot \frac{4}{5} = \frac{8}{15}$.

Figure 3.2

EXAMPLE

5. Using a square, draw a diagram that represents the product $\frac{1}{3} \cdot \frac{2}{5}$.

Solution

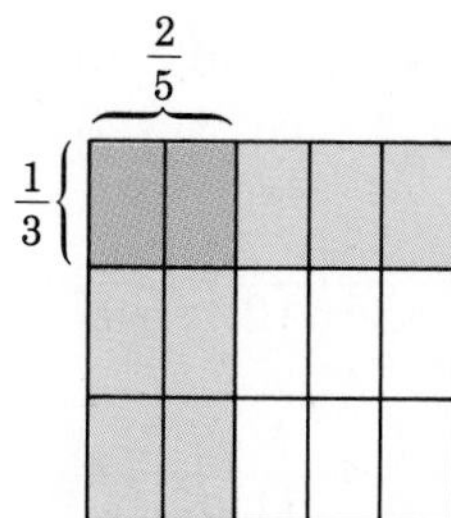

The diagrams are helpful in leading to the following rule for multiplying two fractions.

To multiply fractions:

1. Multiply the numerators.
2. Multiply the denominators.

$$\frac{a}{b}\cdot\frac{c}{d} = \frac{a\cdot c}{b\cdot d}$$

EXAMPLES

6. $\frac{3}{4}\cdot\frac{5}{7} = \frac{3\cdot 5}{4\cdot 7} = \frac{15}{28}$

7. $\frac{1}{5}\cdot\frac{3}{10} = \frac{1\cdot 3}{5\cdot 10} = \frac{3}{50}$

8. $\frac{7}{6}\cdot\frac{11}{2} = \frac{7\cdot 11}{6\cdot 2} = \frac{77}{12}$

Whole numbers can be considered to be fractions with denominator 1. We say that the whole numbers are **equivalent** to their corresponding fractions. Thus,

$0 = \frac{0}{1}$, $1 = \frac{1}{1}$, $2 = \frac{2}{1}$, $3 = \frac{3}{1}$, $4 = \frac{4}{1}$, $5 = \frac{5}{1}$, and so on.

So, in multiplying fractions and whole numbers, rewrite the whole number as a fraction. Also, the rule for multiplying fractions can be extended to multiplication with more than two fractions.

EXAMPLES

9. $\frac{3}{14}\cdot 5 = \frac{3}{14}\cdot\frac{5}{1} = \frac{3\cdot 5}{14\cdot 1} = \frac{15}{14}$

10. $\frac{7}{8}\cdot 0 = \frac{7}{8}\cdot\frac{0}{1} = \frac{7\cdot 0}{8\cdot 1} = \frac{0}{8} = 0$

11. $\frac{1}{4}\cdot\frac{3}{5}\cdot\frac{7}{2} = \frac{1\cdot 3\cdot 7}{4\cdot 5\cdot 2} = \frac{21}{40}$

12. $\frac{5}{11}\cdot\frac{7}{8}\cdot 3\cdot\frac{1}{4} = \frac{5}{11}\cdot\frac{7}{8}\cdot\frac{3}{1}\cdot\frac{1}{4} = \frac{5\cdot 7\cdot 3\cdot 1}{11\cdot 8\cdot 1\cdot 4} = \frac{105}{352}$

The commutative property and associative property for multiplication both apply to fractions. They are stated here for completeness, but there are no related exercises.

COMMUTATIVE PROPERTY OF MULTIPLICATION

If $\frac{a}{b}$ and $\frac{c}{d}$ are fractions, then

$$\frac{a}{b} \cdot \frac{c}{d} = \frac{c}{d} \cdot \frac{a}{b}$$

ASSOCIATIVE PROPERTY OF MULTIPLICATION

If $\frac{a}{b}$, $\frac{c}{d}$, and $\frac{e}{f}$ are fractions, then

$$\frac{a}{b} \cdot \frac{c}{d} \cdot \frac{e}{f} = \frac{a}{b}\left(\frac{c}{d} \cdot \frac{e}{f}\right) = \left(\frac{a}{b} \cdot \frac{c}{d}\right)\frac{e}{f}$$

PRACTICE QUIZ

Find the products.

1. $\frac{1}{4} \cdot \frac{1}{4}$
2. $\frac{3}{5} \cdot \frac{4}{7}$
3. $0 \cdot \frac{5}{6} \cdot \frac{7}{8}$
4. $\frac{1}{2} \cdot \frac{3}{7} \cdot \frac{9}{2}$

ANSWERS

1. $\frac{1}{16}$
2. $\frac{12}{35}$
3. 0
4. $\frac{27}{28}$

Exercises 3.1

Write a fraction that represents the shaded parts in each of the following diagrams.

1. 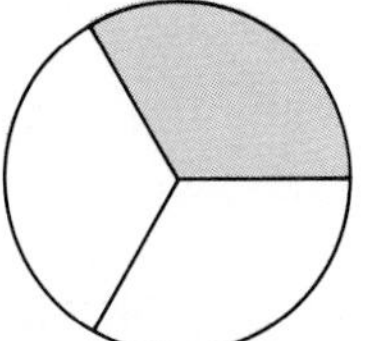

2. 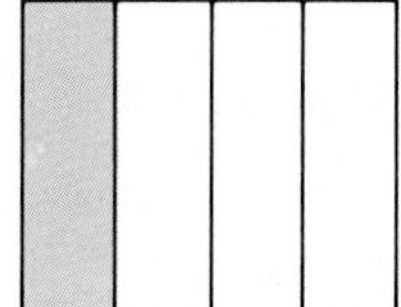

3. 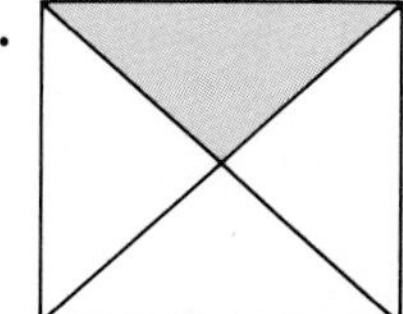

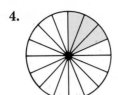

4. **5.** 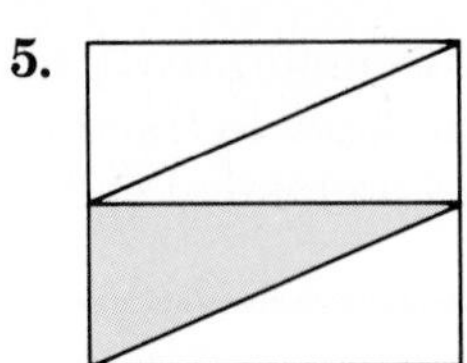 **6.** 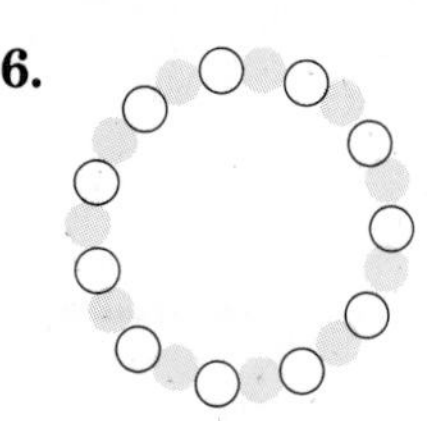

Draw a square similar to Figure 3.1 to illustrate each of the following products.

7. $\frac{5}{6}\cdot\frac{5}{6}$ **8.** $\frac{2}{3}\cdot\frac{1}{5}$ **9.** $\frac{2}{3}\cdot\frac{4}{7}$ **10.** $\frac{5}{6}\cdot\frac{5}{8}$

Find the following products.

11. $\frac{3}{5}\cdot\frac{2}{5}$ **12.** $\frac{3}{4}\cdot\frac{2}{5}$ **13.** $\frac{1}{3}\cdot\frac{1}{3}$ **14.** $\frac{1}{4}\cdot\frac{1}{4}$

15. $\frac{1}{2}\cdot\frac{1}{2}$ **16.** $\frac{1}{8}\cdot\frac{1}{8}$ **17.** $\frac{1}{2}\cdot\frac{3}{4}$ **18.** $\frac{5}{8}\cdot\frac{3}{4}$

19. $\frac{3}{4}\cdot\frac{3}{4}$ **20.** $\frac{5}{6}\cdot\frac{5}{6}$ **21.** $\frac{1}{9}\cdot\frac{4}{9}$ **22.** $\frac{4}{7}\cdot\frac{3}{5}$

23. $\frac{5}{16}\cdot\frac{1}{2}$ **24.** $\frac{7}{16}\cdot\frac{1}{4}$ **25.** $\frac{0}{5}\cdot\frac{7}{6}$ **26.** $\frac{0}{3}\cdot\frac{5}{7}$

27. $\frac{3}{5}\cdot\frac{3}{5}$ **28.** $\frac{4}{9}\cdot\frac{4}{9}$ **29.** $\frac{3}{10}\cdot\frac{3}{10}$ **30.** $\left(\frac{4}{1}\right)\left(\frac{3}{1}\right)$

31. $\left(\frac{2}{1}\right)\left(\frac{5}{1}\right)$ **32.** $\frac{9}{10}\left(\frac{1}{5}\right)$ **33.** $\frac{7}{10}\left(\frac{2}{5}\right)$ **34.** $\left(\frac{3}{5}\right)\frac{1}{10}$

35. $\left(\frac{2}{3}\right)\frac{2}{7}$ **36.** $\frac{5}{6}\cdot\frac{11}{3}$ **37.** $\frac{9}{4}\cdot\frac{11}{5}$ **38.** $\frac{1}{5}\cdot\frac{2}{7}\cdot\frac{3}{11}$

39. $\frac{4}{13}\cdot\frac{2}{5}\cdot\frac{6}{7}$ **40.** $\frac{7}{8}\cdot\frac{7}{9}\cdot\frac{7}{3}$ **41.** $\frac{1}{6}\cdot\frac{1}{10}\cdot\frac{1}{6}$

42. $\frac{9}{100}\cdot 1\cdot 3$ **43.** $\frac{5}{1}\cdot\frac{12}{1}\cdot\frac{14}{1}$ **44.** $\frac{1}{3}\cdot\frac{1}{10}\cdot\frac{7}{3}$

45. $\frac{3}{11}\cdot\frac{0}{8}\cdot\frac{6}{7}$ **46.** $\frac{0}{4}\cdot\frac{3}{8}\cdot\frac{1}{5}$ **47.** $\frac{5}{6}\cdot\frac{5}{11}\cdot\frac{5}{7}\cdot\frac{0}{3}$

48. $\left(\frac{3}{4}\right)\left(\frac{5}{8}\right)\left(\frac{11}{13}\right)$ **49.** $\left(\frac{7}{5}\right)\left(\frac{8}{3}\right)\left(\frac{13}{3}\right)$ **50.** $\left(\frac{1}{10}\right)\left(\frac{3}{5}\right)\left(\frac{3}{10}\right)$

51. Each of the following products has the same value. What is that value?

(a) $\frac{0}{4}\cdot\frac{5}{6}$ (b) $\frac{3}{10}\cdot\frac{0}{8}$ (c) $\frac{1}{10}\cdot\frac{0}{10}$ (d) $\frac{0}{6}\cdot\frac{0}{52}$

52. What is the value, if any, of each of the following numbers?

(a) $\frac{7}{0}$ (b) $\frac{16}{0}$ (c) $\frac{75}{0} \cdot \frac{2}{0}$ (d) $\frac{1}{0} \cdot \frac{0}{2}$

3.2 REDUCING FRACTIONS AND MULTIPLYING FRACTIONS

We know that 1 is the **multiplicative identity** for whole numbers; that is, $a \cdot 1 = a$. The number 1 is also the **multiplicative identity** for fractions since

$$\frac{a}{b} \cdot 1 = \frac{a}{b} \cdot \frac{1}{1} = \frac{a \cdot 1}{b \cdot 1} = \frac{a}{b}$$

MULTIPLICATIVE IDENTITY

For any fraction $\frac{a}{b}$, where $b \neq 0$,

$$\frac{a}{b} \cdot 1 = \frac{a}{b}$$

Now, the number 1 can be written as a fraction with the same counting number as numerator and denominator. In general,

$$1 = \frac{1}{1} = \frac{2}{2} = \frac{3}{3} = \frac{4}{4} = \frac{5}{5} = \cdots = \frac{\boldsymbol{k}}{\boldsymbol{k}} = \cdots$$

where $\boldsymbol{k}$ is any counting number.

Therefore, we have the following important property of fractions.

$$\frac{a}{b} = \frac{a}{b} \cdot 1 = \frac{a}{b} \cdot \frac{\boldsymbol{k}}{\boldsymbol{k}} = \frac{a \cdot \boldsymbol{k}}{b \cdot \boldsymbol{k}} \quad \text{where } \boldsymbol{k} \neq 0$$

This means that a fraction is unchanged if the numerator and denominator are multiplied by the same nonzero number.

If we find a fraction equal to (or equivalent to) a given fraction with a larger denominator, we say that the given fraction has been **raised to higher terms.** So, **to raise a fraction to higher terms,** multiply the numerator and denominator by the same counting number.

EXAMPLES

1. $\frac{3}{4} = \frac{?}{28}$

Solution

We want to raise fourths to twenty-eighths. Try multiplying by $\frac{5}{5}$.

$$\frac{3}{4} = \frac{3}{4} \cdot \frac{5}{5} = \frac{15}{20} \quad \text{WRONG FRACTION}$$

The fraction $\frac{15}{20}$ does equal $\frac{3}{4}$, but $\frac{15}{20}$ has the **wrong denominator.** We want an equal fraction with denominator 28. Since $4 \cdot 7 = 28$, we should use $\frac{7}{7}$ instead of $\frac{5}{5}$.

$$\frac{3}{4} = \frac{3}{4} \cdot \frac{7}{7} = \frac{21}{28} \quad \text{RIGHT FRACTION}$$

Just multiplying by 1 was not good enough. The right form of 1, namely $\frac{7}{7}$, was needed.

2. $\frac{9}{10} = \frac{?}{30}$

Solution

Now we are smarter. Since $10 \cdot 3 = 30$, multiply $\frac{9}{10}$ by $\frac{3}{3}$ to get an equal fraction with denominator 30.

$$\frac{9}{10} = \frac{9}{10} \cdot \frac{3}{3} = \frac{9 \cdot 3}{10 \cdot 3} = \frac{27}{30}$$

Remember: The numerator and denominator must be multiplied by the same number.

3. $\frac{11}{8} = \frac{?}{40}$

Solution

Use $\frac{k}{k} = \frac{5}{5}$ since $8 \cdot 5 = 40$.

$$\frac{11}{8} = \frac{11}{8} \cdot \frac{5}{5} = \frac{11 \cdot 5}{8 \cdot 5} = \frac{55}{40}$$

The improper fraction $\frac{55}{40}$ is a perfectly good answer. We will discuss mixed numbers in Chapter 4.

Changing a fraction to lower terms is the same as **reducing** the fraction. **A fraction is reduced if the numerator and denominator have no common factor other than 1.**

To reduce a fraction to lowest terms:

1. Factor the numerator and denominator into prime factors.
2. Use the fact that $\frac{k}{k} = 1$ and "cancel" or "divide out" common factors.

EXAMPLES

Reduce each fraction to lowest terms.

4. $\frac{21}{35}$

$$\frac{21}{35} = \frac{3 \cdot 7}{5 \cdot 7} = \frac{3}{5} \cdot \frac{7}{7} = \frac{3}{5} \cdot 1 = \frac{3}{5}$$

or $$\frac{21}{35} = \frac{3 \cdot \cancel{7}}{5 \cdot \cancel{7}} = \frac{3}{5}$$ Here we "cancel" or "divide out" the 7's with the understanding that $\frac{7}{7} = 1$.

5. $\frac{72}{90}$

$$\frac{72}{90} = \frac{6 \cdot 12}{9 \cdot 10} = \frac{2 \cdot 3 \cdot 2 \cdot 2 \cdot 3}{3 \cdot 3 \cdot 2 \cdot 5} = \frac{2}{2} \cdot \frac{3}{3} \cdot \frac{3}{3} \cdot \frac{2 \cdot 2}{5} = 1 \cdot 1 \cdot 1 \cdot \frac{4}{5} = \frac{4}{5}$$

or $$\frac{72}{90} = \frac{\cancel{2} \cdot \cancel{3} \cdot 2 \cdot 2 \cdot \cancel{3}}{\cancel{3} \cdot \cancel{3} \cdot \cancel{2} \cdot 5} = \frac{4}{5}$$

6. $\frac{36}{27}$

$$\frac{36}{27} = \frac{2 \cdot 2 \cdot \cancel{3} \cdot \cancel{3}}{3 \cdot \cancel{3} \cdot \cancel{3}} = \frac{4}{3}$$

It is **not necessary** for you to use prime factors. If you "see" any common factor, you can reduce using that factor. However, using prime factors allows you to see all the factors and gives you confidence that the fraction is completely reduced.

$$\frac{36}{27} = \frac{4 \cdot \cancel{9}}{3 \cdot \cancel{9}} = \frac{4}{3}$$

7. $\frac{15}{75}$

$$\frac{15}{75} = \frac{\cancel{3}\cdot\cancel{5}\cdot 1}{\cancel{3}\cdot\cancel{5}\cdot 5} = \frac{1}{5}$$ If all the factors in the numerator (or denominator) cancel, then 1 **must** be used as a factor.

or $$\frac{15}{75} = \frac{15\cdot 1}{15\cdot 5} = \frac{1}{5}$$ The numerator is a factor of the denominator.

Now we can multiply fractions and reduce all in one step by using prime factors.

EXAMPLES

Find each product and reduce.

8. $\frac{18}{35}\cdot\frac{7}{9} = \frac{18\cdot 7}{35\cdot 9} = \frac{2\cdot\cancel{3}\cdot\cancel{3}\cdot\cancel{7}}{5\cdot\cancel{7}\cdot\cancel{3}\cdot\cancel{3}} = \frac{2}{5}$

9. $\frac{9}{10}\cdot\frac{25}{24}\cdot\frac{44}{55} = \frac{9\cdot 25\cdot 44}{10\cdot 24\cdot 55} = \frac{3\cdot\cancel{3}\cdot\cancel{5}\cdot\cancel{5}\cdot\cancel{2}\cdot\cancel{2}\cdot\cancel{11}}{2\cdot\cancel{5}\cdot 2\cdot\cancel{2}\cdot\cancel{2}\cdot\cancel{3}\cdot\cancel{5}\cdot\cancel{11}} = \frac{3}{2\cdot 2} = \frac{3}{4}$

10. $\frac{17}{100}\cdot\frac{25}{34}\cdot 8 = \frac{17\cdot 25\cdot 8}{100\cdot 34\cdot 1} = \frac{\cancel{17}\cdot\cancel{25}\cdot\cancel{4}\cdot\cancel{2}\cdot 1}{\cancel{25}\cdot\cancel{4}\cdot\cancel{2}\cdot\cancel{17}\cdot 1} = \frac{1}{1} = 1$

It is **not necessary** for you to use prime factors if you notice other common factors. However, if you have any difficulty understanding how to multiply and reduce, use prime factors. Then you will be sure that your answer is reduced.

Another technique frequently used is to divide the numerators and denominators by common factors whether they are prime or not. If these factors are easily seen, then this technique is probably faster. But you must be careful and organized. Examples 8, 9, and 10 are shown again using the division technique.

EXAMPLES

8′. $\frac{\overset{2}{\cancel{18}}}{\underset{5}{\cancel{35}}}\cdot\frac{\overset{1}{\cancel{7}}}{\underset{1}{\cancel{9}}} = \frac{2\cdot 1}{5\cdot 1} = \frac{2}{5}$

9 is divided into both 9 and 18.
7 is divided into both 7 and 35.

9′. $\frac{\overset{3}{\cancel{9}}}{\underset{2}{\cancel{10}}}\cdot\frac{\overset{\overset{1}{\cancel{5}}}{\cancel{25}}}{\underset{\underset{2}{\cancel{8}}}{\cancel{24}}}\cdot\frac{\overset{\overset{1}{\cancel{4}}}{\cancel{44}}}{\underset{\underset{1}{\cancel{5}}}{\cancel{55}}} = \frac{3}{4}$

11 is divided into both 44 and 55.
5 is divided into both 5 and 25.
5 is divided into both 5 and 10.
3 is divided into both 9 and 24.
4 is divided into both 4 and 8.

10′. $\frac{\overset{1}{\cancel{17}}}{\underset{\underset{1}{\cancel{4}}}{\cancel{100}}} \cdot \frac{\overset{1}{\cancel{25}}}{\underset{\underset{1}{\cancel{2}}}{\cancel{34}}} \cdot \frac{\overset{\overset{1}{\cancel{2}}}{\cancel{8}}}{1} = \frac{1}{1} = 1$

17 is divided into both 17 and 34.
25 is divided into both 25 and 100.
4 is divided into both 4 and 8.
2 is divided into both 2 and 2.

Use the technique that suits your own abilities and understanding.

PRACTICE QUIZ

Raise to higher terms as indicated.

1. $\frac{3}{5} = \frac{\quad}{100}$

2. $\frac{7}{2} = \frac{\quad}{10}$

Reduce to lowest terms.

3. $\frac{25}{55}$

4. $\frac{34}{51}$

Multiply.

[HINT: Write 6 as $\frac{6}{1}$ and factor before multiplying.]

5. $\frac{17}{100} \cdot \frac{27}{34} \cdot \frac{25}{9} \cdot 6$

ANSWERS

1. $\frac{60}{100}$

2. $\frac{35}{10}$

3. $\frac{5}{11}$

4. $\frac{2}{3}$

5. $\frac{9}{4}$

Exercises 3.2

Raise each fraction to higher terms as indicated.

1. $\frac{3}{4} = \frac{?}{12}$
2. $\frac{3}{5} = \frac{?}{10}$
3. $\frac{2}{3} = \frac{?}{12}$
4. $\frac{1}{2} = \frac{?}{6}$
5. $\frac{6}{7} = \frac{?}{14}$
6. $\frac{5}{8} = \frac{?}{16}$
7. $\frac{5}{9} = \frac{?}{27}$
8. $\frac{3}{4} = \frac{?}{20}$
9. $\frac{1}{2} = \frac{?}{20}$
10. $\frac{1}{4} = \frac{?}{40}$
11. $\frac{3}{8} = \frac{?}{40}$
12. $\frac{1}{6} = \frac{?}{30}$
13. $\frac{2}{3} = \frac{?}{15}$
14. $\frac{1}{6} = \frac{?}{12}$
15. $\frac{3}{5} = \frac{?}{45}$
16. $\frac{2}{3} = \frac{?}{9}$
17. $\frac{10}{11} = \frac{?}{44}$
18. $\frac{5}{11} = \frac{?}{33}$
19. $\frac{5}{21} = \frac{?}{42}$
20. $\frac{3}{7} = \frac{?}{42}$

21. $\frac{5}{7} = \frac{?}{42}$ 22. $\frac{3}{8} = \frac{?}{24}$ 23. $\frac{3}{16} = \frac{?}{80}$ 24. $\frac{1}{16} = \frac{?}{80}$

25. $\frac{7}{2} = \frac{?}{20}$ 26. $\frac{14}{3} = \frac{?}{9}$ 27. $\frac{5}{2} = \frac{?}{20}$ 28. $\frac{12}{5} = \frac{?}{15}$

29. $\frac{3}{1} = \frac{?}{5}$ 30. $\frac{4}{1} = \frac{?}{8}$ 31. $\frac{8}{10} = \frac{?}{100}$ 32. $\frac{9}{10} = \frac{?}{100}$

33. $\frac{6}{10} = \frac{?}{100}$ 34. $\frac{7}{10} = \frac{?}{100}$ 35. $\frac{11}{10} = \frac{?}{60}$ 36. $\frac{3}{100} = \frac{?}{1000}$

37. $\frac{9}{100} = \frac{?}{1000}$ 38. $\frac{9}{10} = \frac{?}{1000}$ 39. $\frac{3}{10} = \frac{?}{1000}$ 40. $\frac{5}{10} = \frac{?}{1000}$

Reduce the following fractions to lowest terms. Just rewrite the fraction if it is already reduced.

41. $\frac{3}{9}$ 42. $\frac{16}{24}$ 43. $\frac{9}{12}$ 44. $\frac{6}{20}$ 45. $\frac{16}{40}$

46. $\frac{24}{30}$ 47. $\frac{14}{36}$ 48. $\frac{5}{11}$ 49. $\frac{0}{25}$ 50. $\frac{75}{100}$

51. $\frac{22}{55}$ 52. $\frac{60}{75}$ 53. $\frac{30}{36}$ 54. $\frac{7}{28}$ 55. $\frac{26}{39}$

56. $\frac{27}{56}$ 57. $\frac{34}{51}$ 58. $\frac{36}{48}$ 59. $\frac{24}{100}$ 60. $\frac{16}{32}$

61. $\frac{30}{45}$ 62. $\frac{28}{42}$ 63. $\frac{12}{35}$ 64. $\frac{66}{84}$ 65. $\frac{14}{63}$

66. $\frac{30}{70}$ 67. $\frac{25}{76}$ 68. $\frac{70}{84}$ 69. $\frac{50}{100}$ 70. $\frac{48}{12}$

71. $\frac{54}{9}$ 72. $\frac{51}{6}$ 73. $\frac{6}{51}$ 74. $\frac{27}{72}$ 75. $\frac{18}{40}$

76. $\frac{144}{156}$ 77. $\frac{150}{135}$ 78. $\frac{121}{165}$ 79. $\frac{140}{112}$ 80. $\frac{96}{108}$

Multiply and reduce each product. [HINT: Factor first so you can reduce as you are multiplying.]

81. $\frac{23}{36} \cdot \frac{20}{46}$ 82. $\frac{7}{8} \cdot \frac{4}{21}$ 83. $\frac{5}{15} \cdot \frac{18}{24}$

84. $\frac{20}{32} \cdot \frac{9}{13} \cdot \frac{26}{7}$ 85. $\frac{69}{15} \cdot \frac{30}{8} \cdot \frac{14}{46}$ 86. $\frac{42}{52} \cdot \frac{27}{22} \cdot \frac{33}{9}$

87. $\frac{3}{4} \cdot 18 \cdot \frac{7}{2} \cdot \frac{22}{54}$ 88. $\frac{9}{10} \cdot \frac{35}{40} \cdot \frac{65}{15}$ 89. $\frac{66}{84} \cdot \frac{12}{5} \cdot \frac{28}{33}$

90. $\frac{24}{100}\cdot\frac{36}{48}\cdot\frac{15}{9}$ **91.** $\frac{17}{10}\cdot\frac{5}{42}\cdot\frac{18}{51}\cdot 4$ **92.** $\frac{75}{8}\cdot\frac{16}{36}\cdot 9\cdot\frac{7}{25}$

93. $5\cdot\frac{3}{10}\cdot\frac{8}{9}\cdot\frac{1}{2}$ **94.** $\frac{2}{5}\cdot\frac{3}{4}\cdot\frac{15}{20}\cdot\frac{1}{18}$ **95.** $6\cdot\frac{2}{3}\cdot\frac{5}{8}\cdot\frac{9}{10}\cdot\frac{7}{12}$

3.3 DIVIDING FRACTIONS

If the product of two fractions is 1, then the fractions are called **reciprocals** of each other. For example,

$\frac{2}{3}$ and $\frac{3}{2}$ are reciprocals because $\frac{2}{3}\cdot\frac{3}{2} = \frac{6}{6} = 1$

$\frac{5}{6}$ and $\frac{6}{5}$ are reciprocals because $\frac{5}{6}\cdot\frac{6}{5} = \frac{30}{30} = 1$

DEFINITION The **reciprocal** of a fraction $\frac{a}{b}$, where $a \neq 0$ and $b \neq 0$, is $\frac{b}{a}$ because

$$\frac{a}{b}\cdot\frac{b}{a} = 1$$

Note: If $a = 0$, then $\frac{0}{b}$ has no reciprocal since $\frac{b}{0}$ is undefined.

To understand how to divide with fractions, remember that division is related to multiplication. With whole numbers we know that

$$18 \div 3 = 6 \quad \text{because} \quad 18 = 6\cdot 3$$

With fractions, the same reasoning gives

$$\frac{2}{3} \div \frac{7}{8} = \square \quad \text{because} \quad \frac{2}{3} = \square\cdot\frac{7}{8}$$

We need to find out what goes in the box. Since division can be indicated with a fraction, we can write

$$\frac{2}{3} \div \frac{7}{8} = \frac{\frac{2}{3}}{\frac{7}{8}}$$

Now, multiplying the numerator (a fraction) and the denominator (another fraction) by the same number will not change the value of the fraction. Multiply by $\frac{8}{7}$, the reciprocal of $\frac{7}{8}$, and see what happens.

$$\frac{2}{3} \div \frac{7}{8} = \frac{\frac{2}{3}}{\frac{7}{8}} = \frac{\frac{2}{3}\cdot\frac{8}{7}}{\frac{7}{8}\cdot\frac{8}{7}} = \frac{\frac{2}{3}\cdot\frac{8}{7}}{1} = \frac{2}{3}\cdot\frac{8}{7}$$

$\left(\text{Note that here } \frac{k}{k} = \frac{\frac{8}{7}}{\frac{8}{7}}.\right)$

A division problem has been changed into a multiplication problem. Thus,

$$\frac{2}{3} \div \frac{7}{8} = \frac{2}{3}\cdot\frac{8}{7} = \frac{16}{21}$$

In general,

$$\frac{a}{b} \div \frac{c}{d} = \frac{a}{b}\cdot\frac{d}{c} \qquad \text{where } b, c, d \neq 0$$

In words: **To divide by any number (except 0), multiply by its reciprocal.**

EXAMPLES

Find each of the following quotients. Reduce to lowest terms whenever possible.

1. $\frac{3}{4} \div \frac{8}{5} = \frac{3}{4}\cdot\frac{5}{8} = \frac{15}{32}$ — The divisor is $\frac{8}{5}$, and we multiply by its reciprocal, $\frac{5}{8}$.

2. $\frac{2}{3} \div \frac{1}{2} = \frac{2}{3}\cdot\frac{2}{1} = \frac{4}{3}$ — The reciprocal of $\frac{1}{2}$ is $\frac{2}{1}$.

3. $\frac{16}{27} \div \frac{8}{9} = \frac{16}{27}\cdot\frac{9}{8} = \frac{\cancel{8}\cdot 2\cdot\cancel{9}}{3\cdot\cancel{9}\cdot\cancel{8}} = \frac{2}{3}$ — The reciprocal of $\frac{8}{9}$ is $\frac{9}{8}$. By factoring, the quotient is reduced to $\frac{2}{3}$.

4. $\frac{5}{16} \div 5 = \frac{5}{16} \div \frac{5}{1} = \frac{\overset{1}{\cancel{5}}}{16}\cdot\frac{1}{\underset{1}{\cancel{5}}} = \frac{1}{16}$ — The reciprocal of 5 is $\frac{1}{5}$.

5. $\frac{9}{10} \div \frac{9}{10} = \frac{\overset{1}{\cancel{9}}}{\underset{1}{\cancel{10}}} \cdot \frac{\overset{1}{\cancel{10}}}{\underset{1}{\cancel{9}}} = \frac{1}{1} = 1$

Here we get the expected result that a number divided by itself is 1.

6. $\frac{7}{5} \div \frac{5}{7} = \frac{7}{5} \cdot \frac{7}{5} = \frac{49}{25}$

We are not dividing a number by itself, and the quotient is not 1.

If you know the product of two numbers and you know one of the numbers, how do you find the other number? The answer is to divide. For example, with whole numbers, suppose the product of two numbers is 24 and one number is 6. To find the other number, divide 24 by 6:

$$24 \div 6 = 4 \quad \text{or} \quad \begin{array}{r} 4 \\ 6\overline{)24} \\ \underline{24} \\ 0 \end{array}$$

The other number is 4.

The following examples show that the same technique applies with fractions.

EXAMPLES

7. The result of multiplying two numbers is $\frac{7}{8}$. If one of the numbers is $\frac{3}{4}$, what is the other number?

Solution

Divide $\frac{7}{8}$ by $\frac{3}{4}$. This is similar to dividing 24 by 6 to find the other factor.

$$\frac{7}{8} \div \frac{3}{4} = \frac{7}{8} \cdot \frac{4}{3} = \frac{7 \cdot \overset{1}{\cancel{4}}}{\underset{2}{\cancel{8}} \cdot 3} = \frac{7}{6}$$

The other number is $\frac{7}{6}$.

Check by multiplying

$$\frac{7}{6} \cdot \frac{3}{4} = \frac{7 \cdot \overset{1}{\cancel{3}}}{\underset{2}{\cancel{6}} \cdot 4} = \frac{7}{8}$$

8. If the product of $\frac{3}{2}$ with another number is $\frac{5}{18}$, what is the other number?

Solution

We know the product of two numbers. So, divide the product by the given number to find the other number.

$$\frac{5}{18} \div \frac{3}{2} = \frac{5}{18} \cdot \frac{2}{3} = \frac{5 \cdot \not{2}}{\not{2} \cdot 9 \cdot 3} = \frac{5}{27}$$

$\frac{5}{27}$ is the other number.

Check by multiplying

$$\frac{3}{2} \cdot \frac{5}{27} = \frac{\not{3} \cdot 5}{2 \cdot \not{3} \cdot 9} = \frac{5}{18}$$

(In rethinking the problem, remember that $\frac{5}{18}$ is the product of two numbers.)

Exercises 3.3

Find each of the following quotients. Reduce whenever possible.

1. $\frac{3}{4} \div \frac{2}{3}$ **2.** $\frac{5}{8} \div \frac{3}{5}$ **3.** $\frac{1}{3} \div \frac{1}{5}$ **4.** $\frac{2}{11} \div \frac{1}{5}$

5. $\frac{2}{7} \div \frac{1}{2}$ **6.** $\frac{7}{16} \div \frac{3}{4}$ **7.** $\frac{9}{16} \div \frac{3}{4}$ **8.** $\frac{5}{12} \div \frac{5}{3}$

9. $\frac{3}{5} \div \frac{27}{25}$ **10.** $\frac{7}{8} \div \frac{2}{3}$ **11.** $\frac{7}{8} \div \frac{3}{2}$ **12.** $\frac{4}{5} \div \frac{2}{3}$

13. $\frac{5}{7} \div \frac{1}{7}$ **14.** $\frac{3}{8} \div \frac{1}{8}$ **15.** $\frac{7}{10} \div \frac{1}{10}$ **16.** $\frac{5}{6} \div \frac{1}{6}$

17. $\frac{2}{15} \div \frac{2}{15}$ **18.** $\frac{7}{11} \div \frac{7}{11}$ **19.** $\frac{3}{4} \div \frac{4}{3}$ **20.** $\frac{9}{10} \div \frac{10}{9}$

21. $\frac{9}{20} \div 3$ **22.** $\frac{14}{20} \div 7$ **23.** $\frac{18}{20} \div 9$ **24.** $\frac{25}{40} \div 10$

25. $\frac{4}{7} \div 2$ **26.** $\frac{6}{11} \div 2$ **27.** $\frac{8}{16} \div 2$ **28.** $\frac{15}{20} \div 3$

29. $\frac{16}{35} \div \frac{1}{7}$ **30.** $\frac{15}{27} \div \frac{1}{9}$ **31.** $\frac{5}{24} \div \frac{3}{8}$ **32.** $\frac{7}{48} \div \frac{14}{16}$

33. $\frac{3}{48} \div \frac{5}{16}$ **34.** $\frac{14}{20} \div \frac{7}{4}$ **35.** $\frac{16}{20} \div \frac{18}{10}$ **36.** $\frac{20}{21} \div \frac{10}{7}$

37. $\frac{13}{4} \div \frac{6}{5}$ **38.** $\frac{12}{13} \div \frac{3}{2}$ **39.** $\frac{16}{35} \div \frac{3}{11}$ **40.** $\frac{19}{24} \div \frac{8}{5}$

41. $\frac{92}{7} \div \frac{46}{11}$ **42.** $14 \div \frac{1}{7}$ **43.** $25 \div \frac{1}{5}$ **44.** $30 \div \frac{1}{10}$

45. $50 \div \frac{1}{2}$

46. The product of $\frac{9}{10}$ with another number is $\frac{5}{3}$. What is the other number?

47. The result of multiplying two numbers is $\frac{31}{3}$. If one of the numbers is $\frac{43}{6}$, what is the other one?

48. The result of multiplying two numbers is 150. If one of the numbers is $\frac{5}{7}$, what is the other number?

49. The product of $\frac{4}{5}$ with another number is 180. What is the other number?

50. The product of $\frac{7}{8}$ and $\frac{5}{10}$ is divided by $\frac{21}{20}$. Find the quotient.

3.4 ADDING FRACTIONS

Figure 3.3 illustrates how the **sum** of the two fractions $\frac{3}{7}$ and $\frac{2}{7}$ might be diagrammed.

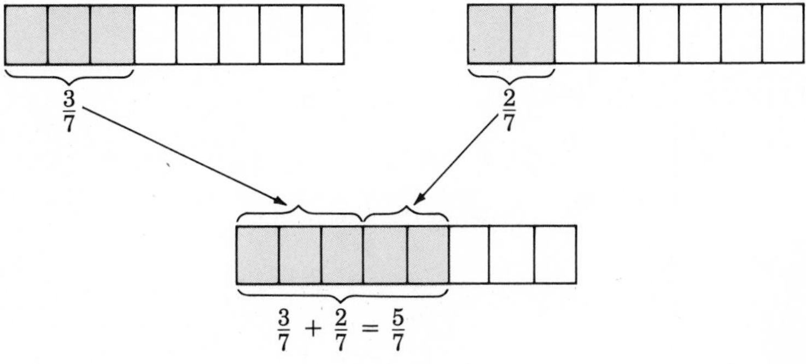

Figure 3.3

The common denominator "names" each fraction. The sum has this common name. Just as 3 apples plus 2 apples gives a total of 5 apples, we have 3 sevenths plus 2 sevenths giving a total of 5 sevenths.

To add two (or more) fractions with the same denominator:

1. Add the numerators.
2. Keep the common denominator.

$$\frac{a}{b} + \frac{c}{b} = \frac{a + c}{b}$$

EXAMPLES

1. $\frac{1}{5} + \frac{3}{5} = \frac{1 + 3}{5} = \frac{4}{5}$

2. $\frac{2}{7} + \frac{3}{7} + \frac{1}{7} = \frac{2 + 3 + 1}{7} = \frac{6}{7}$

You may be able to reduce after adding.

3. $\frac{4}{15} + \frac{6}{15} = \frac{4 + 6}{15} = \frac{10}{15} = \frac{2 \cdot \cancel{5}}{3 \cdot \cancel{5}} = \frac{2}{3}$

4. $\frac{1}{8} + \frac{2}{8} + \frac{7}{8} = \frac{1 + 2 + 7}{8} = \frac{10}{8} = \frac{\cancel{2} \cdot 5}{\cancel{2} \cdot 4} = \frac{5}{4}$

Of course, fractions to be added will not always have the same denominator. In these cases, the smallest common denominator must be found. **The smallest common denominator (SCD) is the least common multiple (LCM) of the denominators.**

EXAMPLES

5. Find the SCD for $\frac{3}{8}$ and $\frac{11}{12}$.

Solution

Using prime factorization:

$$\begin{aligned} 8 &= 2 \cdot 2 \cdot 2 \\ 12 &= 2 \cdot 2 \cdot 3 \\ \text{SCD} &= 2 \cdot 2 \cdot 2 \cdot 3 = 24 \end{aligned}$$

6. Find the SCD for $\frac{5}{21}$ and $\frac{5}{28}$.

Solution

Using prime factorization:

$$\begin{aligned} 21 &= 3 \cdot 7 \\ 28 &= 2 \cdot 2 \cdot 7 \\ \text{SCD} &= 2 \cdot 2 \cdot 3 \cdot 7 = 84 \end{aligned}$$

To add fractions with different denominators:

1. Find the SCD of the denominators.
2. Change each fraction to an equal fraction with that denominator.
3. Add the new fractions.
4. Reduce if possible.

EXAMPLES

7. Find the sum $\frac{1}{2} + \frac{3}{5}$.

Solution

(a) Find the SCD.

$$\text{SCD} = 2 \cdot 5 = 10$$

(b) Find equal fractions with denominator 10.

$$\frac{1}{2} = \frac{1}{2} \cdot \frac{5}{5} = \frac{5}{10}$$

$$\frac{3}{5} = \frac{3}{5} \cdot \frac{2}{2} = \frac{6}{10}$$

(c) Add.

$$\frac{1}{2} + \frac{3}{5} = \frac{5}{10} + \frac{6}{10} = \frac{5+6}{10} = \frac{11}{10}$$

8. Find the sum $\frac{3}{8} + \frac{11}{12}$.

Solution

(a) Find the SCD.

$$\left.\begin{array}{l} 8 = 2 \cdot 2 \cdot 2 \\ 12 = 2 \cdot 2 \cdot 3 \end{array}\right\} \text{SCD} = 2 \cdot 2 \cdot 2 \cdot 3 = 24$$

(b) Find equal fractions with denominator 24.

$$\frac{3}{8} = \frac{3}{8} \cdot \frac{3}{3} = \frac{9}{24}$$ Multiply by $\frac{3}{3}$ since $8 \cdot 3 = 24$.

$$\frac{11}{12} = \frac{11}{12} \cdot \frac{2}{2} = \frac{22}{24}$$ Multiply by $\frac{2}{2}$ since $12 \cdot 2 = 24$.

(c) Add.

$$\frac{5}{8} + \frac{5}{12} = \frac{9}{24} + \frac{22}{24} = \frac{9 + 22}{24} = \frac{31}{24}$$

9. Add $\frac{5}{21} + \frac{5}{28}$ and reduce if possible.

Solution

(a) Find the SCD.

$$\left.\begin{array}{l} 21 = 3 \cdot 7 \\ 28 = 2 \cdot 2 \cdot 7 \end{array}\right\} \text{SCD} = 2 \cdot 2 \cdot 3 \cdot 7 = 84$$

(b) Find equal fractions with denominator 84.

$$\frac{5}{21} = \frac{5}{21} \cdot \frac{4}{4} = \frac{20}{84}$$

$$\frac{5}{28} = \frac{5}{28} \cdot \frac{3}{3} = \frac{15}{84}$$

(c) $\frac{5}{21} + \frac{5}{28} = \frac{20}{84} + \frac{15}{84} = \frac{20 + 15}{84} = \frac{35}{84}$

(d) Now reduce.

$$\frac{35}{84} = \frac{\not{7} \cdot 5}{2 \cdot 2 \cdot 3 \cdot \not{7}} = \frac{5}{12}$$

10. Find the sum $\frac{2}{3} + \frac{1}{6} + \frac{5}{12}$.

Solution

(a) SCD = 12 You can simply observe this or use prime factorizations.

(b) Steps (b), (c), and (d) can be written together in one process.

$$\begin{aligned}\frac{2}{3} + \frac{1}{6} + \frac{5}{12} &= \frac{2}{3}\cdot\frac{4}{4} + \frac{1}{6}\cdot\frac{2}{2} + \frac{5}{12}\\ &= \frac{8}{12} + \frac{2}{12} + \frac{5}{12}\\ &= \frac{15}{12} = \frac{\not{3}\cdot 5}{2\cdot 2\cdot \not{3}} = \frac{5}{4}\end{aligned}$$

Or, the numbers can be written vertically. The process is the same.

$$\begin{aligned}\frac{2}{3} &= \frac{2}{3}\cdot\frac{4}{4} = \frac{8}{12}\\ \frac{1}{6} &= \frac{1}{6}\cdot\frac{2}{2} = \frac{2}{12}\\ \frac{5}{12} &= \frac{5}{12} = \frac{5}{12}\\ \hline & \qquad\quad \frac{15}{12} = \frac{\not{3}\cdot 5}{2\cdot 2\cdot \not{3}} = \frac{5}{4}\end{aligned}$$

11. Add $5 + \frac{7}{10} + \frac{3}{1000}$.

Solution

(a) SCD = 1000 All the denominators are powers of 10, and 1000 is the largest. We can write 5 as $\frac{5}{1}$.

(b) $$\begin{aligned}5 + \frac{7}{10} + \frac{3}{1000} &= \frac{5}{1}\cdot\frac{1000}{1000} + \frac{7}{10}\cdot\frac{100}{100} + \frac{3}{1000}\\ &= \frac{5000}{1000} + \frac{700}{1000} + \frac{3}{1000}\\ &= \frac{5703}{1000}\end{aligned}$$

The following example illustrates a very **COMMON ERROR** that **must** be avoided.

EXAMPLE

12. Find the sum $\frac{3}{2} + \frac{1}{6}$.

Solution

$$\frac{\overset{1}{\cancel{3}}}{2} + \frac{1}{\underset{2}{\cancel{6}}} = \frac{1}{2} + \frac{1}{2} = 1 \quad \text{WRONG}$$

You cannot "cancel" across the + sign.

$$\frac{3}{2} + \frac{1}{6} = \frac{3}{2}\cdot\frac{3}{3} + \frac{1}{6} = \frac{9}{6} + \frac{1}{6} = \frac{10}{6} \quad \text{RIGHT}$$

Now reduce:

$$\frac{10}{6} = \frac{5\cdot\cancel{2}}{3\cdot\cancel{2}} = \frac{5}{3}$$ ("Cancelling" is fine here because 2 is a **factor** in both the numerator and the denominator.)

PRACTICE QUIZ

Find the following sums. Reduce all answers.

1. $\frac{1}{8} + \frac{3}{8} + \frac{2}{8}$
2. $\frac{2}{3} + \frac{5}{8} + \frac{1}{6}$
3. $\frac{7}{10} + \frac{1}{100} + \frac{5}{1000}$

ANSWERS

1. $\frac{3}{4}$
2. $\frac{35}{24}$
3. $\frac{715}{1000} = \frac{143}{200}$

Exercises 3.4

Find the smallest common denominator (SCD) for the fractions in each exercise.

1. $\frac{3}{8}, \frac{5}{16}$ **2.** $\frac{2}{39}, \frac{1}{3}, \frac{4}{13}$ **3.** $\frac{2}{27}, \frac{5}{18}, \frac{1}{6}$

4. $\frac{5}{8}, \frac{1}{12}, \frac{5}{9}$ **5.** $\frac{3}{10}, \frac{1}{100}, \frac{1}{1000}$

Add the following fractions and reduce all answers.

6. $\frac{6}{10} + \frac{4}{10}$ **7.** $\frac{3}{14} + \frac{2}{14}$ **8.** $\frac{1}{20} + \frac{3}{20}$ **9.** $\frac{3}{4} + \frac{3}{4}$

10. $\frac{5}{6} + \frac{4}{6}$ **11.** $\frac{7}{5} + \frac{3}{5}$ **12.** $\frac{11}{15} + \frac{7}{15}$ **13.** $\frac{7}{9} + \frac{8}{9}$

14. $\frac{3}{25} + \frac{12}{25}$ **15.** $\frac{7}{90} + \frac{37}{90} + \frac{21}{90}$ **16.** $\frac{11}{75} + \frac{12}{75} + \frac{62}{75}$

17. $\frac{14}{32} + \frac{7}{32} + \frac{1}{32}$ **18.** $\frac{4}{100} + \frac{35}{100} + \frac{76}{100}$ **19.** $\frac{21}{95} + \frac{33}{95} + \frac{3}{95}$

20. $\frac{1}{200} + \frac{17}{200} + \frac{25}{200}$ **21.** $\frac{1}{12} + \frac{2}{3} + \frac{1}{4}$ **22.** $\frac{3}{8} + \frac{5}{16}$

23. $\frac{2}{5} + \frac{3}{10} + \frac{3}{20}$ **24.** $\frac{3}{4} + \frac{1}{16} + \frac{6}{32}$ **25.** $\frac{2}{7} + \frac{4}{21} + \frac{1}{3}$

26. $\frac{1}{6} + \frac{1}{4} + \frac{1}{3}$ **27.** $\frac{2}{39} + \frac{1}{3} + \frac{4}{13}$ **28.** $\frac{1}{2} + \frac{3}{10} + \frac{4}{5}$

29. $\frac{1}{27} + \frac{4}{18} + \frac{1}{6}$ **30.** $\frac{2}{7} + \frac{3}{20} + \frac{9}{14}$ **31.** $\frac{1}{8} + \frac{1}{12} + \frac{1}{9}$

32. $\frac{2}{5} + \frac{4}{7} + \frac{3}{8}$ **33.** $\frac{2}{3} + \frac{3}{4} + \frac{5}{6}$ **34.** $\frac{1}{5} + \frac{7}{30} + \frac{1}{6}$

35. $\frac{1}{5} + \frac{2}{15} + \frac{1}{6}$ **36.** $\frac{1}{5} + \frac{1}{10} + \frac{1}{4}$ **37.** $\frac{1}{5} + \frac{7}{40} + \frac{1}{4}$

38. $\frac{1}{3} + \frac{5}{12} + \frac{1}{15}$ **39.** $\frac{1}{4} + \frac{1}{20} + \frac{8}{15}$ **40.** $\frac{7}{10} + \frac{3}{25} + \frac{3}{4}$

41. $\frac{5}{8} + \frac{4}{27} + \frac{1}{48}$ **42.** $\frac{3}{16} + \frac{5}{48} + \frac{1}{32}$

43. $\frac{72}{105} + \frac{2}{45} + \frac{15}{21}$ **44.** $\frac{1}{63} + \frac{2}{27} + \frac{1}{45}$

45. $\frac{0}{27} + \frac{0}{16} + \frac{1}{5}$ **46.** $\frac{5}{6} + \frac{0}{100} + \frac{0}{70} + \frac{1}{3}$

47. $\frac{3}{10} + \frac{1}{100} + \frac{7}{1000}$ **48.** $\frac{11}{100} + \frac{15}{10} + \frac{1}{10}$

49. $\frac{17}{1000} + \frac{1}{100} + \frac{1}{10{,}000}$ **50.** $6 + \frac{1}{100} + \frac{3}{10}$

51. $8 + \frac{1}{10} + \frac{9}{100} + \frac{1}{1000}$ **52.** $\frac{1}{10} + \frac{3}{10} + \frac{9}{1000}$

53. $\frac{7}{10} + \frac{5}{100} + \frac{3}{1000}$ **54.** $\frac{1}{2} + \frac{3}{4} + \frac{1}{100}$

55. $\frac{1}{4} + \frac{1}{8} + \frac{7}{100}$ **56.** $\frac{9}{1000} + \frac{7}{1000} + \frac{21}{10{,}000}$

57. $\frac{11}{100} + \frac{1}{2} + \frac{3}{1000}$ **58.** $\frac{3}{4} + \frac{17}{1000} + \frac{13}{10{,}000} + 2$

59. $5 + \frac{1}{10} + \frac{3}{100} + \frac{4}{1000}$

60. $\frac{13}{10{,}000} + \frac{1}{100{,}000} + \frac{21}{1{,}000{,}000}$

61. $\frac{3}{4} + \frac{1}{2} + \frac{5}{12}$

62. $\frac{1}{5} + \frac{2}{15} + \frac{5}{6}$

63. $\frac{7}{8} + \frac{2}{3} + \frac{1}{9}$

64. $\frac{1}{27} + \frac{1}{18} + \frac{4}{9}$

65. $\frac{3}{20} + \frac{1}{100} + \frac{3}{100}$

66. $\frac{13}{100} + \frac{4}{10} + \frac{1}{1000}$

67. $\frac{7}{12} + \frac{1}{9} + \frac{2}{3}$

68. $\frac{1}{3} + \frac{8}{15} + \frac{7}{10}$

69. $\frac{9}{16} + \frac{5}{48} + \frac{3}{32}$

70. $\frac{3}{10} + \frac{1}{20} + \frac{1}{25}$

3.5 SUBTRACTING FRACTIONS

Figure 3.4 shows how the **difference** of the two fractions $\frac{7}{8}$ and $\frac{4}{8}$ might be diagrammed.

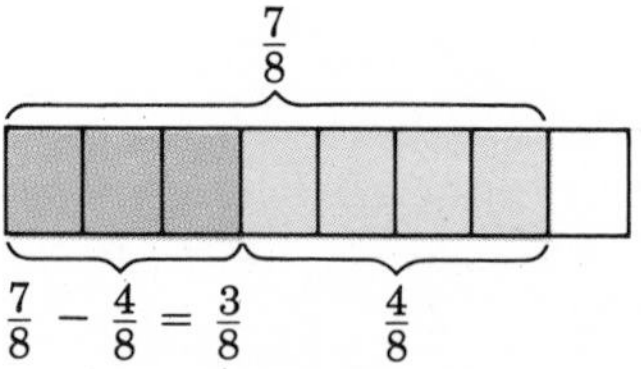

Figure 3.4

From Figure 3.4 we see that $\frac{7}{8} - \frac{4}{8} = \frac{3}{8}$. Just as with addition, the common denominator "names" each fraction. The difference is found by subtracting the numerators and using the common denominator.

To subtract fractions that have the same denominator:

1. Subtract the numerators.
2. Keep the common denominator.

$$\frac{a}{b} - \frac{c}{b} = \frac{a-c}{b}$$

EXAMPLES

1. $\frac{5}{6} - \frac{1}{6} = \frac{5-1}{6} = \frac{4}{6} = \frac{\cancel{2}\cdot 2}{\cancel{2}\cdot 3} = \frac{2}{3}$

2. $\frac{9}{10} - \frac{7}{10} = \frac{9-7}{10} = \frac{2}{10} = \frac{\cancel{2}\cdot 1}{\cancel{2}\cdot 5} = \frac{1}{5}$

3. $\frac{19}{10} - \frac{11}{10} = \frac{19-11}{10} = \frac{8}{10} = \frac{\cancel{2}\cdot 4}{\cancel{2}\cdot 5} = \frac{4}{5}$

4. $\frac{7}{8} - \frac{3}{8} = \frac{7-3}{8} = \frac{4}{8} = \frac{\cancel{4}\cdot 1}{\cancel{4}\cdot 2} = \frac{1}{2}$

To subtract fractions with different denominators:

1. Find the SCD of the denominators.
2. Change each fraction to an equal fraction with that denominator.
3. Subtract the new fractions.
4. Reduce if possible.

EXAMPLES

5. $\frac{9}{10} - \frac{2}{15}$

(a) Find the SCD.

$$\left.\begin{array}{l} 10 = 2\cdot 5 \\ 15 = 3\cdot 5 \end{array}\right\} \text{SCD} = 2\cdot 3\cdot 5 = 30$$

(b) Find equal fractions with denominator 30.

$$\frac{9}{10} = \frac{9}{10}\cdot\frac{3}{3} = \frac{27}{30}$$

$$\frac{2}{15} = \frac{2}{15}\cdot\frac{2}{2} = \frac{4}{30}$$

(c) Subtract.

$$\frac{9}{10} - \frac{2}{15} = \frac{27}{30} - \frac{4}{30} = \frac{27-4}{30} = \frac{23}{30}$$

6. $\dfrac{12}{55} - \dfrac{2}{33}$

Solution

(a) Find the SCD.

$$\left.\begin{aligned}55 &= 5\cdot 11\\ 33 &= 3\cdot 11\end{aligned}\right\} \text{SCD} = 3\cdot 5\cdot 11 = 165$$

(b) Find equal fractions with denominator 165.

$$\frac{12}{55} = \frac{12}{55}\cdot\frac{3}{3} = \frac{36}{165}$$

$$\frac{2}{33} = \frac{2}{33}\cdot\frac{5}{5} = \frac{10}{165}$$

(c) Subtract.

$$\frac{12}{55} - \frac{2}{33} = \frac{36}{165} - \frac{10}{165} = \frac{36-10}{165} = \frac{26}{165}$$

(d) Reduce if possible.

$$\frac{26}{165} = \frac{2\cdot 13}{3\cdot 5\cdot 11} = \frac{26}{165}$$

Cannot be reduced because there are no common factors in the numerator and denominator.

7. $\dfrac{7}{12} - \dfrac{3}{20}$

Solution

(a) $$\left.\begin{aligned}12 &= 2\cdot 2\cdot 3\\ 20 &= 2\cdot 2\cdot 5\end{aligned}\right\} \text{SCD} = 2\cdot 2\cdot 3\cdot 5 = 60$$

(b) Steps (b), (c), and (d) of Example 6 can be written as one process.

$$\frac{7}{12} - \frac{3}{20} = \frac{7}{12}\cdot\frac{5}{5} - \frac{3}{20}\cdot\frac{3}{3} = \frac{35}{60} - \frac{9}{60}$$

$$= \frac{35-9}{60} = \frac{26}{60} = \frac{\not{2}\cdot 13}{\not{2}\cdot 30} = \frac{13}{30}$$

Or, writing the fractions vertically,

$$\begin{aligned}\frac{7}{12} &= \frac{7}{12}\cdot\frac{5}{5} = \frac{35}{60}\\ -\frac{3}{20} &= \frac{3}{20}\cdot\frac{3}{3} = \frac{9}{60}\\ & \qquad\qquad\;\; \frac{26}{60} = \frac{\not{2}\cdot 13}{\not{2}\cdot 30} = \frac{13}{30}\end{aligned}$$

8. $1 - \frac{5}{8}$

Solution

(a) SCD = 8 Since 1 can be written $\frac{1}{1}$, the common denominator is $8 \cdot 1 = 8$.

(b) $1 - \frac{5}{8} = \frac{1}{1} \cdot \frac{8}{8} - \frac{5}{8} = \frac{8}{8} - \frac{5}{8} = \frac{8-5}{8} = \frac{3}{8}$

PRACTICE QUIZ	Find the following differences. Reduce all answers.	ANSWERS
	1. $\frac{5}{9} - \frac{1}{9}$	1. $\frac{4}{9}$
	2. $\frac{7}{6} - \frac{2}{3}$	2. $\frac{1}{2}$
	3. $\frac{7}{10} - \frac{7}{15}$	3. $\frac{7}{30}$
	4. $1 - \frac{3}{4}$	4. $\frac{1}{4}$

Exercises 3.5

Subtract and reduce if possible.

1. $\frac{4}{7} - \frac{1}{7}$ **2.** $\frac{5}{7} - \frac{3}{7}$ **3.** $\frac{9}{10} - \frac{3}{10}$ **4.** $\frac{11}{10} - \frac{7}{10}$

5. $\frac{5}{8} - \frac{1}{8}$ **6.** $\frac{7}{8} - \frac{5}{8}$ **7.** $\frac{11}{12} - \frac{7}{12}$ **8.** $\frac{7}{12} - \frac{3}{12}$

9. $\frac{13}{15} - \frac{4}{15}$ **10.** $\frac{21}{15} - \frac{11}{15}$ **11.** $\frac{5}{6} - \frac{1}{3}$ **12.** $\frac{5}{6} - \frac{1}{2}$

13. $\frac{11}{15} - \frac{3}{10}$ **14.** $\frac{8}{10} - \frac{3}{15}$ **15.** $\frac{3}{4} - \frac{2}{3}$ **16.** $\frac{2}{3} - \frac{1}{4}$

17. $\frac{15}{16} - \frac{21}{32}$ **18.** $\frac{3}{8} - \frac{1}{16}$ **19.** $\frac{5}{4} - \frac{3}{5}$ **20.** $\frac{5}{12} - \frac{1}{6}$

21. $\frac{14}{27} - \frac{7}{18}$ **22.** $\frac{25}{18} - \frac{21}{27}$ **23.** $\frac{8}{45} - \frac{11}{72}$ **24.** $\frac{46}{55} - \frac{10}{33}$

25. $\frac{5}{36} - \frac{1}{45}$ **26.** $\frac{5}{1} - \frac{3}{4}$ **27.** $\frac{4}{1} - \frac{5}{8}$ **28.** $2 - \frac{9}{16}$

29. $1 - \frac{13}{16}$ **30.** $6 - \frac{2}{3}$ **31.** $\frac{9}{10} - \frac{3}{100}$ **32.** $\frac{159}{1000} - \frac{1}{10}$

33. $\frac{76}{100} - \frac{7}{10}$ 34. $\frac{999}{1000} - \frac{99}{100}$ 35. $\frac{54}{100} - \frac{5}{10}$ 36. $\frac{7}{24} - \frac{10}{36}$

37. $\frac{31}{40} - \frac{5}{8}$ 38. $\frac{14}{35} - \frac{12}{30}$ 39. $\frac{20}{35} - \frac{24}{42}$ 40. $\frac{3}{10} - \frac{298}{1000}$

41. $1 - \frac{9}{10}$ 42. $1 - \frac{7}{8}$ 43. $1 - \frac{2}{3}$ 44. $1 - \frac{1}{16}$

45. $1 - \frac{3}{20}$ 46. $1 - \frac{4}{9}$

47. $\begin{array}{r} \frac{7}{8} \\ -\frac{2}{3} \\ \hline \end{array}$ 48. $\begin{array}{r} \frac{14}{15} \\ -\frac{3}{10} \\ \hline \end{array}$ 49. $\begin{array}{r} \frac{1}{10} \\ -\frac{8}{100} \\ \hline \end{array}$ 50. $\begin{array}{r} \frac{3}{100} \\ -\frac{1}{1000} \\ \hline \end{array}$

51. Find the sum of $\frac{1}{4}$ and $\frac{3}{16}$. Subtract $\frac{1}{8}$ from the sum. What is the difference?

52. Find the difference between $\frac{2}{3}$ and $\frac{5}{9}$. Add $\frac{1}{12}$ to the difference. What is the sum?

53. Find the sum of $\frac{3}{4}$ and $\frac{5}{8}$. Multiply the sum by $\frac{2}{3}$. What is the product?

54. Find the product of $\frac{9}{10}$ and $\frac{2}{3}$. Divide the product by $\frac{5}{3}$. What is the quotient?

55. Find the quotient of $\frac{3}{4}$ and $\frac{15}{16}$. Add $\frac{3}{10}$ to the quotient. What is the sum?

3.6 COMPARISONS AND ORDER OF OPERATIONS

Many times we want to compare two (or more) fractions to see which is smaller or larger. Then we can subtract the smaller from the larger, or possibly make some financial decision based on the relative sizes of the fractions. Related word problems will be discussed in detail in later chapters.

To compare two fractions (to find which is larger or smaller):

1. Find the SCD of the denominators.
2. Change each fraction to an equal fraction with that denominator.
3. Compare the numerators.

EXAMPLES

1. Which is larger, $\frac{5}{6}$ or $\frac{7}{8}$? How much larger?

Solution

(a) Find the SCD for 6 and 8.

$$\left.\begin{aligned} 6 &= 2\cdot 3 \\ 8 &= 2\cdot 2\cdot 2 \end{aligned}\right\} \text{SCD} = 2\cdot 2\cdot 2\cdot 3 = 24$$

(b) Find equal fractions with denominator 24.

$$\frac{5}{6} = \frac{5}{6}\cdot\frac{4}{4} = \frac{20}{24} \quad \text{and} \quad \frac{7}{8} = \frac{7}{8}\cdot\frac{3}{3} = \frac{21}{24}$$

(c) $\frac{7}{8}$ is larger than $\frac{5}{6}$ since 21 is larger than 20.

(d) $\frac{7}{8} - \frac{5}{6} = \frac{21}{24} - \frac{20}{24} = \frac{1}{24}$

$\frac{7}{8}$ is larger by $\frac{1}{24}$.

2. Which is larger, $\frac{8}{9}$ or $\frac{11}{12}$? How much larger?

Solution

(a) SCD $= 2\cdot 2\cdot 3\cdot 3 = 36$

(b) $\frac{8}{9} = \frac{8}{9}\cdot\frac{4}{4} = \frac{32}{36}$ and $\frac{11}{12} = \frac{11}{12}\cdot\frac{3}{3} = \frac{33}{36}$

(c) $\frac{11}{12}$ is larger than $\frac{8}{9}$ since 33 is larger than 32.

(d) $\frac{11}{12} - \frac{8}{9} = \frac{33}{36} - \frac{32}{36} = \frac{1}{36}$

$\frac{11}{12}$ is larger by $\frac{1}{36}$.

3. Arrange $\frac{2}{3}, \frac{7}{10}$, and $\frac{9}{15}$ in order, smallest to largest.

Solution

(a) SCD = 30

(b) $\frac{2}{3} = \frac{2}{3} \cdot \frac{10}{10} = \frac{20}{30}$; $\frac{7}{10} = \frac{7}{10} \cdot \frac{3}{3} = \frac{21}{30}$; $\frac{9}{15} = \frac{9}{15} \cdot \frac{2}{2} = \frac{18}{30}$

(c) In order, smallest to largest, $\frac{9}{15}, \frac{2}{3}, \frac{7}{10}$.

An expression with fractions may involve more than one arithmetic operation. To simplify such expressions, we can use the rules for order of operations just as they were discussed in Chapter 2 for whole numbers. Of course, all the rules for fractions must be followed, too. That is, to add or subtract, you need a common denominator; to divide, you multiply by the reciprocal of the divisor.

RULES FOR ORDER OF OPERATIONS

1. First, simplify within grouping symbols, such as parentheses ( ), brackets [ ], or braces { }. Start with the innermost grouping.
2. Second, find any powers indicated by exponents.
3. Third, moving from **left to right,** perform any multiplications or divisions in the order they appear.
4. Fourth, moving from **left to right,** perform any additions or subtractions in the order they appear.

EXAMPLES

4. Evaluate the expression $\frac{1}{2} \div \frac{3}{4} + \frac{5}{6} \cdot \frac{1}{5}$.

Solution

(a) Divide first. $\frac{1}{2} \div \frac{3}{4} + \frac{5}{6} \cdot \frac{1}{5}$

$= \frac{1}{\not{2}} \cdot \frac{\not{4}^{2}}{3} + \frac{5}{6} \cdot \frac{1}{5}$

(b) Now multiply. $= \frac{2}{3} + \frac{\not{5}}{6} \cdot \frac{1}{\not{5}}$

(c) Now add. (SCD = 6)

$$= \frac{2}{3} + \frac{1}{6}$$

$$= \frac{2}{3} \cdot \frac{2}{2} + \frac{1}{6}$$

$$= \frac{4}{6} + \frac{1}{6}$$

$$= \frac{5}{6}$$

5. Evaluate the expression $\left(\frac{3}{4} - \frac{5}{8}\right) \div \left(\frac{15}{16} - \frac{1}{2}\right)$.

Solution

(a) Work inside the parentheses.

$$\left(\frac{3}{4} - \frac{5}{8}\right) \div \left(\frac{15}{16} - \frac{1}{2}\right)$$

$$= \left(\frac{6}{8} - \frac{5}{8}\right) \div \left(\frac{15}{16} - \frac{8}{16}\right)$$

(b) Now divide.

$$= \left(\frac{1}{8}\right) \div \left(\frac{7}{16}\right)$$

$$= \frac{1}{\cancel{8}} \cdot \frac{\overset{2}{\cancel{16}}}{17}$$

$$= \frac{2}{17}$$

6. Evaluate the expression $\frac{1}{2} \cdot \frac{5}{6} + \frac{7}{15} \div 2$.

Solution

$$\frac{1}{2} \cdot \frac{5}{6} + \frac{7}{15} \div 2 = \frac{5}{12} + \frac{7}{15} \div 2$$

$$= \frac{5}{12} + \frac{7}{15} \cdot \frac{1}{2}$$

$$= \frac{5}{12} + \frac{7}{30}$$

$$(\text{SCD} = 60) = \frac{5}{12} \cdot \frac{5}{5} + \frac{7}{30} \cdot \frac{2}{2}$$

$$= \frac{25}{60} + \frac{14}{60}$$

$$= \frac{39}{60} = \frac{\cancel{3} \cdot 13}{\cancel{3} \cdot 20} = \frac{13}{20}$$

7. Evaluate the expression $\frac{9}{10} - \left(\frac{1}{4}\right)^2 + \frac{1}{2}$.

Solution

(a) Use the exponent first.

$$\frac{9}{10} - \left(\frac{1}{4}\right)^2 + \frac{1}{2} = \frac{9}{10} - \frac{1}{16} + \frac{1}{2}$$

(b) Now add and subtract. (SCD = 80)

$$= \frac{9}{10}\cdot\frac{8}{8} - \frac{1}{16}\cdot\frac{5}{5} + \frac{1}{2}\cdot\frac{40}{40}$$

$$= \frac{72}{80} - \frac{5}{80} + \frac{40}{80}$$

$$= \frac{107}{80}$$

Exercises 3.6

Find the larger number of each pair and tell how much larger it is.

1. $\frac{2}{3}, \frac{3}{4}$ **2.** $\frac{5}{6}, \frac{7}{8}$ **3.** $\frac{4}{5}, \frac{17}{20}$ **4.** $\frac{4}{10}, \frac{3}{8}$

5. $\frac{13}{20}, \frac{5}{8}$ 6. $\frac{13}{16}, \frac{21}{25}$ 7. $\frac{14}{35}, \frac{12}{30}$ 8. $\frac{10}{36}, \frac{7}{24}$

9. $\frac{17}{80}, \frac{11}{48}$ **10.** $\frac{37}{100}, \frac{24}{75}$

In each exercise find equivalent fractions with a common denominator. Then write the original numbers in order, smallest to largest.

11. $\frac{2}{3}, \frac{3}{5}, \frac{7}{10}$ **12.** $\frac{8}{9}, \frac{9}{10}, \frac{11}{12}$

13. $\frac{7}{6}, \frac{11}{12}, \frac{19}{20}$ **14.** $\frac{1}{3}, \frac{5}{42}, \frac{3}{7}$

15. $\frac{1}{2}, \frac{1}{3}, \frac{1}{4}$ **16.** $\frac{2}{3}, \frac{3}{4}, \frac{5}{8}$

17. $\frac{7}{9}, \frac{31}{36}, \frac{13}{18}$ **18.** $\frac{17}{12}, \frac{40}{36}, \frac{31}{24}$

19. $\frac{1}{100}, \frac{3}{1000}, \frac{20}{10{,}000}$ **20.** $\frac{32}{100}, \frac{298}{1000}, \frac{3}{10}$

Evaluate each expression using the rules of order of operations.

21. $\frac{1}{2} \div \frac{7}{8} + \frac{1}{7} \cdot \frac{2}{3}$

22. $\frac{3}{5} \cdot \frac{1}{6} + \frac{1}{5} \div 2$

23. $\frac{1}{2} \div \frac{1}{2} + \frac{2}{3} \cdot \frac{2}{3}$

24. $5 - \frac{3}{4} \div 3$

25. $6 - \frac{5}{8} \div 4$

26. $\frac{2}{15} \cdot \frac{1}{4} \div \frac{3}{5} + \frac{1}{25}$

27. $\frac{5}{8} \cdot \frac{1}{10} \div \frac{3}{4} + \frac{1}{6}$

28. $\left(\frac{7}{15} + \frac{8}{21}\right) \div \frac{3}{35}$

29. $\left(\frac{1}{2} - \frac{1}{3}\right) \div \left(\frac{5}{8} + \frac{3}{16}\right)$

30. $\left(\frac{1}{3} + \frac{1}{5}\right) \cdot \left(\frac{3}{4} - \frac{1}{6}\right)$

31. $\left(\frac{1}{2}\right)^2 - \left(\frac{1}{4}\right)^3$

32. $\frac{2}{3} + \frac{3}{4} + \left(\frac{1}{2}\right)^2$

33. $\left(\frac{1}{3}\right)^2 + \left(\frac{1}{6}\right)^2 + \frac{2}{3}$

34. $\frac{1}{2} \div \frac{2}{3} + \left(\frac{1}{3}\right)^2$

35. $\left(\frac{3}{5}\right)^2 \div \left(\frac{1}{3} - \frac{1}{5}\right)$

SUMMARY: CHAPTER 3

DEFINITION

A **rational number** is a number that can be written in the form $\frac{a}{b}$, where a is a whole number and b is a nonzero whole number.

$\frac{a}{b}$ numerator / denominator

Note: The numerator a can be 0, but the denominator b cannot be 0.

Fractions can be used to indicate:

1. Equal parts of a whole.
2. Division.

Improper fractions are fractions in which the numerator is larger than the denominator.

DIVISION BY 0 IS UNDEFINED. NO DENOMINATOR CAN BE 0.

Consider $\frac{5}{0} = \square$. Then $5 = 0 \cdot \square = 0$. But this is impossible; $5 \neq 0$.

Next consider $\frac{0}{0} = \square$. Then $0 = 0 \cdot \square = 0$ for any value of $\square$. This means that $\frac{0}{0}$ could be any number. But in arithmetic an operation such as division cannot give more than one answer.

Thus, $\frac{a}{0}$ **is undefined for any value of** a.

To multiply fractions:

1. Multiply the numerators.
2. Multiply the denominators.

$$\frac{a}{b} \cdot \frac{c}{d} = \frac{a \cdot c}{b \cdot d}$$

COMMUTATIVE PROPERTY OF MULTIPLICATION

If $\frac{a}{b}$ and $\frac{c}{d}$ are fractions, then

$$\frac{a}{b} \cdot \frac{c}{d} = \frac{c}{d} \cdot \frac{a}{b}$$

ASSOCIATIVE PROPERTY OF MULTIPLICATION

If $\frac{a}{b}$, $\frac{c}{d}$, and $\frac{e}{f}$ are fractions, then

$$\frac{a}{b} \cdot \frac{c}{d} \cdot \frac{e}{f} = \frac{a}{b}\left(\frac{c}{d} \cdot \frac{e}{f}\right) = \left(\frac{a}{b} \cdot \frac{c}{d}\right)\frac{e}{f}$$

MULTIPLICATIVE IDENTITY

For any fraction $\frac{a}{b}$, where $b \neq 0$,

$$\frac{a}{b} \cdot 1 = \frac{a}{b}$$

$$\frac{a}{b} = \frac{a}{b} \cdot 1 = \frac{a}{b} \cdot \frac{k}{k} = \frac{a \cdot k}{b \cdot k} \quad \text{where } k \neq 0$$

To reduce a fraction to lowest terms:

1. Factor the numerator and denominator into prime factors.
2. Use the fact that $\frac{k}{k} = 1$ and "cancel" or "divide out" common factors.

DEFINITION

The **reciprocal** of a fraction $\frac{a}{b}$, where $a \neq 0$ and $b \neq 0$, is $\frac{b}{a}$ because

$$\frac{a}{b} \cdot \frac{b}{a} = 1$$

In general,

$$\frac{a}{b} \div \frac{c}{d} = \frac{a}{b} \cdot \frac{d}{c} \quad \text{where } b, c, d \neq 0$$

In words: **To divide by any number (except 0), multiply by its reciprocal.**

To add two (or more) fractions with the same denominator:

1. Add the numerators.
2. Keep the common denominator.

$$\frac{a}{b} + \frac{c}{b} = \frac{a + c}{b}$$

To add fractions with different denominators:

1. Find the SCD of the denominators.
2. Change each fraction to an equal fraction with that denominator.
3. Add the new fractions.
4. Reduce if possible.

To subtract fractions that have the same denominator:

1. Subtract the numerators.
2. Keep the common denominator.

$$\frac{a}{b} - \frac{c}{b} = \frac{a - c}{b}$$

To subtract fractions with different denominators:

1. Find the SCD of the denominators.
2. Change each fraction to an equal fraction with that denominator.
3. Subtract the new fractions.
4. Reduce if possible.

To compare two fractions (to find which is larger or smaller):

1. Find the SCD of the denominators.
2. Change each fraction to an equal fraction with that denominator.
3. Compare the numerators.

RULES FOR ORDER OF OPERATIONS

1. First, simplify within grouping symbols, such as parentheses ( ), brackets [ ], or braces { }. Start with the innermost grouping.
2. Second, find any powers indicated by exponents.
3. Third, moving from **left to right,** perform any multiplications or divisions in the order they appear.
4. Fourth, moving from **left to right,** perform any additions or subtractions in the order they appear.

REVIEW QUESTIONS: CHAPTER 3

1. The denominator of a rational number cannot be ______.

2. $\frac{0}{7} = 0$, but $\frac{7}{0}$ is ______.

3. The reciprocal of $\frac{2}{3}$ is _____, and the reciprocal of $\frac{3}{2}$ is _____.

4. Which property of addition is illustrated by the following statement?

$$\frac{1}{3} + \left(\frac{5}{6} + \frac{1}{2}\right) = \left(\frac{1}{3} + \frac{5}{6}\right) + \frac{1}{2}$$

5. Draw a diagram illustrating the product $\frac{2}{3} \cdot \frac{2}{5}$.

Multiply and reduce all answers.

6. $\frac{1}{3} \cdot \frac{1}{2} \cdot \frac{1}{5}$

7. $\frac{1}{7} \cdot \frac{3}{7}$

8. $\frac{35}{56} \cdot \frac{4}{15} \cdot \frac{5}{10}$

Fill in the missing terms so that each equation is true.

9. $\frac{1}{6} = \frac{?}{12}$

10. $\frac{9}{10} = \frac{?}{60}$

11. $\frac{15}{13} = \frac{?}{65}$

Reduce each fraction to lowest terms.

12. $\frac{15}{30}$

13. $\frac{99}{88}$

14. $\frac{0}{4}$

15. $\frac{150}{120}$

Add or subtract as indicated and reduce all answers.

16. $\frac{3}{7} + \frac{2}{7}$

17. $\frac{5}{6} - \frac{1}{6}$

18. $\frac{5}{8} - \frac{3}{8}$

19. $\frac{1}{12} + \frac{5}{36} + \frac{11}{24}$

20. $\frac{13}{22} - \frac{9}{33}$

21. $\frac{5}{27} + \frac{5}{18}$

22. $1 - \frac{13}{20}$

23. $\frac{3}{4} - \frac{5}{12}$

24. $\begin{array}{r} \frac{2}{3} \\ \frac{1}{8} \\ +\frac{1}{12} \\ \hline \end{array}$

Divide and reduce all answers.

25. $\frac{2}{3} \div 6$

26. $1 \div \frac{3}{5}$

27. $\frac{7}{12} \div \frac{7}{12}$

28. $\frac{15}{16} \div \frac{3}{4}$

29. $\frac{3}{4} \div \frac{15}{16}$

30. Which is larger, $\frac{2}{3}$ or $\frac{4}{5}$? How much larger?

31. Arrange the following fractions in order, smallest to largest:

$$\frac{7}{12}, \frac{5}{9}, \frac{11}{20}$$

Evaluate each expression using the rules of order of operations.

32. $\frac{5}{8} \cdot \frac{3}{10} + \frac{1}{14} \div 2$

33. $\left(\frac{3}{5} - \frac{1}{3}\right) \div \left(\frac{1}{6} + \frac{7}{8}\right)$

34. $\frac{7}{15} + \frac{5}{9} \div \frac{2}{3} - \frac{2}{3}$

35. $\left(\frac{2}{3}\right)^2 - \left(\frac{1}{3}\right)^2 + \frac{1}{18}$

TEST: CHAPTER 3

1. The numbers $\frac{3}{4}$ and $\frac{4}{3}$ are called ______ of each other.

2. A fraction equivalent to $\frac{5}{14}$ with a denominator of 56 is ______.

Reduce to lowest terms.

3. $\frac{45}{75}$

4. $\frac{216}{264}$

For Exercises 5–19, all answers should be reduced to lowest terms.

5. Find the sum of $\frac{7}{10}$ and $\frac{4}{15}$.

6. Find the product of $\frac{5}{9}$ and $\frac{2}{3}$.

7. Find the difference between $\frac{5}{12}$ and $\frac{2}{9}$.

8. Find the quotient when $\frac{6}{7}$ is divided by 3.

Perform the indicated operations.

9. $\frac{4}{9} \cdot \frac{3}{16}$

10. $1 - \frac{11}{15}$

11. $10 \div \frac{2}{5}$

12. $\frac{4}{21} + \frac{5}{21}$

13. $\frac{3}{10} + \frac{1}{8} + \frac{5}{6}$

14. $\frac{4}{5} \div \frac{3}{8}$

15. $\frac{21}{26} \cdot \frac{13}{15} \cdot \frac{5}{7}$

16. $\frac{37}{11} - \frac{15}{11}$

17. $\frac{3}{16} \div \frac{5}{12}$

18. $\frac{3}{10} \div \frac{1}{2} + \frac{7}{8} \cdot \frac{2}{5}$

19. $\frac{9}{10} + \frac{4}{5} \div \frac{2}{7} - \frac{2}{7}$

20. Arrange $\frac{3}{5}$, $\frac{7}{12}$, and $\frac{5}{8}$ in order, smallest to largest.

4 Mixed Numbers

4.1 CHANGING MIXED NUMBERS

A **mixed number** is the sum of a whole number and a fraction. We usually write the whole number and the fraction side by side without the plus sign. For example,

$$3 + \frac{1}{8} = 3\frac{1}{8}$$ Read "three and one-eighth."

$$2 + \frac{3}{4} = 2\frac{3}{4}$$ Read "two and three-fourths."

> To change a mixed number to an improper fraction, add the whole number and the fraction. Remember, the whole number can be written with 1 as the denominator.

EXAMPLES

Change each mixed number to an improper fraction.

1. $3\frac{1}{8} = 3 + \frac{1}{8} = \frac{3}{1}\cdot\frac{8}{8} + \frac{1}{8} = \frac{24}{8} + \frac{1}{8} = \frac{25}{8}$

   The SCD = 8 since the denominator of $\frac{3}{1}$ is 1.

2. $2\frac{3}{4} = 2 + \frac{3}{4} = \frac{2}{1} \cdot \frac{4}{4} + \frac{3}{4} = \frac{8}{4} + \frac{3}{4} = \frac{11}{4}$

The SCD $= 4$ since the denominator of $\frac{2}{1}$ is 1.

3. $10\frac{1}{2} = 10 + \frac{1}{2} = \frac{10}{1} \cdot \frac{2}{2} + \frac{1}{2} = \frac{20}{2} + \frac{1}{2} = \frac{21}{2}$

There is a pattern to changing mixed numbers to improper fractions that leads to a familiar shortcut. Since the denominator of the whole number is always 1, the denominator of the fraction part is always the common denominator. Thus, as you can see in Examples 1, 2, and 3, the whole number is multiplied by the denominator, and the numerator of the fraction is added to this product.

SHORTCUT TO CHANGING MIXED NUMBERS TO FRACTION FORM

1. Multiply the denominator of the fraction part by the whole number.
2. Add the numerator of the fraction part to this product.
3. Write this sum over the denominator of the fraction.

EXAMPLES

Change each mixed number to an improper fraction using the shortcut method.

4. $3\frac{2}{5}$

Solution

(a) Multiply $5 \cdot 3 = 15$ and add 2: $15 + 2 = 17$.

(b) Write 17 over 5:

$$3\frac{2}{5} = \frac{17}{5}$$

5. $9\frac{1}{2}$

Solution

(a) Multiply $2 \cdot 9 = 18$ and add 1: $18 + 1 = 19$.

(b) Write 19 over 2:

$$9\frac{1}{2} = \frac{19}{2}$$

6. $7\frac{3}{10}$

Solution

(a) Multiply $10 \cdot 7 = 70$ and add 3: $70 + 3 = 73$.

(b) Write 73 over 10:

$$7\frac{3}{10} = \frac{73}{10}$$

To reverse the process, change an improper fraction to a mixed number, we use the fact that a fraction can indicate division.

> **To change an improper fraction to a mixed number:**
>
> 1. Divide the numerator by the denominator.
> 2. Write the remainder over the denominator as the fraction part of the mixed number.

EXAMPLES

7. Change $\frac{29}{4}$ to a mixed number.

Solution

$$\begin{array}{r} 7 \\ 4\overline{)29} \\ \underline{28} \\ 1 \end{array} \qquad \frac{29}{4} = 7 + \frac{1}{4} = 7\frac{1}{4}$$

8. Change $\frac{59}{3}$ to a mixed number.

Solution

$$\begin{array}{r} 19 \\ 3\overline{)59} \\ \underline{3} \\ 29 \\ \underline{27} \\ 2 \end{array} \qquad \frac{59}{3} = 19 + \frac{2}{3} = 19\frac{2}{3}$$

> Changing an improper fraction to a mixed number is not the same as reducing it. **Reducing** involves finding common factors in the numerator and denominator. **Changing to a mixed number** involves division of the numerator by the denominator. Common factors are not involved.

The fraction part of a mixed number should be reduced. Either: (1) reduce the improper fraction first, and then change it to a mixed number; or (2) change the improper fraction to a mixed number, and then reduce the fraction part.

EXAMPLE

9. Change $\frac{24}{10}$ to a mixed number with the fraction part reduced.

Solution

(a) Reducing first; then changing to a mixed number:

$$\frac{24}{10} = \frac{\cancel{2}\cdot 12}{\cancel{2}\cdot 5} = \frac{12}{5} \qquad \begin{array}{r} 2 \\ 5\overline{)12} \end{array} \qquad \frac{12}{5} = 2\frac{2}{5}$$

(b) Changing to a mixed number; then reducing:

$$\begin{array}{r} 2 \\ 10\overline{)24} \\ \underline{20} \\ 4 \end{array} \qquad \frac{24}{10} = 2\frac{4}{10} = 2\frac{\cancel{2}\cdot 2}{\cancel{2}\cdot 5} = 2\frac{2}{5}$$

Both procedures give the same result: $2\frac{2}{5}$.

PRACTICE QUIZ		ANSWERS
	1. Reduce $\frac{18}{16}$ to lowest terms.	1. $\frac{9}{8}$
	2. Change $\frac{51}{34}$ to a mixed number.	2. $1\frac{1}{2}$
	3. Change $6\frac{2}{3}$ to an improper fraction.	3. $\frac{20}{3}$

Exercises 4.1

Reduce to lowest terms.

1. $\frac{24}{18}$ **2.** $\frac{25}{10}$ **3.** $\frac{16}{12}$ **4.** $\frac{10}{8}$ **5.** $\frac{39}{26}$

6. $\frac{48}{32}$ **7.** $\frac{35}{25}$ **8.** $\frac{18}{16}$ **9.** $\frac{80}{64}$ **10.** $\frac{75}{60}$

Change to mixed numbers with the fraction part reduced to lowest terms.

11. $\frac{100}{24}$ **12.** $\frac{25}{10}$ **13.** $\frac{16}{12}$ **14.** $\frac{10}{8}$ **15.** $\frac{39}{26}$

16. $\frac{42}{8}$ **17.** $\frac{43}{7}$ **18.** $\frac{34}{16}$ **19.** $\frac{45}{6}$ **20.** $\frac{75}{12}$

21. $\frac{56}{18}$ **22.** $\frac{31}{15}$ **23.** $\frac{36}{12}$ **24.** $\frac{48}{16}$ **25.** $\frac{72}{16}$

26. $\frac{70}{34}$ **27.** $\frac{45}{15}$ **28.** $\frac{60}{36}$ **29.** $\frac{35}{20}$ **30.** $\frac{185}{100}$

Change to fraction form.

31. $4\frac{5}{8}$ **32.** $3\frac{3}{4}$ **33.** $5\frac{1}{15}$ **34.** $1\frac{3}{5}$ **35.** $4\frac{2}{11}$

36. $2\frac{11}{44}$ **37.** $2\frac{9}{27}$ **38.** $4\frac{6}{7}$ **39.** $10\frac{8}{12}$ **40.** $11\frac{3}{8}$

41. $6\frac{8}{10}$ **42.** $14\frac{1}{5}$ **43.** $16\frac{2}{3}$ **44.** $12\frac{4}{8}$ **45.** $20\frac{3}{15}$

46. $9\frac{4}{10}$ **47.** $13\frac{1}{7}$ **48.** $49\frac{0}{12}$ **49.** $17\frac{0}{3}$ **50.** $3\frac{1}{50}$

4.2 ADDING MIXED NUMBERS

Since a mixed number itself represents addition, two or more mixed numbers can be added by adding the whole numbers and the fraction parts separately.

To add mixed numbers:

1. Add the fraction parts.
2. Add the whole numbers.
3. Write the mixed number so that the fraction part is less than 1.

EXAMPLES

Find each sum.

1. $4\frac{2}{7} + 6\frac{3}{7}$

Solution

$$\begin{aligned} 4\frac{2}{7} + 6\frac{3}{7} &= 4 + \frac{2}{7} + 6 + \frac{3}{7} \\ &= (4 + 6) + \left(\frac{2}{7} + \frac{3}{7}\right) \\ &= 10 + \frac{5}{7} \\ &= 10\frac{5}{7} \end{aligned}$$

Or, vertically

$$\begin{array}{r} 4\frac{2}{7} \\ +\ 6\frac{3}{7} \\ \hline 10\frac{5}{7} \end{array}$$

2. $25\frac{1}{6} + 3\frac{7}{18}$

Solution

$$\begin{aligned} 25\frac{1}{6} + 3\frac{7}{18} &= 25 + \frac{1}{6} + 3 + \frac{7}{18} \\ &= (25 + 3) + \left(\frac{1}{6} + \frac{7}{18}\right) \\ &= 28 + \left(\frac{1}{6}\cdot\frac{3}{3} + \frac{7}{18}\right) \quad \text{(The SCD, 18, is needed to add the fraction parts.)} \\ &= 28 + \left(\frac{3}{18} + \frac{7}{18}\right) \\ &= 28\frac{10}{18} = 28\frac{5}{9} \end{aligned}$$

3. $7\frac{2}{3} + 9\frac{4}{5}$

Solution

$$\begin{array}{rcccl} 7\frac{2}{3} & = & 7\,\frac{2}{3}\cdot\frac{5}{5} & = & 7\frac{10}{15} \\ +9\frac{4}{5} & = & 9\,\frac{4}{5}\cdot\frac{3}{3} & = & 9\frac{12}{15} \\ \hline & & & & 16\frac{22}{15} = 16 + 1\frac{7}{15} = 17\frac{7}{15} \end{array}$$

Fraction part is greater than 1. ($\frac{22}{15}$)

Change it to a mixed number. ($1\frac{7}{15}$)

4. $3 + 6\frac{1}{2}$

Solution

$$\begin{aligned} 3 + 6\frac{1}{2} &= 3 + 6 + \frac{1}{2} \\ &= 9 + \frac{1}{2} \\ &= 9\frac{1}{2} \end{aligned}$$

Or vertically,

$$\begin{array}{r} 3 \\ +6\frac{1}{2} \\ \hline 9\frac{1}{2} \end{array}$$

The whole number 3 has no fraction part. Be sure to align the whole numbers.

5. $4\frac{3}{4} + 5 + 7\frac{3}{10}$

Solution

The vertical arrangement is generally easier to use for adding mixed numbers.

$$\begin{array}{rcrcr} 4\frac{3}{4} & = & 4\frac{3}{4}\cdot\frac{5}{5} & = & 4\frac{15}{20} \\ 5 & = & 5 & = & 5 \\ +7\frac{3}{10} & = & 7\frac{3}{10}\cdot\frac{2}{2} & = & 7\frac{6}{20} \\ \hline & & & & 16\frac{21}{20} \end{array}$$

$$16\frac{21}{20} = 16 + 1\frac{1}{20} = 17\frac{1}{20}$$

Fraction part is greater than 1. (↑ under $16\frac{21}{20}$)

Change it to a mixed number. (↑ under $1\frac{1}{20}$)

Exercises 4.2

Find each sum.

1. $10 + 4\frac{5}{7}$

2. $6\frac{1}{2} + 3\frac{1}{2}$

3. $5\frac{1}{3} + 2\frac{2}{3}$

4. $8 + 7\frac{3}{11}$

5. $12 + \frac{3}{4}$

6. $15 + \frac{9}{10}$

7. $3\frac{1}{4} + 5\frac{5}{12}$

8. $5\frac{11}{12} + 3\frac{5}{6}$

9. $21\frac{1}{3} + 13\frac{1}{18}$

10. $9\frac{2}{3} + \frac{5}{6}$

11. $4\frac{1}{2} + 3\frac{1}{6}$

12. $3\frac{1}{4} + 7\frac{1}{8}$

13. $25\frac{1}{10} + 17\frac{1}{4}$

14. $5\frac{1}{7} + 3\frac{1}{3}$

15. $6\frac{5}{12} + 4\frac{1}{3}$

16. $5\frac{3}{10} + 2\frac{1}{14}$

17. $8\frac{2}{9} + 4\frac{1}{27}$

18. $11\frac{3}{4} + 2\frac{5}{16}$

19. $6\frac{4}{9} + 12\frac{1}{15}$

20. $4\frac{1}{6} + 13\frac{9}{10}$

21. $21\frac{3}{4} + 6\frac{3}{4}$

22. $3\frac{5}{8} + 3\frac{5}{8}$

23. $7\frac{3}{5} + 2\frac{1}{8}$

24. $9\frac{1}{8} + 3\frac{7}{12}$

25. $3\frac{1}{3} + 4\frac{1}{4} + 5\frac{1}{5}$

26. $\frac{3}{7} + 2\frac{1}{14} + 2\frac{1}{6}$

27. $20\frac{5}{8} + 42\frac{5}{6}$

28. $25\frac{2}{3} + 1\frac{1}{16}$

29. $32\frac{1}{64} + 4\frac{1}{24} + 17\frac{3}{8}$

30. $3\frac{1}{20} + 7\frac{1}{15} + 2\frac{3}{10}$

31. $\begin{array}{r} 4\frac{1}{3} \\ 8\frac{3}{8} \\ +6\frac{1}{6} \\ \hline \end{array}$

32. $\begin{array}{r} 3\frac{2}{3} \\ 14\frac{1}{10} \\ +\ 5\frac{1}{5} \\ \hline \end{array}$

33. $\begin{array}{r} 13\frac{5}{8} \\ 13\frac{1}{12} \\ +10\frac{1}{4} \\ \hline \end{array}$

34. $\begin{array}{r} 5\frac{1}{8} \\ 1\frac{1}{5} \\ +3\frac{1}{40} \\ \hline \end{array}$

35. $\begin{array}{r} 27\frac{2}{3} \\ 30\frac{5}{8} \\ +31\frac{5}{6} \\ \hline \end{array}$

36. A bus trip is made in three parts. The first part takes $2\frac{1}{3}$ hours, the second part takes $2\frac{1}{2}$ hours, and the third part takes $3\frac{3}{4}$ hours. How many hours does the trip take?

37. A construction company built three sections of highway. One section was $20\frac{7}{10}$ kilometers, the second section was $3\frac{4}{10}$ kilometers, and the third section was $11\frac{6}{10}$ kilometers. What was the total length of highway built?

38. A triangle (three-sided figure) has sides of $42\frac{3}{4}$ feet, $23\frac{1}{2}$ feet, and $22\frac{7}{8}$ feet. What is the perimeter (total distance around) of the triangle?

39. A quadrilateral (four-sided figure) has sides of $3\frac{1}{2}$ inches, $2\frac{1}{4}$ inches, $3\frac{5}{8}$ inches, and $2\frac{3}{4}$ inches. What is the perimeter (total distance around) of the quadrilateral?

40. During one week, Sam played three rounds of golf. The first round took $4\frac{1}{2}$ hours, the second round took $3\frac{3}{4}$ hours, and the third round took $4\frac{3}{4}$ hours. How much time did Sam spend playing golf that week?

4.3 SUBTRACTING MIXED NUMBERS

Subtraction with mixed numbers also involves working with the whole numbers and fraction parts separately.

To subtract mixed numbers:

1. Subtract the fraction parts.
2. Subtract the whole numbers.

EXAMPLES

Find the difference.

1. $4\frac{3}{5} - 1\frac{2}{5}$

Solution

$$\begin{aligned} 4\frac{3}{5} - 1\frac{2}{5} &= (4 - 1) + \left(\frac{3}{5} - \frac{2}{5}\right) \\ &= 3 + \frac{1}{5} \\ &= 3\frac{1}{5} \end{aligned}$$

Or, vertically,

$$\begin{array}{r} 4\frac{3}{5} \\ -1\frac{2}{5} \\ \hline 3\frac{1}{5} \end{array}$$

2. $5\frac{6}{7} - 3\frac{2}{7}$

Solution

$$\begin{array}{r} 5\frac{6}{7} \\ -3\frac{2}{7} \\ \hline 2\frac{4}{7} \end{array}$$

3. $10\frac{3}{5} - 6\frac{3}{20}$

Solution

$$\begin{array}{rcr} 10\frac{3}{5} & = & 10\frac{12}{20} \\ + \ 6\frac{3}{20} & = & 6\frac{3}{20} \\ \hline & & 4\frac{9}{20} \end{array}$$

Sometimes the **fraction part** of the number being subtracted is larger than the fraction part of the first number. By "borrowing" a whole 1 from the whole number, we write the first number as a whole number with an improper fraction part. Then the subtraction of the fraction parts is possible.

If the fraction part being subtracted is larger than the first fraction:

1. "Borrow" a whole 1 from the first whole number.
2. Add this 1 to the first fraction. (This will always give an improper fraction that is larger than the fraction being subtracted.)
3. Now subtract.

EXAMPLES Find the differences.

4. $7\frac{2}{5} - 4\frac{3}{5}$

Solution

$$\begin{array}{r} 7\frac{2}{5} \\ -4\frac{3}{5} \\ \hline \end{array} \quad \frac{3}{5} \text{ is larger than } \frac{2}{5}\text{, so "borrow" 1 from 7.}$$

$$\left(7\frac{2}{5} = 6 + 1 + \frac{2}{5} = 6 + 1\frac{2}{5} = 6\frac{7}{5}\right)$$

$$\begin{array}{rcccr} 7\frac{2}{5} & = & 6 + 1\frac{2}{5} & = & 6\frac{7}{5} \\ -4\frac{3}{5} & = & 4\frac{3}{5} & = & 4\frac{3}{5} \\ \hline & & & & 2\frac{4}{5} \end{array}$$

5. $19\frac{2}{3} - 5\frac{3}{4}$

Solution

$$\begin{array}{rcrcr} 19\frac{2}{3} & = & 19\frac{8}{12} & = & 18\frac{20}{12} \\ -\ 5\frac{3}{4} & = & 5\frac{9}{12} & = & 5\frac{9}{12} \\ \hline & & & & 13\frac{11}{12} \end{array} \qquad 1 + \frac{8}{12} = \frac{12}{12} + \frac{8}{12} = \frac{20}{12}$$

6. $10 - 6\frac{5}{8}$ (Note: The whole number 10 has no fraction part. So, the fraction part is understood to be 0.)

Solution

$$\begin{array}{rcr} 10 & = & 9\frac{8}{8} \\ -\ 6\frac{5}{8} & = & 6\frac{5}{8} \\ \hline & & 3\frac{3}{8} \end{array} \qquad 10 = 9 + 1 = 9 + \frac{8}{8}$$

7. $$\begin{array}{rcrcr} 14\frac{3}{4} & = & 14\frac{15}{20} & = & 13\frac{35}{20} \\ -\ 5\frac{9}{10} & = & 5\frac{18}{20} & = & 5\frac{18}{20} \\ \hline & & & & 8\frac{17}{20} \end{array} \qquad 1 + \frac{15}{20} = \frac{20}{20} + \frac{15}{20} = \frac{35}{20}$$

8. $$\begin{array}{rcrcr} 406\frac{5}{12} & = & 406\frac{25}{60} & = & 405\frac{85}{60} \\ -168\frac{13}{20} & = & 168\frac{39}{60} & = & 168\frac{39}{60} \\ \hline & & & & 237\frac{46}{60} = 237\frac{23}{30} \end{array}$$

Exercises 4.3

Find the differences.

1. $\begin{array}{r} 5\frac{1}{2} \\ -1 \\ \hline \end{array}$

2. $\begin{array}{r} 7\frac{3}{4} \\ -2 \\ \hline \end{array}$

3. $\begin{array}{r} 4\frac{5}{12} \\ -3 \\ \hline \end{array}$

4. $\begin{array}{r} 3\frac{5}{8} \\ -2 \\ \hline \end{array}$

5. $\begin{array}{r} 6\frac{1}{2} \\ -2\frac{1}{2} \\ \hline \end{array}$

6. $\begin{array}{r} 9\frac{1}{4} \\ -5\frac{1}{4} \\ \hline \end{array}$

7. $\begin{array}{r} 5\frac{3}{4} \\ -2\frac{1}{4} \\ \hline \end{array}$

8. $\begin{array}{r} 7\frac{9}{10} \\ -3\frac{3}{10} \\ \hline \end{array}$

9. $\begin{array}{r} 14\frac{5}{8} \\ -11\frac{3}{8} \\ \hline \end{array}$

10. $\begin{array}{r} 20\frac{7}{16} \\ -15\frac{5}{16} \\ \hline \end{array}$

11. $\begin{array}{r} 4\frac{7}{8} \\ -1\frac{1}{4} \\ \hline \end{array}$

12. $\begin{array}{r} 9\frac{5}{16} \\ -2\frac{1}{4} \\ \hline \end{array}$

13. $\begin{array}{r} 5\frac{11}{12} \\ -1\frac{1}{4} \\ \hline \end{array}$

14. $\begin{array}{r} 10\frac{5}{6} \\ -4\frac{2}{3} \\ \hline \end{array}$

15. $\begin{array}{r} 8\frac{5}{6} \\ -2\frac{1}{4} \\ \hline \end{array}$

16. $\begin{array}{r} 15\frac{5}{8} \\ -11\frac{3}{4} \\ \hline \end{array}$

17. $\begin{array}{r} 14\frac{6}{10} \\ -3\frac{4}{5} \\ \hline \end{array}$

18. $\begin{array}{r} 8\frac{3}{32} \\ -4\frac{3}{16} \\ \hline \end{array}$

19. $\begin{array}{r} 12\frac{3}{4} \\ -7\frac{1}{6} \\ \hline \end{array}$

20. $\begin{array}{r} 8\frac{11}{12} \\ -5\frac{9}{10} \\ \hline \end{array}$

21. $4\frac{7}{16} - 3$

22. $5\frac{9}{10} - 2$

23. $7 - 6\frac{2}{3}$

24. $12 - 4\frac{1}{5}$

25. $2 - 1\frac{3}{8}$

26. $75 - 17\frac{5}{6}$

27. $4\frac{9}{16} - 2\frac{7}{8}$

28. $3\frac{7}{10} - 2\frac{5}{6}$

29. $15\frac{11}{16} - 13\frac{7}{8}$

30. $20\frac{3}{6} - 3\frac{4}{8}$

31. $17\frac{3}{12} - 12\frac{2}{8}$

32. $18\frac{2}{7} - 4\frac{1}{3}$

33. $13\frac{5}{8} - 6\frac{11}{20}$

34. $18\frac{7}{8} - 2\frac{2}{3}$

35. $10\frac{3}{10} - 2\frac{1}{2}$

36. $1 - \frac{2}{3}$

37. $1 - \frac{5}{8}$

38. $1 - \frac{3}{4}$

39. $1 - \frac{9}{10}$

40. $1 - \frac{7}{16}$

41. $1 - \frac{12}{15}$

42. $1 - \frac{2}{5}$

43. $1 - \frac{1}{6}$

44. $2 - \frac{1}{2}$

45. $2 - \frac{3}{4}$

46. $93\frac{5}{12} - 36\frac{7}{8}$

47. $206\frac{5}{12} - 69\frac{13}{20}$

48. $182\frac{3}{20} - 104\frac{7}{15}$

49. $250\frac{1}{8} - 84\frac{7}{12}$

50. Sara can paint a room in $3\frac{3}{5}$ hours, and Emily can paint a room of the same size in $4\frac{1}{5}$ hours. How many hours are saved by having Sara paint the room? How many minutes are saved?

51. A teacher graded two sets of test papers. The first set took $3\frac{3}{4}$ hours to grade, and the second set took $2\frac{3}{5}$ hours. How much faster did the teacher grade the second set?

52. Mike takes $1\frac{1}{2}$ hours to clean a pool, and Tom takes $2\frac{1}{3}$ hours to clean the same pool. How much longer does Tom take?

53. A long-distance runner was in training. He ran ten miles in $50\frac{3}{10}$ minutes. Three months later, he ran the same ten miles in $43\frac{7}{10}$ minutes. By how much did his time improve?

54. Certain shares of stock were selling for $43\frac{7}{8}$ dollars per share. One month later, the same stock was selling for $48\frac{1}{2}$ dollars per share. By how much did the stock increase in price?

55. You want to lose 10 pounds in weight. If you weigh 180 pounds now and you lose $3\frac{1}{4}$ pounds in one week and $3\frac{1}{2}$ pounds during the second week, how much weight do you still need to lose?

56. Mr. Johnson originally weighed 240 pounds. During each week of six weeks of dieting, he lost $5\frac{1}{2}$ pounds, $2\frac{3}{4}$ pounds, $4\frac{5}{16}$ pounds, $1\frac{3}{4}$ pounds, $2\frac{5}{8}$ pounds, and $3\frac{1}{4}$ pounds. If he was 35 years old, what did he weigh at the end of the six weeks?

57. A board is 10 feet long. If a piece $6\frac{3}{4}$ feet is cut off, what is the length of the remaining piece?

58. A salesman drove $5\frac{3}{4}$ hours one day and then $6\frac{1}{2}$ hours the next day. How much more time did he spend driving on the second day?

4.4 MULTIPLYING MIXED NUMBERS

The simplest way to multiply mixed numbers is to change them to fraction form and then multiply. Just as with fractions in Chapter 3, you should factor and reduce whenever possible before actually multiplying.

To multiply mixed numbers:

1. Change each number to fraction form.
2. Multiply and reduce these fractions.
3. Change the answer to a mixed number or leave it in fraction form.

EXAMPLES

1. $\left(1\frac{2}{3}\right)\left(2\frac{3}{4}\right) = \left(\frac{5}{3}\right)\left(\frac{11}{4}\right) = \frac{5 \cdot 11}{3 \cdot 4} = \frac{55}{12}$ or $4\frac{7}{12}$

2. $\frac{5}{6} \cdot 3\frac{3}{10} = \frac{5}{6} \cdot \frac{33}{10} = \frac{\cancel{5} \cdot \cancel{3} \cdot 11}{2 \cdot \cancel{3} \cdot 2 \cdot \cancel{5}} = \frac{11}{4}$ or $2\frac{3}{4}$

3. $4\frac{1}{2} \cdot 1\frac{1}{6} \cdot 3\frac{1}{3} = \frac{9}{2} \cdot \frac{7}{6} \cdot \frac{10}{3} = \frac{\cancel{3} \cdot \cancel{3} \cdot 7 \cdot \cancel{2} \cdot 5}{\cancel{2} \cdot 2 \cdot \cancel{3} \cdot \cancel{3}} = \frac{35}{2}$ or $17\frac{1}{2}$

Large mixed numbers can be multiplied in the same way.

EXAMPLE

4. $24\frac{3}{8} \cdot 45\frac{1}{4}$

Solution

(a) Change to improper fractions.

$$\begin{array}{r} 24 \\ \times\ 8 \\ \hline 192 \end{array} \qquad \begin{array}{r} 192 \\ +\ 3 \\ \hline 195 \end{array} \qquad 24\frac{3}{8} = \frac{195}{8}$$

$$\begin{array}{r} 45 \\ \times\ 4 \\ \hline 180 \end{array} \qquad \begin{array}{r} 180 \\ +\ 1 \\ \hline 181 \end{array} \qquad 45\frac{1}{4} = \frac{181}{4}$$

(b) Now multiply.

$$24\frac{3}{8} \cdot 45\frac{1}{4} = \frac{195}{8} \cdot \frac{181}{4} = \frac{35{,}295}{32} = 1102\frac{31}{32}$$

$$\begin{array}{r} 195 \\ \times\ 181 \\ \hline 195 \\ 15\,60 \\ 19\,5 \\ \hline 35{,}295 \end{array} \qquad \begin{array}{r} 1102\frac{31}{32} \\ 32\overline{)35{,}295} \\ 32 \\ \hline 3\,2 \\ 3\,2 \\ \hline 09 \\ 0 \\ \hline 95 \\ 64 \\ \hline 31 \end{array}$$

The product in Example 4 with large mixed numbers can be found by writing the numbers one under the other and then multiplying. But this procedure is not simple because four products must be calculated and then added. The procedure is shown here for comparison, but it is **not** recommended.

$$\begin{array}{rl} 24\frac{3}{8} & \\ 45\frac{1}{4} & \\ \hline \frac{3}{32} & \leftarrow \frac{1}{4}\cdot\frac{3}{8} = \frac{3}{32} \\ 6 & \leftarrow \frac{1}{4}\cdot 24 = 6 \\ 16\frac{7}{8} & \leftarrow 45\cdot\frac{3}{8} = \frac{135}{8} = 16\frac{7}{8} \\ 120 & \\ 960 & \leftarrow 45\cdot 24 \\ \hline 1102\frac{31}{32} & \left(\frac{3}{32} + \frac{7}{8} = \frac{3}{32} + \frac{28}{32} = \frac{31}{32}\right) \end{array}$$

Finding a fractional part **of** a number deserves special attention. This is also common with decimals and percents, as we will discuss later. The key word is **of.** In a phrase such as "$\frac{3}{4}$ of 40," the implication is that you are to **multiply** $\frac{3}{4} \cdot 40$. In effect, **of** means to multiply when used with fractions.

A fraction **of** a number means to **multiply** the number by the fraction.

EXAMPLES

5. Find $\frac{3}{4}$ of 40.

Solution

$$\frac{3}{4}\cdot 40 = \frac{3}{\underset{1}{\cancel{4}}}\cdot\frac{\overset{10}{\cancel{40}}}{1} = 30$$

6. Find $\frac{3}{4}$ of 80.

Solution

$$\frac{3}{4}\cdot 80 = \frac{3}{\underset{1}{\cancel{4}}}\cdot\frac{\overset{20}{\cancel{80}}}{1} = 60$$

7. Find $\frac{2}{3}$ of $\frac{9}{10}$.

Solution

$$\frac{2}{3}\cdot\frac{9}{10} = \frac{\cancel{2}\cdot\cancel{3}\cdot 3}{\cancel{3}\cdot\cancel{2}\cdot 5} = \frac{3}{5}$$

NOTE: Whenever you find a fraction **of** a number (and the fraction is between 0 and 1), the result should always be less than the number. If you get a larger number for an answer, you have made a mistake.

PRACTICE QUIZ	Find the indicated products.	ANSWERS
	1. $4\frac{1}{3}\cdot\frac{2}{13}$	**1.** $\frac{2}{3}$
	2. $5\frac{1}{2}\cdot 3\frac{1}{3}$	**2.** $\frac{55}{3}$ or $18\frac{1}{3}$
	3. Find $\frac{9}{10}$ of 70.	**3.** 63

Exercises 4.4

Find the indicated products.

1. $\left(2\frac{1}{3}\right)\left(3\frac{1}{4}\right)$
2. $\left(1\frac{1}{5}\right)\left(1\frac{1}{7}\right)$
3. $4\frac{1}{2}\left(2\frac{1}{3}\right)$
4. $3\frac{1}{3}\left(2\frac{1}{5}\right)$
5. $6\frac{1}{4}\left(3\frac{3}{5}\right)$
6. $5\frac{1}{3}\left(2\frac{1}{4}\right)$
7. $\left(8\frac{1}{2}\right)\left(3\frac{2}{3}\right)$
8. $\left(9\frac{1}{3}\right)2\frac{1}{7}$
9. $\left(6\frac{2}{7}\right)1\frac{3}{11}$
10. $\left(11\frac{1}{4}\right)1\frac{1}{15}$
11. $6\frac{2}{3}\cdot 4\frac{1}{2}$
12. $4\frac{3}{8}\cdot 2\frac{2}{7}$
13. $9\frac{3}{4}\cdot 2\frac{6}{26}$
14. $7\frac{1}{2}\cdot\frac{2}{15}$
15. $\frac{3}{4}\cdot 1\frac{1}{3}$
16. $3\frac{4}{5}\cdot 2\frac{1}{7}$
17. $12\frac{1}{2}\cdot 2\frac{1}{5}$
18. $9\frac{3}{5}\cdot 1\frac{1}{16}$
19. $6\frac{1}{8}\cdot 3\frac{1}{7}$
20. $5\frac{1}{4}\cdot 11\frac{1}{3}$
21. $\frac{1}{4}\cdot\frac{2}{3}\cdot\frac{6}{7}$
22. $\frac{7}{8}\cdot\frac{24}{25}\cdot\frac{5}{21}$
23. $\frac{3}{16}\cdot\frac{8}{9}\cdot\frac{3}{5}$
24. $\frac{2}{5}\cdot\frac{1}{5}\cdot\frac{4}{7}$
25. $\frac{6}{7}\cdot\frac{2}{11}\cdot\frac{3}{5}$
26. $\left(3\frac{1}{2}\right)\left(2\frac{1}{7}\right)\left(5\frac{1}{4}\right)$
27. $\left(4\frac{3}{8}\right)\left(2\frac{1}{5}\right)\left(1\frac{1}{7}\right)$
28. $\left(6\frac{3}{16}\right)\left(2\frac{1}{11}\right)\left(5\frac{3}{5}\right)$
29. $7\frac{1}{3}\cdot 5\frac{1}{4}\cdot 6\frac{2}{7}$
30. $2\frac{5}{8}\cdot 3\frac{2}{5}\cdot 1\frac{3}{4}$
31. $2\frac{1}{16}\cdot 4\frac{1}{3}\cdot 1\frac{3}{11}$
32. $5\frac{1}{10}\cdot 3\frac{1}{7}\cdot 2\frac{1}{17}$
33. $2\frac{1}{4}\cdot 6\frac{3}{8}\cdot 1\frac{5}{27}$
34. $1\frac{3}{32}\cdot 1\frac{1}{7}\cdot 1\frac{1}{25}$
35. $1\frac{5}{16}\cdot 1\frac{1}{3}\cdot 1\frac{1}{5}$
36. $24\frac{1}{5}\cdot 35\frac{1}{6}$
37. $72\frac{3}{5}\cdot 25\frac{1}{6}$
38. $42\frac{5}{6}\cdot 30\frac{1}{7}$
39. $75\frac{1}{3}\cdot 40\frac{1}{25}$
40. $36\frac{3}{4}\cdot 17\frac{5}{12}$
41. Find $\frac{2}{3}$ of 60.
42. Find $\frac{1}{4}$ of 80.
43. Find $\frac{1}{5}$ of 100.
44. Find $\frac{3}{5}$ of 100.
45. Find $\frac{3}{4}$ of 100.
46. Find $\frac{5}{6}$ of 120.

47. Find $\frac{1}{2}$ of $\frac{5}{8}$.

48. Find $\frac{1}{6}$ of $\frac{3}{4}$.

49. Find $\frac{9}{10}$ of $\frac{15}{21}$.

50. Find $\frac{7}{8}$ of $\frac{8}{10}$.

51. A telephone pole is 32 feet long. If $\frac{5}{16}$ of the pole must be underground and $\frac{11}{16}$ of the pole aboveground, how much of the pole is underground? How much is aboveground?

52. The total distance around a square (its perimeter) is found by multiplying the length of one side by 4. Find the perimeter of a square if the length of one side is $5\frac{1}{16}$ inches.

53. A man driving to work drives $17\frac{7}{10}$ miles one way five days a week. How many miles does he drive each week going to and from work?

54. A length of pipe is $27\frac{3}{4}$ feet. What would be the total length if $36\frac{1}{2}$ of these pipes were laid end to end?

55. A woman reads $\frac{1}{6}$ of a book in 3 hours. If the book contains 540 pages, how many pages will she read in 3 hours? How long will she take to read the entire book?

4.5 DIVIDING MIXED NUMBERS AND ORDER OF OPERATIONS

Division with mixed numbers is the same as division with fractions, as discussed in Section 3.3. Simply change each mixed number to an improper fraction before dividing.

As review,

$$\text{the } \textbf{reciprocal} \text{ of } \quad \frac{c}{d} \quad \text{is} \quad \frac{d}{c}$$

and $\frac{a}{b} \div \frac{c}{d} = \frac{a}{b} \cdot \frac{d}{c}$. That is, **to divide by any number (except 0), multiply by its reciprocal.**

To divide mixed numbers:

1. Change each number to fraction form.
2. Write the reciprocal of the divisor.
3. Multiply.

EXAMPLES

1. $3\frac{1}{4} \div 7\frac{4}{5} = \frac{13}{4} \div \frac{39}{5} = \frac{13}{4} \cdot \frac{5}{39}$

$= \frac{\cancel{13} \cdot 5}{4 \cdot \cancel{13} \cdot 3} = \frac{5}{12}$

Note that the divisor is $\frac{39}{5}$, and we multiply by its reciprocal, $\frac{5}{39}$.

2. $\dfrac{2\frac{1}{4}}{4\frac{1}{2}} = \dfrac{\frac{9}{4}}{\frac{9}{2}} = \frac{\overset{1}{\cancel{9}}}{\underset{2}{\cancel{4}}} \cdot \frac{\overset{1}{\cancel{2}}}{\underset{1}{\cancel{9}}} = \frac{1}{2}$

Note that the divisor is $\frac{9}{2}$, and we multiply by its reciprocal, $\frac{2}{9}$.

3. $6 \div 7\frac{7}{8} = 6 \div \frac{63}{8} = \frac{6}{1} \cdot \frac{8}{63}$

$= \frac{\cancel{3} \cdot 2 \cdot 8}{\cancel{3} \cdot 3 \cdot 7} = \frac{16}{21}$

4. $7\frac{7}{8} \div 6 = \frac{63}{8} \div \frac{6}{1} = \frac{63}{8} \cdot \frac{1}{6}$

$= \frac{\cancel{3} \cdot 3 \cdot 7}{8 \cdot 2 \cdot \cancel{3}} = \frac{21}{16}$ or $1\frac{5}{16}$

The rules for order of operations are restated here for convenience.

RULES FOR ORDER OF OPERATIONS

1. First, simplify within grouping symbols, such as parentheses ( ), brackets [ ], or braces { }. Start with the innermost grouping.
2. Second, find any powers indicated by exponents.
3. Third, moving from **left to right,** perform any multiplications or divisions in the order they appear.
4. Fourth, moving from **left to right,** perform any additions or subtractions in the order they appear.

EXAMPLE

5. Use the rules for order of operations to simplify the following expression.

Solution

$$2\frac{1}{2}\cdot 1\frac{1}{6} + 7 \div \frac{3}{4} = \frac{5}{2}\cdot\frac{7}{6} + \frac{7}{1}\cdot\frac{4}{3}$$ Multiply and divide from left to right.

$$= \frac{35}{12} + \frac{28}{3}$$ Now add.

$$= \frac{35}{12} + \frac{28}{3}\cdot\frac{4}{4}$$

$$= \frac{35}{12} + \frac{112}{12}$$

$$= \frac{147}{12} = 12\frac{3}{12} = 12\frac{1}{4}$$

Or, you could write:

$$2\frac{1}{2}\cdot 1\frac{1}{6} = \frac{5}{2}\cdot\frac{7}{6} = \frac{35}{12} = 2\frac{11}{12}$$ multiplying

$$7 \div \frac{3}{4} = \frac{7}{1}\cdot\frac{4}{3} = \frac{28}{3} = 9\frac{1}{3}$$ dividing

$$\begin{array}{rcr} 2\frac{11}{12} & = & 2\frac{11}{12} \\ +9\frac{1}{3} & = & 9\frac{4}{12} \\ \hline & & 11\frac{15}{12} = 12\frac{3}{12} = 12\frac{1}{4} \end{array}$$ adding the results

Sometimes expressions are given in the form of **complex fractions,** fractions in which the numerator or denominator or both are themselves fractions. The fraction bar is a symbol of inclusion as well as a division symbol.

To simplify a complex fraction:

1. Simplify the numerator to a single fraction.
2. Simplify the denominator to a single fraction.
3. Divide the numerator by the denominator.

EXAMPLES

6. Simplify the complex fraction $\dfrac{\frac{3}{4}+\frac{1}{2}}{1-\frac{1}{3}}$.

Solution

Simplify the numerator and denominator separately, then divide.

$$\frac{3}{4}+\frac{1}{2}=\frac{3}{4}+\frac{2}{4}=\frac{5}{4} \quad \text{numerator}$$

$$1-\frac{1}{3}=\frac{3}{3}-\frac{1}{3}=\frac{2}{3} \quad \text{denominator}$$

So,

$$\frac{\frac{3}{4}+\frac{1}{2}}{1-\frac{1}{3}}=\frac{\frac{5}{4}}{\frac{2}{3}}=\frac{5}{4}\div\frac{2}{3}=\frac{5}{4}\cdot\frac{3}{2}=\frac{15}{8}=1\frac{7}{8}$$

7. Simplify $\dfrac{2\frac{1}{3}}{\frac{1}{4}+\frac{1}{3}}$.

Solution

$$2\frac{1}{3}=\frac{7}{3} \quad \text{Change } 2\tfrac{1}{3} \text{ to an improper fraction.}$$

$$\frac{1}{4}+\frac{1}{3}=\frac{1}{4}\cdot\frac{3}{3}+\frac{1}{3}\cdot\frac{4}{4}=\frac{3}{12}+\frac{4}{12}=\frac{7}{12} \quad \text{denominator}$$

So,

$$\frac{2\frac{1}{3}}{\frac{1}{4}+\frac{1}{3}}=\frac{\frac{7}{3}}{\frac{7}{12}}=\frac{7}{3}\cdot\frac{12}{7}=\frac{\cancel{7}\cdot\cancel{3}\cdot 4}{\cancel{3}\cdot\cancel{7}\cdot 1}=\frac{4}{1}=4$$

As with whole numbers, the **average** of mixed numbers and fractions can be found by adding the numbers, then dividing the sum by the number of numbers that were added.

EXAMPLE

8. Find the average of $1\frac{1}{2}$, $\frac{3}{4}$, and $2\frac{5}{8}$.

Solution

$$\begin{aligned} 1\frac{1}{2} &= 1\frac{1}{2}\cdot\frac{4}{4} = 1\frac{4}{8} \\ \frac{3}{4} &= \frac{3}{4}\cdot\frac{2}{2} = \frac{6}{8} \\ +2\frac{5}{8} &= 2\frac{5}{8} \phantom{\cdot\frac{2}{2}} = 2\frac{5}{8} \\ \hline & 3\frac{15}{8} = 3 + 1\frac{7}{8} = 4\frac{7}{8} \end{aligned}$$

$$4\frac{7}{8} \div 3 = \frac{39}{8} \div \frac{3}{1} = \frac{39}{8}\cdot\frac{1}{3} = \frac{\cancel{3}\cdot 13}{8\cdot\cancel{3}} = \frac{13}{8} = 1\frac{5}{8}$$

The average is $1\frac{5}{8}$.

In some situations we know the product of two numbers and one of the numbers. To find the other number, the product should be divided by the given number.

EXAMPLES

9. A certain mathematics class currently has 30 students in attendance. This represents $\frac{2}{3}$ of the original number of students in the class. How many students started the class?

Solution

$$\frac{2}{3}\cdot ? = 30$$

To find the missing number, divide 30 by $\frac{2}{3}$.

$$30 \div \frac{2}{3} = \frac{\overset{15}{\cancel{30}}}{1}\cdot\frac{3}{\underset{1}{\cancel{2}}} = 45$$

45 students started the class.

10. The product of $2\frac{1}{3}$ with another number is $5\frac{1}{6}$. What is the other number?

Solution

$$2\frac{1}{3} \cdot ? = 5\frac{1}{6}$$

To find the missing number, *divide* $5\frac{1}{6}$ by $2\frac{1}{3}$.

$$5\frac{1}{6} \div 2\frac{1}{3} = \frac{31}{6} \div \frac{7}{3} = \frac{31}{\underset{2}{\cancel{6}}} \cdot \frac{\overset{1}{\cancel{3}}}{7} = \frac{31}{14} \quad \text{or} \quad 2\frac{3}{14}$$

The other number is $2\frac{3}{14}$.

Check

$$2\frac{1}{3} \cdot 2\frac{3}{14} = \frac{\cancel{7}}{3} \cdot \frac{31}{\underset{2}{\cancel{14}}} = \frac{31}{6} = 5\frac{1}{6}$$

Exercises 4.5

Find the indicated quotients.

1. $\frac{2}{21} \div \frac{2}{7}$
2. $\frac{9}{32} \div \frac{5}{8}$
3. $\frac{5}{12} \div \frac{3}{4}$
4. $\frac{6}{17} \div \frac{6}{17}$
5. $\frac{5}{6} \div 3\frac{1}{4}$
6. $\frac{7}{8} \div 7\frac{1}{2}$
7. $\frac{29}{50} \div 3\frac{1}{10}$
8. $4\frac{1}{5} \div 3\frac{1}{3}$
9. $2\frac{1}{17} \div 1\frac{1}{4}$
10. $5\frac{1}{6} \div 3\frac{1}{4}$
11. $2\frac{2}{49} \div 3\frac{1}{14}$
12. $6\frac{5}{6} \div 2$
13. $4\frac{1}{5} \div 3$
14. $4\frac{5}{8} \div 4$
15. $6\frac{5}{6} \div \frac{1}{2}$
16. $4\frac{5}{8} \div \frac{1}{4}$
17. $4\frac{1}{5} \div \frac{1}{3}$
18. $1\frac{1}{32} \div 3\frac{2}{3}$
19. $7\frac{5}{11} \div 4\frac{1}{10}$
20. $13\frac{1}{7} \div 4\frac{2}{11}$

Evaluate the following expressions using the rules for order of operations.

21. $\frac{3}{5} \cdot \frac{1}{6} + \frac{1}{5} \div 2$
22. $\frac{1}{2} \div \frac{1}{2} + 1 - \frac{2}{3} \cdot 3$

23. $3\frac{1}{2} \cdot 5\frac{1}{3} + \frac{5}{12} \div \frac{15}{16}$

24. $2\frac{1}{4} + 1\frac{1}{5} + 2 \div \frac{20}{21}$

25. $\frac{5}{8} - \frac{1}{3} \cdot \frac{2}{5} + 6\frac{1}{10}$

26. $1\frac{1}{6} \cdot 1\frac{2}{19} \div \frac{7}{8} + \frac{1}{38}$

27. $\frac{3}{10} + \frac{5}{6} \div \frac{1}{4} \cdot \frac{1}{8} - \frac{7}{60}$

28. $5\frac{1}{7} \div (2 + 1)$

29. $\left(2 - \frac{1}{3}\right) \div \left(1 - \frac{1}{3}\right)$

30. $\left(2\frac{4}{9} + 1\frac{1}{18}\right) \div \left(1\frac{2}{9} - \frac{1}{6}\right)$

31. $\dfrac{2 + \frac{1}{5}}{1 + \frac{1}{4}}$

32. $\dfrac{\frac{2}{3} + \frac{1}{5}}{4\frac{1}{2}}$

33. $\dfrac{4 + \frac{1}{3}}{6 + \frac{1}{4}}$

34. $\dfrac{2 - \frac{1}{3}}{1 - \frac{1}{3}}$

35. $\dfrac{\frac{2}{3} - \frac{1}{4}}{\frac{3}{5} - \frac{1}{4}}$

36. $\dfrac{\frac{5}{6} - \frac{2}{3}}{\frac{5}{8} - \frac{1}{16}}$

37. $\dfrac{1}{\frac{1}{12} + \frac{1}{6}}$

38. $\dfrac{1}{\frac{1}{5} + \frac{2}{15}}$

39. $\dfrac{\frac{3}{10} + \frac{1}{6}}{3}$

40. $\dfrac{\frac{7}{12} + \frac{1}{15}}{5}$

41. $\dfrac{7\frac{1}{3} + 2\frac{1}{5}}{6\frac{1}{9} + 2}$

42. $\dfrac{5\frac{2}{3} - 1\frac{1}{6}}{3\frac{1}{2} + 3\frac{1}{6}}$

43. $\dfrac{6\frac{1}{100} + 5\frac{3}{100}}{2\frac{1}{2} + 3\frac{1}{10}}$

44. $\dfrac{4\frac{7}{10} - 2\frac{9}{10}}{5\frac{1}{100}}$

45. Find the average of the numbers $\frac{7}{8}$, $\frac{9}{10}$, and $1\frac{3}{4}$.

46. Find the average of the numbers $\frac{5}{6}$, $\frac{1}{15}$, and $\frac{17}{30}$.

47. The product of $\frac{9}{10}$ with another number is $1\frac{2}{3}$. What is the other number?

48. The result of multiplying two numbers is $10\frac{1}{3}$. If one of the numbers is $7\frac{1}{6}$, what is the other number?

49. An airplane is carrying 150 passengers. This is $\frac{6}{7}$ of its capacity. What is the capacity of the airplane?

50. The sale price of a coat is \$135. This is $\frac{3}{4}$ of the original price. What was the original price?

51. If the product of $5\frac{1}{2}$ and $2\frac{1}{4}$ is added to the quotient of $\frac{9}{10}$ and $\frac{3}{4}$, what is the sum?

52. If the quotient of $\frac{5}{8}$ and $\frac{1}{2}$ is subtracted from the product of $2\frac{1}{4}$ and $3\frac{1}{5}$, what is the difference?

53. If $\frac{9}{10}$ of 70 is divided by $\frac{3}{4}$ of 10, what is the quotient?

54. If $\frac{2}{3}$ of $4\frac{1}{4}$ is added to $\frac{5}{8}$ of $6\frac{1}{3}$, what is the sum?

SUMMARY: CHAPTER 4

To change a mixed number to an improper fraction, add the whole number and the fraction. Remember, the whole number can be written with 1 as the denominator.

SHORTCUT TO CHANGING MIXED NUMBERS TO FRACTION FORM

1. Multiply the denominator of the fraction part by the whole number.
2. Add the numerator of the fraction part to this product.
3. Write this sum over the denominator of the fraction.

To change an improper fraction to a mixed number:

1. Divide the numerator by the denominator.
2. Write the remainder over the denominator as the fraction part of the mixed number.

To add mixed numbers:

1. Add the fractions parts.
2. Add the whole numbers.
3. Write the mixed number so that the fraction part is less than 1.

To subtract mixed numbers:

1. Subtract the fractions parts.
2. Subtract the whole numbers.

If the fraction part being subtracted is larger than the first fraction:

1. "Borrow" a whole 1 from the first whole number.
2. Add this 1 to the first fraction.
3. Now subtract.

To multiply mixed numbers:

1. Change each number to fraction form.
2. Multiply and reduce these fractions.
3. Change the answer to a mixed number or leave it in fraction form.

A fraction **of** a number means to **multiply** the number by the fraction.

To divide mixed numbers:

1. Change each number to fraction form.
2. Write the reciprocal of the divisor.
3. Multiply.

RULES FOR ORDER OF OPERATIONS

1. First, simplify within grouping symbols, such as parentheses ( ), brackets [ ], or braces { }. Start with the innermost grouping.
2. Second, find any powers indicated by exponents.
3. Third, moving from **left to right,** perform any multiplications or divisions in the order they appear.
4. Fourth, moving from **left to right,** perform any additions or subtractions in the order they appear.

REVIEW QUESTIONS: CHAPTER 4

Change to mixed numbers with the fraction part reduced.

1. $\frac{53}{8}$ **2.** $\frac{91}{13}$ **3.** $\frac{342}{100}$

Change to fraction form.

4. $5\frac{1}{10}$ **5.** $2\frac{11}{12}$ **6.** $13\frac{2}{5}$

Perform the indicated operations. Reduce all fractions to lowest terms.

7. $27\frac{1}{4} + 3\frac{1}{2}$ **8.** $5\frac{2}{5} - 4\frac{2}{3}$ **9.** $4\frac{5}{7} \cdot 2\frac{6}{11}$

10. $\frac{5}{6} \div 3\frac{3}{4}$ **11.** $4\frac{5}{8} + 2\frac{3}{4}$ **12.** $\frac{5}{12}\left(6\frac{3}{10}\right)\left(7\frac{1}{9}\right)$

13. $7\frac{1}{11}\left(2\frac{3}{4}\right)\left(5\frac{1}{3}\right)$ **14.** $2\frac{4}{5} \div 4$ **15.** $12\frac{5}{6} - 6\frac{1}{4}$

16. $4\frac{7}{8} - 3\frac{1}{4}$

17.
$$\begin{array}{r} 14\frac{5}{12} \\ -\ 3\frac{3}{4} \\ \hline \end{array}$$

18.
$$\begin{array}{r} 42\frac{7}{9} \\ 53\frac{4}{15} \\ +24\frac{9}{10} \\ \hline \end{array}$$

19. $4\frac{2}{7} \div 3\frac{3}{5} + 4\frac{1}{6} \cdot 2\frac{4}{5}$ **20.** $2\frac{3}{10} - 5\frac{3}{5} \div 4\frac{2}{3} + 2\frac{1}{6}$

Simplify.

21. $\dfrac{\dfrac{1}{5}+\dfrac{1}{6}}{2\frac{1}{3}}$

22. $\dfrac{\dfrac{7}{8}-\dfrac{3}{16}}{\dfrac{1}{3}-\dfrac{1}{4}}$

23. Find the average of $\frac{3}{4}$, $\frac{5}{8}$, and $\frac{9}{10}$.

24. Find the average of $1\frac{3}{10}$, $2\frac{1}{5}$, and $1\frac{3}{4}$.

25. Find $\frac{2}{3}$ of 96.

26. Find $\frac{2}{5}$ of $6\frac{1}{4}$.

27. The product of $20\frac{2}{3}$ with some other number is $6\frac{1}{2}$. What is the other number?

28. While dieting, Ken lost $4\frac{1}{3}$ pounds the first week, $1\frac{3}{4}$ pounds the second week, and $2\frac{1}{6}$ pounds the third week. What was his average weekly weight loss?

29. The sale price of a television is 375 dollars. This is $\frac{3}{5}$ of the original price. What was the original price?

30. The sum of $12\frac{5}{8}$ and some other number is $20\frac{1}{4}$. What is the other number?

31. Rachel wants to make matching dresses for her three daughters. The pattern requires $1\frac{7}{8}$ yards of material for a size 2 dress, $2\frac{1}{4}$ yards for a size 5, and $3\frac{1}{2}$ yards for a size 8. What is the least amount of material Rachel should buy?

32. Jason won 7500 dollars in the state lottery. If $\frac{1}{5}$ of his prize money was withheld for taxes, how much money did he actually receive?

33. Michael is building a fence $4\frac{1}{2}$ feet high. It is recommended that $\frac{1}{4}$ of the length of each fence post should be underground. How long must each fence post be?

TEST: CHAPTER 4

Reduce all fractions to lowest terms.

1. Change $\frac{34}{7}$ to a mixed number.

2. Change $7\frac{3}{8}$ to fraction form.

3. Find the sum of $19\frac{3}{7}$ and $23\frac{9}{14}$.

4. Find the difference between $18\frac{2}{5}$ and $7\frac{7}{10}$.

5. Find the product of $4\frac{3}{8}$ and $2\frac{2}{5}$.

6. Find the quotient of 3 and $4\frac{1}{5}$.

Perform the indicated operations.

7. $\begin{array}{r} 6\frac{5}{14} \\ +\ \frac{16}{21} \\ \hline \end{array}$

8. $\begin{array}{r} 9 \\ -7\frac{9}{11} \\ \hline \end{array}$

9. $4\frac{2}{3} + 3\frac{3}{4}$

10. $4\frac{3}{8} - 2\frac{5}{6}$

11. $4\frac{1}{5} + 2\frac{7}{8} + 7\frac{1}{20}$

12. $7\frac{2}{9} - 6\frac{3}{4}$

13. $\left(2\frac{2}{11}\right)\left(4\frac{1}{8}\right)$

14. $3\frac{1}{9} \div 5\frac{1}{4}$

15. $\left(1\frac{1}{6}\right)\left(2\frac{3}{7}\right)\left(5\frac{5}{8}\right)$

16. $2\frac{2}{7} \cdot 5\frac{3}{5} + \frac{3}{5} \div 2$

17. Simplify $\dfrac{1\frac{1}{9} + \frac{5}{18}}{4 - 2\frac{2}{3}}$

18. Find $\frac{4}{9}$ of $6\frac{3}{7}$.

19. Find the average of $2\frac{3}{4}$, $3\frac{2}{5}$, and $3\frac{3}{10}$.

20. The number $5\frac{1}{4}$ is the product of $\frac{2}{3}$ with some other number. What is the other number?

CUMULATIVE TEST I: CHAPTERS 1–4

1. Write 203,021 in word form.
2. Complete the following sentence: 11(4)(2) = 11(8) is an example of the ______ property of ______.
3. Find the sum of 876 and 129.
4. Add: 2056 + 92 + 825.
5. Find the difference of 3004 and 796.
6. Round off 139,726 to the nearest thousand.
7. On Monday morning, Kirk had $342 in his checking account. That day, he deposited $600 in his account and wrote checks for the following amounts: $96, $150, and $475. What is the new balance in his checking account on Tuesday morning?
8. Find the product of 68 and 79.
9. Multiply: 204(3500).
10. Complete the following: 3(10 + 7) = 30 + 21 is an example of the ______ property.
11. Find the quotient of 221 and 17.
12. Divide the following and express any remainder as a whole number: 14,180 ÷ 28.
13. Vikki bought two pairs of shoes for $17 per pair, a blouse for $28, a pair of pants for $19, and a sweater for $24. What was the average cost per item of clothing?
14. Evaluate: 2^5
15. Evaluate: $17 + (11 \cdot 6 + 3^2) \div 3 \cdot 10$
16. Which of the following numbers will divide 72,450 exactly?

    2, 3, 4, 5, 9, 10
17. Complete the following: There are ______ even prime numbers greater than 3.
18. Find the prime factorization of 360.

Find the LCM for each of the following groups of numbers.

19. 42 and 105
20. 12, 45, and 50

Complete each of the following statements.

21. A fraction equivalent to $\frac{11}{13}$ with a denominator of 78 is ______.

22. The division $15 \div 32$ can be written as the fraction ______.

23. The reciprocal of 4 is ______.

24. Arrange the following fractions in order from smallest to largest: $\frac{3}{5}$, $\frac{5}{7}$, and $\frac{17}{35}$.

Perform the indicated operations. Reduce all fractions to lowest terms.

25. $\frac{3}{8} \div 6$

26. $\frac{5}{9}\left(\frac{3}{10}\right)\left(\frac{2}{3}\right)$

27. $\frac{7}{8} - \frac{5}{12}$

28. $\frac{7}{10} + \frac{3}{5} + \frac{1}{6}$

29. $37\frac{4}{9} + 10\frac{7}{12}$

30. $3\frac{4}{5} - 2\frac{5}{7}$

31. $3\frac{2}{3}\left(2\frac{5}{11}\right)$

32. $3\frac{3}{8} \div 2\frac{1}{4}$

33. $3\frac{11}{14} + 2\frac{5}{7} + 4\frac{3}{4}$

34. $4\left(5\frac{2}{5}\right)\left(4\frac{1}{6}\right)$

35. Evaluate: $1\frac{1}{3} + 6\frac{4}{5} \div 2\frac{1}{8} - 1\frac{7}{10}$.

36. Find the average of the following numbers:

$$1\frac{2}{5},\ \frac{3}{4},\ \frac{1}{10},\ \text{and}\ \frac{7}{8}.$$

37. If a car has an 18-gallon gas tank, how many gallons of gas will it take to fill the tank when it is $\frac{3}{8}$ full?

38. $\dfrac{1\frac{3}{10} + 2\frac{3}{5}}{2\frac{4}{5} - 1\frac{1}{2}}$

39. Find $\frac{4}{9}$ of $5\frac{2}{5}$.

40. David needed to gain weight to join the football team. When he began his training, he weighed $135\frac{3}{4}$ pounds. Six weeks later, he weighed in at $157\frac{1}{2}$ pounds. What was his average weekly weight gain (in pounds)?

5 Decimal Numbers

5.1 READING AND WRITING DECIMAL NUMBERS

The powers of ten are

$$1, \quad 10, \quad 100, \quad 1000, \quad 10{,}000, \quad 100{,}000, \quad \text{and so on}$$

Fractions and mixed numbers with a power of ten in the denominator are called **decimal numbers.** Examples of decimal numbers in fraction form or mixed number form are:

$$\frac{3}{10}, \quad \frac{5}{100}, \quad \frac{19}{10{,}000}, \quad 6\frac{1}{10}, \quad \frac{8}{1}, \quad \text{and} \quad 12\frac{23}{100}$$

The common notation for decimal numbers uses a decimal point and an extension of the place value system. We write whole numbers to the left of the decimal point and fractions to the right of the decimal point. The value of each place is shown in Figure 5.1.

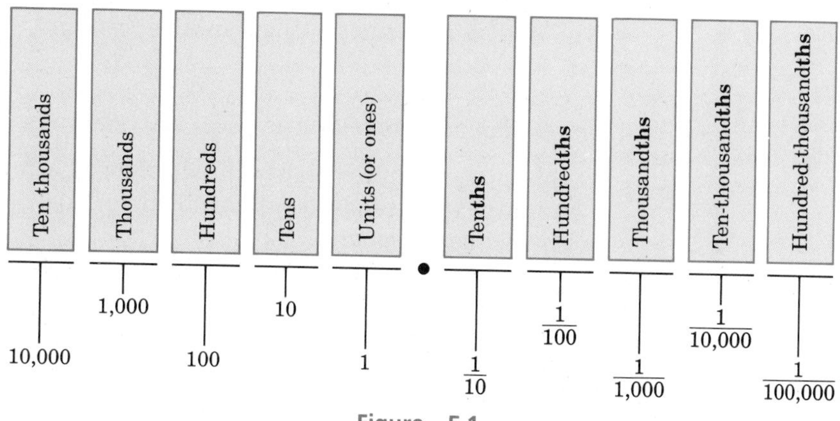

Figure 5.1

NOTE: th is used to indicate the fraction parts. You should memorize the value of each position.

> **To read or write a decimal number:**
> 1. Read (or write) the whole number as before.
> 2. Read (or write) **and** in place of the decimal point.
> 3. Read (or write) the fraction part as a whole number with the name of the place of the last digit.

EXAMPLES

1. 4 8 . 6 The same as $48\frac{6}{10}$.

   forty-eight **and** six tenths

   And indicates the decimal point and the digit 6 is in the tenths position.

2. 5 . 3 9 8 The same as $5\frac{398}{1000}$.

   five **and** three hundred ninety-eight thousandths

   And indicates the decimal point and the digit 8 is in the thousandths position.

3. 1 2 . 0 0 7 5 The same as $12\frac{75}{10{,}000}$.

   twelve **and** seventy-five ten-thousandths

   The digit 5 is in the ten-thousandths position.

4. In reading a fraction such as $\frac{183}{1000}$, you read the numerator as a whole number (one hundred eighty-three), then attach the name of the denominator (thousandths). Remember to follow the same procedure with decimal numbers.

   0.183 is read "one hundred eighty-three thousandths"

   (one hundred eighty-three: as a numerator; thousandths: as a denominator)

SPECIAL NOTES

1. The **th** at the end of a word indicates a fraction part (a part to the right of the decimal point).

   eight hundred = 800
   eight hundred**ths** = 0.08

2. The hyphen (-) indicates one word.

   eight hundred thousand = 800,000
   eight hundred-thousand**ths** = 0.00008

EXAMPLES

5. Write $13\frac{506}{1000}$ in decimal notation.

   13.506

6. Write fifteen thousandths in decimal notation.

   0.015

   Note that the digit 5 is in the thousandths position.

7. Write sixty-two and forty-three hundredths in decimal notation.

   62.43

8. Write six hundred and five thousandths in decimal notation.

   600.005

9. Write six hundred five thousandths in decimal notation.

   0.605

   Note carefully how the use of **and** in Example 8 gives a completely different result from the result in Example 9.

As a safety measure when writing a check, the amount is written in number form and in word form to avoid problems of spelling and poor penmanship. (See Example 10.)

EXAMPLE 10.

MARY KELLEY
123 ELM STREET 123-4569
FRANKLIN, COLORADO 80001

No. **125**

70–1876
711

January 1 19 87

Pay to the order of Management Systems $ 550 00

Five hundred fifty and no/100 Dollars

SURBURBAN BANK
FRANKLIN, COLORADO
80001

Mary Kelley

⑈035446⑆021⑆679⑈

PRACTICE QUIZ		ANSWERS
	1. Write 20.7 in words.	**1.** twenty and seven tenths
	2. Write 18.051 in words.	**2.** eighteen and fifty-one thousandths
	3. Write $4\frac{6}{100}$ in decimal notation.	**3.** 4.06
	4. Write eight hundred and three tenths in decimal notation.	**4.** 800.3

Exercises 5.1

Write the following mixed numbers in decimal notation.

1. $37\frac{498}{1000}$ **2.** $18\frac{76}{100}$ **3.** $4\frac{11}{100}$ **4.** $56\frac{3}{100}$

5. $87\frac{3}{1000}$ **6.** $95\frac{2}{10}$ **7.** $62\frac{7}{10}$ **8.** $100\frac{25}{100}$

9. $100\frac{38}{100}$ **10.** $250\frac{623}{1000}$

Write the following decimal numbers in mixed number form.

11. 82.56 **12.** 93.07 **13.** 10.576 **14.** 100.6

15. 65.003

Write the following numbers in decimal notation.

16. three tenths **17.** fourteen thousandths

18. seventeen hundredths

19. six and twenty-eight hundredths

20. sixty and twenty-eight thousandths

21. seventy-two and three hundred ninety-two thousandths

22. eight hundred fifty and thirty-six ten-thousandths

23. seven hundred and seventy-seven hundredths

24. eight thousand four hundred ninety-two and two hundred sixty-three thousandths

25. six hundred thousand, five hundred and four hundred two thousandths

Write the following decimal numbers in words.

26. 0.5 **27.** 0.93 **28.** 5.06 **29.** 32.58

30. 71.06 **31.** 35.078 **32.** 7.003 **33.** 18.102

34. 50.008 **35.** 607.607 **36.** 593.86 **37.** 593.860

38. 4700.617 **39.** 5000.005 **40.** 603.0065 **41.** 900.4638

Write sample checks for the amounts indicated.

42. $356.45

MARY KELLEY
123 ELM STREET 123-4569
FRANKLIN, COLORADO 80001

No. 125

70–1876
711

January 1 19 87

Pay to the order of Management Systems $ ____________

______________________________________ Dollars

SURBURBAN BANK
FRANKLIN, COLORADO
80001

Mary Kelley

⑈035446⑆021⑆679⑈

43. \$651.50

MARY KELLEY
123 ELM STREET 123-4569
FRANKLIN, COLORADO 80001

No. 125

70–1876
711

January 1 19 *87*

Pay to the order of *Management Systems* \$ ______

______ Dollars

SURBURBAN BANK
FRANKLIN, COLORADO
80001

Mary Kelley

⑈035446⑆021⑆679⑈

44. \$2506.64

MARY KELLEY
123 ELM STREET 123-4569
FRANKLIN, COLORADO 80001

No. 125

70–1876
711

January 1 19 *87*

Pay to the order of *Management Systems* \$ ______

______ Dollars

SURBURBAN BANK
FRANKLIN, COLORADO
80001

Mary Kelley

⑈035446⑆021⑆679⑈

5.2 ROUNDING OFF DECIMAL NUMBERS

Measuring devices made by humans give only approximate measurements. (See Figure 5.2.) The units can be large, such as miles and kilometers, or small, such as inches and centimeters. But there are always smaller units, such as eighths of an inch and millimeters. If a recipe calls for 1.5 cups of flour and you put in 1.52 cups of flour or 1.47 cups of flour, is this acceptable? Will the result be tasty?

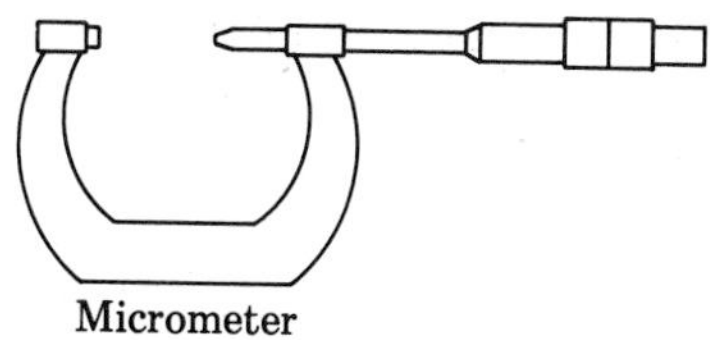

(a) The micrometer is marked to give approximate measures of circular objects.

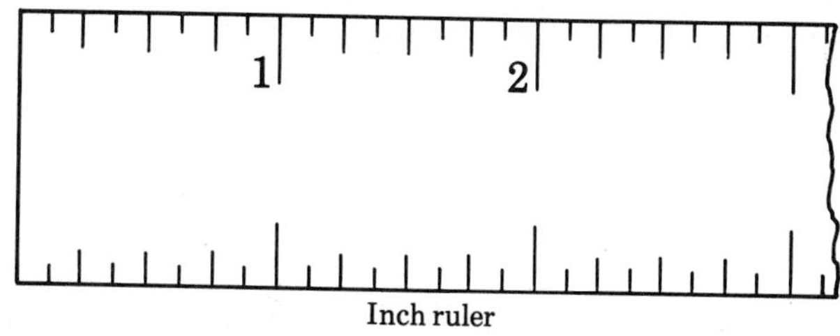

(b) The ruler is marked to give approximate measures of lengths in fourths and eighths of an inch.

Figure 5.2

Rounding Off: Rounding off a given number means to find another number close to the given number. The desired place of accuracy must be stated.

EXAMPLES

1. Round off 5.76 to the nearest tenth.

Solution

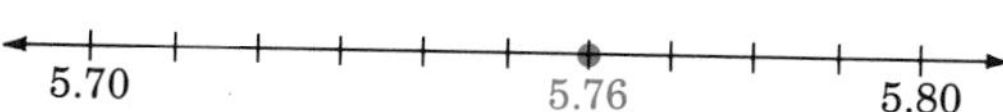

We can see from the number line that 5.76 is closer to 5.80. So, 5.76 rounds off to 5.80 or 5.8 (to the nearest tenth).

2. Round off 0.153 to the nearest hundredth.

Solution

0.153 is between 0.150 and 0.160.

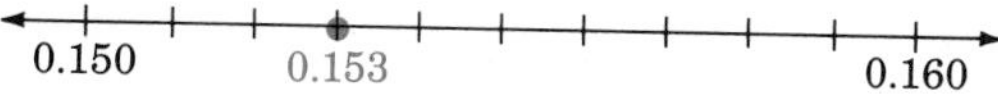

Looking at the numbers on the number line, we can see that 0.153 is closer to 0.150. So, 0.153 rounds off to 0.150 or 0.15 (to the nearest hundredth).

The number line is a good aid for illustrating the rounding-off concept. However, the following set of rules is more useful. You should be aware that there are different sets of rules for working with approximate numbers in various situations. These rules are discussed in shop classes, engineering classes, surveying classes, and so on. They will not be discussed here.

RULES FOR ROUNDING OFF DECIMAL NUMBERS

1. Look at the single digit just to the right of the place of desired accuracy.
2. If this digit is 5 or greater, make the digit in the desired place of accuracy one larger and replace all digits to the right with zeros.
3. If this digit is less than 5, leave the digit in the desired place of accuracy as it is and replace all digits to the right with zeros.

NOTE: This rule is used in many situations, but not all. For example, if the price of an item in a store involves a fraction of a cent, the merchant will always round off to the next higher cent. If three cans of beans cost \$1.00, then one can of beans will cost 34¢, not $33\frac{1}{3}$¢ or 33¢.

EXAMPLES

3. Round off 6.749 to the nearest tenth.

Solution

(a) 6.7 4 9

7 is in the tenths position. The next digit is 4.

(b) Since 4 is less than 5, leave the 7 and replace 4 and 9 with 0's.

(c) 6.749 rounds off to 6.700 or 6.7 to the nearest tenth.

NOTE: 6.700 and 6.7 are both correct. We have a choice of dropping trailing 0's to the right of the decimal point. Usually the trailing 0's will be dropped to indicate the position of accuracy.

4. Round off 3.45196 to the nearest thousandth.

Solution

(a) 3.4 5 1 9 6

1 is in the thousandths position. The next digit is 9.

(b) Since 9 is greater than 5, change 1 to 2 and replace 9 and 6 with 0's.

(c) 3.45196 rounds off to 3.45200 or 3.452 to the nearest thousandth.

5. Round off 4.9838 to the nearest tenth.

Solution

(a) 4.9 8 3 8

9 is in the tenths position. 8 is the next digit.

(b) Since 8 is greater than 5, make 9 one larger and replace 8 and 3 and 8 with 0's. (Making 9 one larger gives 10 which affects the digit 4, too.)

(c) 4.9838 rounds off to 5.0000 or 5.0 to the nearest tenth.

NOTE: The 0 in the tenths position is written to indicate the position of rounding off.

6. Round off 17.835 to the nearest hundredth.

Solution

(a) 1 7.8 3 5

hundredths position next digit

(b) Since the next digit is 5, make 3 one larger, change 3 to 4, and replace 5 with 0.

(c) 17.835 rounds off to 17.840 or 17.84 to the nearest hundredth.

7. Round off 8472 to the nearest hundred.

Solution

(a) The decimal point is understood to be to the right of 2. In whole numbers, the decimal point is optional. If there is no decimal point, its position is understood to be to the right of the rightmost digit.

(b) The digit in the hundreds position is 4.

(c) The next digit is 7.

(d) Since 7 is greater than 5, change the 4 to 5 and replace 7 and 2 with 0's.

(e) So, 8472 rounds off to 8500 (to the nearest hundred).

SPECIAL NOTE: The 0's must **not** be dropped in a whole number. They can only be dropped if they are trailing 0's to the right of the decimal point.

PRACTICE QUIZ	Round off as indicated.	ANSWERS
	1. 572.3 (nearest ten)	**1.** 570
	2. 6.749 (nearest tenth)	**2.** 6.7
	3. 7558 (nearest thousand)	**3.** 8000
	4. 0.07921 (nearest thousandth)	**4.** 0.079

Exercises 5.2

Fill in the blanks to correctly complete each statement.

1. Round off 34.76 to the nearest tenth.

(a) The digit in the tenths position is ______.

(b) The next digit is ______.

(c) Since ______ is greater than 5, change ______ to ______ and replace 6 with 0.

(d) So, 34.76 rounds off to ______ to the nearest tenth.

2. Round off 6.832 to the nearest hundredth.

(a) The digit in the hundredths position is ______.

(b) The next digit is ______.

(c) Since ______ is less than 5, leave ______ as it is and replace ______ with 0.

(d) 6.832 rounds off to ______ to the nearest hundredth.

3. Round off 1.00643 to the nearest ten-thousandth.

(a) The digit in the ten-thousandths position is ______.

(b) The next digit is ______.

(c) Since ______ is less than 5, leave ______ as it is and replace ______ with 0.

(d) 1.00643 rounds off to ______ to the nearest ______.

4. Round off 6275.38 to the nearest ten.

(a) The digit in the tens position is ______.

(b) The next digit is ______.

(c) Since ______ is equal to 5, change ______ to ______ and replace 5, 3, and 8 with ______.

(d) 6275.38 rounds off to ______ to the nearest ten.

Round off each of the following decimal numbers as indicated.

To the nearest tenth:

5. 89.015 **6.** 7.555 **7.** 18.009 **8.** 37.666

9. 14.3338 **10.** 0.036

To the nearest hundredth:

11. 0.385 **12.** 0.296 **13.** 5.722 **14.** 8.987

15. 6.996 **16.** 13.1346 **17.** 0.0782 **18.** 6.0035

19. 5.7092 **20.** 2.8347

To the nearest thousandth:

21. 0.0672 **22.** 0.05550 **23.** 0.6338 **24.** 7.6666

25. 32.4785 **26.** 9.4302 **27.** 17.36371 **28.** 4.44449

29. 0.00191 **30.** 20.76962

To the nearest whole number (or nearest unit):

31. 479.23 **32.** 6.8 **33.** 17.5 **34.** 19.999

35. 382.48 **36.** 649.66 **37.** 439.78 **38.** 701.413

39. 6333.11 **40.** 8122.825

To the nearest ten:

41. 5163. **42.** 6475 **43.** 495 **44.** 572.5

45. 998.5 **46.** 378.92 **47.** 5476.2 **48.** 76,523.1

49. 92,540.9 **50.** 7007.7

To the nearest thousand:

51. 7398 **52.** 62,275 **53.** 47,823.4 **54.** 103,499

55. 217,480.2 **56.** 9872.5 **57.** 379,500 **58.** 4,500,762

59. 7,305,438 **60.** 573,333.3

61. 0.0005783 (nearest hundred-thousandth)

62. 0.5449 (nearest hundredth)

63. 473.8 (nearest ten)

64. 5.00632 (nearest thousandth)

65. 473.8 (nearest hundred)

66. 5750 (nearest thousand)

67. 3.2296 (nearest thousandth)

68. 15.548 (nearest tenth)

69. 78,419 (nearest ten thousand)

70. 78,419 (nearest ten)

5.3 ADDING AND SUBTRACTING DECIMAL NUMBERS

Decimal numbers can be written in an expanded form, such as

$$\begin{aligned} 5.237 &= 5 + 2\left(\frac{1}{10}\right) + 3\left(\frac{1}{100}\right) + 7\left(\frac{1}{1000}\right) \\ &= 5 + \frac{2}{10} + \frac{3}{100} + \frac{7}{1000} \end{aligned}$$

Thus, to add 5.237 + 6.15, we can write

$$\begin{aligned} &5 + \frac{2}{10} + \frac{3}{100} + \frac{7}{1000} + 6 + \frac{1}{10} + \frac{5}{100} \\ &= (5 + 6) + \left(\frac{2}{10} + \frac{1}{10}\right) + \left(\frac{3}{100} + \frac{5}{100}\right) + \frac{7}{1000} \end{aligned}$$

This procedure can be accomplished in a much easier way by writing the decimal numbers one under the other and keeping the decimal points in line. In this way, the whole numbers will be added properly, tenths added to tenths, hundredths to hundredths, and so on. The decimal point in the sum is in line with the other decimal points.

$$\text{Add} \quad \begin{array}{r} 5.237 \\ \underline{6.150} \\ 11.387 \end{array}$$

Zeros may be written to the right of the last digit in the fraction part to help keep the digits in the correct line. This will not change the value of any number.

To add decimal numbers:

1. Write the numbers one under the other.
2. Keep the decimal points in line.
3. Keep digits with the same position value in line (zeros may be filled in as aids).
4. Add the numbers, placing the decimal point in the answer in line with the other decimal points.

EXAMPLES

1. Find the sum 6.3 + 5.42 + 14.07.

Solution

$$\begin{array}{r} 6.30 \\ 5.42 \\ +14.07 \\ \hline 25.79 \end{array}$$

← 0 may be filled in to help keep the digits in line.

2. Find the sum 9 + 4.86 + 37.479 + 0.6.

Solution

$$\begin{array}{r} 9.000 \\ 4.860 \\ 37.479 \\ +\ 0.600 \\ \hline 51.939 \end{array}$$

The decimal point is understood to be to the right of 9, as 9.0.

0's are filled in to help keep the digits in line.

3. Add.

$$\begin{array}{r} 56.2 \\ 85.75 \\ +29.001 \\ \hline \end{array}$$

You can write

$$\begin{array}{r} 56.200 \\ 85.750 \\ +29.001 \\ \hline 170.951 \end{array}$$

To subtract decimal numbers:

1. Write the numbers one under the other.
2. Line up the decimal points.
3. Keep digits with the same position value in line (zeros may be filled in).
4. Subtract, placing the decimal point in the answer in line with the other decimal points.

EXAMPLES

4. Find the difference 16.715 − 4.823.

Solution

$$\begin{array}{r} 16.715 \\ -\ \ 4.823 \\ \hline 11.892 \end{array}$$

5. Find the difference 21.2 − 13.716.

Solution

$$\begin{array}{rl} 21.200 & \leftarrow \text{fill in 0's.} \\ -13.716 & \\ \hline 7.484 & \end{array}$$

6. Find the difference 17 − 0.5618.

Solution

$$\begin{array}{rl} 17.0000 & \leftarrow \text{fill in 0's.} \\ -\ \ 0.5618 & \\ \hline 16.4382 & \end{array}$$

7. Mrs. Finn went to the local store and bought a pair of shoes for $42.50, a blouse for $25.60, and a shirt for $37.55. How much did she spend? (Tax was included in the prices.)

Solution

$$\begin{array}{r} \$\ 42.50 \\ 25.60 \\ +\ 37.55 \\ \hline \$105.65 \end{array}$$

She spent $105.65.

8. Joe decided he needed some new fishing equipment. He bought a new rod for $55, a rod (on sale) for $22.50, and some fishing line for $2.70. If tax totaled $4.82, how much change did he receive from a $100 bill?

Solution

(a) Find the total of his expenses including tax.

$$\begin{array}{r} \$55.00 \\ 22.50 \\ 2.70 \\ +\ \ 4.82 \\ \hline \$85.02 \end{array}$$

(b) Subtract the answer in part (a) from $100.

$$\begin{array}{r} \$100.00 \\ -\ \ \ \ 85.02 \\ \hline \$\ 14.98 \end{array}$$

His change was $14.98.

PRACTICE QUIZ	Find each indicated sum or difference.	ANSWERS
	1. 46.2 + 3.07 + 2.6	**1.** 51.87
	2. 9 + 5.6 + 0.58	**2.** 15.18
	3. 6.4 − 3.7	**3.** 2.7
	4. 18 − 0.4384	**4.** 17.5616

Exercises 5.3

Find each of the indicated sums.

1. 0.6 + 0.4 + 1.3 **2.** 5 + 6.1 + 0.4 **3.** 0.59 + 6.91 + 0.05

4. 3.488 + 16.593 + 25.002 **5.** 37.02 + 25 + 6.4 + 3.89

6. 4.0086 + 0.034 + 0.6 + 0.05 **7.** 43.766 + 9.33 + 17 + 206

8. 52.3 + 6 + 21.01 + 4.005 **9.** 2.051 + 0.2006 + 5.4 + 37

10. 5 + 2.37 + 463 + 10.88

11. $\begin{array}{r} 47.3 \\ 42.03 \\ +29.003 \\ \hline \end{array}$ **12.** $\begin{array}{r} 1.007 \\ 20.063 \\ +\ 0.49 \\ \hline \end{array}$ **13.** $\begin{array}{l} 4.128 \\ 0.02 \\ +3. \\ \hline \end{array}$ **14.** $\begin{array}{r} 5.0015 \\ 2.443 \\ +0.0469 \\ \hline \end{array}$

15. $\begin{array}{r} 75.2 \\ 3.682 \\ +14.995 \\ \hline \end{array}$ **16.** $\begin{array}{r} 107.39 \\ 5.061 \\ 23.54 \\ +\ 64.9801 \\ \hline \end{array}$ **17.** $\begin{array}{r} 34.967 \\ 50.6 \\ 8.562 \\ +\ 9.3 \\ \hline \end{array}$ **18.** $\begin{array}{r} 4.156 \\ 3.7 \\ 25.682 \\ +13.405 \\ \hline \end{array}$

19. $\begin{array}{r} 74. \\ 3.529 \\ 52.62 \\ +\ 7.001 \\ \hline \end{array}$ **20.** $\begin{array}{r} 983.4 \\ 47.518 \\ 805.411 \\ +300.766 \\ \hline \end{array}$

Find each of the indicated differences.

21. 5.2 − 3.76 **22.** 17.83 − 8.9 **23.** 29.5 − 13.61

24. 1.0057 − 0.03 **25.** 78.015 − 13.068

26. $\begin{array}{r} 22.418 \\ -17.523 \\ \hline \end{array}$ **27.** $\begin{array}{l} 4.8 \\ -0.0026 \\ \hline \end{array}$ **28.** $\begin{array}{r} 31.009 \\ -\ 0.534 \\ \hline \end{array}$ **29.** $\begin{array}{l} 4. \\ -1.0566 \\ \hline \end{array}$

30. $\begin{array}{r} 40.718 \\ -\ 6.532 \\ \hline \end{array}$

31. Mrs. Johnson bought the following items at a department store: dress, \$47.25; shoes, \$35.75; purse, \$12.50. How much did she spend? What was her change if she gave the clerk a \$100 bill? (Tax was included in the prices.)

32. Brian got a haircut for \$10.00 (including a shampoo) and a shave for \$3.50. If he tipped the barber \$2.00, how much change did he receive from a \$20 bill?

33. The inside radius of a pipe is 2.38 inches, and the outside radius is 2.63 inches as shown in the figure. What is the thickness of the pipe? (Note: The radius of a circle is the distance from the center of the circle to a point on the circle.)

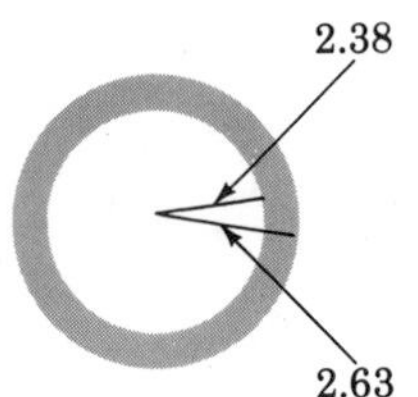

34. Mr. Johnson bought the following items at a department store: slacks, \$32.50; shoes, \$43.75; shirt, \$18.60. How much did he spend? What was his change if he gave the clerk a \$100 bill? (Tax was included in the prices.)

35. An architect's scale drawing shows a rectangular lot 2.38 inches on one side and 3.76 inches on the other side. What was the perimeter (distance around) of the rectangle on the drawing?

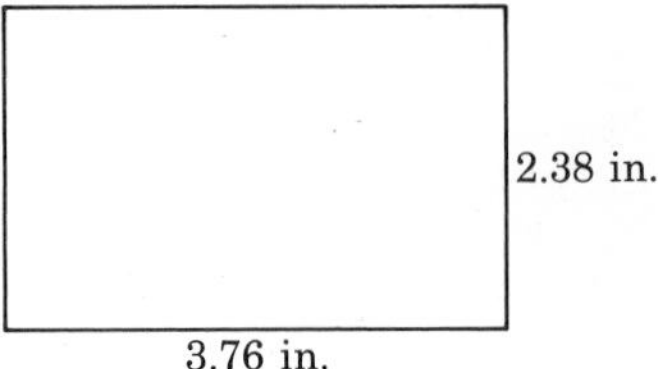

5.4 MULTIPLYING DECIMAL NUMBERS

To illustrate how to place the decimal point in a product, several products are shown in both fraction notation and decimal notation. Remember that decimals are fractions.

FRACTIONS	*DECIMALS*
$\frac{1}{10} \cdot \frac{1}{100} = \frac{1}{1000}$	$\begin{array}{r} .1 \\ .01 \\ \hline .001 \end{array}$ 3 places (or thousandths)

FRACTIONS	*DECIMALS*	
$\frac{3}{10} \cdot \frac{5}{100} = \frac{15}{1000}$	.3 .05 .015	3 places (or thousandths) 3 places (or thousandths)
$\frac{6}{100} \cdot \frac{4}{1000} = \frac{24}{100{,}000}$	.004 .06 .00024	5 places (or hundred-thousandths) 5 places (or hundred-thousandths)

As you probably noted in the examples just shown, **there is no need to keep the decimal points lined up for multiplication.** The following rule explains how to multiply two decimal numbers.

> **To multiply decimal numbers:**
>
> 1. Multiply the two numbers as if they were whole numbers.
> 2. Count the total number of places to the right of the decimal points in both numbers being multiplied.
> 3. This sum is the number of places to the right of the decimal point in the product.

EXAMPLES

1. Multiply 2.432 × 5.1.

Solution

```
   2.432   ←3 places ⎫
 ×   5.1   ←1 place  ⎭ total of 4 places
   2432
 12 160
 12.4032   ←4 places in the product
```

2. Multiply 4.35 × 12.6.

Solution

```
   4.35   ←2 places ⎫
 × 12.6   ←1 place  ⎭ total of 3 places
  2 610
  8 70
 43 5
 54.810   ←3 places in the product
```

3. Multiply (0.046)(0.007).

Solution

$$\begin{array}{r} 0.046 \\ \times\ 0.007 \\ \hline 0.000322 \end{array}$$

← 3 places (0.046), ← 3 places (0.007) } total of 6 places

← 6 places in the product

This means that three 0's had to be inserted between the 3 and the decimal point.

We multiplied whole numbers by powers of ten in Section 1.6 by placing zeros to the right of the number. This is the same as moving the decimal point.

To multiply a decimal number by a power of 10:

1. Move the decimal point to the **right.**
2. Move it the same number of places as the number of 0's in the power of 10.

Multiplication by 10 moves the decimal point one place **to the right.**
Multiplication by 100 moves the decimal point two places **to the right.**
Multiplication by 1000 moves the decimal point three places **to the right,**
and so on.

EXAMPLES

4. $10(9.23) = 92.3$ — Move decimal point 1 place to the right.

5. $100(9.23) = 923.$ — Move decimal point 2 places to the right.

6. $1000(0.8642) = 864.2$ — Move decimal point 3 places to the right.

7. $1000(7.5) = 7500.$ — Move decimal point 3 places to the right. Note that two 0's had to be inserted.

Exponents can also be used to indicate the power of ten.

EXAMPLES

8. $10^2(4.9631) = 496.31$ The exponent tells how many places to move the decimal point.

9. $10^3(4.9631) = 4963.1$

Word problems can involve several operations with decimal numbers. The words do not usually say directly to add, subtract, or multiply. You must reason from experience what to do with the numbers.

EXAMPLE

10. You can buy a car for \$7500 cash, or you can make a down payment of \$1500 and then pay \$566.67 each month for twelve months. How much can you save by paying cash?

Solution

(a) Find the amount paid in monthly payments.

```
 $ 566.67
 ×     12
  1133 34
  5666 7
 $6800.04   paid in monthly payments
```

(b) Find the total amount paid by adding the down payment to the answer in part (a).

```
 $ 1500.00
 + 6800.04
 $ 8300.04   total paid
```

(c) Find the savings by subtracting \$7500 from the answer in part (b).

```
 $ 8300.04
 - 7500.00
 $  800.04   savings by paying cash
```

PRACTICE QUIZ

Find each of the indicated products.

1. (.8)(.2)
2. (5.6)(.04)
3. $10^4(3.781)$

ANSWERS

1. 0.16
2. 0.224
3. 37810.

Exercises 5.4

Find each of the indicated products.

1. (0.6)(0.7)
2. (0.3)(0.8)
3. 5(1.8)
4. 9(3.4)
5. 8(2.7)
6. 4(9.6)
7. 1.4(0.3)
8. 1.5(0.6)
9. (0.2)(0.02)
10. (0.3)(0.03)
11. 5.4(0.02)
12. 7.3(0.01)
13. 0.23×0.12
14. 0.15×0.15
15. 8.1×0.006
16. 7.1×0.008
17. 0.06×0.01
18. 0.25×0.01
19. 3(0.125)
20. 4(0.375)
21. 1.6(0.875)
22. 5.3(0.75)
23. 6.9(0.25)
24. 4.8(0.25)
25. 0.83(6.1)
26. 0.27(0.24)
27. 0.16(0.5)
28. 0.28(0.5)
29. 3.29(0.01)
30. 5.78(0.02)
31. $\begin{array}{r} 0.005 \\ \times 0.009 \\ \hline \end{array}$
32. $\begin{array}{r} 0.006 \\ \times 0.004 \\ \hline \end{array}$
33. $\begin{array}{r} 0.137 \\ \times\ 0.06 \\ \hline \end{array}$
34. $\begin{array}{r} 0.106 \\ \times\ 0.09 \\ \hline \end{array}$
35. $\begin{array}{r} 1.07 \\ \times\ 0.5 \\ \hline \end{array}$
36. $\begin{array}{r} 5.08 \\ \times\ 0.4 \\ \hline \end{array}$
37. $\begin{array}{r} 0.0106 \\ \times\ 0.087 \\ \hline \end{array}$
38. $\begin{array}{r} 0.0213 \\ \times\ 0.065 \\ \hline \end{array}$
39. $\begin{array}{r} 83.105 \\ \times\ 0.111 \\ \hline \end{array}$
40. $\begin{array}{r} 17.002 \\ \times\ 0.101 \\ \hline \end{array}$
41. $\begin{array}{r} 86.1 \\ \times 0.057 \\ \hline \end{array}$
42. $\begin{array}{r} 7.83 \\ \times 0.18 \\ \hline \end{array}$
43. $\begin{array}{r} 95.62 \\ \times\ 0.57 \\ \hline \end{array}$
44. $\begin{array}{r} 6.002 \\ \times\ 0.57 \\ \hline \end{array}$
45. $\begin{array}{r} 8.034 \\ \times\ 0.29 \\ \hline \end{array}$
46. 100(3.46)
47. 100(20.57)
48. 100(7.82)
49. 100(6.93)
50. 100(16.1)
51. 100(38.2)
52. 10(0.435)
53. 10(0.719)
54. 10(1.86)
55. 1000(4.1782)
56. $10^3(0.38)$
57. $10^3(0.47)$
58. $10^4(0.005)$
59. $10^4(0.00615)$
60. $10^4(7.4)$
61. If an architect makes a drawing to scale so that 1 inch represents 6.75 feet, what distance is represented by 5.5 inches?
62. To buy a car, a man can pay $2036.50 cash, or he can put $400 down and make 18 monthly payments of $104.30. How much does he save by paying cash?

63. An automobile dealer makes $150.70 on each used car he sells and $425.30 on each new car he sells. How much did he make in a month if he sold 11 used and 6 new cars?

64. Suppose a tax assessor figures the tax at 0.07 times the assessed value of a home. If the assessed value is figured at a rate of 0.32 times the market value, what taxes are paid on a home with a market value of $136,500?

65. If the sale price of a new refrigerator is $583 and sales tax is figured at 0.06 times the price, what is the total amount paid for the refrigerator?

66. If you drive south at 57.6 miles per hour for 3 hours, then west at 52.4 miles per hour for 4 hours, how far have you driven in the 7 hours? (Assume that you started at least 300 miles east of the Pacific Ocean and 200 miles north of the Gulf of Mexico.)

67. Multiply the numbers 2.456 and 3.16, then round off the product to the nearest tenth. Next, round off each of the numbers to the nearest tenth first and then multiply and round off this product to the nearest tenth. Did you get the same answer?

68. If you paid $2500 as a down payment on a new car and thirty-six monthly payments of $275.50, how much did you pay for the car?

69. If you were paid a salary of $350 per week and $13.75 for each hour you worked over 40 hours, how much would you make if you worked 45 hours in one week?

70. If you bought a magazine for $2.50, a candy bar for $.65, a milk shake for $1.75, french fries for $.85, and you had to pay a tax of 0.06 times the total, how much change would you get from a $10 bill?

5.5 DIVIDING DECIMAL NUMBERS

Division with whole numbers gives a quotient and possibly a remainder.

```
      27   quotient
  35)959
     70
     259
     245
      14   remainder
```

Now that we have decimal numbers, we can continue to divide and get a decimal quotient.

$$\begin{array}{r} 27.4 \\ 35\overline{)959.0} \\ \underline{70} \\ 259 \\ \underline{245} \\ 14\,0 \\ \underline{14\,0} \\ 0 \end{array}$$

quotient is a decimal number

If the divisor is a decimal number, multiply both the divisor and dividend by a power of 10 to make the divisor a whole number. For example, we can write

$$4.9\overline{)51.45} \quad \text{as} \quad \frac{51.45}{4.9} \times \frac{10}{10} = \frac{514.5}{49}$$

Thus, $4.9\overline{)51.45}$ is the same as $49\overline{)514.5}$

To divide decimal numbers:

1. Move the decimal point in the divisor so it is a whole number.
2. Move the decimal point in the dividend the same number of places to the right.
3. Place the decimal point in the quotient directly above the new decimal point in the dividend.
4. Divide just as with whole numbers.

NOTES:

1. In moving the decimal point, you are multiplying by a power of 10.
2. **Be sure to place the decimal point in the quotient before actually dividing.**

EXAMPLES

1. $51.45 \div 4.9$.

Solution

(a) Write down the numbers.

$$4.9\overline{)51.45}$$

(b) Move the decimal points so that the divisor is a whole number.

$$4.9.\overline{)51.4.5}$$ (decimal point in quotient)

Move each decimal point 1 place. This makes the whole number 49 the divisor.

(c) Proceed to divide as with whole numbers.

```
      10.5
49.)514.5
    49
     24
      0
     24 5
     24 5
        0
```

2. 5.1 ÷ 1.36.

Solution

(a) Write down the numbers.

$$1.36\overline{)5.1}$$

(b) Move the decimal points so the divisor is a whole number. Add 0's in the dividend if needed.

$$1.36.\overline{)5.10.00}$$ (decimal point in quotient)

Add 0's as needed.

Move each decimal point 2 places.

(c) Divide.

```
        3.75
136.)510.00
     408
     102 0
      95 2
       6 80
       6 80
          0
```

3. 6.3252 ÷ 6.3.

Solution

```
        1.004
6.3.)6.3.252
      6 3
      0 2
        0
        25
         0
        252
        252
          0
```

Note: There **must** be a digit to the right of the decimal point in the quotient above every digit to the right of the decimal point in the dividend.

In the examples so far, the remainder has eventually been 0. This will not always be so. We must decide ahead of time how accurate the quotient is to be.

When the remainder is not zero:

1. Decide first how many decimal places are to be in the quotient.
2. Divide until the quotient is one digit past the place of desired accuracy.
3. Using this last digit, round off the quotient to the desired place of accuracy.

EXAMPLES

4. Find 8.24 ÷ 2.9 to the nearest tenth.

Solution

Divide until the quotient is in hundredths (one more place than tenths), then round off to tenths.

```
        2.84 ≈ 2.8     ← read approximately
2.9.)8.2.40
      5 8
      2 4 4
      2 3 2
        1 20
        1 16
           4
```

$8.24 \div 2.9 \approx 2.8$ accurate to the nearest tenth.

5. Find 17 ÷ 3.3 to the nearest thousandth.

Solution

Divide until the quotient is in ten-thousandths; then round off to thousandths.

```
           5.1515 ≈ 5.152
   3.3.)17.0.0000
        16 5           add as many 0's as needed
           5 0
           3 3
           1 70
           1 65
              50
              33
              170
              165
                5
```

$17 \div 3.3 \approx 5.152$ accurate to thousandths.

> **To divide a decimal number by a power of 10:**
>
> 1. Move the decimal point to the **left.**
> 2. Move it the same number of places as the number of 0's in the power of 10.
>
> Division by 10 moves the decimal point **one** place **to the left.**
> Division by 100 moves the decimal point **two** places **to the left.**
> Division by 1000 moves the decimal point **three** places **to the left,** and so on.

EXAMPLES

6. $4.16 \div 100 = \dfrac{4.16}{100} = 0.0416$

7. $782 \div 10 = \dfrac{782}{10} = 78.2$

8. $\dfrac{593.3}{1000} = 0.5933$

9. $\dfrac{186.4}{100} = 1.864$

As a general comment for understanding work with powers of ten:

1. Multiplying by a power of ten will make a number larger, so move the decimal point to the right.
2. Dividing by a power of ten will make a number smaller, so move the decimal point to the left.

PRACTICE QUIZ	Find each of the indicated quotients.	ANSWERS
	1. $4\overline{)1.83}$ (nearest hundredth)	**1.** 0.46
	2. $.06\overline{)43.721}$ (nearest thousandth)	**2.** 728.683
	3. $\frac{42.31}{1000}$	**3.** 0.04231
	4. $10^3(42.31)$	**4.** 42,310

Exercises 5.5

Find each quotient to the nearest tenth.

1. $4.68 \div 2$ **2.** $1.71 \div 3$ **3.** $4.95 \div 0.5$

4. $1.62 \div 0.9$ **5.** $0.064 \div 0.8$ **6.** $0.63 \div 0.7$

7. $82.24 \div 0.04$ **8.** $16.02 \div 0.03$ **9.** $48 \div 2.4$

10. $28 \div 5.6$ **11.** $8.7\overline{)45.62}$ **12.** $5.3\overline{)26.32}$

13. $9.4\overline{)6.538}$ **14.** $4.6\overline{)5}$ **15.** $7.05\overline{)0.4977}$

16. $0.37\overline{)4.683}$ **17.** $0.23\overline{)65.226}$ **18.** $1.62\overline{)34}$

19. $1.33\overline{)75}$

Find each quotient to the nearest hundredth.

20. $24\overline{)0.1463}$ **21.** $1.23\overline{)14.91129}$ **22.** $0.075\overline{)0.42753}$

23. $2.7\overline{)2.583}$ **24.** $23\overline{)62.949}$ **25.** $9\overline{)2}$

26. $26\overline{)5.729}$ **27.** $13\overline{)65.476}$ **28.** $3.181\overline{)6}$

Find each quotient to the nearest thousandth.

29. $.023\overline{)0.71}$ **30.** $6.9\overline{)29.3}$ **31.** $85.3\overline{)24.31}$

32. $2.57\overline{)0.4961}$ **33.** $13\overline{)1.029}$ **34.** $14\overline{)4.073}$

35. $16.2\overline{)0.11623}$ **36.** $25.7\overline{)6.27}$ **37.** $0.23\overline{)45.221}$

Divide as indicated.

38. $78.4 \div 100$ **39.** $50.36 \div 100$ **40.** $45.621 \div 1000$

41. $18.6 \div 1000$ **42.** $\dfrac{167}{10}$ **43.** $\dfrac{138.1}{10}$

44. $\dfrac{1.54}{10{,}000}$ **45.** $\dfrac{169.9}{10{,}000}$

Multiply as indicated.

46. 10(167) **47.** 10(138.1) **48.** 10,000(1.54)

49. 10,000(169.9)

50. Find the average of the numbers 86.7, 49.2, and 75.4, correct to the nearest tenth.

51. If a car averages 24.6 miles per gallon, how many miles will it go on 18 gallons of gas?

52. If a motorcycle averages 32.4 miles per gallon, how many miles will it go on 7 gallons of gas?

53. If a car travels 300 miles on 16 gallons of gas, how many miles does it travel per gallon?

54. If a bicyclist rode 250.6 miles in 13.2 hours, what was her average speed (to the nearest tenth)?

55. A quarter section of beef can be bought cheaper than the same amount of meat purchased a few pounds at a time. What is the cost per pound if 150 pounds cost $187.50?

56. If you drive 9.5 hours at an average speed of 52.2 miles per hour, how far will you drive?

57. If new tires cost $56.50 per tire and tax is figured at 0.06 times the cost of each tire, what will you pay for 4 new tires?

58. If you bought 10 books for a total price of $225 plus tax at 0.06 times the price, what average amount did you pay per book including tax?

59. If the total price of a stereo was $312.70 including tax at 0.06 times the list price, you can find the list price by dividing the total price by 1.06. What was the list price?

60. If the interest paid on a 30-year mortgage for a home loan of $60,000 is going to be $189,570, what will the monthly payments be on the loan and interest? (This does not include insurance or taxes.)

5.6 DECIMALS AND FRACTIONS

A decimal number can be written in fraction form by writing a fraction with

1. Numerator: a whole number with all the digits of the decimal number.
2. Denominator: the power of ten that names the rightmost digit.

For example,

$$0.25 = \frac{25}{100} \quad \text{and} \quad 0.025 = \frac{25}{1000}$$

5 is in the hundredths position   5 is in the thousandths position.

We may or may not want to reduce some fractions. If we choose to reduce, factoring works just as it did in Chapter 4.

EXAMPLES

Change each decimal number to fraction form and reduce if possible.

1. $0.25 = \frac{25}{100} = \frac{\cancel{5} \cdot \cancel{5} \cdot 1}{2 \cdot \cancel{5} \cdot 2 \cdot \cancel{5}} = \frac{1}{4}$

   ↑ hundredths

2. $0.32 = \frac{32}{100} = \frac{\cancel{4} \cdot 8}{\cancel{4} \cdot 25} = \frac{8}{25}$

   ↑ hundredths

3. $0.131 = \frac{131}{1000}$

   ↑ thousandths

4. $0.075 = \frac{75}{1000} = \frac{\cancel{25} \cdot 3}{\cancel{25} \cdot 40} = \frac{3}{40}$

   ↑ thousandths

5. $2.6 = \frac{26}{10} = \frac{\cancel{2} \cdot 13}{\cancel{2} \cdot 5} = \frac{13}{5}$

   ↑ tenths

   or, as a mixed number,

$$2.6 = 2\frac{6}{10} = 2\frac{3}{5}$$

6. $1.42 = \frac{142}{100} = \frac{\not{2} \cdot 71}{\not{2} \cdot 50} = \frac{71}{50}$

($\uparrow$ hundredths)

or, as a mixed number,

$$1.42 = 1\frac{42}{100} = 1\frac{21}{50}$$

> To change a fraction into a decimal number, divide the numerator by the denominator.
>
> 1. If the remainder is 0, the decimal is said to be **terminating.**
> 2. If the remainder is not 0, the decimal is said to be **nonterminating.**

The following examples illustrate fractions that convert to terminating decimals.

EXAMPLES Change each fraction to a decimal number.

7. $\frac{3}{8}$

$$\begin{array}{r} .375 \\ 8\overline{)3.000} \\ \underline{2\,4} \\ 60 \\ \underline{56} \\ 40 \\ \underline{40} \\ 0 \end{array}$$

$\frac{3}{8} = .375$

8. $\frac{3}{4}$

$$\begin{array}{r} .75 \\ 4\overline{)3.00} \\ \underline{2\,8} \\ 20 \\ \underline{20} \\ 0 \end{array}$$

$\frac{3}{4} = .75$

9. $\frac{4}{5}$

$$\begin{array}{r} .8 \\ 5\overline{)4.0} \\ \underline{4\,0} \\ 0 \end{array}$$

$\frac{4}{5} = .8$

Nonterminating decimals can be **repeating** or **nonrepeating.** A repeating nonterminating decimal has a repeating pattern to its digits. Every fraction that has a whole number numerator and denominator will be either a terminating decimal or a repeating decimal. Nonrepeating decimals will be discussed in detail in later courses.

The following examples illustrate nonterminating repeating decimals.

EXAMPLES

10. $\frac{1}{3}$

$$\begin{array}{r} .333 \\ 3\overline{)1.000} \\ \underline{9} \\ 10 \\ \underline{9} \\ 10 \\ \underline{9} \\ 1 \end{array}$$

The 3 will repeat without end.

Continuing to divide will give a remainder of 1 each time.

We write $\frac{1}{3} = 0.333\ldots$ The three dots mean "and so on" or to continue without stopping.

11. $\frac{7}{12}$

$$\begin{array}{r} .5833 \\ 12\overline{)7.0000} \\ \underline{6\,0} \\ 1\,00 \\ \underline{96} \\ 40 \\ \underline{36} \\ 40 \\ \underline{36} \\ 4 \end{array}$$

The 3 will repeat without end.

Continuing to divide will give a remainder of 4 each time.

We write $\frac{7}{12} = 0.58333\ldots.$

12. $\frac{1}{7}$

$$\begin{array}{r} .142857 \\ 7\overline{)1.000000} \\ \underline{7} \\ 30 \\ \underline{28} \\ 20 \\ \underline{14} \\ 60 \\ \underline{56} \\ 40 \\ \underline{35} \\ 50 \\ \underline{49} \\ 1 \end{array}$$

The six digits will repeat in the the same pattern without end.

The remainder will repeat in sequence 1, 3, 2, 6, 4, 5, 1, and so on. Therefore, the digits in the quotient will also repeat.

We write $\frac{1}{7} = 0.142857142857142857\ldots.$

Another way to write repeating decimals is to write a **bar** over the repeating digits. Thus, in Examples 10, 11, and 12, we can write

$$\frac{1}{3} = 0.\overline{3} \quad \text{and} \quad \frac{7}{12} = 0.58\overline{3} \quad \text{and} \quad \frac{1}{7} = 0.\overline{142857}$$

We may choose to round off the quotient to some decimal place just as was done with division in Section 5.5. Perform the division one place past the desired round-off position.

EXAMPLES

Find the decimal representation of each fraction to the nearest hundredth.

13. $\frac{5}{11}$

$$\begin{array}{r} .454 \\ 11\overline{)5.000} \\ \underline{4\,4} \\ 60 \\ \underline{55} \\ 50 \\ \underline{44} \end{array} \approx .45 \quad \text{(nearest hundredth)}$$

14. $\frac{5}{6}$

$$\begin{array}{r} .833 \\ 6\overline{)5.000} \\ \underline{4\,8} \\ 20 \\ \underline{18} \\ 20 \\ \underline{18} \end{array} \approx .83 \quad \text{(nearest hundredth)}$$

Exercises 5.6

Change each decimal to fraction form. Do not reduce.

1. 0.9 **2.** 0.3 **3.** 0.5 **4.** 0.8

5. 0.62 **6.** 0.38 **7.** 0.57 **8.** 0.41

9. 0.526 **10.** 0.625 **11.** 0.016 **12.** 0.012

13. 5.1 **14.** 7.2 **15.** 8.15 **16.** 6.35

Change each decimal to fraction form (or mixed number form) and reduce if possible.

17. 0.125 **18.** 0.36 **19.** 0.18 **20.** 0.375

21. 0.225 **22.** 0.455 **23.** 0.17 **24.** 0.029

25. 3.2 **26.** 1.25 **27.** 6.25 **28.** 2.75

Change each fraction to decimal form. If the decimal is nonterminating, write it using the bar notation over the repeating pattern of digits.

29. $\frac{2}{3}$ 30. $\frac{5}{16}$ 31. $\frac{7}{11}$ 32. $\frac{3}{11}$

33. $\frac{11}{16}$ 34. $\frac{9}{16}$ 35. $\frac{3}{7}$ 36. $\frac{5}{7}$

37. $\frac{1}{6}$ 38. $\frac{5}{18}$ 39. $\frac{5}{9}$ 40. $\frac{2}{9}$

Change each fraction to decimal form rounded off to the nearest thousandth.

41. $\frac{7}{24}$ 42. $\frac{16}{33}$ 43. $\frac{5}{12}$ 44. $\frac{13}{16}$

45. $\frac{1}{32}$ 46. $\frac{1}{14}$ 47. $\frac{16}{13}$ 48. $\frac{20}{9}$

49. $\frac{30}{21}$ 50. $\frac{40}{3}$

SUMMARY: CHAPTER 5

DEFINITION A **decimal number** is a rational number that has a power of ten as its denominator.

To read or write a decimal number:

1. Read (or write) the whole number as before.
2. Read (or write) **and** in place of the decimal point.
3. Read (or write) the fraction part as a whole number with the name of the place of the last digit.

SPECIAL NOTES

1. The **th** at the end of a word indicates a fraction part (a part to the right of the decimal point).
2. The hyphen (-) indicates one word.

To **round off** a number means to find another number close to the original number.

RULE FOR ROUNDING OFF DECIMAL NUMBERS

1. Look at the single digit just to the right of the digit that is in the place of desired accuracy.
2. If this digit is 5 or greater, make the digit in the desired place of accuracy one larger and replace all digits to the right with zeros.
3. If this digit is less than 5, leave the digit that is in the place of desired accuracy as it is and replace all digits to the right with zeros.

To add decimal numbers:

1. Write the numbers one under the other.
2. Keep the decimal points in line.
3. Keep digits with the same position value in line (zeros may be filled in as aids).
4. Add the numbers, placing the decimal point in the answer in line with the other decimal points.

To subtract decimal numbers:

1. Write the numbers one under the other.
2. Line up the decimal points.
3. Keep digits with the same position value in line (zeros may be filled in).
4. Subtract, placing the decimal point in the answer in line with the other decimal points.

To multiply decimal numbers:

1. Multiply the two numbers as if they were whole numbers.
2. Count the total number of places to the right of the decimal points in both numbers being multiplied.
3. This sum is the number of places to the right of the decimal point in the product.

To multiply a decimal number by a power of 10:

1. Move the decimal point to the **right.**
2. Move it the same number of places as the number of 0's in the power of 10.

To divide decimal numbers:

1. Move the decimal point in the divisor so it is a whole number.
2. Move the decimal point in the dividend the same number of places to the right.
3. Place the decimal point in the quotient directly above the new decimal point in the dividend.
4. Divide just as with whole numbers.

To divide a decimal number by a power of 10:

1. Move the decimal point to the **left.**
2. Move it the same number of places as the number of 0's in the power of 10.

To change a fraction into a decimal number, divide the numerator by the denominator.

(a) If the remainder is 0, the decimal is said to be **terminating.**

(b) If the remainder is not 0, the decimal is said to be **nonterminating.**

REVIEW QUESTIONS: CHAPTER 5

1. A decimal number is a rational number that has a power of ______ as its denominator.

Write the following decimal numbers in words.

2. 0.4 **3.** 7.08 **4.** 92.137 **5.** 18.5526

Write the following decimal numbers in mixed number form.

6. 81.47 **7.** 100.03 **8.** 9.592 **9.** 200.5

Write the following numbers in decimal notation.

10. two and seventeen hundredths

11. eighty-four and seventy-five thousandths

12. three thousand three and three thousandths

Round off as indicated.

13. 5863 (nearest hundred)

14. 7.649 (nearest tenth)

15. 0.0385 (nearest thousandth)

16. 2.069876 (nearest hundred-thousandth)

Add or subtract as indicated.

17. 5.4 + 7.34 + 14.08

18. 3 + 7.86 + 52.891 + 0.4

19. 34.967 + 40.8 + 9.451 + 8.2

20. 32.5 − 14.71

21. 16.92 − 7.9

22. 5 − 1.0377

23. Add $\begin{array}{r} 78.6 \\ 9.683 \\ +15.989 \\ \hline \end{array}$

24. Subtract $\begin{array}{r} 42.008 \\ -19.3 \\ \hline \end{array}$

Multiply.

25. (.8)(.9) **26.** (.2)(.1) **27.** (.02)(.32)

28. 100(2.35) **29.** 10(.17632) **30.** $10^3(5.9641)$

31. $\begin{array}{r} 2.4 \\ \times\ .05 \\ \hline \end{array}$ **32.** $\begin{array}{r} 1.08 \\ \times\ .16 \\ \hline \end{array}$ **33.** $\begin{array}{r} 36.5 \\ \times\ 4.7 \\ \hline \end{array}$

Divide. (Round off to the nearest hundredth.)

34. $4\overline{)2.83}$ **35.** $.06\overline{)52.832}$ **36.** $1.003\overline{)200.6}$

Divide by moving the decimal point the correct number of places.

37. $\dfrac{296.1}{100}$ **38.** $\dfrac{5.67}{10^3}$ **39.** $\dfrac{19.435}{10}$

40. Find the average (to the nearest tenth) of 16.5, 23.4, and 30.7.

Change to fraction (or mixed number) form and reduce if possible.

41. 0.07 **42.** 2.025 **43.** 0.015

Change to decimal form. If the decimal is nonterminating, write it using a bar over the repeating digits.

44. $\frac{1}{3}$ **45.** $\frac{5}{8}$ **46.** $2\frac{4}{9}$

Change to decimal form rounded off to the nearest thousandth.

47. $\frac{15}{17}$ **48.** $\frac{99}{101}$

TEST: CHAPTER 5

1. Write 0.036 as a fraction and reduce to lowest terms.

2. Write $\frac{37}{400}$ as a decimal. **3.** Write 300.03 in word form.

4. Write in decimal notation: five thousand and thirty-two hundredths.

Round off as indicated.

5. 193.182 to the nearest hundredth

6. 193.182 to the nearest hundred

7. 193.182 to the nearest tenth

Perform the indicated operations. Round off all quotients to the nearest hundredths.

8. 52.536 + 46.849 **9.** 19 − 3.08

10. 10 + 12.3 + 19.47 **11.** 0.103 − 0.07921

12. 7.3 + 0.92 + 21.307 **13.** 50.872 − 36.938

14. (0.56)(0.025) **15.** 364 ÷ 0.052

16. $\begin{array}{r} 41.62 \\ \times 0.134 \\ \hline \end{array}$ **17.** $17.1\overline{)51.82}$

18. $10^3(13.85)$ **19.** $21 \div 10^4$

20. Jeremy used 21 gallons of gas to drive 419 miles. What was his average miles per gallon (to the nearest tenth)?

6 Ratio and Proportion

6.1 RATIO AND PRICE PER UNIT

We know that a fraction can be used to:

1. Indicate a part of a whole

$$\frac{3}{8} \quad \text{means} \quad \frac{\text{3 pieces of cherry pie}}{\text{8 pieces in the whole pie}}$$

2. Indicate division

$$\frac{3}{8} \quad \text{means} \quad 3 \div 8 \quad \text{or} \quad \begin{array}{r} .375 \\ 8\overline{)3.000} \\ \underline{2\,4} \\ 60 \\ \underline{56} \\ 40 \\ \underline{40} \\ 0 \end{array}$$

A third use of fractions is to compare two numbers. For example,

$$\frac{6}{2} \quad \text{might mean} \quad \frac{\text{6 minutes}}{\text{2 hours}} \quad \text{or} \quad \frac{\text{6 apples}}{\text{2 oranges}}$$

In these comparisons, we must know the units. Such a comparison is called a **ratio.**

DEFINITION A **ratio** is a comparison of two quantities. The ratio of a to b can be written $\frac{a}{b}$ or $a : b$ or a to b

To avoid confusion, the units in a ratio should be written down or otherwise explained in the problem. If possible, the units should be the same. Generally, the ratio should also be reduced.

EXAMPLES

1. During baseball season, major league players' batting averages are in the newspapers. What does a batting average of .300 indicate?

   Solution

   A batting average is a ratio of hits to times at bat. Thus, .300 means

   $$.300 = \frac{300 \text{ hits}}{1000 \text{ times at bat}}$$

   Reducing gives

   $$.300 = \frac{300}{1000} = \frac{3 \cdot \cancel{100}}{10 \cdot \cancel{100}} = \frac{3 \text{ hits}}{10 \text{ times at bat}}$$

   This means that you can expect this player to get 3 hits every 10 times he comes to bat. (A more detailed explanation would be discussed in courses in probability and/or statistics.)

2. Write the comparison of 2 feet to 5 yards as a ratio.

   Solution

   (a) We can write $\frac{2 \text{ feet}}{5 \text{ yards}}$.

   (b) Changing to common units gives us a different look at the same ratio. Since there are 3 feet in a yard, 5 yards = 15 feet. So,

   $$\frac{2 \text{ feet}}{5 \text{ yards}} = \frac{2 \text{ feet}}{15 \text{ feet}} \quad \text{or} \quad 2 : 15 \quad \text{or} \quad 2 \text{ to } 15$$

3. What is the reduced ratio of 300 centimeters (cm) to 2 meters (m)? Centimeters and meters are units used in the metric system and will be discussed in detail in Chapter 10.

   Solution

   There are 100 centimeters in a meter. The ratio is

   $$\frac{300 \text{ cm}}{2 \text{ m}} = \frac{300 \text{ cm}}{200 \text{ cm}} = \frac{3}{2}$$

   The reduced ratio can also be written 3 : 2 or 3 to 2.

4. What is the ratio of 8 ounces (oz) to 1 pound (lb)?

Solution

There are 16 ounces in 1 pound. So, the ratio is

$$\frac{8 \text{ oz}}{1 \text{ lb}} = \frac{8 \text{ oz}}{16 \text{ oz}} = \frac{1}{2} \quad \text{or} \quad 1:2 \quad \text{or} \quad 1 \text{ to } 2$$

When you buy groceries, you want to get the most for your money. When the same item is in two (or more) different sized packages, you want the better (or best) buy. By using a ratio of price to units, you can determine the **price per unit** or **unit price.**

To find the **price per unit,** divide the cost by the number of units.

EXAMPLES

5. A 12-ounce can of beans is priced at 80¢ while an 18-ounce can of the same beans is $1.10. Which is the better buy?

Solution

We write the ratio of price to unit and divide.

(a) $\frac{80¢}{12 \text{ oz}} = \frac{6.7¢}{1 \text{ oz}}$ or 6.7¢ per ounce

$$\begin{array}{r} 6.66 \\ 12\overline{)80.00} \\ \underline{72} \\ 8\,0 \\ \underline{7\,2} \\ 80 \\ \underline{72} \end{array}$$

(b) $\frac{110¢}{18 \text{ oz}} = \frac{6.2¢}{1 \text{ oz}}$ or 6.2¢ per ounce

$$\begin{array}{r} 6.11 \\ 18\overline{)110.00} \\ \underline{108} \\ 2\,0 \\ \underline{1\,8} \\ 20 \\ \underline{18} \end{array}$$

In this case, the larger can (18 ounces for $1.10) is the better buy since the price per ounce is smaller.

NOTE: $\frac{110}{18} = 6.11$ would round off to 6.1, but with money, we take the next larger digit. So, 6.2 is the reported number.

6. Pancake syrup comes in three different sized bottles: 36 fluid ounces for \$3.29, 24 fluid ounces for \$2.49, and 12 fluid ounces for \$1.59. Find the price per fluid ounce for each size of bottle and tell which is the best buy.

Solution

(a) $\dfrac{\$3.29}{36\text{ oz}} = \dfrac{329¢}{36\text{ oz}} = \dfrac{9.2¢}{1\text{ oz}} = 9.2¢/\text{oz}$

(b) $\dfrac{\$2.49}{24\text{ oz}} = \dfrac{249¢}{24\text{ oz}} = \dfrac{10.4¢}{1\text{ oz}} = 10.4¢/\text{oz}$

(c) $\dfrac{\$1.59}{12\text{ oz}} = \dfrac{159¢}{12\text{ oz}} = \dfrac{13.3¢}{1\text{ oz}} = 13.3¢/\text{oz}$

The largest container (36 fluid ounces) is the best buy. In actual practice, people who do not use much pancake syrup may want to buy the smallest bottle. Even though they pay more per unit, they do not end up throwing any into the trash, which would be more expensive in the long run.

Exercises 6.1

Write the following comparisons as ratios reduced to lowest terms. Use common units in the numerator and denominator whenever possible. Problems 6–8 use metric measures. (See Section 9.1 as a reference.)

1. 1 dime to 3 nickels
2. 30 chairs to 25 students
3. 2 yards to 5 feet
4. 18 inches to 2 feet
5. 3 bookshelves to 18 feet of lumber
6. 10 centimeters to 1 decimeter
7. 100 centimeters to 1 meter
8. 1000 millimeters to 1 meter
9. \$525 to 100 stocks in the stock market
10. 38 miles to 2 gallons of gas
11. 5 minutes to 2 hours
12. 5 ounces to 2 pounds

13. 4 quarters to 1 dollar
14. 5 quarters to 3 dollars
15. 6 ounces to 3 pounds
16. 125 hits to 500 times at bat
17. 12 inches to 2 feet
18. 80 miles to 4 gallons of gas
19. 1800 feet to 1 second (speed of sound)
20. 300,000,000 meters to 1 second (speed of light)

Find the unit price of each of the following items and tell which is the better (or best) buy.

21. boxed rice
    16 oz (1 lb) at 89¢
    32 oz (2 lb) at $1.69

22. noodles
    42 oz at $2.99
    14 oz at $1.39

23. sour cream
    8 oz ($\frac{1}{2}$ pt) at 69¢
    11 oz at $1.29

24. coffee
    48 oz (3 lb) at $7.29
    13 oz at $2.45

25. instant coffee
    8 oz at $3.49
    2 oz at $1.29

26. instant coffee
    12 oz at $4.49
    8 oz at $3.69

27. frozen orange juice
    16 fl oz at $1.69
    12 fl oz at 99¢
    6 fl oz at 69¢

28. tortilla chips
    11 oz at $2.32
    16 oz at $2.74
    7.5 oz at $1.72

29. boxed dry milk
    9.6 oz at $1.49
    25.6 oz at $3.59

30. cookies
    16 oz at $1.99
    20 oz at $2.59

31. saltine crackers
    8 oz at 99¢
    16 oz at $1.19

32. peanut butter
    18 oz at $1.85
    28 oz at $2.69

33. aluminum foil
    200 sq ft at $4.19
    75 sq ft at $1.59
    25 sq ft at 69¢

34. liquid dish soap
    32 oz at $2.29
    22 oz at $1.69
    12 oz at $1.09

35. sliced bologna
    8 oz at $1.09
    12 oz at $1.59

36. sliced ham
    16 oz at $4.69
    8 oz at $2.59

37. laundry detergent
9 lb 3 oz (147 oz) at $5.99
72 oz at $3.99
42 oz at $2.39
17 oz at $1.23

38. peanut butter
12 oz at $1.29
18 oz at $1.89
28 oz at $2.79
40 oz at $3.89

39. bottled bleach
1 gal (128 fl oz) at $1.14
$\frac{1}{2}$ gal (64 fl oz) at 85¢
1 qt (32 fl oz) at 59¢

40. jelly
18 oz at $1.29
32 oz at $1.69
48 oz at $2.49

41. boxed doughnuts
14 oz at $1.49
9 oz at $1.09
5 oz at 69¢

42. mustard
8 oz at 65¢
12 oz at $1.09
24 oz at $1.19

43. mayonnaise
16 oz at $1.23
32 oz at $1.69

44. salad dressing
8 oz at $1.09
16 oz at $1.69

45. mustard
8 oz at 69¢
16 oz at $1.09

46. catsup
14 oz at 83¢
32 oz at 99¢

47. chili with beans
40 oz at $2.29
15 oz at 89¢

48. dill pickles
32 oz at $1.59
16 oz at $1.19

49. chili beans
30 oz at 89¢
15.5 oz at 59¢

50. baked beans
31 oz at 79¢
16 oz at 53¢

6.2 PROPORTIONS

The statement $\frac{6}{8} = \frac{3}{4}$ is an equation that says that the two ratios $\frac{6}{8}$ and $\frac{3}{4}$ are equal. Is this true? Is the equation $\frac{5}{8} = \frac{7}{10}$ true? These equations are called **proportions,** and we need some method to determine whether a proportion is true or false.

DEFINITION A **proportion** is a statement that two ratios are equal. In symbols,

$$\frac{a}{b} = \frac{c}{d} \quad \text{is a proportion.}$$

In a proportion, the numbers are called **terms** and are identified as follows.

A proportion has four terms:

$$\frac{a}{b} = \frac{c}{d}$$

first term (a), second term (b), third term (c), fourth term (d)

Extremes: Terms 1 and 4 (a and d)

Means: Terms 2 and 3 (b and c)

EXAMPLES

1. $\frac{9}{10} = \frac{18}{20}$ is a proportion.

   9 and 20 are the extremes.
   10 and 18 are the means.

2. $\frac{8.4}{4.2} = \frac{10.2}{5.1}$ is a proportion.

   8.4 and 5.1 are the extremes.
   4.2 and 10.2 are the means.

A proportion is true if the product of the extremes equals the product of the means.

$\frac{a}{b} = \frac{c}{d}$ if and only if $a \cdot d = b \cdot c$, where $b \neq 0$ and $d \neq 0$.

NOTE: The terms can be decimal numbers, fractions, mixed numbers, or whole numbers. The rules are the same.

EXAMPLES

Determine whether the following proportions are true or false.

3. $\frac{9}{13} = \frac{4.5}{6.5}$

Solution

$$\begin{array}{r} 6.5 \\ \times\ 9 \\ \hline 58.5 \end{array} \qquad \begin{array}{r} 4.5 \\ \times\ 13 \\ \hline 13\ 5 \\ 45\ \ \\ \hline 58.5 \end{array}$$

Since $\underbrace{9(6.5)}_{\text{extremes}} = \underbrace{13(4.5)}_{\text{means}}$, the proportion is true.

4. $\frac{5}{8} = \frac{7}{10}$

Solution

$$5 \cdot 10 = 50 \quad \text{and} \quad 8 \cdot 7 = 56$$

Since $50 \neq 56$, the proportion is false.

5. $\frac{1}{4} : \frac{2}{3} = \frac{9}{10} : \frac{12}{5}$ The colon can be used to indicate ratios.

Solution

The proportion can be written $\dfrac{\frac{1}{4}}{\frac{2}{3}} = \dfrac{\frac{9}{10}}{\frac{12}{5}}$.

The extremes are $\frac{1}{4}$ and $\frac{12}{5}$.

The means are $\frac{2}{3}$ and $\frac{9}{10}$.

$$\frac{1}{4} \cdot \frac{12}{5} = \frac{1 \cdot \not{4} \cdot 3}{\not{4} \cdot 5} = \frac{3}{5} \quad \text{and} \quad \frac{2}{3} \cdot \frac{9}{10} = \frac{\not{2} \cdot \not{3} \cdot 3}{\not{3} \cdot \not{2} \cdot 5} = \frac{3}{5}$$

Since the product of the extremes equals the product of the means $\left(\frac{3}{5} = \frac{3}{5}\right)$, the proportion is true.

Exercises 6.2

List the means and extremes for each of the following proportions.

1. $\frac{4}{5} = \frac{20}{25}$ **2.** $\frac{4}{9} = \frac{12}{27}$ **3.** $\frac{\frac{1}{2}}{8} = \frac{3}{48}$ **4.** $\frac{8}{12} = \frac{6}{9}$

Determine whether each proportion is true or false.

5. $\frac{5}{6} = \frac{10}{12}$ **6.** $\frac{2}{7} = \frac{5}{17}$ **7.** $\frac{7}{21} = \frac{4}{12}$ **8.** $\frac{6}{15} = \frac{2}{5}$

9. $\frac{5}{8} = \frac{12}{17}$ **10.** $\frac{12}{15} = \frac{20}{25}$ **11.** $\frac{5}{3} = \frac{15}{9}$ **12.** $\frac{6}{8} = \frac{15}{20}$

13. $\frac{2}{5} = \frac{4}{10}$ **14.** $\frac{3}{5} = \frac{60}{100}$ **15.** $\frac{125}{1000} = \frac{1}{8}$ **16.** $\frac{3}{8} = \frac{375}{1000}$

17. $\frac{1}{4} = \frac{25}{100}$ **18.** $\frac{7}{8} = \frac{875}{1000}$ **19.** $\frac{3}{16} = \frac{9}{48}$ **20.** $\frac{2}{3} = \frac{66}{100}$

21. $\frac{1}{3} = \frac{33}{100}$ **22.** $\frac{14}{6} = \frac{21}{8}$ **23.** $\frac{4}{9} = \frac{7}{12}$ **24.** $\frac{19}{16} = \frac{20}{17}$

25. $\frac{3}{6} = \frac{4}{8}$ **26.** $\frac{12}{18} = \frac{14}{21}$ **27.** $\frac{5}{6} = \frac{7}{8}$ **28.** $\frac{7.5}{10} = \frac{3}{4}$

29. $\frac{6.2}{3.1} = \frac{10.2}{5.1}$ **30.** $\frac{8\frac{1}{2}}{2\frac{1}{3}} = \frac{4\frac{1}{4}}{1\frac{1}{6}}$ **31.** $\frac{6\frac{1}{5}}{1\frac{1}{7}} = \frac{3\frac{1}{10}}{\frac{8}{14}}$

32. $\frac{6}{24} = \frac{10}{48}$ **33.** $\frac{7}{16} = \frac{3\frac{1}{2}}{8}$ **34.** $\frac{10}{17} = \frac{5}{8\frac{1}{2}}$

35. $3 : 6 = 5 : 10$ **36.** $6 : 16 = 9 : 24$ **37.** $210 : 7 = 20 : \frac{2}{3}$

38. $3.75 : 3 = 7.5 : 6$ **39.** $3 : 5 = 60 : 100$ **40.** $12 : 1.09 = 36 : 3.27$

41. $2\frac{1}{2} : \frac{3}{4} = 1\frac{1}{2} : \frac{2}{3}$ **42.** $1\frac{1}{4} : 1\frac{1}{2} = \frac{1}{4} : \frac{1}{2}$ **43.** $6 : 1.56 = 2 : 0.52$

44. $3\frac{1}{5} : 1 = 3\frac{3}{5} : 1\frac{2}{5}$ **45.** $8.5 : 6.5 = 4.5 : 3.5$

6.3 SOLVING PROPORTIONS

As we will see in Section 6.4, proportions are useful in solving certain types of word problems. In these applications, there will be some unknown quantity that can be represented as one term in a proportion. Finding the value of this unknown term is called **solving the proportion.**

To find the unknown term in a proportion:

1. Write the proportion representing the unknown term with some letter, such as x, y, w, A, B, etc.
2. Write an equation that sets the product of the extremes equal to the product of the means.
3. Divide both sides of the equation by the number multiplying the unknown. (This number is called the **coefficient** of the unknown.)
4. The resulting equation gives the missing value for the unknown.

IMPORTANT NOTE: Write one equation under the other, just as in the examples.

EXAMPLES

1. Find x if $\frac{3}{6} = \frac{5}{x}$.

Solution

(a) Write the proportion: $\frac{3}{6} = \frac{5}{x}$

(b) Write product of extremes equal to product of means: $3 \cdot x = 6 \cdot 5$

(c) Divide both sides by 3, the coefficient of x: $\frac{3 \cdot x}{3} = \frac{30}{3}$

(d) Reduce to find the solution: $\frac{\cancel{3} \cdot x}{\cancel{3}} = \frac{\overset{10}{\cancel{30}}}{\cancel{3}}$

$x = 10$

2. Find y if $\frac{6}{16} = \frac{y}{24}$.

Solution

(a) Write the proportion: $\frac{6}{16} = \frac{y}{24}$

(b) Write product of extremes equal to product of means: $6 \cdot 24 = 16 \cdot y$

(c) Divide both sides by 16, the coefficient of y:

$$\frac{6 \cdot 24}{16} = \frac{16 \cdot y}{16}$$

(d) Reduce to find the missing value:

$$\frac{\overset{3}{\cancel{6}} \cdot \overset{3}{\cancel{24}}}{\underset{\cancel{2}}{\cancel{16}}} = \frac{\cancel{16} \cdot y}{\cancel{16}}$$

$$9 = y$$

Second Solution

(a) Write the proportion:

$$\frac{6}{16} = \frac{y}{24}$$

(b) Reduce the fraction $\frac{6}{16}$ to $\frac{3}{8}$:

$$\frac{3}{8} = \frac{y}{24}$$

(c) Solve as before:

$$3 \cdot 24 = 8 \cdot y$$

$$\frac{72}{8} = \frac{\cancel{8} \cdot y}{\cancel{8}}$$

$$9 = y$$

3. Find w if $\frac{w}{7} = \frac{20}{\frac{2}{3}}$.

Solution

(a) Write the proportion:

$$\frac{w}{7} = \frac{20}{\frac{2}{3}}$$

(b) Set product of extremes equal to product of means:

$$w \cdot \frac{2}{3} = 7 \cdot 20$$

(c) Divide both sides by $\frac{2}{3}$ $\left(\text{or multiply by } \frac{3}{2}, \text{ the reciprocal of } \frac{2}{3}\right)$:

$$\frac{w \cdot \frac{2}{3}}{\frac{2}{3}} = \frac{7 \cdot 20}{\frac{2}{3}} \quad \text{or} \quad w \cdot \frac{2}{3} \cdot \frac{3}{2} = 7 \cdot 20 \cdot \frac{3}{2}$$

(d) Reduce. Remember that to divide by a fraction, you multiply by its reciprocal:

$$\frac{w \cdot \frac{2}{\cancel{3}}}{\frac{2}{\cancel{3}}} = \frac{7 \cdot 20}{\frac{2}{3}} \quad \text{or} \quad w \frac{\cancel{2}}{\cancel{3}} \cdot \frac{\cancel{3}}{\cancel{2}} = 7 \cdot \overset{10}{\cancel{20}} \cdot \frac{3}{\cancel{2}}$$

$$w = 7 \cdot \overset{10}{\cancel{20}} \cdot \frac{3}{\cancel{2}} = 210 \qquad w = 210$$

4. Find A if $\frac{A}{9} = \frac{7.5}{6}$.

Solution

$$\frac{A}{9} = \frac{7.5}{6}$$

$$6 \cdot A = 9(7.5)$$

$$\frac{\cancel{6} \cdot A}{\cancel{6}} = \frac{67.5}{6}$$

$$A = 11.25$$

$$\begin{array}{r} 11.25 \\ 6\overline{)67.50} \\ \underline{6} \\ 07 \\ \underline{6} \\ 15 \\ \underline{12} \\ 30 \\ \underline{30} \end{array}$$

The unknown can be a decimal or fraction or mixed number as well as a whole number.

IMPORTANT NOTE AGAIN: Remember to write one equation under the other, just as in the examples. Even in Example 4, where the division is written to the right, the equations are aligned.

5. Find x if $\frac{x}{15} = \frac{5}{24}$

Solution

$$\frac{x}{15} = \frac{5}{24}$$

$$24 \cdot x = 5 \cdot 15$$

$$\frac{\cancel{24} \cdot x}{\cancel{24}} = \frac{75}{24}$$

$$x = \frac{\cancel{3} \cdot 5 \cdot 5}{\cancel{3} \cdot 8} = \frac{25}{8}$$

$$x = 3\frac{1}{8}$$

Here the unknown term is written as a mixed number. You might have chosen to divide $(25 \div 8)$ and get a decimal answer of 3.125. Both answers are correct.

PRACTICE QUIZ	Solve the following proportions.	ANSWERS
	1. $\frac{3}{5} = \frac{R}{100}$	1. $R = 60$
	2. $\frac{2\frac{1}{2}}{6} = \frac{3}{y}$	2. $y = \frac{36}{5}$ or $7\frac{1}{5}$
	3. $\frac{3}{2} = \frac{B}{4.5}$	3. 6.75

Exercises 6.3

Be sure to study the examples carefully step by step before you do these exercises.

Solve the following proportions.

1. $\frac{3}{6} = \frac{6}{x}$ **2.** $\frac{7}{21} = \frac{y}{6}$ **3.** $\frac{5}{7} = \frac{z}{28}$ **4.** $\frac{4}{10} = \frac{5}{x}$

5. $\frac{8}{B} = \frac{6}{30}$ **6.** $\frac{7}{B} = \frac{5}{15}$ **7.** $\frac{1}{2} = \frac{x}{100}$ **8.** $\frac{3}{4} = \frac{x}{100}$

9. $\frac{A}{3} = \frac{7}{2}$ **10.** $\frac{x}{100} = \frac{1}{20}$ **11.** $\frac{3}{5} = \frac{60}{D}$ **12.** $\frac{3}{16} = \frac{9}{x}$

13. $\frac{\frac{1}{2}}{x} = \frac{5}{10}$ **14.** $\frac{\frac{2}{3}}{3} = \frac{y}{127}$ **15.** $\frac{\frac{1}{3}}{x} = \frac{5}{9}$ **16.** $\frac{\frac{3}{4}}{7} = \frac{3}{z}$

17. $\frac{\frac{1}{8}}{6} = \frac{\frac{1}{2}}{w}$ **18.** $\frac{\frac{1}{6}}{5} = \frac{5}{w}$ **19.** $\frac{1}{4} = \frac{1\frac{1}{2}}{y}$ **20.** $\frac{1}{5} = \frac{x}{2\frac{1}{2}}$

21. $\frac{1}{5} = \frac{x}{7\frac{1}{2}}$ **22.** $\frac{2}{5} = \frac{R}{100}$ **23.** $\frac{3}{5} = \frac{R}{100}$ **24.** $\frac{A}{4} = \frac{75}{100}$

25. $\frac{A}{4} = \frac{50}{100}$ **26.** $\frac{20}{B} = \frac{1}{4}$ **27.** $\frac{30}{B} = \frac{25}{100}$ **28.** $\frac{A}{20} = \frac{15}{100}$

29. $\frac{1}{3} = \frac{R}{100}$ **30.** $\frac{2}{3} = \frac{R}{100}$ **31.** $\frac{9}{x} = \frac{4\frac{1}{2}}{11}$ **32.** $\frac{y}{6} = \frac{2\frac{1}{2}}{12}$

33. $\frac{x}{4} = \frac{1\frac{1}{4}}{5}$ **34.** $\frac{5}{x} = \frac{2\frac{1}{4}}{27}$ **35.** $\frac{x}{3} = \frac{16}{3\frac{1}{5}}$ **36.** $\frac{6.2}{5} = \frac{x}{15}$

37. $\frac{3.5}{2.6} = \frac{10.5}{B}$ **38.** $\frac{4.1}{3.2} = \frac{x}{6.4}$ **39.** $\frac{7.8}{1.3} = \frac{x}{0.26}$ **40.** $\frac{7.2}{y} = \frac{4.8}{14.4}$

41. $\frac{150}{300} = \frac{R}{100}$ **42.** $\frac{19.2}{96} = \frac{R}{100}$ **43.** $\frac{12}{B} = \frac{25}{100}$

44. $\frac{13.5}{B} = \frac{15}{100}$ **45.** $\frac{A}{42} = \frac{65}{100}$

6.4 SOLVING WORD PROBLEMS USING PROPORTIONS

To solve a word problem using proportions:

1. Read the problem carefully (several times if necessary).
2. Decide what is unknown (what you are trying to find) and assign some letter to represent it.
3. Set up a proportion using Pattern A or Pattern B. (Both patterns will give the same answer.)

   Pattern A: Each ratio has different units but they are in the same order. For example,

   $$\frac{500 \text{ miles}}{20 \text{ gallons}} = \frac{x \text{ miles}}{30 \text{ gallons}}$$

   Pattern B: Each ratio has the same units, the numerators correspond, and the denominators correspond. For example,

   $$\frac{500 \text{ miles}}{x \text{ miles}} = \frac{20 \text{ gallons}}{30 \text{ gallons}}$$

   (500 miles corresponds to 20 gallons and x miles corresponds to 30 gallons.)
4. Solve the proportion. $x = 750$ miles

EXAMPLES

1. You drove your car 500 miles and used 20 gallons of gasoline. How many miles would you expect to drive using 300 gallons of gasoline?

Solution

(a) Let x represent the unknown miles.

(b) Set up a proportion using either Pattern A or Pattern B. Label the numerators and denominators to be sure the units are in the same order.

$$\frac{500 \text{ miles}}{20 \text{ gallons}} = \frac{x \text{ miles}}{30 \text{ gallons}}$$

Pattern A Each ratio has different units, but numerators are the same units and denominators are the same units.

(c) Solve the proportion.

$$\begin{aligned} \frac{500}{20} &= \frac{x}{30} \\ 500 \cdot 30 &= 20 \cdot x \\ \frac{15{,}000}{20} &= \frac{\cancel{20} \cdot x}{\cancel{20}} \\ 750 &= x \end{aligned}$$

You would expect to drive 750 miles on 30 gallons of gas.

Note: *Any* of the following proportions would give the same answer.

$$\frac{20 \text{ gallons}}{500 \text{ miles}} = \frac{30 \text{ gallons}}{x \text{ miles}}$$

Pattern A Each ratio has different units, but numerators are the same units and denominators are the same units.

or

$$\frac{500 \text{ miles}}{x \text{ miles}} = \frac{20 \text{ gallons}}{30 \text{ gallons}}$$

Pattern B Each ratio has the same units, and numerators correspond and denominators correspond.

or

$$\frac{x \text{ miles}}{500 \text{ miles}} = \frac{30 \text{ gallons}}{20 \text{ gallons}}$$

Pattern B Each ratio has the same units, and numerators correspond and denominators correspond.

2. An architect draws the plans for a building using a scale of $\frac{1}{2}$ inch to represent 10 feet. How many feet would 6 inches represent?

Solution

(a) Let y represent the unknown feet.

(b) Set up a proportion labeling the numerators and denominators.

$$\frac{\frac{1}{2}\text{ inch}}{6\text{ inches}} = \frac{y\text{ feet}}{10\text{ feet}} \quad \text{WRONG}$$

This proportion is **WRONG** because $\frac{1}{2}$ inch does *not* correspond to y feet. $\frac{1}{2}$ inch corresponds to 10 feet.

$$\frac{\frac{1}{2}\text{ inch}}{6\text{ inches}} = \frac{10\text{ feet}}{y\text{ feet}} \quad \text{RIGHT}$$

(c) Solve the **RIGHT** proportion.

$$\frac{\frac{1}{2}}{6} = \frac{10}{y}$$

$$\frac{1}{2} \cdot y = 6 \cdot 10$$

$$\frac{\cancel{\frac{1}{2}} \cdot y}{\cancel{\frac{1}{2}}} = \frac{60}{\frac{1}{2}}$$

$$y = 120$$

6 inches would represent 120 feet.

3. A recommended mixture of weed killer is 3 cupfuls for 2 gallons of water. How many cupfuls should be mixed with 5 gallons of water?

Solution

(a) Let x = unknown number of cupfuls of weed killer.

(b) Set up a proportion.

$$\frac{x\text{ cupfuls}}{5\text{ gallons}} = \frac{3\text{ cupfuls}}{2\text{ gallons}}$$

(c) Solve the proportion.

$$\frac{x}{5} = \frac{3}{2}$$

$$2 \cdot x = 5 \cdot 3$$

$$\frac{\cancel{2} \cdot x}{\cancel{2}} = \frac{15}{2}$$

$$x = 7\frac{1}{2}$$

$7\frac{1}{2}$ cupfuls of weed killer should be mixed with 5 gallons of water.

Nurses and doctors work with proportions when prescribing medicine and giving injections. Medical texts write proportions in the form

$$2 : 40 :: x : 100$$

instead of

$$\frac{2}{40} = \frac{x}{100}$$

The double colon :: is used in place of an equal (=) sign. Using either notation, the solution is found by setting the product of the extremes equal to the product of the means and solving the equation for the unknown quantity.

EXAMPLE

4. Solve the proportion

2 ounces : 40 grams :: x ounces : 100 grams.

Solution

means

$$2 : 40 :: x : 100$$

extremes

$$2 \cdot 100 = 40 \cdot x$$

$$\frac{200}{40} = \frac{\cancel{40} \cdot x}{\cancel{40}}$$

$$5 = x$$

The solution is $x = 5$ ounces.

Exercises 6.4

Solve the following word problems using proportions.

1. A map maker uses a scale of 2 inches to represent 30 miles. How many miles do 3 inches represent?
2. If gasoline sells for $0.78 per gallon, what will 10 gallons cost?
3. If gasoline sells for $0.78 per gallon, how many gallons can be bought with $8.58?
4. If the odds on a horse are $5 to win on a $2 bet, how much can be won with a $5 bet?
5. An investor thinks she should make $12 for every $100 she invests. How much would she expect to make on a $1500 investment?
6. If one dozen (12) eggs cost $1.09, what would three dozen eggs cost?
7. The price of a certain fabric is $1.75 per yard. How many yards can be bought with $35 (not including tax)?
8. An artist figures she can paint 3 portraits every two weeks. At this rate, how long will it take her to paint 18 portraits?
9. Two units of a certain gas weigh 175 grams. What is the weight of 5 units of this gas?
10. A baseball team bought 8 bats for $96. What would they pay for 10 bats?
11. A store owner expects to make a profit of $2 on an item that sells for $10. How much profit will he expect to make on an item that sells for $60?
12. A saleswoman makes $8 for every $100 worth of the product she sells. What will she make if she sells $5000 worth of the product?
13. If property taxes are figured at $1.50 for every $100 in evaluation, what taxes will be paid on a home valued at $85,000?

14. A condominium owner pays property taxes of $2000 per year. If taxes are figured at a rate of $1.25 for every $100 in value, what is the value of his condominium?
15. Sales tax is figured at 6¢ for every $1.00 of merchandise purchased. What was the purchase price on an item that had a sales tax of $2.04?
16. An architect drew plans for a city park using a scale of $\frac{1}{4}$ inch to represent 25 feet. How many feet would 2 inches represent?
17. A building 14 stories high casts a shadow of 30 feet at a certain time

of day. What is the length of the shadow of a 20-story building at the same time of day in the same city?

18. Two numbers are in the ratio of 4 to 3. The number 10 is in that same ratio to a fourth number. What is the fourth number?

19. A car is traveling at 45 miles per hour. Its speed is increased by 3 miles per hour every 2 seconds. By how much will its speed increase in 5 seconds? How fast will the car be traveling?

20. A truck is traveling at 55 miles per hour and the driver brakes to slow down 2 miles per hour every 3 seconds. How long will it take the truck to slow to a speed of 45 miles per hour?

21. A salesman figured he drove 560 miles every two weeks. How far would he drive in three months (12 weeks)?

22. Driving steadily, a woman made a trip of 200 miles in $4\frac{1}{2}$ hours. How long would she take to drive 500 miles at the same rate of speed?

23. If you can drive 286 miles in $5\frac{1}{2}$ hours, how long will it take you to drive 468 miles at the same rate of speed?

24. What will 21 gallons of gasoline cost if gasoline costs $1.25 per gallon?

25. If diesel fuel costs $1.10 per gallon, how much diesel fuel will $24.53 buy?

26. An electric fan makes 180 revolutions per minute. How many revolutions will the fan make if it runs for 24 hours?

27. An investor made $144 in one year on a $1000 investment. What would she have earned if her investment had been $4500?

28. A typist can type 8 pages of manuscript in 56 minutes. How long will this typist take to type 300 pages?

29. On a map, $1\frac{1}{2}$ inches represent 40 miles. How many inches represent 50 miles?

30. In the metric system, there are 2.54 centimeters in 1 inch. How many centimeters are there in 1 foot?

31. If 40 pounds of fertilizer are used on 2400 square feet of lawn, how many pounds of fertilizer are needed for a lawn of 5400 square feet?

32. An English teacher must read and grade 27 essays. If the teacher takes 20 minutes to read and grade 3 essays, how much time will he need to grade all 27 essays?

33. If 2 cups of flour are needed to make 12 biscuits, how much flour will be needed to make 9 of the same kind of biscuits?

34. If 2 cups of flour are needed to make 12 biscuits, how many of the same kind of biscuits can be made with 3 cups of flour?

35. There are one thousand grams in one kilogram. How many grams are there in four and seven tenths kilograms?

The following exercises are examples of proportions in medicine. Be sure to label your answers. (The abbreviations are from the metric system.)

36. 1 liter : 1000 mL :: x liter : 5000 mL

37. 1 kg : 1000 g :: x kg : 2700 g

38. 1 mg : 1000 mcg :: x mg : 4 mcg

39. 1 g : 1000 mg :: 0.5 g : x mg

40. 1 dram : 60 grains :: x dram : 90 grains

41. $\frac{\text{1 ounce}}{\text{8 drams}} = \frac{\text{0.5 ounce}}{x}$

42. $\frac{\text{1 dram}}{\text{60 minims}} = \frac{x}{\text{180 minims}}$

43. $\frac{\text{1 ounce}}{\text{480 minims}} = \frac{\text{4 ounces}}{x}$

44. $\frac{\text{1 g}}{\text{15 grains}} = \frac{x}{\text{60 grains}}$

45. $\frac{\text{1 grain}}{\text{60 mg}} = \frac{x}{\text{500 mg}}$

46. $\frac{\text{1 ounce}}{\text{30 g}} = \frac{\text{2 ounces}}{x}$

47. $\frac{\text{1 mL}}{\text{15 minims}} = \frac{\text{5 mL}}{x}$

48. $\frac{\text{1 tsp}}{\text{4 mL}} = \frac{x}{\text{12 mL}}$

49. $\frac{\text{1 ounce}}{\text{30 mL}} = \frac{x}{\text{15 mL}}$

50. $\frac{\text{1 pint}}{\text{500 mL}} = \frac{\text{1.5 pints}}{x}$

SUMMARY: CHAPTER 6

DEFINITION

A **ratio** is a comparison of two quantities. The ratio of a to b can be written

$$\frac{a}{b} \quad \text{or} \quad a : b \quad \text{or} \quad a \text{ to } b$$

To find the **price per unit,** divide the cost by the number of units.

DEFINITION A **proportion** is a statement that two ratios are equal. In symbols,

$$\frac{a}{b} = \frac{c}{d} \quad \text{is a proportion.}$$

A proportion has four terms:

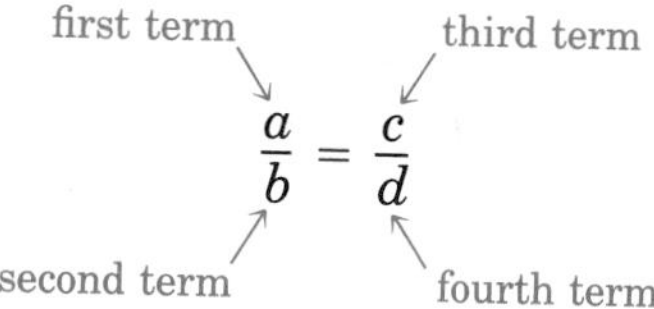

Extremes: Terms 1 and 4 (a and d)

Means: Terms 2 and 3 (b and c)

A proportion is true if the product of the extremes equals the product of the means.

$$\frac{a}{b} = \frac{c}{d} \quad \text{if and only if} \quad a \cdot d = b \cdot c,$$

where $b \neq 0$ and $d \neq 0$

NOTE: The terms can be decimal numbers, fractions, mixed numbers, or whole numbers. The rules are the same.

To find the unknown term in a proportion:

1. Write the proportion representing the unknown term with some letter, such as x, y, w, A, B, etc.
2. Write an equation that sets the product of the extremes equal to the product of the means.
3. Divide both sides of the equation by the number multiplying the unknown. (This number is called the **coefficient** of the unknown.)
4. The resulting equation gives the missing value for the unknown.

REVIEW QUESTIONS: CHAPTER 6

Write the following comparisons as ratios. Reduce to lowest terms and use common units whenever possible.

1. 2 dimes to 5 nickels
2. 20 inches to 1 yard
3. 10.16 centimeters to 4 inches
4. 51 miles to 3 gallons of gas
5. 18 hours to 2 days
6. 26 girls to 39 boys

Find the unit prices and tell which is the better buy.

7. Packaged cheese: 6 oz at 79¢
   "Deli" cheese: 1 lb (16 oz) at $1.99
8. 1 qt (32 fl oz) milk at 53¢
   1 gal (128 fl oz) milk at $2.02
9. 1 lb 2 oz loaf of bread at 85¢
   $1\frac{1}{2}$ lb loaf of bread at $1.15
10. 6 oz bologna at $1.29
    8 oz bologna at $1.49

Name the means and the extremes for each proportion.

11. $\frac{4}{5} = \frac{16}{20}$
12. $\frac{\frac{1}{6}}{3} = \frac{\frac{1}{9}}{2}$
13. $3 : 7 = 0.75 : 1.75$

Determine whether the following proportions are true or false.

14. $\frac{3}{5} = \frac{9}{15}$
15. $\frac{15}{20} = \frac{18}{24}$
16. $\frac{6.5}{14} = \frac{8}{16}$

Solve the following proportions. Reduce all fractions.

17. $\frac{10}{12} = \frac{x}{6}$
18. $\frac{1.7}{5.1} = \frac{100}{y}$
19. $\frac{7\frac{1}{2}}{3\frac{1}{3}} = \frac{w}{2\frac{1}{4}}$
20. $a : 7 = \frac{1}{3} : 5$

21. A motorcycle averages 42.8 miles per gallon of gas. How many miles can the motorcycle travel on 3.5 gallons of gas?

22. Find the difference between sixty-four and five hundred thirty-six ten-thousandths and fifty-nine and three thousand six hundred eighty-one ten-thousandths.

23. On a certain map, 1 inch represents 35.5 miles. What distance is represented by 4.7 inches?

24. A part-time clerk earned \$420 the first month on a new job. This was 0.6 of what he had anticipated. How much had he anticipated making?

25. If a machine produces 5000 hairpins in 2 hours, how many will it produce in two 8-hour days?

26. An automobile was slowing down at the rate of 5 miles per hour (mph) for every 3 seconds. If the automobile was going 65 mph when it began to slow down, how fast was it going at the end of 12 seconds?

27. An architect draws house plans using a scale of $\frac{3}{4}$ inch to represent 10 feet. How many feet are represented by 2 inches?

28. If you can drive 200 miles in $4\frac{1}{2}$ hours, how far could you drive (at the same rate) in 6 hours?

29. The ratio of kilometers to miles is about 8 to 5. Find the rate in kilometers per hour that is equivalent to 40 miles per hour.

30. In a certain hospital, 55 out of every 100 children born are boys. In a year when 1040 children are born in the hospital, how many are girls?

TEST: CHAPTER 6

Complete each statement.

1. The fraction $\frac{7}{4}$ can represent the ratio ______ to ______.

2. In the proportion $\frac{7}{8} = \frac{63}{72}$, the extremes are ______ and ______.

3. Is the proportion $2\frac{1}{3} : 7 = 3\frac{1}{2} : 10.5$ true or false?

Write the following comparisons as ratios. Reduce to lowest terms and use common units whenever possible.

4. 2 weeks to 35 days

5. 24 seconds to 1 hour

6. 6 nickels to 3 quarters

Find the unit prices and tell which is the better buy.

7. $\frac{1}{2}$ pint (8 oz) cottage cheese at 54¢ or
1 quart (32 oz) cottage cheese at \$2.05

Solve the following proportions. Reduce all fractions.

8. $\frac{9}{17} = \frac{x}{51}$

9. $\frac{3}{5} = \frac{10}{y}$

10. $\frac{50}{z} = \frac{0.5}{0.75}$

11. $\frac{x}{\frac{3}{8}} = \frac{\frac{1}{4}}{9}$

12. $\frac{x}{1.53} = \frac{2.2}{3}$

13. $\frac{2.25}{y} = \frac{1.5}{13}$

14. $x : 5 = 3\frac{3}{5} : 7$

15. $\frac{1}{2}$ is to 25 as x is to 10.

16. The scale on a street map indicates that $1\frac{1}{4}$ inches represent $\frac{1}{2}$ mile. How many miles do 3 inches represent?

17. A new type of truck averaged 29.2 miles per gallon on a test track. How many miles did it travel on the track if it used 0.83 gallon of gas?

18. You borrowed \$500 for 6 months from a loan company and paid \$45 in interest. How much interest would you pay if you borrowed \$600 for a year at the same interest rate?

19. A lighthouse light revolves once every 30 seconds. How many times will it revolve in 12 hours?

20. In a certain voting precinct, 7 out of every 15 voters are registered as Democrats. In a recent local election, 795 votes were tallied at this precinct. How many Democrats would have been expected to vote?

7 Percent

7.1 UNDERSTANDING PERCENT

The word **percent** comes from the Latin *per centum* meaning "per hundred." So, **percent means hundredths** or **the ratio of a number to 100.** The symbol % is called the **percent sign.** You can think of it as a rearrangement of the digits for 100 in the form 0/0. For example,

$$\frac{50}{100} = 50\% \qquad \text{and} \qquad \frac{27}{100} = 27\%$$

In Figure 7.1 there are 100 small squares and 27 are shaded. Thus, $\frac{27}{100}$ or 27% of the large square is shaded.

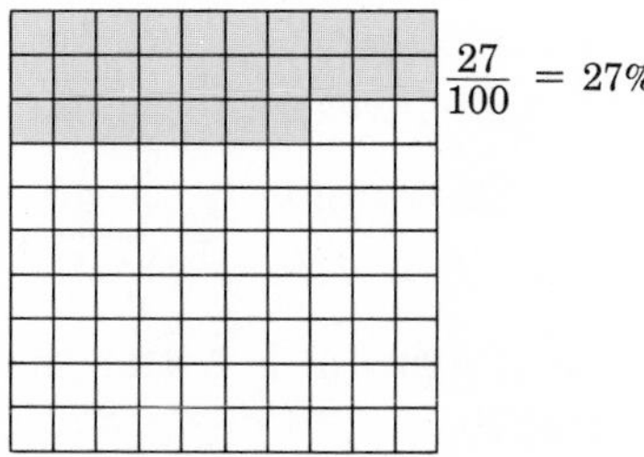

Figure 7.1

EXAMPLES

Each fraction is changed to a percent.

1. $\frac{30}{100} = 30\%$ Write the numerator and the % sign. Remember % means hundredths.

2. $\frac{45}{100} = 45\%$

3. $\frac{25}{100} = 25\%$

4. $\frac{6.3}{100} = 6.3\%$

5. $\frac{18\frac{1}{2}}{100} = 18\frac{1}{2}\%$

6. $\frac{200}{100} = 200\%$ If a number is larger than 1, then it is more than 100%.

Percent is commonly used in connection with taxes, discount, commission, batting averages, profit, and loss. **Percent of profit** is the ratio of money made to money invested.

EXAMPLES

7. Calculate the percent of profit for both (a) and (b) and tell which is the better investment.

   (a) \$150 made as profit by investing \$300

   (b) \$200 made as profit by investing \$500

   Solution

   (a) $\frac{\$150 \text{ profit}}{\$300 \text{ invested}} = \frac{\cancel{3} \cdot 50}{\cancel{3} \cdot 100} = \frac{50}{100} = 50\%$

   (b) $\frac{\$200 \text{ profit}}{\$500 \text{ invested}} = \frac{\cancel{5} \cdot 40}{\cancel{5} \cdot 100} = \frac{40}{100} = 40\%$

   Investment (a) is better than investment (b) because 50% is larger than 40%. Obviously, \$200 profit is more than \$150 profit. But the money risked (\$500) is also greater. The use of percent gives an effective method for comparison because both ratios have the same denominator (100).

8. Which is the better investment?

(a) make $40 profit by investing $200

(b) make $75 profit by investing $300

Solution

Write each ratio as hundredths and compare the percents.

(a) $\dfrac{\$40 \text{ profit}}{\$200 \text{ invested}} = \dfrac{\not{2} \cdot 20}{\not{2} \cdot 100} = \dfrac{20}{100} = 20\%$

(b) $\dfrac{\$75 \text{ profit}}{\$300 \text{ invested}} = \dfrac{\not{3} \cdot 25}{\not{3} \cdot 100} = \dfrac{25}{100} = 25\%$

Investment (b) is better since 25% is larger than 20%.

Exercises 7.1

What percent of each square is shaded?

1. 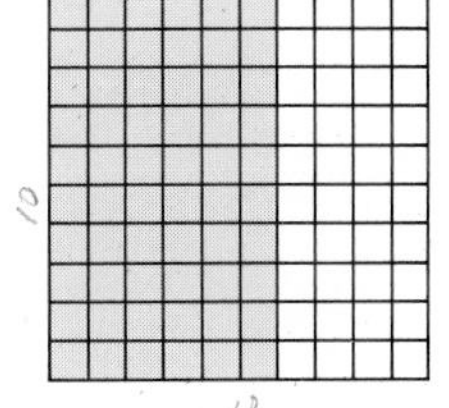

2. 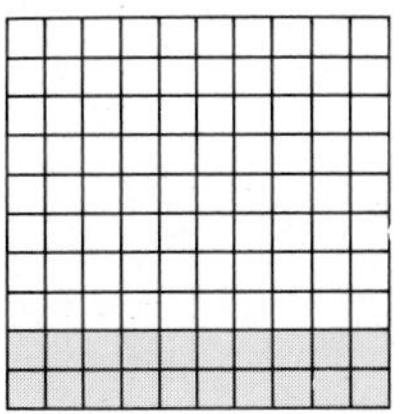

3. 

4. 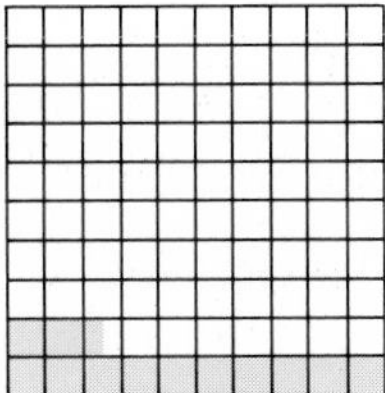

5. 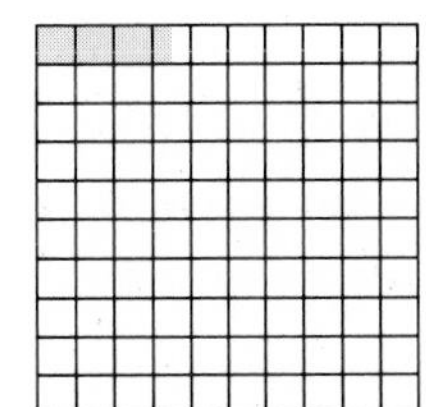

6. 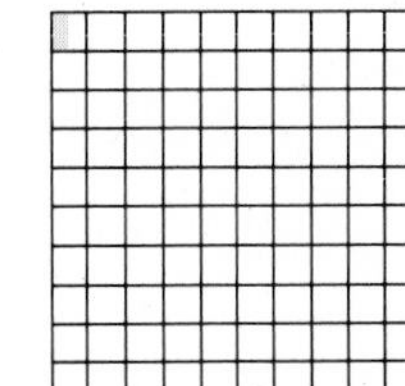

7. 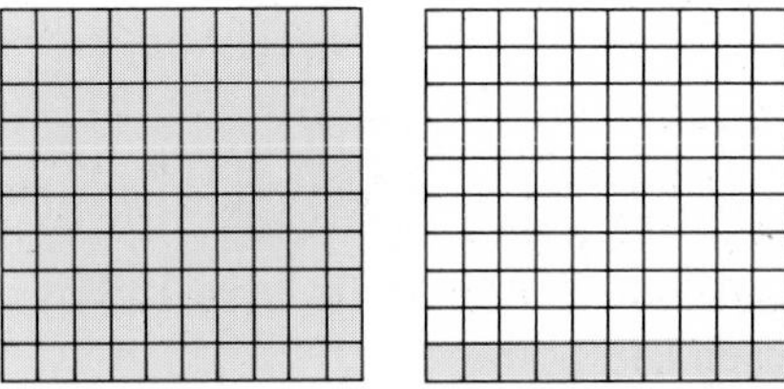

8. 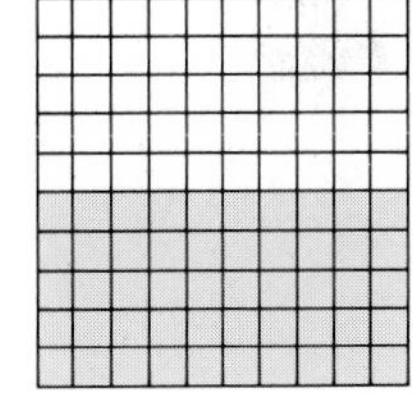

9. 
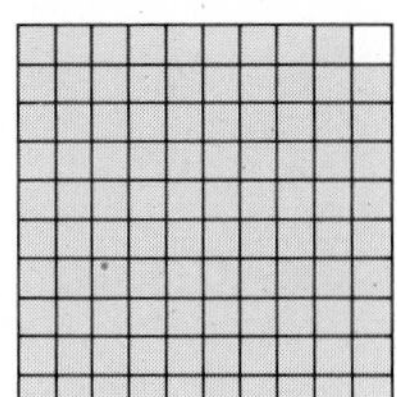

10.
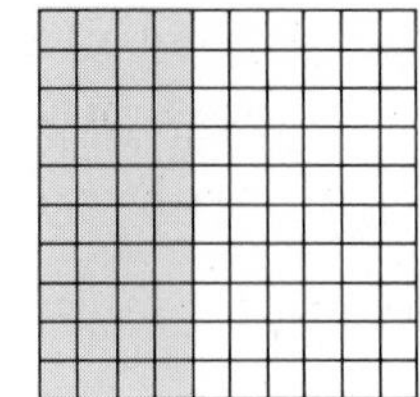

Change the following fractions to percents.

11. $\frac{30}{100}$ **12.** $\frac{20}{100}$ **13.** $\frac{40}{100}$ **14.** $\frac{50}{100}$ **15.** $\frac{7}{100}$

16. $\frac{8}{100}$ **17.** $\frac{90}{100}$ **18.** $\frac{15}{100}$ **19.** $\frac{25}{100}$ **20.** $\frac{35}{100}$

21. $\frac{45}{100}$ **22.** $\frac{65}{100}$ **23.** $\frac{75}{100}$ **24.** $\frac{42}{100}$ **25.** $\frac{53}{100}$

26. $\frac{68}{100}$ **27.** $\frac{77}{100}$ **28.** $\frac{48}{100}$ **29.** $\frac{125}{100}$ **30.** $\frac{110}{100}$

31. $\frac{150}{100}$ **32.** $\frac{175}{100}$ **33.** $\frac{200}{100}$ **34.** $\frac{250}{100}$ **35.** $\frac{236}{100}$

36. $\frac{120}{100}$ **37.** $\frac{16.3}{100}$ **38.** $\frac{27.2}{100}$ **39.** $\frac{13.4}{100}$ **40.** $\frac{38.6}{100}$

41. $\frac{20.25}{100}$ **42.** $\frac{93.5}{100}$ **43.** $\frac{0.5}{100}$ **44.** $\frac{1.5}{100}$ **45.** $\frac{0.25}{100}$

46. $\frac{3\frac{1}{2}}{100}$ **47.** $\frac{10\frac{1}{4}}{100}$ **48.** $\frac{1\frac{1}{4}}{100}$ **49.** $\frac{24\frac{1}{2}}{100}$ **50.** $\frac{17\frac{3}{4}}{100}$

In each problem, write the ratio of profit to investment as hundredths and compare the percents. Tell which investment is better, (a) or (b).

51. (a) a profit of \$36 on a \$200 investment
(b) a profit of \$51 on a \$300 investment

52. (a) a profit of \$40 on a \$400 investment
(b) a profit of \$60 on a \$500 investment

53. (a) a profit of \$70 on a \$700 investment
(b) a profit of \$30 on a \$300 investment

54. (a) a profit of \$150 on a \$1500 investment
(b) a profit of \$200 on a \$2000 investment

55. (a) a profit of \$300 on a \$2000 investment
(b) a profit of \$360 on a \$3000 investment

7.2 DECIMALS AND PERCENTS

We know that percent means hundredths. So, by changing a decimal to a fraction with denominator 100, we effectively change the decimal to a percent. For example,

$$0.42 = \frac{42}{100} = 42\%$$

$$0.76 = \frac{76}{100} = 76\%$$

Not all decimals are hundredths. In these cases, we would proceed as follows:

$$0.253 = \frac{0.253 \times 100}{100} = \frac{25.3}{100} = 25.3\%$$

That is, by multiplying and dividing by 100, we can change the form of the decimal without changing its value. Again,

$$0.802 = \frac{0.802 \times 100}{100} = \frac{80.2}{100} = 80.2\%$$

By noting the way the decimal point is moved, the change can be done directly using the following rule.

To change a decimal to a percent:

1. Move the decimal point two places to the right.
2. Write the % sign.

EXAMPLES

Change the following decimals to percents.

1. $0.67 = 67\%$

   decimal point moved two places to the right; % sign added

2. $0.324 = 32.4\%$

   decimal point moved two places to the right; % sign added

3. $0.005 = 0.5\%$

4. $0.0025 = 0.25\%$

5. $1.5 = 1.50 = 150\%$

   a number larger than 1 (↑ 1.5); is more than 100% (↑ 150%)

6. $2.35 = 235\%$

7. $0.03 = 3\%$

8. $0.3 = 30\%$

To change percents to decimals, we simply reverse the procedure for changing decimals to percents.

To change a percent to a decimal:

1. Move the decimal point two places to the left.
2. Drop the % sign.

EXAMPLES

Change the following percents to decimals.

9. $56\% = 0.56$

   understood decimal point (↑ 56%); decimal point moved two places left; % sign dropped

10. $38\% = 0.38$

11. $18.5\% = 0.185$

12. $15\frac{1}{4}\% = 0.15\frac{1}{4}$

    or $15\frac{1}{4}\% = 15.25\% = 0.1525$

13. $0.6\% = 0.006$

14. $0.23\% = 0.0023$

15. 240% = 2.40

16. 100% = 1.00 = 1

17. 101% = 1.01

18. 5% = 0.05

PRACTICE QUIZ	Change from decimals to percents.	ANSWERS
	1. 0.34	**1.** 34%
	2. 1.75	**2.** 175%
	Change from percents to decimals.	
	3. 200%	**3.** 2.00
	4. 1.5%	**4.** 0.015

Exercises 7.2

Change the following decimals to percents.

1. 0.02	**2.** 0.09	**3.** 0.1	**4.** 0.5	**5.** 0.7
6. 0.9	**7.** 0.36	**8.** 0.52	**9.** 0.83	**10.** 0.75
11. 0.25	**12.** 0.30	**13.** 0.40	**14.** 0.65	**15.** 0.025
16. 0.035	**17.** 0.046	**18.** 0.055	**19.** 0.003	**20.** 0.004
21. 1.10	**22.** 1.30	**23.** 1.25	**24.** 1.75	**25.** 2
26. 1.08	**27.** 1.05	**28.** 1.5	**29.** 2.3	**30.** 2.15

Change the following percents to decimals.

31. 2%	**32.** 7%	**33.** 10%	**34.** 18%
35. 15%	**36.** 20%	**37.** 25%	**38.** 30%
39. 35%	**40.** 80%	**41.** 10.1%	**42.** 11.5%
43. 13.2%	**44.** 17.3%	**45.** $5\frac{1}{4}\%$	**46.** $6\frac{1}{2}\%$
47. $13\frac{3}{4}\%$	**48.** $15\frac{1}{4}\%$	**49.** $20\frac{1}{4}\%$	**50.** $18\frac{1}{2}\%$
51. 0.25%	**52.** 1.25%	**53.** 0.17%	**54.** 0.50%

55. 125% **56.** 150% **57.** 130% **58.** 120%

59. 222% **60.** 215%

61. Suppose sales tax is figured at 6%. Change 6% to a decimal.

62. The interest rate on a loan is 15%. Change 15% to a decimal.

63. The sales commission in a retail store is figured at $8\frac{1}{2}\%$ for the clerk. Change $8\frac{1}{2}\%$ to a decimal.

64. The discount during a special sale on dresses is 30%. Change 30% to a decimal.

65. The rate of profit based on sales price is 45%. Change 45% to a decimal.

66. A bookstore figures its profit to be \$0.20 for every \$1.00 in sales. Change 0.20 to a percent.

67. In calculating his sales commission, Mr. John multiplies by the decimal 0.12. Change 0.12 to a percent.

68. To calculate what your maximum house payment should be, a banker multiplied your income by 0.28. Change 0.28 to a percent.

69. The discount you earn by paying cash is found by multiplying the amount of your purchase by 0.02. Change 0.02 to a percent.

70. Suppose the state license fee is figured by multiplying the cost of your car by 0.005. Change 0.005 to a percent.

7.3 FRACTIONS AND PERCENTS (CALCULATORS OPTIONAL)

If a fraction has denominator 100, we can change it to a percent by writing the numerator and adding the % sign (Section 7.1). If the denominator is a factor of 100 (2, 4, 5, 10, 20, 25, or 50), we can write it in an equivalent form with denominator 100, and then change it to a percent. For example,

$$\frac{3}{4} = \frac{3}{4} \cdot \frac{25}{25} = \frac{75}{100} = 75\%$$

$$\frac{1}{2} = \frac{1}{2} \cdot \frac{50}{50} = \frac{50}{100} = 50\%$$

$$\frac{4}{5} = \frac{4}{5} \cdot \frac{20}{20} = \frac{80}{100} = 80\%$$

However, most fractions do not have denominators that are factors of 100. A more general approach (easily applied with calculators) is to change the fraction to decimal form (Section 5.6), then change the decimal to a percent (Section 7.2).

> **To change a fraction to a percent:**
>
> 1. Change the fraction to a decimal (divide the denominator into the numerator).
> 2. Change the decimal to a percent.

EXAMPLES

1. Change $\frac{3}{4}$ to a percent.

Solution

(a) Divide:

$$\begin{array}{r} .75 \\ 4\overline{)3.00} \\ \underline{2\,8} \\ 20 \\ \underline{20} \\ 0 \end{array}$$

(b) Change .75 to a percent: $\frac{3}{4} = 0.75 = 75\%$

2. Change $\frac{5}{8}$ to a percent.

Solution

(a) Divide:

$$\begin{array}{r} .625 \\ 8\overline{)5.000} \\ \underline{4\,8} \\ 20 \\ \underline{16} \\ 40 \\ \underline{40} \\ 0 \end{array}$$

(b) Change .625 to a percent: $\frac{5}{8} = 0.625 = 62.5\%$

3. Change $\frac{18}{20}$ to a percent.

Solution

(a) Divide:

$$\begin{array}{r} .9 \\ 20\overline{)18.0} \\ \underline{18\ 0} \\ 0 \end{array}$$

(b) $\frac{18}{20} = .9 = 90\%$

4. Change $2\frac{1}{4}$ to a percent.

Solution

(a) $2\frac{1}{4} = \frac{9}{4}$

$$\begin{array}{r} 2.25 \\ 4\overline{)9.00} \\ \underline{8} \\ 1\ 0 \\ \underline{8} \\ 20 \\ \underline{20} \\ 0 \end{array}$$

(b) $2\frac{1}{4} = \frac{9}{4} = 2.25 = 225\%$

ROUNDING OFF DECIMAL QUOTIENTS

Decimal quotients that are exact with four decimal places or less will be written with four decimal places; otherwise, decimal quotients will be rounded off to the third decimal place (thousandths).

USING A CALCULATOR

If your instructor agrees that this is the appropriate time, you may choose to use a calculator to do the long division when changing a fraction to a decimal. Most calculators give answers accurate to 8 digits, so you will need to round off your answers according to the preceding statement about **Rounding Off Decimal Quotients.**

EXAMPLES

5. $\frac{1}{3} = 0.3333333$ using a calculator.

Rounding off the decimal quotient:

$\frac{1}{3} = 0.333 = 33.3\%$ The answer is rounded off and not exact.

Without a calculator, we divide and use fractions:

$$\begin{array}{r} .33\frac{1}{3} \\ 3\overline{)1.00} \\ \underline{9} \\ 10 \\ \underline{9} \\ 1 \end{array}$$

$\frac{1}{3} = .33\frac{1}{3} = 33\frac{1}{3}\%$

$33\frac{1}{3}\%$ is exact and 33.3% is rounded off.

Both answers are acceptable, but be aware that 33.3% is a rounded-off answer.

6. $\frac{2}{3} = 0.6666666$ using a calculator.

Rounding off the decimal quotient:

$\frac{2}{3} = 0.667 = 66.7\%$ Remember, the answer is rounded off.

Without a calculator, we divide and use fractions:

$$\begin{array}{r} .66\frac{2}{3} \\ 3\overline{)2.00} \\ \underline{1\,8} \\ 20 \\ \underline{18} \\ 2 \end{array}$$

$\frac{2}{3} = .66\frac{2}{3} = 66\frac{2}{3}\%$

Both $66\frac{2}{3}\%$ and 66.7% are acceptable. Remember that 66.7% is a rounded-off answer.

7. $\frac{1}{7} = 0.1428571 = 14.3\%$ using a calculator.

Long division:

$$\begin{array}{r} .14\frac{2}{7} \\ 7\overline{)1.00} \\ \underline{7} \\ 30 \\ \underline{28} \\ 2 \end{array} = 14\frac{2}{7}\%$$

or:

$$\begin{array}{r} .1428 \\ 7\overline{)1.0000} \\ \underline{7} \\ 30 \\ \underline{28} \\ 20 \\ \underline{14} \\ 60 \\ \underline{56} \\ 4 \end{array} = 14.3\%$$

by rounding off the decimal quotient

To change a percent to a fraction or mixed number:

1. Write the percent as a fraction with denominator 100 and drop the % sign.
2. Reduce the fraction.

EXAMPLES

8. Change 60% to a fraction.

$$60\% = \frac{60}{100} = \frac{3 \cdot \cancel{20}}{5 \cdot \cancel{20}} = \frac{3}{5}$$

9. Change 18% to a fraction.

$$18\% = \frac{18}{100} = \frac{9 \cdot \cancel{2}}{50 \cdot \cancel{2}} = \frac{9}{50}$$

10. Change $7\frac{1}{4}\%$ to a fraction.

$$7\frac{1}{4}\% = \frac{7\frac{1}{4}}{100} = \frac{\frac{29}{4}}{100} = \frac{29}{4} \cdot \frac{1}{100} = \frac{29}{400}$$

11. Change 130% to a mixed number.

$$130\% = \frac{130}{100} = \frac{13 \cdot \not{10}}{10 \cdot \not{10}} = \frac{13}{10} = 1\frac{3}{10}$$

A COMMON MISUNDERSTANDING

The fractions $\frac{1}{4}$ and $\frac{1}{2}$ are often confused with the percents $\frac{1}{4}\%$ and $\frac{1}{2}\%$. The differences can be emphasized by changing the numbers to decimals.

$$\frac{1}{4} = 0.25 \quad \text{and} \quad \frac{1}{4}\% = 0.25\% = 0.0025$$

$$0.25 \neq 0.0025$$

Similarly,

$$\frac{1}{2} = 0.50 \quad \text{and} \quad \frac{1}{2}\% = 0.5\% = 0.005$$

$$0.50 \neq 0.005$$

You can think of $\frac{1}{4}$ as being one-fourth of a dollar (a quarter) and $\frac{1}{4}\%$ as being one-fourth of a penny. $\frac{1}{2}$ can be thought of as one-half of a dollar and $\frac{1}{2}\%$ as one-half of a penny.

PRACTICE QUIZ

Change to percents	ANSWERS
1. $\frac{3}{20}$	**1.** 15%
2. $\frac{3}{8}$	**2.** 37.5%
Change to fractions.	
3. 35%	**3.** $\frac{7}{20}$
4. 40%	**4.** $\frac{2}{5}$

Exercises 7.3

Change the following numbers to percents. A calculator may be used if your instructor thinks its use is appropriate at this time.

1. $\frac{3}{100}$ **2.** $\frac{16}{100}$ **3.** $\frac{7}{100}$ **4.** $\frac{29}{100}$ **5.** $\frac{1}{2}$ **6.** $\frac{3}{4}$

7. $\frac{1}{4}$ **8.** $\frac{1}{20}$ **9.** $\frac{11}{20}$ **10.** $\frac{7}{10}$ **11.** $\frac{3}{10}$ **12.** $\frac{3}{4}$

13. $\frac{1}{5}$ **14.** $\frac{2}{5}$ **15.** $\frac{4}{5}$ **16.** $\frac{1}{50}$ **17.** $\frac{13}{50}$ **18.** $\frac{1}{25}$

19. $\frac{12}{25}$ **20.** $\frac{24}{25}$ **21.** $\frac{1}{8}$ **22.** $\frac{5}{8}$ **23.** $\frac{7}{8}$ **24.** $\frac{1}{9}$

25. $\frac{5}{9}$ **26.** $\frac{2}{7}$ **27.** $\frac{3}{7}$ **28.** $\frac{5}{6}$ **29.** $\frac{7}{11}$ **30.** $\frac{5}{11}$

31. $1\frac{1}{14}$ **32.** $1\frac{1}{6}$ **33.** $1\frac{1}{20}$ **34.** $1\frac{1}{4}$ **35.** $1\frac{3}{4}$ **36.** $1\frac{1}{5}$

37. $1\frac{3}{8}$ **38.** $2\frac{1}{2}$ **39.** $2\frac{1}{10}$ **40.** $2\frac{1}{15}$

Change the following percents to fractions or mixed numbers.

41. 10% **42.** 5% **43.** 15% **44.** 17% **45.** 25%

46. 30% **47.** 50% **48.** $12\frac{1}{2}\%$ **49.** $37\frac{1}{2}\%$ **50.** $16\frac{2}{3}\%$

51. $33\frac{1}{3}\%$ **52.** $66\frac{2}{3}\%$ **53.** 33% **54.** $\frac{1}{2}\%$ **55.** $\frac{1}{4}\%$

56. 1% **57.** 100% **58.** 125% **59.** 120% **60.** 150%

61. 0.3% **62.** 2.5% **63.** 62.5% **64.** 0.2% **65.** 0.75%

7.4 TYPES OF PERCENT PROBLEMS (CALCULATORS OPTIONAL)

Consider the sentence

35% of 80 is 28.

We want to translate the sentence into an equation as follows:

35%	of	80	is	28
↓	↓	↓	↓	↓
Rate or Percent	times	Base	=	Amount or Percentage

This basic relationship holds for the three types of problems related to percent.

> **R** = Rate or Percent (as a decimal or fraction)
> **B** = Base (number we are finding a percent of)
> **A** = Amount or Percentage (a part of the Base)
> "of" means times (multiply)
> "is" means =
> The relationship between A, B, and R is
>
> $$\mathbf{R \times B = A}$$

The equation

$$R \times B = A$$

has three quantities in it. Finding the value of one quantity when the other two are known corresponds to one of the three **types of percent problems.**

> There are three basic types of problems using percent.
>
> **Type 1** Finding a percent of a number:
>
> What is 45% of 70?
>
> $$\begin{array}{ccccc} \downarrow & \downarrow & \downarrow & \downarrow & \downarrow \\ A & = & 0.45 & \times & 70 \\ A & = & R & \times & B \end{array}$$
>
> **Type 2** Finding a number knowing that a percent of that number is a certain amount:
>
> 30% of what number is 18?
>
> $$\begin{array}{ccccc} \downarrow & \downarrow & \downarrow & \downarrow & \downarrow \\ 0.30 & \times & B & = & 18 \\ R & \times & B & = & A \end{array}$$
>
> **Type 3** Finding the percent of a number represented by a certain amount:
>
> What percent of 84 is 16.8?
>
> $$\begin{array}{ccccc} \downarrow & \downarrow & \downarrow & \downarrow & \downarrow \\ R & \times & 84 & = & 16.8 \\ R & \times & B & = & A \end{array}$$

Remember,

(a) "of" means to multiply;

(b) "is" means =; and

(c) The percent is changed to decimal or fraction form.

EXAMPLES

PROBLEM TYPE 1

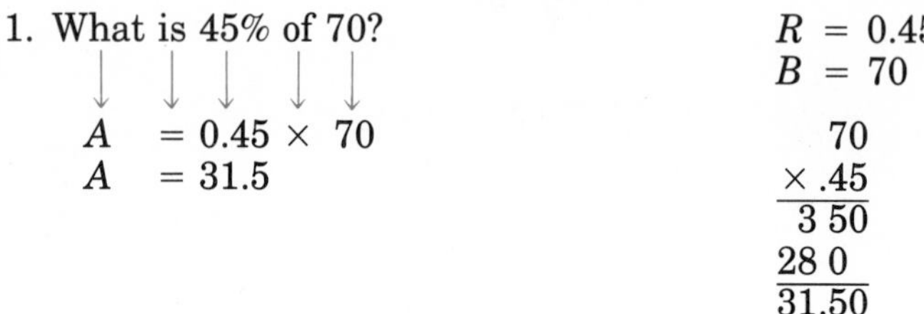

1. What is 45% of 70?

$A = 0.45 \times 70$
$A = 31.5$

$R = 0.45$
$B = 70$

$$\begin{array}{r} 70 \\ \times .45 \\ \hline 3\,50 \\ 28\,0 \\ \hline 31.50 \end{array}$$

So, 31.5 is 45% of 70.

The multiplication can be done with a calculator or to the side as shown here.

2. 18% of 200 is what?

$0.18 \times 200 = A$
$36 = A$

$R = 0.18$
$B = 200$

$$\begin{array}{r} .18 \\ \times \quad 200 \\ \hline 36.00 \end{array}$$

So, 18% of 200 is 36.

Problems of Type 1 are the most common and the easiest to solve. The rate (R) and base (B) are known. The unknown amount (A) is found by multiplying the rate times the base.

EXAMPLES

PROBLEM TYPE 2

3. 30% of what number is 18?

$.30 \times B = 18$

$R = .30$
$A = 18$

Here the **coefficient** of B is .30 and both sides of the equation are to be divided by .30. This is the same procedure we used in Chapter 6 when solving proportions.

$.30 \times B = 18$

$\frac{.30 \times B}{.30} = \frac{18}{.30}$

$B = 60$

$$\begin{array}{r} 60. \\ .30\overline{)18.00} \\ 18\,0 \\ \hline 00 \\ 0 \\ \hline \end{array}$$

So, 30% of 60 is 18.

4. 82% of ___ is 246?

$$.82 \times B = 246$$

$R = .82$
$A = 246$

$$.82 \times B = 246$$

$$\frac{.82 \times B}{.82} = \frac{246}{.82}$$

$$B = 300$$

$$.82\overline{)246.00} = 300.$$

So, 82% of 300 is 246.

EXAMPLES

PROBLEM TYPE 3

5. What percent of 84 is 16.8?

$$R \times 84 = 16.8$$

$B = 84$
$A = 16.8$

$$R \times 84 = 16.8$$

$$\frac{R \times 84}{84} = \frac{16.8}{84}$$

$$R = .2 = 20\%$$

$$84\overline{)16.8} = .2$$

So, 20% of 84 is 16.8.

6. ___% of 85 is 27.2.

$$R \times 85 = 27.2$$

$$R \times 85 = 27.2$$

$$\frac{R \times 85}{85} = \frac{27.2}{85}$$

$$R = .32 = 32\%$$

$$85\overline{)27.20} = .32$$

So, 32% of 85 is 27.2.

The multiplication or division shown in these examples can easily be done with a calculator. But you should always write the equation down first so you know that you are performing the correct operation and you know what quantity is unknown.

When working with applications, the wording of the problem is unlikely to be just as shown in Examples 1–6. The following examples show some alternative wording, but each problem is still one of the three types.

EXAMPLES

7. Find 65% of 42.

Solution

Here the Rate = 65%
and the Base = 42.
This is a TYPE 1 problem:

65% of 42 is ___.

$$.65 \times 42 = A$$
$$27.3 = A$$

$$\begin{array}{r} .65 \\ \times\ 42 \\ \hline 1.30 \\ 26.0 \\ \hline 27.30 \end{array}$$

8. What percent of 125 gives a result of 31.25?

Solution

Here the Rate is unknown.
The Base = 125 and Amount = 31.25.
This is a TYPE 3 problem:

___% of 125 is 31.25.

$$R \times 125 = 31.25$$

$$\frac{R \times \cancel{125}}{\cancel{125}} = \frac{31.25}{125}$$

$$R = .25 = 25\%$$

$$\begin{array}{r} .25 \\ 125\overline{)31.25} \\ 25\ 0 \\ \hline 6\ 25 \\ 6\ 25 \\ \hline \end{array}$$

Some percents are so common that their decimal and fraction equivalents should be memorized. Their fractional values are particularly easy to work with, and many times calculations can be done mentally.

COMMON PERCENT-DECIMAL-FRACTION EQUIVALENTS

$100\% = 1.00 = 1$	$33\frac{1}{3}\% = .33\frac{1}{3} = \frac{1}{3}$	$12\frac{1}{2}\% = .125 = \frac{1}{8}$
$25\% = .25 = \frac{1}{4}$	$66\frac{2}{3}\% = .66\frac{2}{3} = \frac{2}{3}$	$37\frac{1}{2}\% = .375 = \frac{3}{8}$
$50\% = .50 = \frac{1}{2}$		$62\frac{1}{2}\% = .625 = \frac{5}{8}$
$75\% = .75 = \frac{3}{4}$		$87\frac{1}{2}\% = .875 = \frac{7}{8}$

EXAMPLES

These examples show how using fraction equivalents for percents can simplify your work.

9. What is 25% of 24?

$$A = \frac{1}{\not{4}} \times \overset{6}{\not{24}} = 6$$

TYPE 1 problem

$25\% = \frac{1}{4}$

10. Find 75% of 36.

$$A = \frac{3}{\not{4}} \times \overset{9}{\not{36}} = 27$$

TYPE 1 problem

$75\% = \frac{3}{4}$

11. What is the value of $33\frac{1}{3}\%$ of 72?

$$A = \frac{1}{\not{3}} \times \overset{24}{\not{72}} = 24$$

TYPE 1 problem

$33\frac{1}{3}\% = \frac{1}{3}$

12. $37\frac{1}{2}\%$ of what number is 300?

$$\frac{3}{8} \times B = 300$$

TYPE 2 problem

$$\frac{\not{8}}{\not{3}} \times \frac{\not{3}}{\not{8}} \times B = \frac{8}{\not{3}} \times \overset{100}{\not{300}}$$

$\frac{8}{3}$ is the reciprocal of $\frac{3}{8}$.

$$B = 800$$

Remember that a percent is just another form of a fraction and that, when you find a percent of a number, the amount will be:

(a) smaller than the number if the percent is less than 100%;

(b) greater than the number if the percent is more than 100%.

PRACTICE QUIZ

Solve each problem for the unknown quantity.

1. 5% of 70 is ______.
2. ______% of 80 is 9.6.
3. 15% of ______ is 13.5.

ANSWERS

1. 3.5
2. 12%
3. 90

Exercises 7.4

Solve each problem for the unknown quantity. (A calculator may be used as an aid.)

1. 10% of 70 is ______.
2. 5% of 62 is ______.
3. 15% of 60 is ______.
4. 25% of 72 is ______.
5. 75% of 12 is ______.
6. 60% of 30 is ______.
7. 100% of 36 is ______.
8. 80% of 50 is ______.
9. 2% of ______ is 3.
10. 20% of ______ is 17.
11. 3% of ______ is 21.
12. 30% of ______ is 21.
13. 100% of ______ is 75.
14. 50% of ______ is 42.
15. 150% of ______ is 63.
16. 110% of ______ is 330.
17. ______% of 60 is 90.
18. ______% of 150 is 60.
19. ______% of 75 is 15.
20. ______% of 12 is 4.
21. ______% of 34 is 17.
22. ______% of 30 is 6.
23. ______% of 48 is 16.
24. ______% of 100 is 35.

Each of the following problems is one of the three types discussed, with slightly changed wording. Remember that *B* is the number you are finding the percent of.

25. ______ is 50% of 25.
26. ______ is 31% of 76.
27. 22 is 20% of ______.
28. 86 is 100% of ______.
29. 13 is ______% of 10.
30. 15 is ______% of 10.
31. 24 is $33\frac{1}{3}$% of ______.
32. 92.1 is 15% of ______.
33. 119.6 is 23% of ______.
34. 9.5 is 25% of ______.
35. 36 is ______% of 18.
36. 60 is ______% of 40.
37. ______ is 96% of 17.
38. ______ is 84% of 32.
39. ______ is 18% of 325.
40. ______ is 28% of 460.
41. Find 18% of 244.
42. Find 15.2% of 75.
43. Find 120% of 60.
44. What percent of 32 gives a result of 8?
45. 100 is 125% of what number?

Use fractions to solve the following problems. Do the work mentally if you can after you have set up the related equation.

46. Find 50% of 32.

47. Find $66\frac{2}{3}\%$ of 60.

48. What is $12\frac{1}{2}\%$ of 80?

49. What is $62\frac{1}{2}\%$ of 16?

50. $33\frac{1}{3}\%$ of 75 is ______.

51. 25% of 150 is ______.

52. 75% of what number is 21?

53. 50% of what number is 35?

54. $37\frac{1}{2}\%$ of ______ is 61.2.

55. 100% of ______ is 76.3.

7.5 APPLICATIONS WITH PERCENT: DISCOUNT, COMMISSION, SALES TAX, OTHERS (CALCULATORS RECOMMENDED)

You should be able to operate with decimals, round off decimals, change percents to decimals, and set up and solve the three basic types of percent problems. Since this section is concerned with the application of skills already learned, the author recommends the use of a calculator. A calculator is not meant to replace necessary skills and understanding but to enhance these abilities by providing answers rapidly and accurately. You must know from your own experience, knowledge, and understanding what numbers to work with and what operations to use.

GENERAL PLAN FOR SOLVING PERCENT PROBLEMS

1. Read the problem carefully.
2. Decide what is unknown (A, B, or R).
3. Write down the values you do know for A, B, or R. (You must know two of them.)
4. Set up and solve the equation $R \times B = A$.
5. Check to see that the answer is reasonable. Rework the problem if the answer does not make sense to you.

In this section you may do any of the multiplying, dividing, adding, or subtracting with a calculator. Be sure to set up the equation $R \times B = A$ whenever possible so that you will know you are doing the right operations.

When solving problems related to money, round off answers to the next higher cent.

EXAMPLES

1. A microwave oven is on sale at a 30% discount. What is the discount if the original price is marked at $275? What is the sale price?

Solution

There are two problems here:

(a) find the discount (a percent problem),
and (b) find the sale price.

(a) Find the discount.

30% of $275 is ______. TYPE 1 problem

$$.30 \times 275 = A$$
$$82.50 = A$$

$$\begin{array}{r} 275 \\ \times \quad .30 \\ \hline \$82.50 \end{array} \quad \text{discount}$$

The discount is $82.50.

(b) Find the sale price. The problem does not say specifically how to do this. We know that a discount is subtracted from the original price.

$$\begin{array}{rl} \$275.00 & \text{original price} \\ -\ 82.50 & \text{discount} \\ \hline \$192.50 & \text{sale price} \end{array}$$

The sale price is $192.50.

2. If sales tax is figured at 6%, what would be the final cost of the microwave oven in Example 1?

Solution

(a) Find the amount of the sales tax.

6% of $192.50 is ______. TYPE 1 problem

$$.06 \times 192.50 = A$$
$$11.55 = A$$

$$\begin{array}{r} \$192.50 \\ \times \quad .06 \\ \hline \$11.5500 \end{array}$$

The sales tax is $11.55.

(b) Add this tax to the sale price to get the final cost. (This method of computing sales tax and final cost is common in your daily life.)

$$\begin{array}{rl} \$192.50 & \text{sale price} \\ +\ 11.55 & \text{sales tax} \\ \hline \$204.05 & \text{final cost} \end{array}$$

The final cost of the microwave oven is $204.05.

3. An auto dealer paid \$7566 for a large order of a special part. This was not the original price. He received a 3% discount off the original price because he paid cash. What was the original price?

Solution

Do **not** take 3% of \$7566. We know that if 3% of something is gone, then 97% remains (100% − 3% = 97%). That is, \$7566 represents 97% of the original price.

97% of ______ is \$7566.

$.97 \times B = 7566$ TYPE 2 problem

$$\frac{.97 \times B}{.97} = \frac{7566}{.97}$$

$B = 7800$

```
        7800.
.97)7566.00
    679
     776
     776
       0 0
         0
        00
         0
```

The original price was \$7800.

If you want to check this result, take 3% of \$7800 and subtract this amount from \$7800. You should get \$7566. (Do **not** take 3% of \$7566.)

4. A saleswoman earns a salary of \$900 a month plus a commission of 8% on whatever she sells after she has sold \$6500 in merchandise. What did she earn the month she sold \$11,800 in merchandise?

Solution

(a) First find the amount of her commission. She earns 8% of what she sells over \$6500.

```
  $11,800
 −  6,500
  $ 5,300   base of commission
```

8% of \$5300 is ______. TYPE 1 problem

$.08 \times 5300 = A$

$424 = A$

```
   5300
 ×  .08
 424.00   commission
```

(b) Her monthly pay is her salary plus her commission.

```
  $900.00   salary
 + 424.00   commission
 $1324.00   pay
```

She earned \$1324.

Percent of profit was introduced as a **ratio** of profit to investment in Section 7.1. Another common use of percent of profit is related to merchandise where profit is the difference between selling price and cost.

Profit: The difference between selling price and cost

(profit = selling price − cost)

Percent of Profit: There are two approaches to this:

1. Percent of profit based on cost is the ratio of profit to cost:

$$\frac{\text{profit}}{\text{cost}} = \%\text{ profit based on cost}$$

2. Percent of profit based on selling price is the ratio of profit to selling price:

$$\frac{\text{profit}}{\text{selling price}} = \%\text{ profit based on selling price}$$

EXAMPLE

5. A company manufactures light fixtures that cost \$21 each to produce and are sold for \$28 each. What is the profit on each fixture? What is the percent of profit based on cost? What is the percent of profit based on selling price?

Solution

$$\begin{array}{rl} \$28 & \text{selling price} \\ -\ 21 & \text{cost} \\ \hline \$\ 7 & \text{profit} \end{array}$$

The profit on each light fixture is \$7. Using ratios, then changing the fractions to percents,

$$\frac{\$7 \text{ profit}}{\$21 \text{ cost}} = \frac{1}{3} = .33\frac{1}{3} = 33\frac{1}{3}\%\text{ profit based on cost}$$

$$\frac{\$7 \text{ profit}}{\$28 \text{ selling price}} = \frac{1}{4} = .25 = 25\%\text{ profit based on selling price}$$

Percent of profit **based on cost** is higher than percent of profit **based on selling price.** The business community reports whichever percent serves its purpose better. Your responsibility as an investor or consumer is to know which percent is reported and what it means to you.

EXAMPLE

6. Women's coats were on sale for \$250. This was a discount of \$100 from the original selling price. What was the original selling price? What was the rate of discount? If the coats cost the store owner \$200, what was his percent of profit based on cost? What was his percent of profit based on the actual selling price?

Solution

(a) Find the original selling price.

\$250	sale price
+ 100	discount
\$350	original selling price

(b) Find the rate of discount.

__% of \$350 is \$100. TYPE 3 problem

$$R \times 350 = 100$$

$$\frac{R \times 350}{350} = \frac{100}{350}$$

$$R = .2857 \quad \text{(rounded off)}$$

$$R = 28.57\% \quad \text{rate of discount}$$

(c) Find the profit first, then find each percent of profit.

\$250.00	actual selling price
− 200.00	cost
\$ 50.00	profit

$$\frac{\$50 \text{ profit}}{\$200 \text{ cost}} = \frac{1}{4} = .25 = 25\% \text{ profit based on cost}$$

$$\frac{\$50 \text{ profit}}{\$250 \text{ selling price}} = \frac{1}{5} = .20 = 20\% \text{ profit based on selling price}$$

Remember, even though you may do the actual calculations with a calculator, be sure to write the steps of the problem and the results in a neat, organized form. The form used in the examples is a good model.

Exercises 7.5

Some of the following problems involve several calculations. Write down the known information and the basic set-up to work each problem. Calculators are recommended for these exercises.

1. An appliance salesman works on a commission of 8%. What was his income for the month he sold $15,000 worth of appliances?
2. A car salesman earns a commission of 7% on each car he sells. How much did he earn on the sale of a car for $4950?
3. If sales tax is figured at 6%, how much tax was paid to buy three textbooks priced at $25.00, $15.50, and $34.95?
4. A new briefcase was priced at $275.00. If it was to be marked down 30%, what would be the discount?
5. A realtor works on a 6% commission. What is his commission on a house he sold for $95,000?
6. A store owner received a 3% discount from the manufacturer because she bought $6500 worth of dresses. What was the amount of the discount? What did she pay for the dresses?
7. A sales clerk receives a monthly salary of $500 plus a commission of 6% on all sales over $2500. What did the clerk earn the month she sold $6000 in merchandise?
8. If a salesman works on a 10% commission only, how much will he have to sell to earn $1800 in one month?
9. A computer programmer was told she would be given a bonus of 5% of any money her programs could save the company. How much would she have to save the company to earn a bonus of $600?
10. The property taxes on a house were $750. What was the tax rate if the house was valued at $25,000?
11. Towels were on sale at a discount of 30%. If the sale price was $3.01, what was the original price?
12. Sheets were marked $12.50 and pillow cases were marked $4.50. What is the sale price of each item if each item is discounted 25% from the marked price?
13. One shoe salesman worked on a straight 9% commission. His friend worked on a salary of $300 per month plus a 5% commission. How much did each salesman make during the month in which each sold $4500 worth of shoes?
14. A student missed 3 problems on a math test and was given a grade of 85%. If all the problems were of equal value, how many problems were on the test?

15. In one season, a basketball player missed 15% of his free throws. How many free throws did he make if he attempted 180 free throws?

16. A basketball player made 120 of 300 shots she attempted. What percent of her shots did she make? What percent did she miss?

17. The discount on a fur coat was $150. This was a 20% discount. What was the original selling price of the coat? What was the sale price? What was paid for the coat if a 6% sales tax was added to the sale price?

18. If sales tax is figured at 6%, what was the tax on a purchase of $30.20? What was paid for the purchase?

19. Golf clubs were marked on sale for $320. This was a discount of 20% off the original selling price. What was the original selling price? The clubs cost the golf pro $240. What was his profit? What was his percent of profit based on cost? What was his percent of profit based on the sale price?

20. The discount on men's suits was $50, and they were on sale for $200. What was the original selling price? What was the rate of discount? If the suits cost the store owner $150 each, what was his percent of profit based on cost? What was his percent of profit based on the selling price?

21. In one year, Mr. James earned $15,000. He spent $4800 on rent, $5250 on food, and $1800 on taxes. What percent of his income did he spend on each of those items?

22. The author of a book was told she would have to cut the number of pages by 12% in order for the book to sell at a popular price and still show a profit. What percent of the pages were in the final form? If the book contained 220 pages in final form, how many pages were in it originally? How many pages were cut?

23. In order to get more subscribers, a book club offered three books whose total selling price was originally $17.55 for $7.02. What was the amount of the discount? Based on the original selling price, what was the rate of discount on these three books?

24. The cost of a television set to a store owner was $350, and he sold the set for $490. What was his profit? What was his percent of profit based on cost? What was his percent of profit based on selling price?

25. A car dealer bought a used car for $1500. He marked up the price so he would make a profit of 25% based on his cost. What was the selling price? If the customer paid 8% of the selling price in taxes and fees, what did the customer pay for the car?

26. A man weighed 200 pounds. He lost 20 pounds in three months. What percent did he lose? Then he gained back 20 pounds two

months later. What percent did he gain? The loss and the gain are the same, but the two percents are different. Why?

27. An auto supply store received a shipment of auto parts together with the bill for $845.30. Some of the parts were not as ordered, however, and were returned at once. The value of the parts returned was $175.50. The terms of the billing provided the store with a 2% discount if it paid cash within two weeks. What did the store finally pay for the parts it kept if it paid cash within two weeks?

28. Suppose you sell your home for $100,000 and you owe the Savings and Loan $60,000 on the first trust deed. You pay a real estate agent 6% of the selling price and other fees and taxes totaling $1200. How much cash do you have after the sale? (You may pay income taxes later unless you buy a new home.)

29. You purchase a new home for $98,000. The bank will loan you 80% of the purchase price. How much cash do you need if loan fees are 2% of the loan and other fees total $850?

30. Margorie enrolled in freshman calculus. She had the choice of buying the text in hardback form for $26 or in paperback for $19.50. Tax is figured at 6% of the selling price. If the bookstore buys back hardback books for 50% of the selling price and paperback books for 30% of the selling price, which book is the more economical buy for Margorie if she sells her book back to the bookstore at the end of the semester? How much will she save?

SUMMARY: CHAPTER 7

Percent means hundredths.

To change a decimal to a percent:

1. Move the decimal point two places to the right.
2. Write the % sign.

To change a percent to a decimal:

1. Move the decimal point two places to the left.
2. Drop the % sign.

To change a fraction to a percent:

1. Change the fraction to a decimal (divide the denominator into the numerator).
2. Change the decimal to a percent.

To change a percent to a fraction or mixed number:

1. Write the percent as a fraction with denominator 100 and drop the % sign.
2. Reduce the fraction.

BASIC RELATIONSHIPS FOR PERCENT PROBLEMS

$\boldsymbol{R}$ = Rate or Percent (as a decimal or fraction)
$\boldsymbol{B}$ = Base (number we are finding a percent of)
$\boldsymbol{A}$ = Amount or Percentage (a part of the Base)
"of" means times (multiply)
"is" means =

The relationship between A, B, and R is

$$\boldsymbol{R \times B = A}$$

Profit: The difference between selling price and cost

(profit = selling price − cost)

Percent of Profit: There are two approaches to this:

1. Percent of profit based on cost is the ratio of profit to cost:

$$\frac{\text{profit}}{\text{cost}} = \%\text{ profit based on cost}$$

2. Percent of profit based on selling price is the ratio of profit to selling price:

$$\frac{\text{profit}}{\text{selling price}} = \%\text{ profit based on selling price}$$

REVIEW QUESTIONS: CHAPTER 7

1. Percent means ______.

Change the following fractions to percents.

2. $\frac{85}{100}$ **3.** $\frac{18}{100}$ **4.** $\frac{37}{100}$ **5.** $\frac{16\frac{1}{2}}{100}$

6. $\frac{15.2}{100}$ **7.** $\frac{115}{100}$

Change the following decimals to percents.

8. 0.06 **9.** 0.3 **10.** 0.67

11. 0.027 **12.** 3 **13.** 1.2

Change the following percents to decimals.

14. 35% **15.** 4% **16.** 0.25%

17. $\frac{1}{4}\%$ **18.** 7.1% **19.** 132%

Change the following numbers to percents.

20. $\frac{6}{10}$ **21.** $\frac{3}{20}$ **22.** $\frac{4}{25}$

23. $\frac{3}{8}$ **24.** $\frac{5}{12}$ **25.** $1\frac{4}{15}$

Change the following percents to fractions or mixed numbers.

26. 14% **27.** 40% **28.** 66%

29. $12\frac{1}{2}\%$ **30.** 400% **31.** $33\frac{1}{2}\%$

Solve each problem for the unknown quantity.

32. 30% of 52 is ______. **33.** 15% of 17 is ______.

34. 3% of ______ is 7. **35.** 42% of ______ is 18.

36. ______% of 36 is 7.2. **37.** ______% of 48 is 16.

38. 75 is ______% of 300. **39.** ______ is 6% of 18.25.

40. 5 is 10% of ______.

41. 14 if $5\frac{1}{2}$% of ______.

42. ______ is $6\frac{1}{2}$% of 15.

43. 62 is ______% of 31.

44. A shirt was marked 25% off. What would you pay for the shirt if the original price was $15 and you had to pay 6% sales tax?

45. Men's topcoats were on sale for $180. This was a discount of $30 from the original price. If the store owner paid $120 for the coats, what was his percent of profit based on his coat? Based on the selling price?

46. A student received a grade of 75% on a statistics test. If there were 32 problems on the test, all of equal value, how many problems did the student miss?

47. A salesman works on a 9% commission on his sales over $10,000 each month, plus a base salary of $600 per month. How much did he make the month he sold $25,000 in merchandise?

48. The property taxes on a house were $1800. What was the tax rate if the house was valued at $150,000?

49. Mary's allowance each week was $15. She saved for 6 weeks, then she spent $5 on a movie, $35 on clothes, and $20 on a gift for her parents' anniversary. What percent of her savings did she spend on each item?

50. The discount on a new car was $1500, including a rebate from the company. What was the original price of the car if the discount was 15% of the original price? What would be paid for the car if taxes and license fees totaled $650?

TEST: CHAPTER 7

Change the following numbers to percents.

1. $\frac{4.5}{100}$
2. 0.036
3. 5
4. $\frac{9}{20}$
5. $1\frac{3}{8}$

Change the following percents to decimals.

6. 8%
7. 123%
8. $12\frac{1}{4}\%$

Change the following percents to fractions or mixed numbers. Reduce to lowest terms.

9. 350%
10. $9\frac{3}{4}\%$
11. 6.2%

Solve each problem for the unknown quantity.

12. 22% of 60 is ______.
13. 4.15 is ______% of 83.
14. 25% of ______ is 18.
15. 18% of 3000 is ______.
16. 97 is 100% of ______.
17. 5.6 is ______% of 16.
18. A new refrigerator was marked at $750. For paying cash, the customer was allowed a 5% discount. What did the customer pay, including 6% sales tax, if she paid cash?
19. The bookstore sold used English books for $15. If the bookstore had paid $10 for the books, what was the percent of profit based on cost? Based on selling price?
20. A realtor works on a 6% commission. She must give 30% of her commission to the company she works for. What did she make on a house she sold for $85,000?

8 Applications with Calculators

8.1 SIMPLE INTEREST

You are encouraged to use a calculator for all the problems in this chapter. Many of the problems have several steps, and while a calculator is not necessary, it will save you a great deal of time. **Write down the results of intermediate steps in a neat, organized manner, labeling these results whenever possible.** [HINT: Ask a friend to trace the steps in your calculations after you have written down all your work.]

Accuracy to three decimal places will be sufficient, but you should avoid rounding off if you can. Be aware that some answers will differ slightly if you round off at different times in a calculation involving multiplication and/or division. For example, to calculate $\$500 \times 0.09 \times \frac{2}{3}$, you might write:

$$\$500 \times 0.15 \times \frac{2}{3} = \frac{500 \times 0.15 \times 2}{3} = \frac{150}{3} = \$50$$

Or, using a decimal and a calculator, $\frac{2}{3} = 0.667$ (to three decimal places):

$$\$500 \times 0.15 \times \frac{2}{3} = 500 \times 0.15 \times 0.667 = \$50.03$$

Even though \$50 is the correct answer, \$50.03 must also be accepted if calculators are allowed. The 3¢ error is called a **round-off error.** You must understand that slight errors can and do occur with the use of calculators and rounded-off decimals.

Interest is money paid for the use of money. The money that is invested or borrowed is called the **principal.** The **rate** is the **percent of interest** and is almost always stated as **an annual (yearly) rate.**

Interest is either paid or earned, depending on whether you are the borrower or the lender. In either case, the calculations involved are the same. The purpose of this section is to explain how interest is calculated. The interest rates vary from year to year and from one area of the world to another, but the concept of interest is the same everywhere.

There are two kinds of interest, **simple interest** and **compound interest.** Compound interest involves interest paid on interest and will be discussed in Section 8.2. Many loans are based on simple interest. Only one payment (including the principal and the interest) is made at the end of the term of the loan. No monthly payments are made.

FORMULA FOR CALCULATING SIMPLE INTEREST

$$I = P \times R \times T$$

where

I = Interest
P = Principal, the amount invested or borrowed
R = Rate, the percent stated as an annual (yearly) rate
T = Time, in years or fraction of a year

NOTE: We will use 360 days in one year (30 days in a month), a common practice in business and banking although, with the advent of large computers, many lending institutions now base their calculations on 365 days per year and pay or charge interest on a daily basis.

EXAMPLES

1. You want to borrow \$2000 at 14% interest for one year. How much interest would you pay?

Solution

$$I = P \times R \times T$$
$$P = \$2000$$
$$R = 14\% = 0.14$$
$$T = 1 \text{ year}$$
$$I = \$2000 \times 0.14 \times 1 = \$280$$

You would pay \$280 interest.

2. You decide that you might need the $2000 for only 90 days. How much interest would you pay? (The annual interest rate would still be stated as 14%.)

Solution

$$I = P \times R \times T$$
$$P = \$2000$$
$$R = 14\% = 0.14$$
$$T = 90 \text{ days} = \frac{90}{360} \text{ year} = \frac{1}{4} \text{ year}$$
$$I = \$2000 \times 0.14 \times \frac{1}{4} = \frac{280}{4} = \$70$$

or
$$I = \$2000 \times 0.14 \times 0.25 = \$70$$

You would pay $70 interest if you borrowed the money for 90 days.

3. John loaned $500 to a friend for 6 months at an interest rate of 12%. How much will his friend pay him at the end of the 6 months?

Solution

John will be paid interest plus the original principal. Find the interest and then add the principal.

$$I = P \times R \times T$$
$$P = \$500$$
$$R = 12\% = 0.12$$
$$T = 6 \text{ months} = \frac{6}{12} \text{ year} = \frac{1}{2} \text{ year}$$
$$I = \$500 \times 0.12 \times \frac{1}{2} = \frac{60}{2} = \$30$$
$$\text{Principal} + \text{Interest} = \$500 + \$30 = \$530$$

John will be paid $530.

The equation

$$I = P \times R \times T$$

has four quantities represented. If any three are known, the fourth can be found by solving the related equation, just as with proportions (Chapter 6) and percents (Chapter 7).

EXAMPLES

4. How much (what principal) would you need to invest if you wanted to make \$100 in interest in 30 days and your investment would return 12% interest?

Solution

Here the principal is unknown, while the interest (\$100), the rate of interest (12%), and the time (30 days) are all known.

$$I = \$100 \qquad R = 0.12 \qquad T = \frac{30}{360} = \frac{1}{12}$$

Using the formula and substituting the known values,

$$I = P \times R \times T$$

$$100 = P \times 0.12 \times \frac{1}{12}$$

$$100 = P \times 0.01$$

$$\frac{100}{0.01} = \frac{P \times \cancel{0.01}}{\cancel{0.01}}$$

$$10{,}000 = P$$

Or, using a fraction for R,

$$100 = P \times \frac{\overset{1}{\cancel{12}}}{100} \times \frac{1}{\underset{1}{\cancel{12}}}$$

$$100 = P \times \frac{1}{100}$$

$$\frac{100}{1} \times 100 = P \times \frac{1}{\cancel{100}} \times \frac{\cancel{100}}{1}$$

$$10{,}000 = P$$

You would need a principal of \$10,000 invested at a rate of 12% for 30 days to make \$100 interest.

5. What interest rate would you be paying if you borrowed \$1000 for 6 months and paid \$60 in interest?

Solution

Here the rate is unknown.

$$I = \$60 \qquad P = \$1000 \qquad T = \frac{6}{12} = \frac{1}{2}$$

Using the formula and substituting the known values,

$$I = P \times R \times T$$

$$60 = 1000 \times R \times \frac{1}{2}$$

$$60 = \overset{500}{\cancel{1000}} \times \frac{1}{\cancel{2}} \times R$$

$$\frac{60}{500} = \frac{\cancel{500} \times R}{\cancel{500}}$$

$$0.12 = R$$

$$\begin{array}{r} .12 \\ 500\overline{)60.00} \\ \underline{50\ 0} \\ 10\ 00 \\ \underline{10\ 00} \end{array}$$

You would be paying 12% interest.

6. Jose wants to borrow \$1500 at 15% and is willing to pay \$300 in interest. How long can he keep the money?

Solution

Here the time is unknown.

$$I = \$300 \qquad P = \$1500 \qquad R = 0.15 = \frac{15}{100}$$

Substituting these known values in the formula,

$$I = P \times R \times T$$

$$300 = 1500 \times \frac{15}{100} \times T$$

$$300 = \overset{15}{\cancel{1500}} \times \frac{15}{\cancel{100}} \times T$$

$$\frac{300}{225} = \frac{\cancel{225} \times T}{\cancel{225}}$$

$$T = \frac{300}{225} = \frac{2 \cdot 2 \cdot \cancel{3} \cdot \cancel{25}}{3 \cdot \cancel{3} \cdot \cancel{25}} = \frac{4}{3}\text{ yr}$$

$$T = 1\frac{1}{3}\text{ yr or 1 yr 4 mo}$$

Jose can borrow the money for $1\frac{1}{3}$ years.

Two important facts related to all these examples are:

1. Only one formula is necessary for working with simple interest, $I = P \times R \times T$.
2. You must be able to use the skills with fractions, decimals, and percents learned in earlier chapters.

Exercises 8.1

1. What is the simple interest on \$500 at 10% for one year?

2. What is the simple interest on \$2000 at 12% for one year?

3. How much interest would be paid on a loan of \$1000 at 15% for 6 months?

4. How much interest would be paid on a loan of \$3000 at 12% for 8 months?

5. You invested \$2000 at 13% for 60 days. How much interest did your money earn?

6. Stacey loaned her uncle \$1500 at 10% interest for 9 months. How much interest did she earn?

7. What principal will earn \$50 in interest if it is invested at 8% for 90 days?

8. What principal will earn \$75 in interest if it is invested for 60 days at 9%?

9. How long will it take for \$1000 invested at 10% to earn \$50 simple interest?

10. What length of time will it take to earn \$500 simple interest if \$2000 is invested at 15%?

11. What will be the interest earned in one year on a savings account of \$800 if the bank pays 6% interest?

12. If interest is paid at 10% for one year, what will a principal of \$600 earn?

13. If a principal of \$900 is invested at a rate of 14% for 90 days, what will be the interest earned?

14. A loan of \$5000 is made at 11% for a period of 6 months. How much interest is paid?

15. If you borrow \$750 for 30 days at 18%, how much interest will you pay?

16. How much interest is paid on a 60-day loan of \$500 at 12%?

17. Find the simple interest paid on a savings account of \$1800 for 120 days at 8%.

18. A savings account of \$2300 is left for 90 days drawing interest at a rate of 7%. How much interest is earned? What is the amount in the account at the end of 90 days?

19. Every 6 months a stock pays 10% dividends (interest on investment). What will be the earnings of \$14,600 invested for 6 months?

20. One thousand dollars worth of merchandise is charged at a local department store for 60 days at 18% interest. How much is owed at the end of 60 days?

21. You buy an oven on sale from $500 to $450, but you don't pay the bill for 60 days and are charged interest at a rate of 18%. How much do you pay for the oven by waiting 60 days to pay? How much did you save by buying the oven on sale?

22. A friend borrows $500 from you for a period of 8 months and pays you interest at 6%. How much interest are you paid? Suppose you ask 8% instead. Then how much interest are you paid?

23. How much would you have to invest at 8% for 60 days to earn interest of $500?

24. How many days must you leave $1000 in a savings account at $5\frac{1}{2}$% to earn $11.00?

25. What is the rate of interest charged if a loan of $2500 for 90 days is paid off with $2562.50?

26. Determine the missing item in each row.

PRINCIPAL	RATE	TIME	INTEREST
$ 400	16%	90 days	$ (a)
$ (b)	15%	120 days	$ 5.00
$ 560	12%	(c)	$ 5.60
$2700	(d)	40 days	$25.50

27. Determine the missing item in each row.

PRINCIPAL	RATE	TIME	INTEREST
$ 500	18%	30 days	$ (a)
$ 500	18%	(b)	$15.00
$ 500	(c)	90 days	$22.50
$ (d)	18%	30 days	$ 1.50

28. If you have a savings account of $25,000 drawing interest at 8%, how much interest will you earn in 6 months? How long must you leave the money in the account to earn $1500?

29. You have accumulated \$50,000, and you want to live on the interest each year. If you need \$800 a month to live on, what interest rate must you earn on your \$50,000? If you are earning $13\frac{1}{2}\%$, payable each month, what will your monthly interest be?

30. Your \$2500 savings account draws interest at $5\frac{1}{2}\%$. How many days will it take for you to earn \$68.75? If the interest rate is then raised to 6%, what will your money earn in the next 6 months?

31. A bank decides to loan \$5 million dollars to a contractor to build some houses. How much interest will the bank earn in one year if the interest rate is $14\frac{1}{4}\%$?

32. A credit card company has \$20 million dollars loaned to its customers at 18% interest. How much interest will it earn in one month?

33. A small airline company borrowed \$7.5 million dollars to buy some new airplanes. The loan rate was 15% and the airline paid \$562,500 in interest. What was the length of time of the loan?

34. A department store keeps \$15 million in merchandise in stock. If the store pays interest at 9% on a bank loan for this stock, how much interest will the store pay in 3 months' time?

35. Determine the missing item in each row.

PRINCIPAL	RATE	TIME	INTEREST
\$1000	$10\frac{1}{2}\%$	60 days	\$ (a)
\$ 800	$13\frac{1}{2}\%$	(b)	\$ 18.00
\$2000	(c)	9 months	\$172.50
\$ (d)	$7\frac{1}{2}\%$	1 year	\$ 85.00

8.2 COMPOUND INTEREST

Interest paid on interest is called **compound interest.** We calculate the simple interest over each period of time **with a new principal for each calculation.** The new principal is the previous principal **plus** the earned interest.

To calculate compound interest:

1. Using $I = P \times R \times T$, calculate the simple interest where T is period of time for compounding. For example,

$$T = 1 \text{ for compounding annually;}$$

$$T = \frac{1}{2} \text{ for compounding semiannually;}$$

$$T = \frac{1}{4} \text{ for compounding quarterly;}$$

$$T = \frac{1}{12} \text{ for compounding monthly;}$$

$$T = \frac{1}{360} \text{ for compounding daily.}$$

2. Add this interest to the principal to create a new value for the principal.
3. Continue Steps 1 and 2 until the entire interest period is covered.

NOTE: Most savings and loan associations and banks pay interest compounded daily. A computer or a set of tables is needed to do such interest calculations over a period of time. Since the purpose here is to teach the concept of compound interest, the problems will involve compounding annually, semiannually, quarterly, or monthly, but not daily.

EXAMPLES

1. You deposit \$1000 in an account that pays 8% interest compounded quarterly (every 3 months). How much interest would you earn in one year?

Solution

Calculate the simple interest four times with $T = \frac{1}{4}$. Add this interest to the previous principal before each calculation. You will be calculating interest paid on interest.

$$I = P \times R \times T$$

(a) $I = \$1000 \times 0.08 \times \frac{1}{4}$ $\quad P = \$1000$

$= 80 \times \frac{1}{4} = \$20$ $\quad$ interest

(b) $I = \$1020 \times 0.08 \times \frac{1}{4}$ $P = \$1000 + \$20 = \$1020$

$= 81.60 \times \frac{1}{4} = \$20.40$ interest

(c) $I = \$1040.40 \times 0.08 \times \frac{1}{4}$ $P = \$1020 + \$20.40 = \$1040.40$

$= 83.23 \times \frac{1}{4} = \$20.81$ interest

(d) $I = \$1061.21 \times 0.08 \times \frac{1}{4}$ $P = \$1040.40 + \$20.81 = \$1061.21$

$= 84.90 \times \frac{1}{4} = \$21.23$ interest

$$\begin{array}{r} \$20.00 \\ 20.40 \\ 20.81 \\ \underline{21.23} \\ \$82.44 \end{array}$$ total interest

2. In Example 1, how much more interest did you earn by having your interest compounded quarterly for one year than if it had just been calculated as simple interest for one year?

Solution

Simple interest for one year would be

$$I = \$1000 \times 0.08 \times 1 = \$80$$

$$\begin{array}{rl} \$82.44 & \text{compound interest} \\ -\ \underline{80.00} & \text{simple interest} \\ \$\ 2.44 & \text{more by compounding quarterly} \end{array}$$

3. \$5000 is deposited in a savings account, and interest is compounded semiannually at 10%. How much interest will be earned in one year?

Solution

There are two calculations with $T = \frac{1}{2}$ because interest is accumulated every six months.

(a) $I = \$5000 \times 0.10 \times \frac{1}{2}$ $\quad P = \$5000$

$= \$250$

(b) $I = \$5250 \times 0.10 \times \frac{1}{2}$ $\quad P = \$5000 + \$250 = \$5250$

$= \$262.50$

$$\begin{array}{r} \$250.00 \\ -\ \ 262.50 \\ \hline \$512.50 \end{array}$$ interest in one year

4. If an account is compounded monthly at 12%, how much interest will $2000 earn in three months?

Solution

Use $T = \frac{1}{12}$ and calculate the interest three times.

(a) $I = \$2000 \times 0.12 \times \frac{1}{12}$ $\quad P = \$2000$

$= \$20$

(b) $I = \$2020 \times 0.12 \times \frac{1}{12}$ $\quad P = \$2000 + \$20 = \$2020$

$= \$20.20$

(c) $I = \$2040.20 \times 0.12 \times \frac{1}{12}$ $\quad P = \$2020 + \$20.20 = \$2040.20$

$= \$20.41$

$$\begin{array}{r} \$20.00 \\ 20.20 \\ +\ \ 20.41 \\ \hline \$60.61 \end{array}$$ interest in three months

5. Suppose your income is $1000 per month (or $12,000 per year), and you will receive a cost of living raise each year. If inflation is at 9% each year, in how many years will you be making $24,000?

Solution

Compound annually at 9% until the principal (base salary) plus interest (raise) totals $24,000.

In table form (inflation at 9%),

YEAR	BASE	+	RAISE	=	TOTAL
1	$12,000.00		$ 0		$12,000.00
2	12,000.00		1080.00		13,080.00
3	13,080.00		1177.20		14,257.20
4	14,257.20		1283.15		15,540.35
5	15,540.35		1398.63		16,938.98
6	16,938.98		1524.51		18,463.49
7	18,463.49		1661.71		20,125.20
8	20,125.20		1811.27		21,936.47
9	21,936.47		1974.28		23,910.75

In 9 years, you would have almost doubled your salary, but your relative purchasing power would be the same as it was 9 years before. (Actually, you will be in a higher income tax bracket, and you will not be as well off as you were 9 years before.)

6. Bill invested $5000 at 12% interest to be compounded quarterly. How much interest will he earn in 9 months?

Solution

Use $T = \frac{1}{4}$ and calculate three times to cover 9 months since 9 months $= \frac{3}{4}$ year.

(a) $I = \overset{50}{\cancel{5000}} \times \frac{\overset{3}{\cancel{12}}}{\cancel{100}} \times \frac{1}{\cancel{4}} = \$150$

Here $\frac{12}{100}$ is used for 12% to show how the calculations can be done with fractions.

(b) $I = \overset{51.50}{\cancel{5150}} \times \frac{\overset{3}{\cancel{12}}}{\cancel{100}} \times \frac{1}{\cancel{4}} = \$154.50$

Notice how division by 100 moves the decimal point two places to the left.

(c) $I = \overset{53.0050}{\cancel{5300.50}} \times \frac{\overset{3}{\cancel{12}}}{\cancel{100}} \times \frac{1}{\cancel{4}} = 159.0150$

$= \$159.02$

$$\begin{array}{r} \$150.00 \\ 154.50 \\ \underline{159.02} \\ \$436.52 \end{array}$$ interest in nine months $\left(\frac{3}{4} \text{ of a year}\right)$

Exercises 8.2

1. If a bank compounds interest quarterly at 12% (on a certificate deposit), what will your $13,000 deposit be worth in 6 months? In one year?

2. An amount of $9000 is deposited in a savings and loan account, and interest is compounded monthly at 10%. What will be the balance of the account in 6 months?

3. If an account is compounded quarterly at a rate of 6%, what will be the interest earned on $5000 in one year? What will be the total amount in the account? How much more interest is earned in the first year because compounding is done quarterly rather than annually?

4. How much interest will be earned on a savings account of $4000 compounded semiannually at 6% in 4 years? What will be the balance of the account? If the saver took out the interest every 6 months for spending money, what would be the balance in 2 years?

5. How much interest will be earned on a savings account of $3000 in 2 years if interest is compounded annually at 6%? If interest is compounded semiannually? If interest is compounded quarterly?

6. If interest is calculated at 10% and compounded quarterly, what will be the value of $15,000 in $2\frac{1}{2}$ years?

7. You borrowed $4000 and agreed to make equal payments of $1000 each plus interest over the next 4 years. Interest is at a rate of 8% based only on what you owe. How much interest did you pay? How much interest would you have paid if you had not made the annual payments and paid only the $4000 plus interest compounded annually at the end of 4 years?

8. You borrow $600 from a friend and agree to pay 12% interest and pay in 60 days. But you can only pay $100 plus interest. Your friend agrees to let you pay $100 plus interest every 60 days until the debt is paid. How long will it take you to repay your friend? How much interest will you pay?

9. What will be the value of a $15,000 savings account at the end of 3 years if interest is calculated at 10% compounded annually? Suppose the interest is calculated at 5% compounded semiannually. Is the value the same? If not, what is the difference? Suppose the semiannual compounding is at 5% for 6 years. Now will the value be the same?

10. Calculate the interest earned in one year on $10,000 compounded monthly at 14%. What is the difference between this and simple interest at 14% for one year?

11. Calculate the interest you would pay on a loan of $6500 for one year if the interest were compounded every 3 months at 18% and you made no monthly payments. If you made payments of $1000 plus interest every 3 months and the balance plus interest at the end of the year, how much interest would you pay?

12. If a savings account of $10,000 draws interest at a rate of 10% compounded annually, how many years will it take for the $10,000 to double in value? (Make a table similar to the one in Example 5.)

13. If a savings account of $10,000 draws interest at a rate of 6% compounded annually, how many years will it take for the $10,000 to double in value? (Make a table similar to the one in Example 5.)

14. If a savings account of $10,000 draws interest at a rate of 12% compounded annually, how many years will it take for the $10,000 to double in value? (Make a table similar to the one in Example 5.)

15. Suppose $5000 has been invested in a savings account that is compounded annually at 8%. A table showing data for six years is given. Answer the following questions using this table.

TABLE OF INTEREST COMPOUNDED AT 8%

YEAR	PRINCIPAL	+	INTEREST	=	TOTAL
1	$5000.00		$400.00		$5400.00
2	5400.00		432.00		5832.00
3	5832.00		466.56		6298.56
4	6298.56		503.89		6802.45
5	6802.45		544.20		7346.65
6	7346.65		587.74		7934.39

(a) How much interest was earned during the third year?

(b) How much interest was earned during the first two years?

(c) How much interest was earned during the third and fourth years together?

(d) Why is the answer to part (c) larger than the answer to part (b)?

(e) How much interest accumulated in six years?

(f) How much interest would be earned during the seventh year?

8.3 BALANCING A CHECKING ACCOUNT

Many people do not balance their checking accounts simply because they were never told how. Others trust the bank's calculations over their own, but this is a poor practice because, for a variety of reasons, "computers do make errors." Also, because the bank's statement comes only once a month, you should know your current balance so your account will not be overdrawn. Overdrawn accounts pay a penalty and can lead to a bad credit rating.

Like many applications with mathematics, balancing a checking account follows a pattern of steps. The procedure and related ideas are given in the following lists.

The bank or the savings and loan company sends you a statement of your checking account each month. This statement contains a record of:

1. The beginning balance;
2. A list of all checks paid by the bank;
3. A list of deposits you made;
4. Any interest paid to you (some checking accounts do pay interest);
5. Any service charge by the bank; and
6. The closing balance.

Your checkbook register contains a record of:

1. All the checks you have written; and
2. The current balance.

Your current balance will not agree with the closing balance on the bank statement because:

1. You have not recorded the interest.
2. You have not recorded the service charge.
3. The bank does not have a record of all the checks you have written. (Several checks will be *outstanding* because they have not been received yet for payment by the bank by the date given on the bank statement.)

To balance your checking account (sometimes called **reconciling** the bank statement with your checkbook register):

1. Go through your checkbook register and the bank statement and put a check mark (✔) by each check paid and deposit recorded on the bank statement.
2. In your checkbook register:
   (a) Add any interest paid to your current balance.
   (b) Subtract any service charge from this balance.
   This number is your **true balance.**
3. Find the total of all the outstanding checks (no ✔ mark) in your checkbook register (checks not yet received by the bank.)
4. On a reconciliation sheet:
   (a) Enter the balance from the bank statement.
   (b) Add any deposits you have made that are not recorded (no ✔ mark) on the bank statement.
   (c) Subtract the total of your outstanding checks (found in Step 3). This number should agree with your **true balance.** Now your checking account is balanced.

If your checking account does not balance:

1. Go over your arithmetic in the balancing procedure.
2. Go over your arithmetic check by check in your checkbook register.
3. Make an appointment with bank personnel to find the reasons for any other errors.

EXAMPLES

1.

YOUR CHECKBOOK REGISTER

Check No.	Date	Transaction Description	Payment (−)	(✓)	Deposit (+)	Balance
			Balance brought forward			312.12
102	2-3	Painted Lady (cosmetics)	14.80	✓		−14.80 297.32
103	2-4	Nu-Grocers (groceries)	76.51			−76.51 220.81
104	2-9	Wood Lumber (plywood)	53.70	✓		−53.70 167.11
	2-16	Deposit ( )		✓	100.00	+100.00 267.11
105	2-21	Metal Box (nails)	10.14	✓		−10.14 256.97
	2-26	Deposit ( )		✓	250.00	+250.00 506.97
		Bank Interest ( )		✓		—— 506.97
		Service Charge ( )	3.00	✓		−3.00 503.97
		( )				——
		( )				——
					True Balance	503.97

BANK STATEMENT
Checking Account Activity

Transaction Description	Amount	(✓)	Running Balance	Date
Beginning Balance		✓	312.12	2-1
Check # 102	14.80	✓	297.32	2-4
Check # 104	53.70	✓	243.62	2-5
Deposit	100.00	✓	343.62	2-16
Check #105	10.14	✓	333.48	2-24
Service Charge	3.00	✓	330.45	2-28
Ending Balance				

RECONCILIATION SHEET

A. First, mark ✓ beside each check and deposit listed in both your checkbook register and on the bank statement.
B. Second, in your checkbook register, add any interest paid and subtract any service charge listed on the bank statement.
C. Third, find the total of all outstanding checks.

Outstanding Checks

No.	Amount
103	76.51
Total	76.51

Statement Balance	330.48
Add deposits not credited	+250.00
Total	580.48
Subtract total amount of checks outstanding	−76.51
True Balance	503.97

2.

YOUR CHECKBOOK REGISTER

Check No.	Date	Transaction Description	Payment (−)	(✓)	Deposit (+)	Balance
			Balance brought forward			505.21
152	1-3	U-Auto-Lease (car lease)	198.60	✓		−198.60 306.61
153	1-10	ABC Power Co. (electric bill)	75.00			−75.00 231.61
154	1-10	Fly-By-Nite Airway (plane ticket)	112.00	✓		−112.00 119.61
	1-15	Deposit ( )		✓	500.00	+500.00 619.61
155	1-15	Safe-T Drugs (medicine)	5.60	✓		−5.60 614.01
156	1-17	Milady Togs (dress)	49.80	✓		−49.80 564.21
157	2-03	U-Auto Lease (car lease)	198.60			−198.60 365.61
		Bank Interest ( )		✓		—— 365.61
		Service Charge ( )	4.00	✓		−4.00 361.61
		( )				——
					True Balance	361.61

BANK STATEMENT
Checking Account Activity

Transaction Description	Amount	(✓)	Running Balance	Date
Beginning Balance		✓	505.21	1-01
Check #152	198.60	✓	306.61	1-06
Check #154	112.00	✓	194.61	1-14
Deposit	500.00	✓	694.61	1-15
Check #155	5.60	✓	689.01	1-18
Check #156	49.80	✓	639.21	1-20
Service Charge	4.00	✓	635.21	1-31
Ending Balance			635.21	

RECONCILIATION SHEET

A. First, mark ✓ beside each check and deposit listed in both your checkbook register and on the bank statement.
B. Second, in your checkbook register, add any interest paid and subtract any service charge listed on the bank statement.
C. Third, find the total of all outstanding checks.

Outstanding Checks

No.	Amount
153	75.00
157	198.60
Total	273.60

Statement Balance	635.21
Add deposits not credited	+ ——
Total	635.21
Subtract total amount of checks outstanding	−273.60
True Balance	361.61

Exercises 8.3

For each of the following problems, you are given a copy of a checkbook register, the corresponding bank statement, and a reconciliation sheet. You are to find the true balance of the account on both the checkbook register and on the reconciliation sheet as shown in Examples 1 and 2. Follow the directions on the reconciliation sheet.

1.

YOUR CHECKBOOK REGISTER

Check No.	Date	Transaction Description	Payment (−)	(✓)	Deposit (+)	Balance
			Balance brought forward			-0-
	7-15	Deposit ( )			700.00	+700.00 700.00
1	7-15	Quiet Town Apt. (rent/deposit)	520.00			-520.00 180.00
2	7-15	Pa Bell Telephone (phone installation)	32.16			- 32.16 147.84
3	7-15	XYZ Power Co. (gas/elect. hookup)	46.49			- 46.49 101.35
4	7-16	Foodway Stores (groceries)	51.90			- 51.90 49.45
	7-20	Deposit ( )			350.00	+350.00 399.45
5	7-23	Comfy Furniture (sofa, chair)	300.50			-300.50 98.95
	8-1	Deposit ( )			350.00	+350.00 448.95
		Interest ( )			2.50	+ 2.50
		Service ( )	2.00			- 2.00
			True Balance			

BANK STATEMENT
Checking Account Activity

Transaction Description	Amount	(✓)	Running Balance	Date
Beginning Balance	700.00		700.00	7-15
Check #1	520.00		180.00	7-16
Check #4	51.90		128.10	7-17
Check #2	32.16		95.94	7-18
Check #3	46.49		49.45	7-18
Deposit	350.00		399.45	7-20
Interest	2.50		401.95	7-31
Service Charge	2.00		399.95	7-31
Ending Balance			399.95	

RECONCILIATION SHEET

A. First, mark ✓ beside each check and deposit listed in both your checkbook register and on the bank statement.

B. Second, in your checkbook register, add any interest paid and subtract any service charge listed on the bank statement.

C. Third, find the total of all outstanding checks.

Outstanding Checks	
No.	Amount
Total	______

Statement Balance	______
Add deposits not credited	+ ______
Total	______
Subtract total amount of checks outstanding	− ______
True Balance	______

2.

YOUR CHECKBOOK REGISTER

Check No.	Date	Transaction Description	Payment (−)	(✓)	Deposit (+)	Balance
			Balance brought forward			1610.39
1234	12-7	Pearl City (pearl ring)	524.00			-524.00 1086.39
1235	12-7	Comp-U-Tate (home computer)	801.60			-801.60 284.79
1236	12-8	Sportz Hutz (skis)	206.25			-206.25 78.54
1237	12-8	Guild Card Shop (christmas cards)	25.50			-25.50 53.04
	12-10	Deposit ( )			1000.00	+1000.00 1053.04
1238	12-14	Toys-R-We (stuffed panda)	80.41			-80.41
1239	12-24	Meat Markette (turkey)	18.39			-18.39
1240	12-24	Poodle Shoppe (pedigreed puppy)	300.00			-300.00
1241	12-31	Homey Sav. & Loan (mortgage payment)	600.00			-600.00
		Service Charge ( )	-0-			———
			True Balance			

BANK STATEMENT
Checking Account Activity

Transaction Description	Amount	(✓)	Running Balance	Date
Beginning Balance			1610.39	12-01
Check #1234	524.00		1086.39	12-08
Check #1236	206.25		880.14	12-09
Check #1237	25.50		854.64	12-09
Deposit	1000.00		1854.64	12-10
Check #1235	801.60		1053.04	12-11
Check #1238	80.41		972.63	12-15
Check #1239	18.39		954.24	12-27
Check #1240	300.00		654.24	12-28
Ending Balance			654.24	

RECONCILIATION SHEET

A. First, mark ✓ beside each check and deposit listed in both your checkbook register and on the bank statement.

B. Second, in your checkbook register, add any interest paid and subtract any service charge listed on the bank statement.

C. Third, find the total of all outstanding checks.

Outstanding Checks	
No.	Amount
Total	______

Statement Balance	______
Add deposits not credited	+ ______
Total	______
Subtract total amount of checks outstanding	− ______
True Balance	______

3.

YOUR CHECKBOOK REGISTER

Check No.	Date	Transaction Description	Payment (−)	(✓)	Deposit (+)	Balance
			Balance brought forward			756.14
271	6-15	Parts, Parts, Parts (spark plugs)	12.72			-12.72 743.42
272	6-24	Firerock Tire Co. (2 tires)	121.40			———
273	6-30	Gus' Gas Station (tune-up)	75.68			———
	7-1	Deposit ( )			250.00	———
274	7-1	Prudent Ins. Co. (car insurance)	300.00			———
		Service Charge ( )				———
		( )				———
		( )				———
		( )				———
		( )				———
			True Balance			

BANK STATEMENT
Checking Account Activity

Transaction Description	Amount	(✓)	Running Balance	Date
Beginning Balance			756.14	6-01
Check #271	12.72		743.42	6-16
Check #272	121.40		622.02	6-26
Service Charge	1.00		621.02	6-30
Ending Balance			621.02	

RECONCILIATION SHEET

A. First, mark ✓ beside each check and deposit listed in both your checkbook register and on the bank statement.

B. Second, in your checkbook register, add any interest paid and subtract any service charge listed on the bank statement.

C. Third, find the total of all outstanding checks.

Outstanding Checks	
No.	Amount
Total	______

Statement Balance	______
Add deposits not credited	+ ______
Total	______
Subtract total amount of checks outstanding	− ______
True Balance	______

4.

YOUR CHECKBOOK REGISTER

Check No.	Date	Transaction Description	Payment (−)	(✓)	Deposit (+)	Balance
			Balance brought forward			12.14
419	1-2	Postmaster (stamps)	10.00			-10.00 2.14
	1-3	Deposit ( )			525.50	
420	1-17	E. Hamel, DDS (dental check-up)	63.50			
421	1-26	Cash ( )	100.00			
422	1-31	Up-Top Apartments (rent)	350.00			
		Service Charge ( )				
		( )				
		( )				
		( )				
		( )				
			True Balance			

BANK STATEMENT
Checking Account Activity

Transaction Description	Amount	(✓)	Running Balance	Date
Beginning Balance			12.14	1-01
Deposit	525.50		537.64	1-03
Check #419	10.00		527.64	1-03
Check #420	63.50		464.14	1-20
Check #421	100.00		364.14	1-26
Service Charge	2.00		362.14	1-31
Ending Balance			362.14	

RECONCILIATION SHEET

A. First, mark ✓ beside each check and deposit listed in both your checkbook register and on the bank statement.

B. Second, in your checkbook register, add any interest paid and subtract any service charge listed on the bank statement.

C. Third, find the total of all outstanding checks.

Outstanding Checks No.	Amount
Total	

Statement Balance	
Add deposits not credited	+
Total	
Subtract total amount of checks outstanding	−
True Balance	

5.

YOUR CHECKBOOK REGISTER

Check No.	Date	Transaction Description	Payment (−)	(✓)	Deposit (+)	Balance
			Balance brought forward			967.22
772	4-13	C.P. Hay (accountant)	85.00			
	4-14	Deposit ( )			1200.00	
773	4-14	E.Z. Pharmacy (aspirin)	4.71			
774	4-15	I.R.S. (income tax)	2,000.00			
775	4-30	Heavy Finance Co. (loan payment)	52.50			
	5-1	Deposit ( )			600.00	
		Interest ( )				
		Service Charge ( )				
		( )				
		( )				
			True Balance			

BANK STATEMENT
Checking Account Activity

Transaction Description	Amount	(✓)	Running Balance	Date
Beginning Balance			967.22	4-01
Deposit	1200.00		2167.22	4-14
Check #772	85.00		2082.22	4-15
Check #773	4.71		2077.51	4-15
Interest	2.82		2080.33	4-30
Service Charge	4.00		2076.33	4-30
Ending Balance			2076.33	

RECONCILIATION SHEET

A. First, mark ✓ beside each check and deposit listed in both your checkbook register and on the bank statement.

B. Second, in your checkbook register, add any interest paid and subtract any service charge listed on the bank statement.

C. Third, find the total of all outstanding checks.

Outstanding Checks No.	Amount
Total	

Statement Balance	
Add deposits not credited	+
Total	
Subtract total amount of checks outstanding	−
True Balance	

6.

YOUR CHECKBOOK REGISTER

Check No.	Date	Transaction Description	Payment (−)	(✓)	Deposit (+)	Balance
			Balance brought forward			1403.49
86	9-1	Now Stationers (school supplies)	17.12			
87	9-2	Young-At-Heart (clothes)	192.50			
88	9-4	H.S.U. Bookstore (books)	56.28			
89	9-7	Regent's Office (tuition)	380.00			
90	9-7	Off-Campus Apts. (rent)	240.00			
91	9-27	State Telephone Co. (phone bill)	24.62			
92	9-30	Up-N-Up Foods (groceries)	47.80			
		Service Charge ( )				
		( )				
		( )				
			True Balance			

BANK STATEMENT
Checking Account Activity

Transaction Description	Amount	(✓)	Running Balance	Date
Beginning Balance			1403.49	9-01
Check #86	17.12		1386.37	9-02
Check #87	192.50		1193.87	9-05
Check #88	56.28		1137.59	9-05
Check #90	240.00		897.59	9-10
Service Charge	4.00		893.59	9-30
Ending Balance			893.59	

RECONCILIATION SHEET

A. First, mark ✓ beside each check and deposit listed in both your checkbook register and on the bank statement.

B. Second, in your checkbook register, add any interest paid and subtract any service charge listed on the bank statement.

C. Third, find the total of all outstanding checks.

Outstanding Checks No.	Amount
Total	

Statement Balance	
Add deposits not credited	+
Total	
Subtract total amount of checks outstanding	−
True Balance	

7.

YOUR CHECKBOOK REGISTER

Check No.	Date	Transaction Description	Payment (−)	(✓)	Deposit (+)	Balance
			Balance brought forward			602.82
14	6-20	Aisle Bridal (flowers)	102.40			
	6-22	Deposit ( )			1000.00	
15	6-24	Tuxedo Junction (tux)	55.65			
16	6-28	D. Lohengrin (organist)	25.00			
17	6-28	D-Lux Limo (limo rental)	75.00			
18	6-30	K.K. Katering (food caterer)	700.00			
19	7-1	Luv-Lee Stationers (thank-you cards)	15.20			
		Service Charge ( )				
		( )				
		( )				
			True Balance			

BANK STATEMENT
Checking Account Activity

Transaction Description	Amount	(✓)	Running Balance	Date
Beginning Balance			602.82	6-01
Deposit	1000.00		1602.82	6-22
Check #14	102.40		1500.42	6-22
Check #15	55.65		1444.77	6-26
Service Charge	1.00		1443.77	6-30
Ending Balance			1443.77	

RECONCILIATION SHEET

A. First, mark ✓ beside each check and deposit listed in both your checkbook register and on the bank statement.

B. Second, in your checkbook register, add any interest paid and subtract any service charge listed on the bank statement.

C. Third, find the total of all outstanding checks.

Outstanding Checks No.	Amount
Total	

Statement Balance	
Add deposits not credited	+
Total	
Subtract total amount of checks outstanding	−
True Balance	

8.

YOUR CHECKBOOK REGISTER

Check No.	Date	Transaction Description	Payment (−)	(✓)	Deposit (+)	Balance
			Balance brought forward			278.32
326	8-12	J. J. Jones (birthday check)	40.00			
	8-15	Deposit ( )			500.00	
327	8-15	N. O. Payne, M.D. (physical exam)	260.00			
328	8-15	Local Waterworks (water/trash)	27.40			
	8-22	Deposit ( )	400.12			
329	8-27	Time-O-Life (magazine subscrip)	12.50			
330	8-29	Biggar Mortgage (house payment)	750.00			
		Interest ( )				
		Service Charge ( )				
		( )				
		True Balance				

BANK STATEMENT
Checking Account Activity

Transaction Description	Amount	(✓)	Running Balance	Date
Beginning Balance			278.32	8-01
Deposit	500.00		778.32	8-15
Check #326	40.00		738.32	8-15
Check #327	260.00		478.32	8-17
Deposit	400.12		878.44	8-22
Check #328	27.40		851.04	8-22
Interest	1.82		852.86	8-31
Service Charge	4.00		848.86	8-31
Ending Balance			848.86	

RECONCILIATION SHEET

A. First, mark ✓ beside each check and deposit listed in both your checkbook register and on the bank statement.

B. Second, in your checkbook register, add any interest paid and subtract any service charge listed on the bank statement.

C. Third, find the total of all outstanding checks.

Outstanding Checks No.	Amount
Total	

Statement Balance	
Add deposits not credited	+
Total	
Subtract total amount of checks outstanding	−
True Balance	

9.

YOUR CHECKBOOK REGISTER

Check No.	Date	Transaction Description	Payment (−)	(✓)	Deposit (+)	Balance
			Balance brought forward			147.02
203	2-3	Food Stoppe (groceries)	26.90			
204	2-8	Ekhon Oil (gasoline bill)	71.45			
205	2-14	Rose's Roses (flowers)	25.00			
206	2-14	S.M.R.U. (alumni dues)	20.00			
	2-15	Deposit ( )			600.00	
207	2-26	SRO (theater tickets)	52.50			
208	2-28	MPG Mtg. (house payment)	500.00			
		Service Charge ( )				
		( )				
		( )				
		True Balance				

BANK STATEMENT
Checking Account Activity

Transaction Description	Amount	(✓)	Running Balance	Date
Beginning Balance			147.02	2-01
Check #203	26.90		120.12	2-04
Check #204	71.45		48.67	2-14
Deposit	600.00		648.67	2-15
Check #205	25.00		623.67	2-15
Service Charge	3.00		620.67	2-28
Ending Balance			620.67	

RECONCILIATION SHEET

A. First, mark ✓ beside each check and deposit listed in both your checkbook register and on the bank statement.

B. Second, in your checkbook register, add any interest paid and subtract any service charge listed on the bank statement.

C. Third, find the total of all outstanding checks.

Outstanding Checks No.	Amount
Total	

Statement Balance	
Add deposits not credited	+
Total	
Subtract total amount of checks outstanding	−
True Balance	

10.

YOUR CHECKBOOK REGISTER

Check No.	Date	Transaction Description	Payment (−)	(✓)	Deposit (+)	Balance
			Balance brought forward			4071.82
996	10-1	Red-E Credit (loan payment)	200.75			
997	10-10	United Ways (charity donation)	25.00			
998	10-21	MacIntosh Farms (barrel of apples)	42.20			
999	10-26	Fun Haus (costume rental)	35.00			
1000	10-28	Yum Yum Shoppe (Halloween candy)	12.14			
1001	10-29	B-Sharp, Inc. (piano tuners)	20.00			
1002	10-30	Food-2-Goo! (party platter)	78.50			
1003	10-31	Cash ( )	300.00			
1004	10-31	Principal S&L (house payment)	1250.60			
		Interest ( )				
			True Balance			

BANK STATEMENT
Checking Account Activity

Transaction Description	Amount	(✓)	Running Balance	Date
Beginning Balance			4071.82	10-01
Check #996	200.75		3871.07	10-05
Check #998	42.20		3828.87	10-23
Check #999	35.00		3793.87	10-30
Check #1003	300.00		3493.87	10-31
Interest	16.29		3510.16	10-31
Ending Balance			3510.16	

RECONCILIATION SHEET

A. First, mark ✓ beside each check and deposit listed in both your checkbook register and on the bank statement.

B. Second, in your checkbook register, add any interest paid and subtract any service charge listed on the bank statement.

C. Third, find the total of all outstanding checks.

Outstanding Checks

No.	Amount
Total	

Statement Balance	
Add deposits not credited	+
Total	
Subtract total amount of checks outstanding	−
True Balance	

8.4 READING GRAPHS

Graphs are "pictures" of numerical information. Graphs appear almost daily in newspapers and magazines and frequently in textbooks and corporate reports. Communication with well-drawn graphs can be accurate, effective, and fast. Most computers can be programmed to draw graphs, and anyone whose work involves a computer in any way will probably be expected to understand graphs.

Four types of graphs are:

1. **Bar Graphs:** To emphasize comparative amounts
2. **Circle Graphs:** To help in understanding percents or parts of a whole
3. **Line Graphs:** To indicate tendencies or trends over a period of time
4. **Pictographs:** To emphasize the topic being related as well as quantities

> All graphs should:
>
> 1. Be clearly labeled. 2. Be easy to read. 3. Have titles.

The following examples illustrate the four types of graphs listed. There are several questions related to each graph to test your understanding. Some questions can be answered directly from the graph while others involve some calculations.

These examples are given in a form that asks the student to complete some of the answers to the questions. This style is particularly useful either as a classroom activity or to lead individual students in step-by-step understanding. The answers are given at the bottom of the page.

COMPLETION EXAMPLE 1

Figure 8.1 shows a bar graph. Note that the scale on the left and the data on the bottom (months) are clearly labeled and the graph itself has a title. The questions following can be easily answered by looking at the graph.

(a) What were the sales in January? $100,000
(b) During what month were sales lowest? March
(c) During what month were sales highest? February and June
(d) What were the sales during each of the highest sales months?

____

(e) What were the sales in April? ________________

The following questions will take some calculations after reading the graph.

(f) What was the amount of decrease in sales between February and March?

	\$150,000	February sales
−	50,000	March sales
	\$100,000	decrease in sales

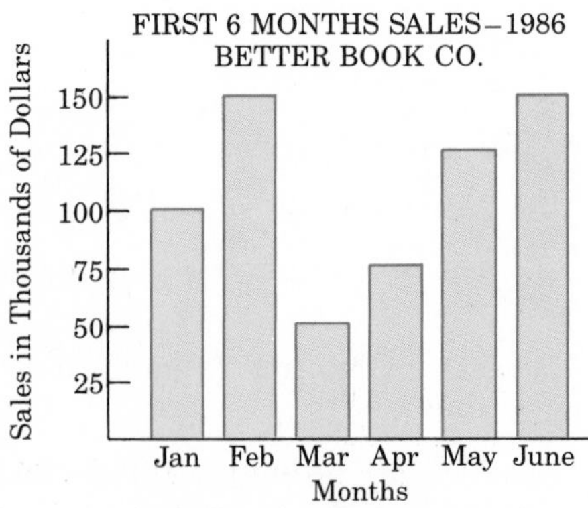

Figure 8.1 Bar graph

(g) What was the percent of decrease?

$$\frac{100{,}000 \text{ decrease}}{150{,}000 \text{ February sales}} = \frac{2}{3} = .666 = 67\% \text{ decrease}$$

COMPLETION EXAMPLE 2

Figure 8.2 shows percents budgeted for various items in a home for one year. Suppose a family has an annual income of $15,000. Using Figure 8.2, calculate how much will be allocated to each item indicated in the graph.

Solution

	ITEM	*AMOUNT*
Housing	0.25 × $15,000 =	$ 3750.00
Food	0.20 × $15,000 =	3000.00
Taxes	0.05 × $15,000 =	750.00
Clothing	0.07 × $15,000 =	1050.00
Savings	0.10 × $15,000 =	________
Education	0.15 × $15,000 =	________
Entertainment	____________ =	________
Transportation & Maintenance	____________ =	________

What is the total of all the amounts? ____________

HOME BUDGET FOR 1 YEAR

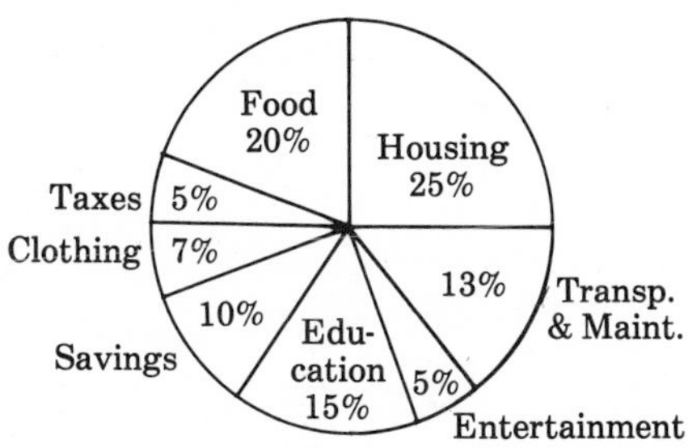

Figure 8.2 Circle graph

Completion Example Answers

1. (d) $150,000
   (e) $75,000
2. Savings $1500.00
   Education $2250.00
   Entertainment 0.05 × 15,000 = $750.00
   Transportation & Maintenance 0.13 × 15,000 = $1950.00
   Total of all amounts is 100% of $15,000 which is $15,000.

COMPLETION EXAMPLE 3

Figure 8.3 is a line graph that shows the relationships between daily high and low temperatures. From the graph you can see that temperatures tended to rise during the week but fell sharply on Saturday.

What was the lowest high temperature? 66°
On what day did this occur? Sunday
What was the highest low temperature? 70°
On what day did this occur? Friday

Find the average difference between the daily high and low temperatures for the week shown.

Solution

First find the differences; then average these differences.

Sunday	66 – 60 =	6°
Monday	70 – 62 =	8°
Tuesday	76 – 66 =	10°
Wednesday	72 – 66 =	______
Thursday	80 – 66 =	______
Friday	80 – 70 =	______
Saturday	74 – 62 =	______
		______ total of differences
	7)	______ average difference

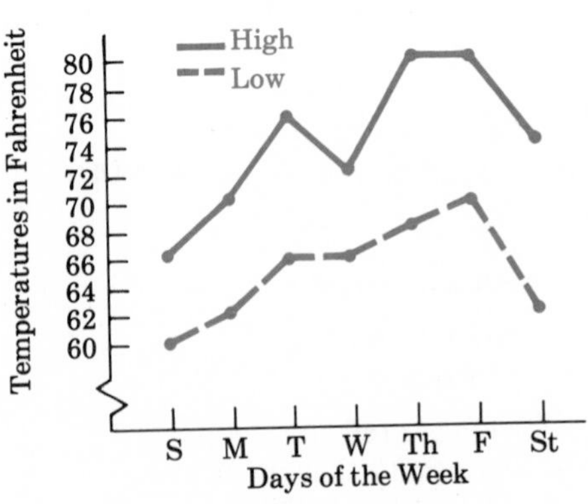

Figure 8.3 Line graph

COMPLETION EXAMPLE 4

Figure 8.4 shows a pictograph of new home construction in five counties in 1980. Answer the following questions.

(a) Which county had the most number of new homes built?

____

How many? _______________

(b) Which county had the least number of new homes built?

____

How many? _______________

(c) What was the difference between new home construction in Counties A and D? _______________

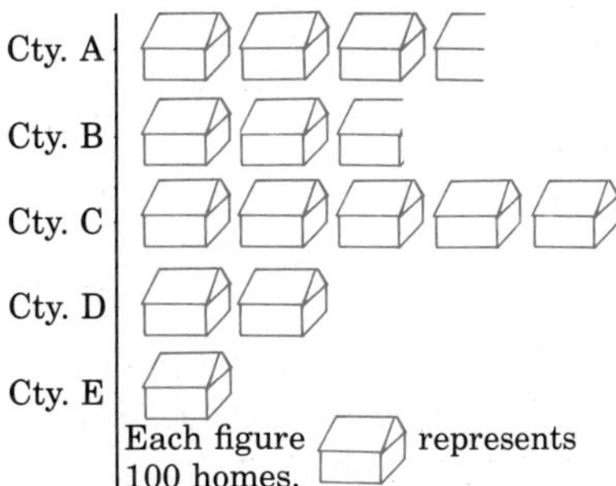

Figure 8.4 Pictograph

Completion Example Answers

3.

Sunday	6°
Monday	8°
Tuesday	10°
Wednesday	6°
Thursday	14°
Friday	10°
Saturday	12°
	66° total of differences

$$\begin{array}{r} 9.4^\circ \\ 7\overline{)66.0} \\ \underline{63} \\ 3\,0 \\ \underline{2\,8} \\ 2 \end{array}$$ average difference

4. (a) County C, 500 homes
(b) County E, 100 homes
(c) 350 − 200 = 150 homes difference

Exercises 8.4

After you have carefully studied the four graphs in Examples 1–4 and have answered the related questions, answer the question related to each of the graphs in Problems 1–12. Some questions can be answered directly from the graphs while others will require you to make some calculations.

1. (a) Which field of study has the largest number of declared majors?
(b) Which field of study has the smallest number of declared majors?
(c) How many declared majors are indicated in the entire graph?
(d) What percent are computer science majors?

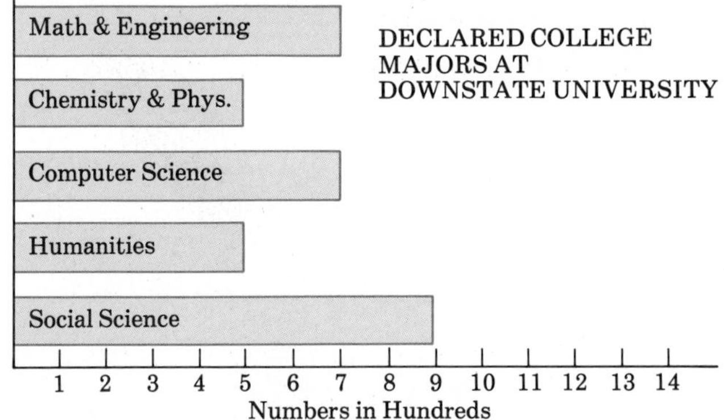

2. The school budget shown is based on a total budget of $12,500,000.
(a) What amount will be spent on each category?
(b) How much more will be spent on teachers' salaries than on administration salaries?
(c) What percent will be spent on items other than salaries?
(d) How much will be spent on these items?

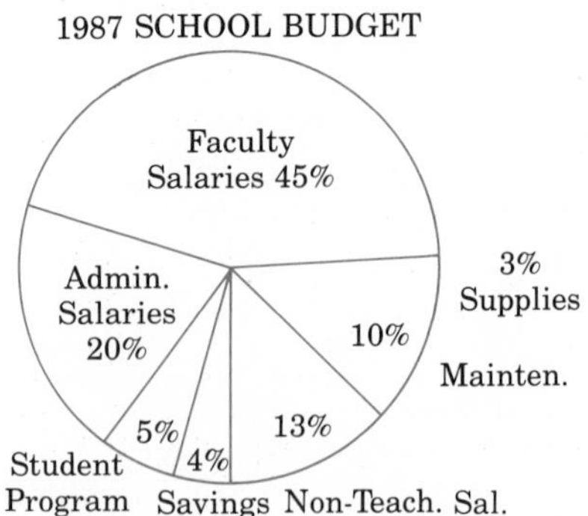

3. (a) In what year was the rainfall least?
   (b) What was the most rainfall in a year?
   (c) What year?
   (d) What was the average rainfall over the 6-year period?

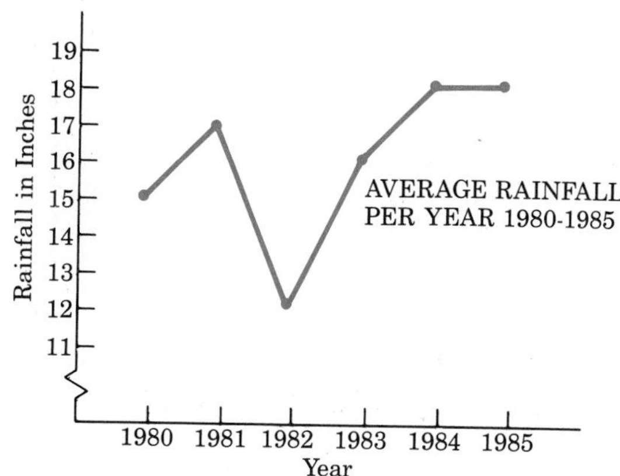

4. (a) Which farm showed the most corn production?
   (b) How much was this?
   (c) What percent of the total of the production of the four farms was this?

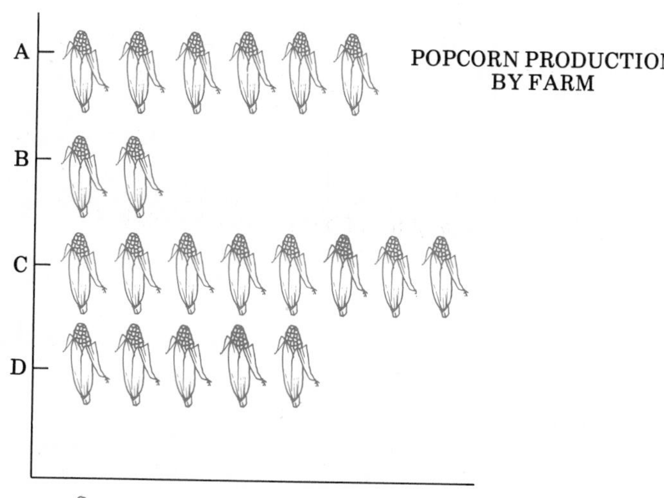

5. Station XYZ is off the air from 2 A.M. to 6 A.M., so there are only 20 hours of daily programming. Sports are not shown here because they are considered special events. In the 20-hour period shown, how much time (in minutes) is devoted to each category?

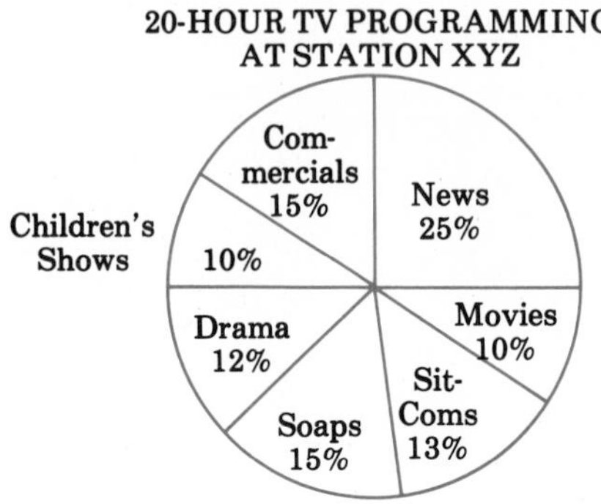

6. (a) If on Monday morning you had 100 shares in each of the three stocks shown (oil, steel, wheat), and you held the stock all week, in which stock would you have lost money?
   (b) How much?
   (c) In which stock would you have gained money?
   (d) How much?
   (e) In which stock could you have made the most money if you had sold at the best time?
   (f) How much?
   (g) In which stock could you have lost the most money if you had sold at the worst time?
   (h) How much?

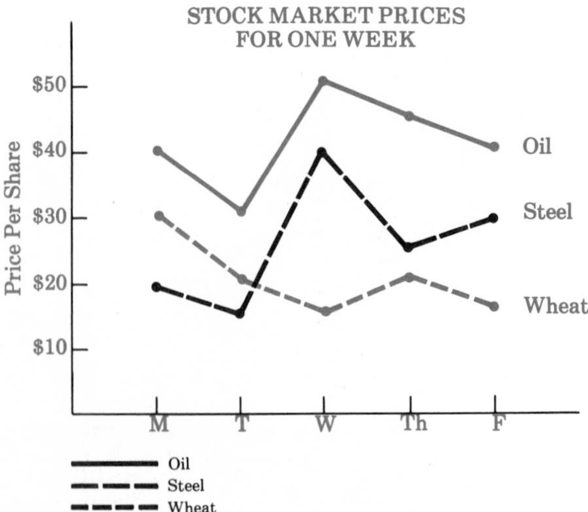

7. (a) Which city had the greatest number of marathon runners?
   (b) How many?
   (c) What percent was this of the total number in the race?
   (d) Which city had the greatest number of women runners?
   (e) How many?
   (f) What percent of the total number of women runners was this?

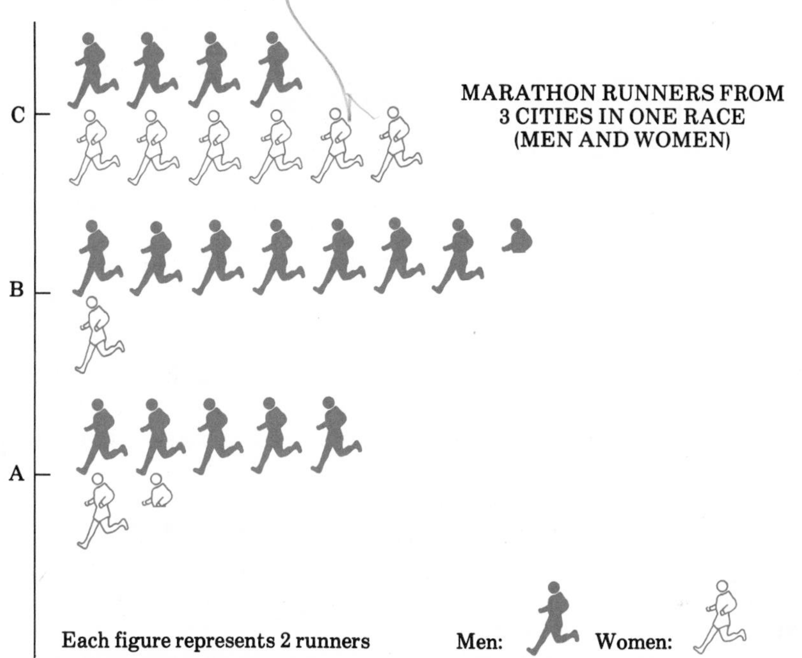

8. (a) What was the total number of vehicles that crossed this intersection in the 2 weeks?
   (b) Which day averaged the highest number?
   (c) How many?
   (d) What percent of the total intersection traffic was counted on Sundays?
   (e) On Mondays?

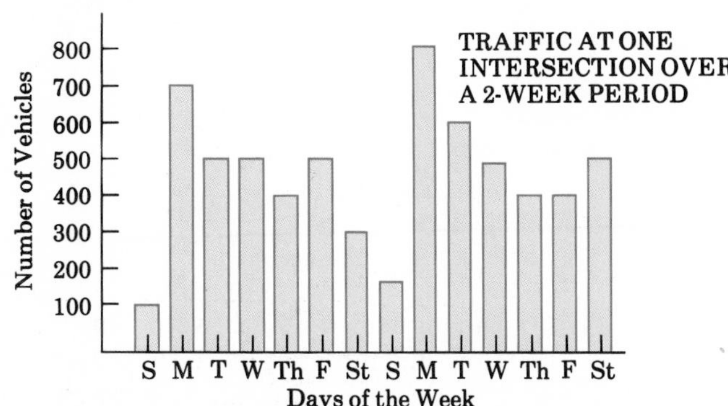

9. (a) During what month or months in 1985 were interest rates highest?
   (b) Lowest?
   (c) What was the average of the interest rates over the entire year?

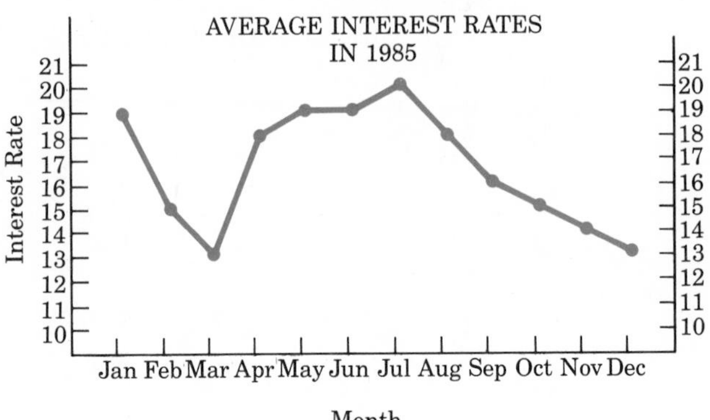

10. Assume that all five students (Al, Bob, Ron, Sue, Ann) had graduated with comparable grades from the same high school.
   (a) What would be a major difficulty in putting the two graphs shown here together on one graph?
   (b) Who worked the most hours per week?
   (c) Who had the lowest college GPA?
   (d) If Bob spent 30 hours per week studying for his classes, what percent of his total work week (part-time work plus study time) did he spend studying?
   (e) Ann also spent 30 hours per week studying. What percent of her work week did she spend studying?
   (f) Does there seem to be any relationship between college GPA and job hours?

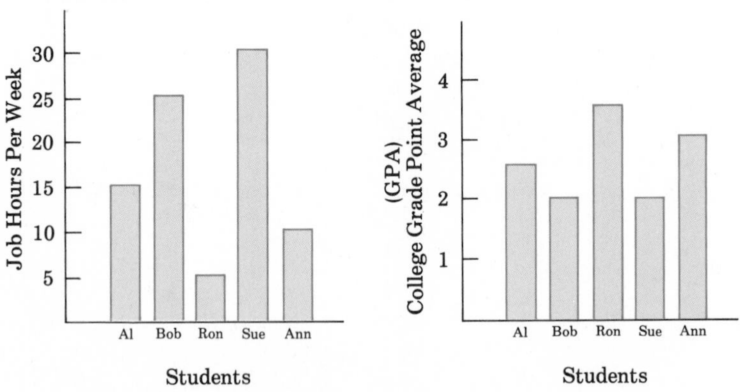

11. (a) During which month were domestic sales highest?
   (b) How much higher were they than for the lowest month?
   (c) What was the difference in import sales for the same two months?

(d) What was the difference between domestic and import sales in March?
(e) What percent of sales were imports in December 1983?
(f) In December 1984? (Answers will be approximate.)

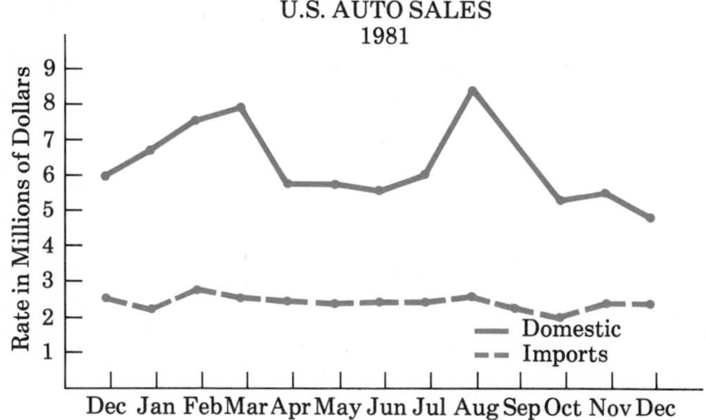

12. (a) What percent of workers were in each of the four areas in 1860?
(b) In 1980?
(c) Which area of work seems to have had the most stable percent of workers between 1860 and 1980?
(d) What is the difference between the highest and lowest percents for this area?
(e) Which area has had the most growth?
(f) What was its lowest percent and when?
(g) What was its highest percent and when?
(h) Which area has had the most decline?

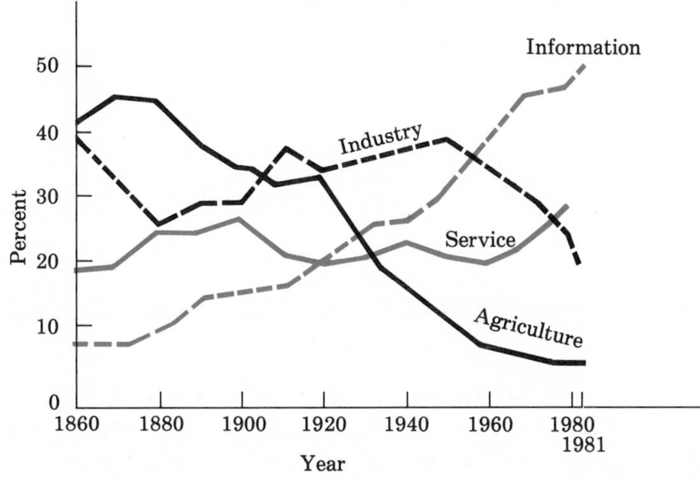

8.5 STATISTICS (MEAN, MEDIAN, MODE, AND RANGE)

Statistics is basically the study of numerical characteristics of large amounts of numerical values indicating some type of information. In this section, we will discuss only four measures (or four statistics) that are easily found or calculated: mean, median, mode, and range. Other measures, such as standard deviation and variance (both indicate "how spread out" the information is), correlation coefficients (Is there a correlation between your schooling and your life-time earnings?), and z-scores (How does your score on a math exam compare to your score on a history exam?) are discussed in detail in statistics courses. You would need two semesters of algebra to understand these topics, so keep working hard.

The following terms and their corresponding definitions are necessary for understanding the problems in this section. They are given in a list form for easy reference.

TERMS USED IN THE STUDY OF STATISTICS

Data: Value(s) measuring some characteristic of interest (We will consider only numerical data.)

Statistic: A single number describing some characteristic of the data.

Mean: The arithmetic average of the data (Add all the data and divide by the number of data items.)

Median: The middle data item (Arrange the data in order and pick out the middle item.)

Mode: The single data item that appears the most number of times (Some data may have more than one mode. We will leave the discussion of such a situation to a course in statistics. In this text, if data has a mode, there will be only one.)

Range: The difference between the largest and smallest data items

In Examples 1–4, answer the questions about statistics using the data from Group A and Group B.

GROUP A: Annual Income for 8 Families

$18,000; $12,000; $15,000; $17,000; $35,000; $70,000; $15,000; $20,000

GROUP B: Grade Point Averages (GPA) for 11 Students

2.0; 2.0; 1.9; 3.1; 3.5; 2.9; 2.5; 3.6; 2.0; 2.4; 3.4

EXAMPLES

1. Find the mean income for the families in Group A.

Solution

Find the sum of the 8 incomes and divide by 8.

```
$ 18,000              $25,250
  12,000            8)202,000
  15,000              16
  17,000               42
  35,000               40
  70,000                20
  15,000                16
  20,000                 40
$202,000                 40
                          00
                           0
                           0
```

You may want to use a calculator to do this arithmetic. The mean annual income is $25,250.

2. Find the median income for Group A and the median GPA for Group B.

Solution

To find the median:

1. Arrange the data in order.
2. If there is an **even** number of items, the median is the average of the two middle items.
3. If there is an **odd** number of items, the median is the middle item.

Arrange both sets of data in order.

GROUP A

$12,000; $15,000; $15,000; $17,000; $18,000; $20,000; $35,000; $70,000

GROUP B

1.9; 2.0; 2.0; 2.0; 2.4; 2.5; 2.9; 3.1; 3.4; 3.5; 3.6

For Group A, the median is the average of the 4th and 5th items because there is an even number (8) of items.

$$\text{median} = \frac{\$17{,}000 + \$18{,}000}{2} = \frac{35{,}000}{2} = \$17{,}500$$

For Group B, the median is the 6th item because, with an **odd** number of 11 items, the 6th item is the middle item.

$$\text{median} = 2.5$$

3. Find the mode and the range for both Group A and Group B.

Solution

The mode is the most frequent item. From the arranged data in Example 2, we can see that:

for Group A, the mode is \$15,000
for Group B, the mode is 2.0

The range is the difference between the largest and smallest items:

$$\text{Group A range} = \$70{,}000 - \$12{,}000 = \$58{,}000$$
$$\text{Group B range} = 3.6 - 1.9 = 1.7$$

4. COMMENTARY Of the four statistics mentioned in this section, the mean and median are most commonly used. Many people feel that the mean (or arithmetic average) is relied on too much in reporting central tendencies for data such as income, housing costs, and taxes where a few very high items can *distort* the picture of a central tendency. As you can see in Group A data, the median of \$17,500 is probably more representative of the data than the mean of \$25,250. This is because the one high income of \$70,000 raises the mean considerably.

   When you read an article in a magazine or newspaper that reports means or medians, you should now have a better understanding of the implications.

Exercises 8.5

For each of the following problems, find (a) the mean, (b) the median, (c) the mode, and (d) the range of the given data.

1. Ten math students had the following scores on a final exam:

75, 83, 93, 65, 85,
85, 88, 90, 55, 71

2. Joe did the following number of sit-ups each morning for a week:

25, 52, 48, 42, 38, 58, 52

3. Fifteen college students reported the following hours of sleep the night before an exam:

4, 6, 6, 7, 6.5, 6.5, 7.5, 8.5
5, 6, 4.5, 5.5, 9, 3, 8

4. The local high school basketball team scored the following points per game during their 20-game season:

85, 60, 62, 70, 75, 52, 88, 50, 80, 72,
90, 85, 85, 93, 70, 75, 68, 73, 65, 82

5. Stacey went to six different repair shops to get the following estimates to repair her car. (The accident was not her fault; her car was parked at the time.)

\$425, \$525, \$325, \$300, \$500, \$325

6. Mike kept track of his golf scores for twelve rounds of eighteen holes each. His scores were:

85, 90, 82, 85, 87, 80,
78, 82, 88, 82, 86, 81

7. The local weather station recorded the following daily high temperatures for one month:

75, 76, 76, 78, 85, 82, 85, 88, 90, 90,
88, 95, 96, 92, 88, 88, 80, 80, 78, 80,
78, 76, 77, 75, 75, 74, 70, 70, 72, 73

8. The Big City fire department reported the following mileage for tires used on their nine fire trucks:

14,000; 14,000; 11,000; 15,000;
9,000; 14,000; 12,000; 10,000; 9,000

9. The city planning department issued the following numbers of building permits over a three-week period:

17, 19, 18, 35, 30, 29, 23, 14,
18, 16, 20, 18, 18, 25, 30

10. Police radar measured the following speeds of thirty-five cars on one street:

28, 24, 22, 38, 40, 25, 24, 35, 25,
23, 22, 50, 31, 37, 45, 28, 30, 30,
30, 25, 35, 32, 45, 52, 24, 26, 18,
20, 30, 32, 33, 48, 58, 30, 25

11. On a one-day fishing trip, Mr. and Mrs. Johnson recorded the following lengths of fish they caught (measured in inches):

14.3; 13.6; 10.5; 15.5; 20.1;
10.9; 12.4; 25.0; 30.2; 32.5

12. A machine puts out parts measured in thickness to the nearest hundredth of an inch. One hundred parts were measured and the results are tallied in the following chart:

THICKNESS MEASURED	0.80	0.83	0.84	0.85	0.87
NUMBER OF PARTS	22	41	14	20	3

SUMMARY: CHAPTER 8

Interest is money paid for the use of money. The **principal** is the money that is invested or borrowed. The **rate** is the **percent of interest** and is generally an annual rate.

Simple interest is interest paid on principal only. **Compound interest** is interest paid on interest.

FORMULA FOR CALCULATING SIMPLE INTEREST

$I = P \times R \times T$ where I = interest
P = principal
R = rate
T = time (in years)

To calculate compound interest:

1. Using $I = P \times R \times T$, calculate the simple interest where T is period of time for compounding. For example,

$$T = 1 \text{ for compounding annually;}$$

$$T = \frac{1}{2} \text{ for compounding semiannually;}$$

$$\text{T} = \frac{1}{4} \text{ for compounding quarterly;}$$

$$T = \frac{1}{12} \text{ for compounding monthly;}$$

$$T = \frac{1}{360} \text{ for compounding daily.}$$

2. Add this interest to the principal to create a new value for the principal.
3. Continue Steps 1 and 2 until the entire interest period is covered.

To balance your checking account (sometimes called **reconciling** the bank statement with your checkbook register):

1. Go through your checkbook register and the bank statement and put a check mark (✓) by each check paid and deposit recorded on the bank statement.
2. In your checkbook register:
   (a) Add any interest paid to your current balance.
   (b) Subtract any service charge from this balance.
   This number is your **true balance.**
3. Find the total of all the outstanding checks (no ✓ mark) in your checkbook register (checks not yet received by the bank).
4. On a reconciliation sheet:
   (a) Enter the balance from the bank statement.
   (b) Add any deposits you have made that are not recorded (no ✓ mark) on the bank statement.
   (c) Subtract the total of your outstanding checks (found in Step 3). This number should agree with your **true balance.** Now your checking account is balanced.

Four types of graphs are:

1. **Bar Graphs:** To emphasize comparative amounts
2. **Circle Graphs:** To help in understanding percents or parts of a whole
3. **Line Graphs:** To indicate tendencies or trends over a period of time
4. **Pictographs:** To emphasize the topic being related as well as quantities

TERMS USED IN THE STUDY OF STATISTICS

Data: Value(s) measuring some characteristic of interest (We will consider only numerical data.)

Statistic: A single number describing some characteristic of the data

Mean: The arithmetic average of the data (Add all the data and divide by the number of data items.)

Median: The middle data item (Arrange the data in order and pick out the middle item.)

Mode: The single data item that appears the most number of times (Some data may have more than one mode. We will leave the discussion of such a situation to a course in statistics. In this text, if data has a mode, there will be only one.)

Range: The difference between the largest and smallest data items

REVIEW QUESTIONS: CHAPTER 8

1. What will be the interest earned in one year on a savings account of \$1500 if the bank pays $6\frac{1}{2}\%$ interest?

2. If a stock pays 12% dividends every 6 months, what will be the dividend paid on an investment of \$13,450 in 6 months?

3. If a principal of \$1000 is invested at a rate of 9% for 30 days, what will be the interest earned?

4. You made \$50 on an investment at 10% for 3 months. What principal did you invest?

5. Determine the missing items in each row.

PRINCIPAL	RATE	TIME	INTEREST
$ 200	12%	180 days	?
$ 300	18%	?	$ 81
$1000	?	1 year	$ 85
?	9%	18 months	$270

6. How much interest will be earned on a savings account of $6500 in 2 years if interest is compounded annually at 6%? If interest is compounded semiannually? If interest is compounded quarterly?

7. Make a table based on an initial principal of $10,000 and an inflation rate of 8% that indicates the principal, interest, and total for 9 years. (See Example 5, Section 8.2.)

Refer to the circle graph for Exercises 8–11.

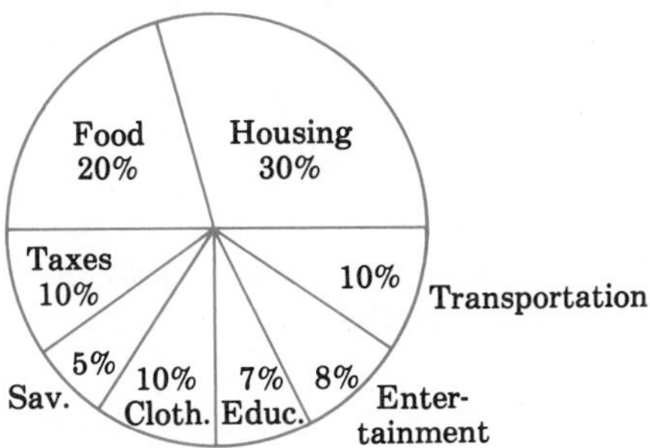

8. The circle graph shows a home budget. What amount will be spent on each category if the family income is $35,000?

9. How much more will the family spend for food than for clothing if their income is increased to $40,000 (to the nearest dollar)?

10. What fractional part of the family income is spent for food, housing, **and** taxes combined?

11. How much will the family spend for food, housing, **and** transportation combined if their income is reduced to $30,000 (to the nearest dollar)?

Refer to the following test scores for Exercises 12–15:

94, 86, 92, 70, 88, 91,
88, 70, 89, 88, 96, 92

12. Find the mean score.

13. Find the mode (if any).
14. Find the median score.
15. Find the range of the scores.

TEST: CHAPTER 8

1. You loan \$2500 to a friend for one year at 13% simple interest. What will he pay you at the end of the year?
2. You made \$112 interest on an investment at 7% for 90 days. What principal did you invest?
3. Referring to Exercise 2, what would the rate of simple interest have been if you had earned \$144 interest with the same principal for the same length of time?
4. How many years will it take (to the nearest year) for \$20,000 invested at a rate of $12\frac{1}{2}\%$ simple interest to double?
5. Calculate the interest you would receive in 6 months on a savings account of \$1200 at 7% if the interest were compounded monthly.
6. Referring to Exercise 5, what is the difference between this amount and the amount you would earn if simple interest were calculated for the 6 months?

Refer to the circle graph for Exercises 7–9.

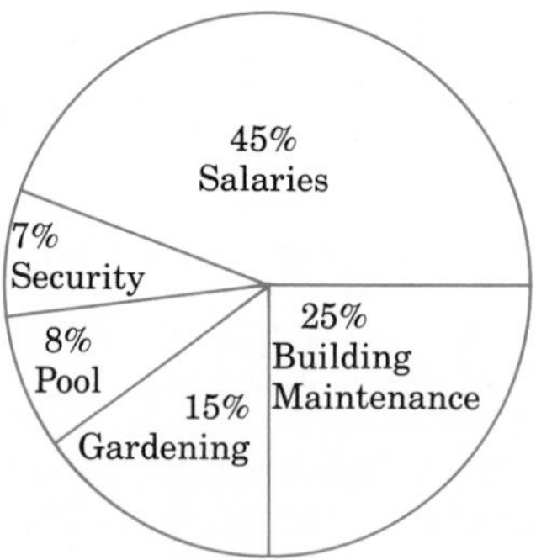

7. The budget for an apartment complex is shown in the graph. What fractional part of the budget is spent for salaries **and** security combined?
8. How much will be spent for building maintenance in one year if the budget is \$250,000 per month?

9. How much will be spent for salaries **and** security combined in 6 months if the monthly budget is $150,000?

Refer to the bar graph for Exercises 10–12.

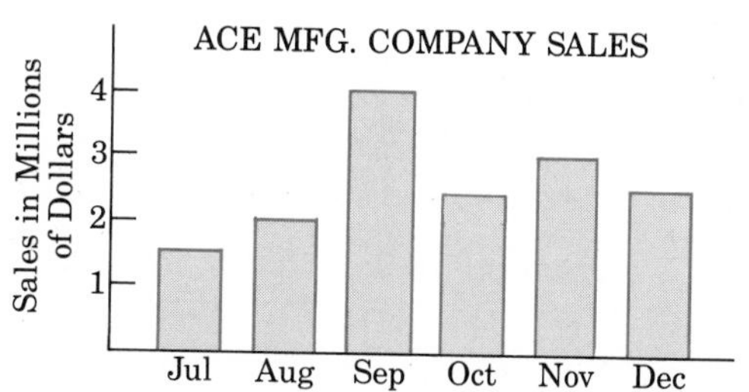

10. What were the total sales for the 6-month period?

11. What percent of the total sales for the 6 months were the July sales (to the nearest percent)?

12. What was the growth percent between August and September?

Find (a) the mean, (b) the median, (c) the mode (if any), and (d) the range for the groups of data in Exercises 13 and 14.

13. The number of hours of television viewing per day for a certain group:

3, 2, 2, 0, 1,
4, 1, 2, 3, 2

14. The number of centimeters of precipitation during one 6-month period in a certain area:

2.54, 10.16, 7.62, 20.32, 12.70, 7.62

CUMULATIVE TEST II: CHAPTERS 1–8

All fractions should be reduced to lowest terms.

1. Write the following in standard notation: two hundred thousand, sixteen.
2. Write the following in decimal notation: three hundred and four thousandths.
3. Round off 16.996 to the nearest hundredth.
4. Find the decimal equivalent to $\frac{14}{35}$.
5. Find the decimal equivalent to $\frac{21}{40}$.
6. Write $\frac{9}{5}$ as a percent.
7. Write $1\frac{1}{2}\%$ as a decimal.

Perform the indicated operations for Exercises 8–19.

8. $\frac{2}{15} + \frac{11}{15} + \frac{7}{15}$
9. $4 - \frac{3}{11}$
10. $\frac{2}{5} \cdot \frac{1}{3} \cdot \frac{4}{7}$
11. $2\frac{4}{15} + 3\frac{1}{6} + 4\frac{7}{10}$
12. $70\frac{1}{4} - 23\frac{5}{6}$
13. $4\frac{5}{7} \cdot 2\frac{6}{11}$
14. $6 \div 3\frac{1}{3}$
15. (700)(8000)
16. 403 − 4.012
17. 71 + 0.354 + 4.39
18. (0.27)(0.043)
19. 27.404 ÷ 0.34
20. Evaluate $(36 \div 3^2 \cdot 2) + 12 \div 4 - 2^2$.
21. Use the tests for divisibility to determine if 732 can be divided exactly by 2, 3, 4, 5, 9, 10.
22. Find the prime factorization of 396.
23. Find the LCM of 14, 21, and 30.

Complete each of the following (Exercises 24–28).

24. Division by ______ is undefined.

25. In the proportion $\frac{7}{8} = \frac{140}{160}$, the extremes are ______ and ______.

26. 15% of ______ is 7.5.

27. $9\frac{1}{4}$% of 200 is ______.

28. 65 is ______% of 26.

29. Solve for x: $\dfrac{1\frac{2}{3}}{x} = \dfrac{10}{2\frac{1}{4}}$

30. In a certain company, three out of every five employees are male. How many female employees are there out of the 490 people working for this company?

31. The sum of two numbers is 521. If one of the numbers is 196, what is the other number?

32. Milt has the following scores on his first three math tests: 75, 87, and 79. To earn a B grade for the course, he must have at least an 80 average. What is the least score that he can get on his fourth test to maintain a B average?

33. An investment pays $6\frac{1}{4}$% simple interest. What is the interest on $4800 invested for 8 months?

34. How much is put into a savings account that pays $5\frac{1}{2}$% simple interest if after 100 days, the interest earned is $69.30?

35. Find the simple interest rate if an investment of $6200 earns $1395 in $2\frac{1}{2}$ years.

36. How long must $3000 be left in an account paying 11% simple interest to triple in value (to the nearest year)?

37. Five thousand dollars is placed in an account paying 8% interest, compounded quarterly. How much is in the account at the end of 9 months (to the nearest cent)?

Refer to the following test scores for Exercises 38–40:

27, 36, 45, 72, 63, 36, 27, 18, 36, 90

38. Find the mean score.

39. Find the median score.

40. Find the range of the scores.

9 Measurement: The Metric System

9.1 LENGTH AND PERIMETER

About 90% of the people in the world use the metric system of measurement. The United States is the only major industrialized country still committed to the U.S. Customary System (formerly called the English System). Even in the United States, the metric system has been used for years in such fields as medicine, science, and military activities.

The **meter** is the basic unit of length in the metric system. Smaller and larger units are named by putting a prefix in front of the basic unit, for example, **centi**meter and **kilo**meter. The prefixes* we will use are, from largest to smallest unit size,

kilo
hecto
deka
deci
centi
milli

Some examples of metric lengths are shown in Figure 9.1.

*Other prefixes that indicate extremely small units are micro, nano, pico, femto, and atto. Prefixes that indicate extremely large units are mega, giga, and tera. These prefixes will not be used in this text.

one millimeter (mm): About the width of the wire in a paper clip

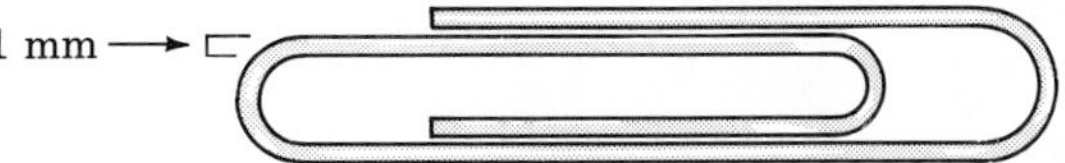

one centimeter (cm): About the width of a paper clip

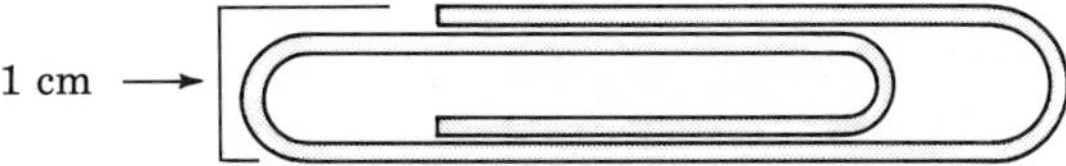

one meter (m): Just over 39 inches, or slightly longer than a yard
one kilometer (km): About 0.62 of a mile

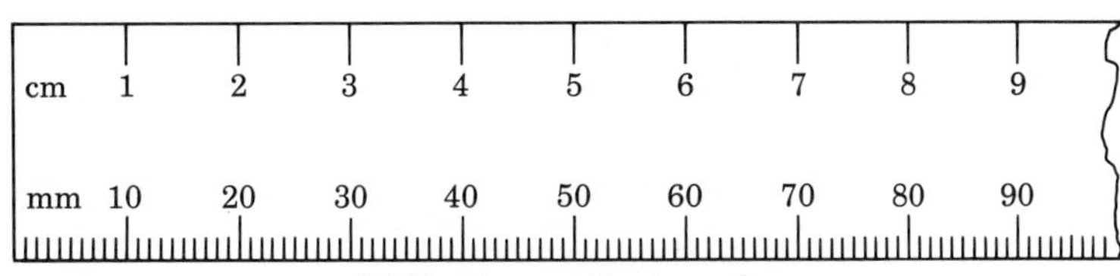

Millimeter- centimeter ruler

Figure 9.1

Table 9.1 contains the metric prefixes and their values. **These prefixes must be memorized, and memorized in order.** Table 9.2 lists the measures of length, their respective relationships to the meter, and their abbreviations.

Table 9.1 Metric Prefixes and Their Values

PREFIX	VALUE	
milli	0.001	—thousandths
centi	0.01	—hundredths
deci	0.1	—tenths
basic unit	1	—ones
deka	10	—tens
hecto	100	—hundreds
kilo	1000	—thousands

Table 9.2 Measures of Length

1 **milli**meter	(mm)	=	0.001 meter
1 **centi**meter	(cm)	=	0.01 meter
1 **deci**meter	(dm)	=	0.1 meter
1 meter	(m)	=	1.0 meter
1 **deka**meter	(dam)	=	10 meters
1 **hecto**meter	(hm)	=	100 meters
1 **kilo**meter	(km)	=	1000 meters

As you can tell from studying Tables 9.1 and 9.2, the metric units of length are related to each other by powers of 10. That is, you simply multiply by 10 to get the equivalent measure in the next higher unit.

The following method of using a chart makes conversion from one metric unit to another quite easy.

A chart can be used to change from one unit of length to another.

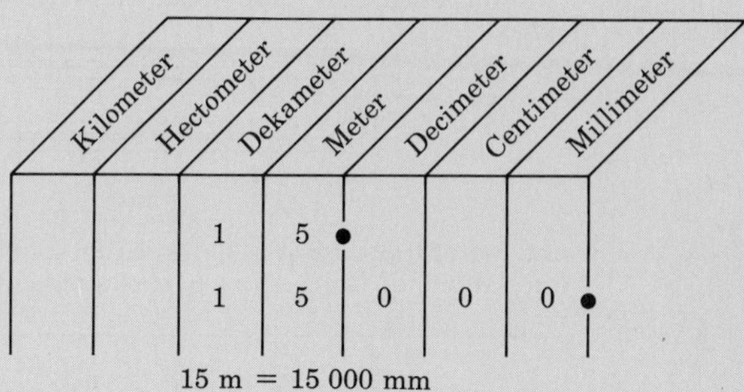

1. List each unit across the top. Memorize the unit prefixes in order.
2. Enter the given number so that each digit is in one column and the decimal point is on the given unit line.
3. Move the decimal point to the desired unit line.
4. Fill in the spaces with 0's using one digit per column.

SPECIAL NOTE: In the metric system,

1. A 0 is written to the left of the decimal point if there is no whole number part. (0.25 m)
2. No commas are used in writing numbers. If a number has more than four digits (left or right of the decimal point), the digits are grouped in threes from the decimal point with a space between the groups. (14 000 m)

The use of the chart shows how the following equivalent measures can be found. Note that there are no periods in metric abbreviations.

EXAMPLES

1. 5 m = 0.005 km (If there is no whole number part, 0 is written.)

2. 3 m = 300 cm

3. 16 cm = 0.16 m

4. 7.6 cm = 76 mm

5. 8 km = 800 dam

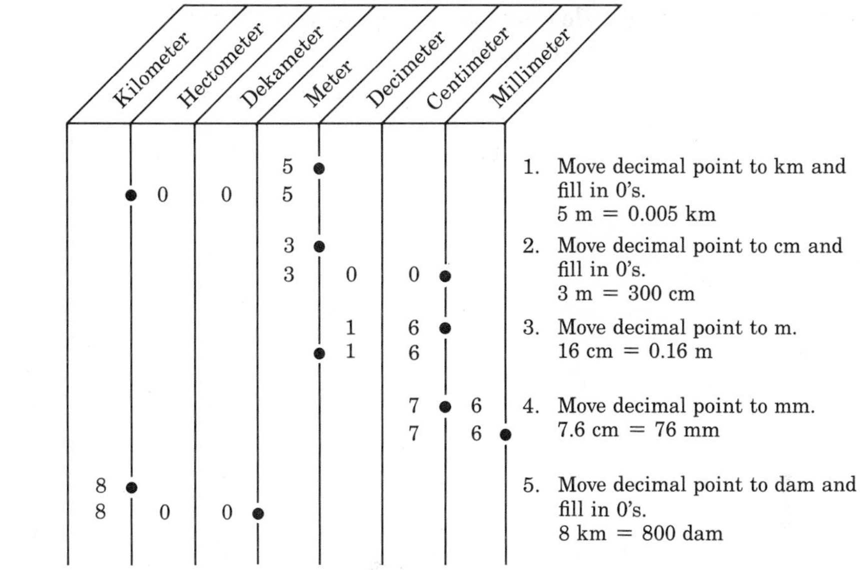

Kilometer
Hectometer
Dekameter
Meter
Decimeter
Centimeter
Millimeter
1. Move decimal point to km and fill in 0's.
5 m = 0.005 km
2. Move decimal point to cm and fill in 0's.
3 m = 300 cm
3. Move decimal point to m.
16 cm = 0.16 m
4. Move decimal point to mm.
7.6 cm = 76 mm
5. Move decimal point to dam and fill in 0's.
8 km = 800 dam

Make your own chart on a piece of paper and see if you agree with the results in Example 6–10.

EXAMPLES

6. 4 m = 4000 mm 7. 3.1 cm = 31 mm 8. 50 cm = 0.5 m

9. 18 km = 18 000 m 10. 2.5 km = 2500 m

After some practice, you may not need a chart to help in changing units. The following technique can be used; however, the chart method is highly recommended.

CHANGING METRIC MEASURES WITHOUT A CHART

1. To change to a measure to:

   One unit smaller, multiply by 10. 3 cm = 30 mm
   Two units smaller, multiply by 100. 5 m = 500 cm
   Three units smaller, multiply by 1000. 14 m = 14 000 mm

   and so on.

2. To change to a measure to:

   One unit larger, divide by 10. 50 cm = 5 dm
   Two units larger, divide by 100. 50 cm = 0.5 m
   Three units larger, divide by 1000. 13 mm = 0.013 m

   and so on.

EXAMPLES

SMALLER UNITS ⟶ LARGER NUMBERS

11. 42 m = 420 dm = 4200 cm = 42 000 mm

12. 17.3 m = 173 dm = 1730 cm = 17 300 mm

LARGER UNITS ⟶ SMALLER NUMBERS

13. 6 m = 0.6 dam = 0.06 hm = 0.006 km

14. 112 m = 11.2 dam = 1.12 hm = 0.112 km

Geometric figures and the related formulas for perimeter provide useful applications of measurements of length.

Perimeter: Total distance around a plane geometric figure

Circumference: Perimeter of a circle

Radius: The distance from the center of a circle to a point on the circle

Diameter: Distance from one point on a circle to another point on the circle measured through the center

Six geometric figures and the formulas for finding their perimeters are shown here.

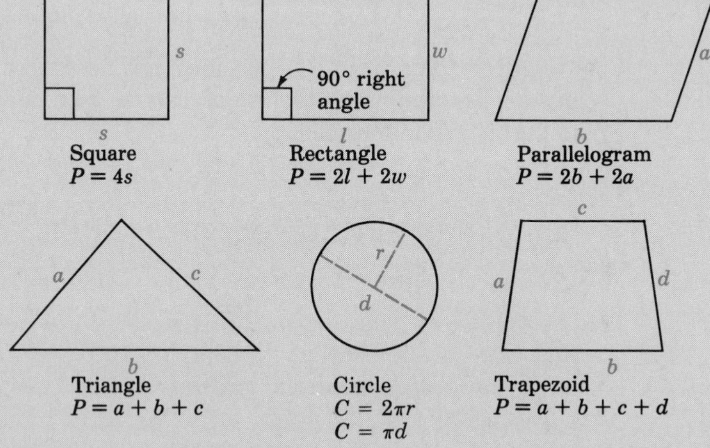

NOTE: π is the symbol used for the constant number 3.1415926535 . . . . This constant is an infinite decimal with no pattern to its digits. For our purposes, we will use $\pi \approx 3.14$, but you should be aware that 3.14 is only an approximation to π.

EXAMPLES

15. Find the perimeter of a rectangle with length 20 cm and width 15 cm.

Solution

Sketch the figure first.

$$P = 2l + 2w$$
$$P = 2 \cdot 20 + 2 \cdot 15$$
$$= 40 + 30 = 70 \text{ cm}$$

$w = 15$ cm

$l = 20$ cm

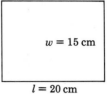

The perimeter is 70 cm.

16. Find the circumference of a circle with diameter 3 m.

Solution

Sketch the figure first.

$$C = \pi d$$
$$C = 3.14(3) = 9.42 \text{ m}$$

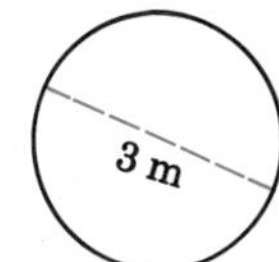

The circumference is 9.42 m.

17. Find the perimeter of a triangle with sides of 4 cm, 0.7 dm, and 80 mm. Write your answer in millimeters.

Solution

Sketch the figure and change all the units to millimeters.

$$P = a + b + c$$
$$P = 40 + 70 + 80$$
$$= 190 \text{ mm}$$

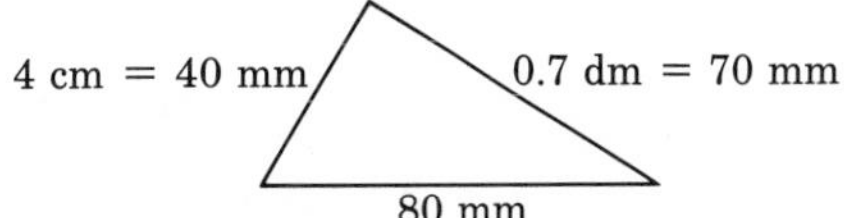

The perimeter is 190 mm.

PRACTICE QUIZ

Draw a metric chart and change the following units as indicated.

	ANSWERS
1. 4 m = ______ mm	**1.** 4000 mm
2. 3.1 cm = ______ mm	**2.** 31 mm
3. 50 cm = ______ m	**3.** 0.5 m
4. 18 km = ______ m	**4.** 18 000 m

Exercises 9.1

1. Write the six metric prefixes discussed in this section in order from largest to smallest.

Change the following units as indicated. Use the chart method until you are accustomed to the metric system.

2. 1 m = _____ cm
3. 5 m = _____ cm
4. 12 m = _____ cm
5. 6 m = _____ cm
6. 2 m = _____ mm
7. 0.3 m = _____ mm
8. 0.7 m = _____ mm
9. 1.4 m = _____ mm
10. 1.6 cm = _____ mm
11. 1.8 cm = _____ mm
12. 25 cm = _____ mm
13. 35 cm = _____ mm
14. 4 m = _____ dm
15. 16 m = _____ dm
16. 7 dm = _____ cm
17. 21 dm = _____ cm
18. 3 km = _____ m
19. 5 km = _____ m
20. 5.28 km = _____ m
21. 6.4 km = _____ m
22. 11 mm = _____ cm
23. 26 mm = _____ cm
24. 72 mm = _____ cm
25. 48 mm = _____ cm
26. 6 mm = _____ dm
27. 12 mm = _____ dm
28. 20 mm = _____ m
29. 30 mm = _____ m
30. 145 mm = _____ m
31. 256 mm = _____ m
32. 25 cm = _____ m
33. 32 cm = _____ m
34. 150 cm = _____ m
35. 170 cm = _____ m
36. 3000 m = _____ km
37. 2400 m = _____ km
38. 500 m = _____ km
39. 400 m = _____ km
40. 3.45 m = _____ cm
41. 4.62 m = _____ cm
42. 6.3 cm = _____ m
43. 5.2 cm = _____ m
44. 3.25 m = _____ mm
45. 6.41 m = _____ mm
46. 3 mm = _____ cm
47. 5 mm = _____ cm
48. 32 mm = _____ m
49. 57 mm = _____ m
50. 20 000 m = _____ km
51. 35 000 m = _____ km

52. 1.5 km = ______ m

53. 2.3 km = ______ m

54. 0.5 m = ______ km

55. Match each formula for perimeter to its corresponding geometric figure.

______ (a) square	(1) $P = 2l + 2w$
______ (b) parallelogram	(2) $P = 4s$
______ (c) circle	(3) $P = 2b + 2a$
______ (d) rectangle	(4) $C = 2\pi r$
______ (e) trapezoid	(5) $P = a + b + c$
______ (f) triangle	(6) $P = a + b + c + d$

56. Find the perimeter of a triangle with sides of 4 cm, 8.3 cm, and 6.1 cm.

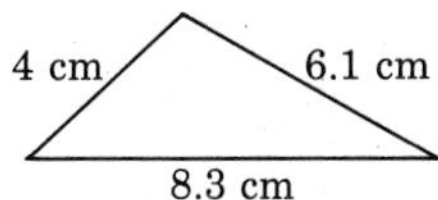

57. Find the perimeter of a rectangle with length 35 mm and width 17 mm.

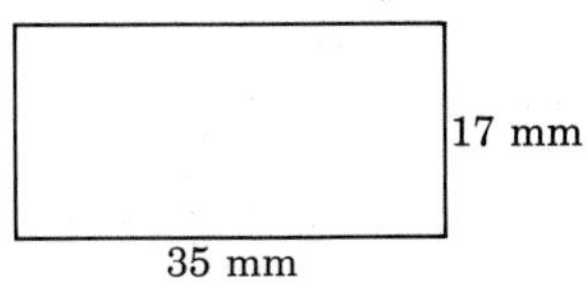

58. Find the perimeter of a square with sides of 13.3 m.

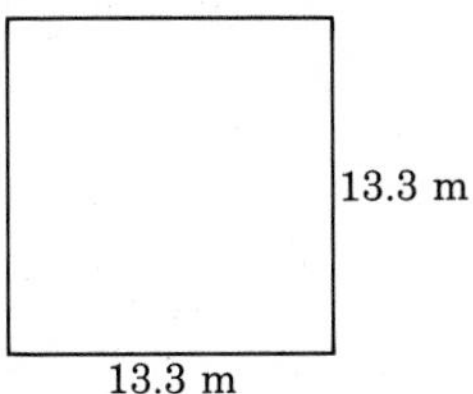

59. Find the circumference of a circle with radius 5 cm. (Use $\pi \approx 3.14$.)

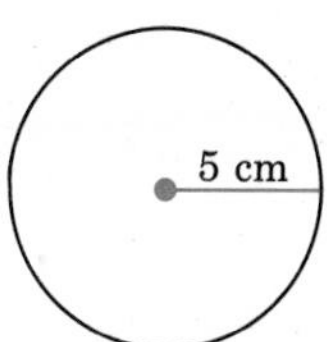

60. Find the perimeter of a parallelogram with one side 43 cm and another side 20 mm. Write your answer in millimeters.

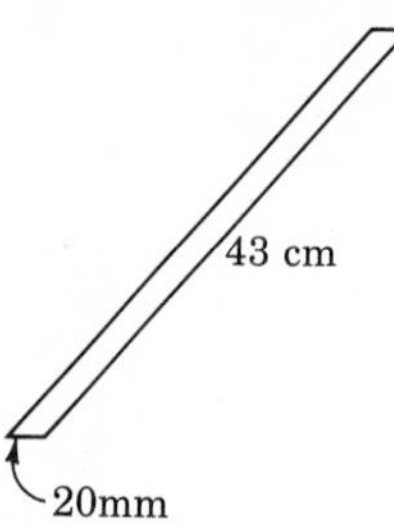

61. Find the circumference of a circle with diameter 6.2 cm. (Use $\pi \approx 3.14$.)

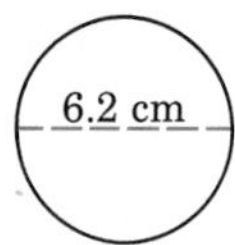

62. Find the perimeter of a rectangle with length 50 m and width 50 dm. Write your answer in meters.

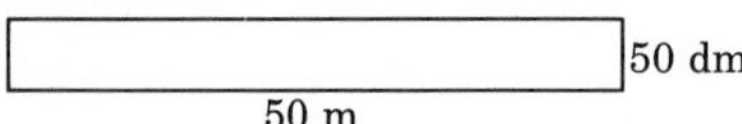

63. Find the perimeter of a triangle with sides of 5 cm, 55 mm, and 0.3 dm. Write your answers in centimeters.

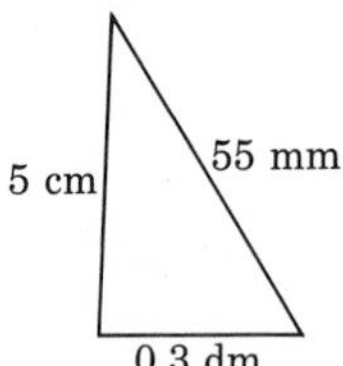

64. Find the circumference of a circle in meters if its radius is 70 cm. (Use $\pi \approx 3.14$.)

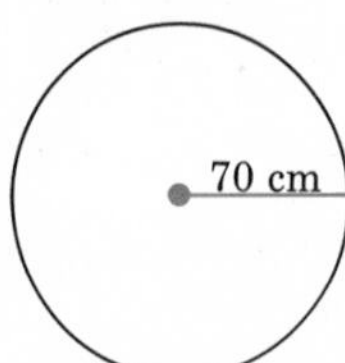

65. Find the perimeter of a square in meters if one side is 4 km long.

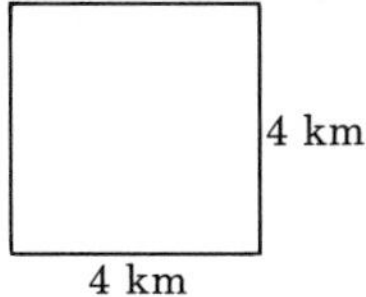

9.2 AREA

Area is a measure of the interior, or enclosure, of a surface. For example, the two rectangles in Figure 9.2 have different areas because they have different amounts of interior space, or different amounts of space are enclosed by the sides of the figures.

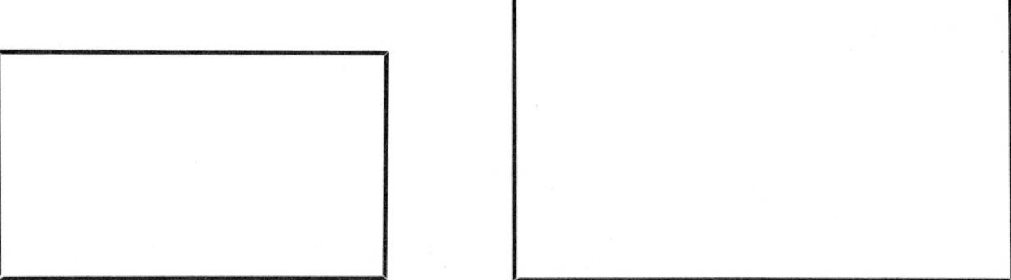

These two rectangles have different **areas**

Figure 9.2

Area is measured in square units. A square that is 1 centimeter long on each side is said to have an area of 1 square centimeter, or the area is 1 cm^2. A rectangle that is 7 cm on one side and 4 cm on the other side encloses 28 squares that have area 1 cm^2. So the rectangle is said to have an area of 28 square centimeters or 28 cm^2. (See Figure 9.3.)

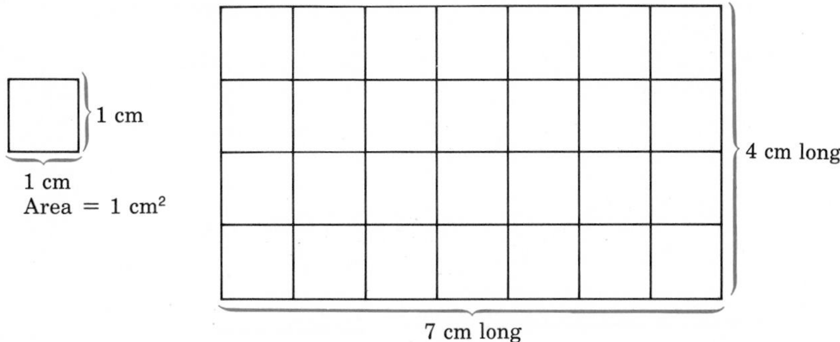

Area = 7 cm × 4 cm = 28 cm^2

There are 28 squares that are each 1 cm^2 in the large rectangle

Figure 9.3

Table 9.3 shows area measures useful for relatively small areas. For example, the area of the floor of your classroom could be measured in square meters for carpeting, and the area of this page of paper could be measured in square centimeters. Other measures, listed in Table 9.4, are used for measuring land.

Table 9.3 Measures of Small Area

		$1\ cm^2$	$= 100\ mm^2$
	$1\ dm^2 =$	$100\ cm^2$	$= 10\ 000\ mm^2$
$1\ m^2$	$= 100\ dm^2$	$= 10\ 000\ cm^2$	$= 1\ 000\ 000\ mm^2$

[NOTE: Each smaller unit of area is 100 times the previous unit of area–**not** just 10 times.]

EXAMPLES

1. A square 1 centimeter on a side encloses 100 square millimeters.

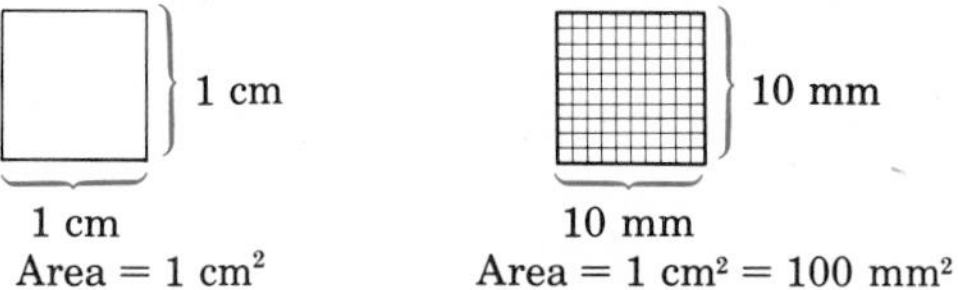

2. A square 1 decimeter (10 cm) on a side encloses 1 square decimeter. ($1\ dm^2 = 100\ cm^2 = 10\ 000\ mm^2$)

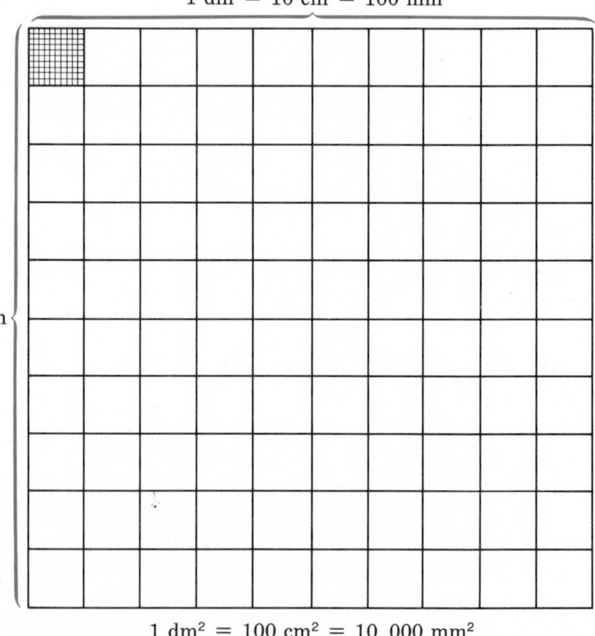

A chart similar to that used in Section 9.1 can be used to change measures of area. **The key difference is that there must be two digits in each column.** This corresponds to multiplication of the previous unit of area by 100.

A chart can be used to change from one unit of area to another.

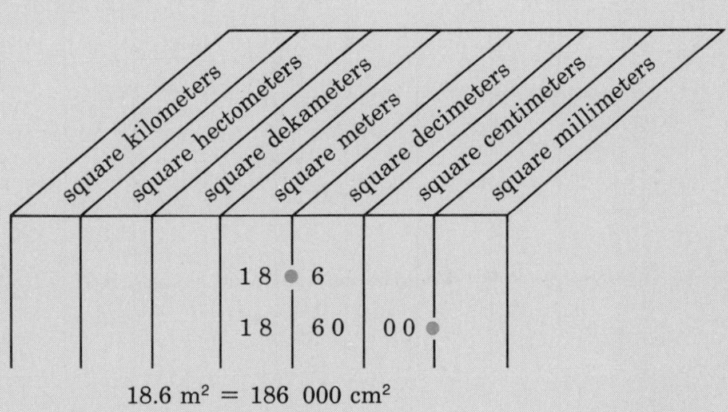

$18.6\ m^2 = 186\ 000\ cm^2$

1. List each area unit across the top. (Abbreviations will do.)
2. Enter the given number so that there are **two** digits in each column with the decimal point on the given unit line.
3. Move the decimal point to the desired unit line.
4. Fill in the spaces with 0's using two digits per column.

The use of the chart on page 302 shows how the following equivalent measures can be found.

EXAMPLES

3. $5\ cm^2 = 500\ mm^2$
4. $3\ dm^2 = 30\ 000\ mm^2$
5. $1.4\ m^2 = 14\ 000\ cm^2$
6. $1.4\ m^2 = 0.014\ dam^2$
7. $8\ km^2 = 8\ 000\ 000\ m^2$

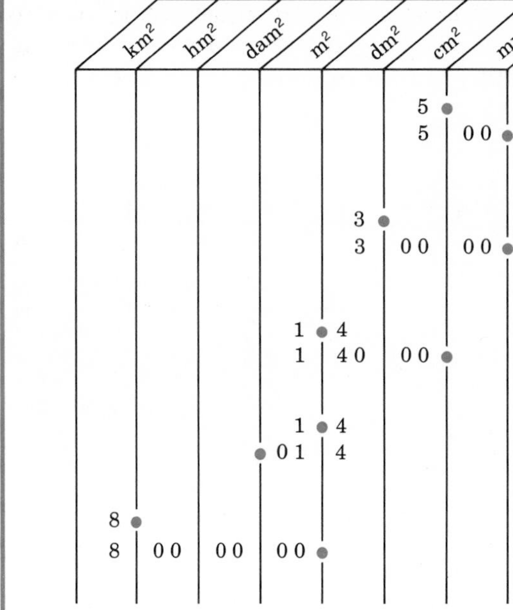

3. Move decimal point to mm^2 and fill in 0's.
   5 cm^2 = 500 mm^2

4. Move decimal point to mm^2 and fill in 0's.
   3 dm^2 = 30 000 mm^2

5. Move decimal point to cm^2 and fill in 0's.
   1.4 m^2 = 14 000 cm^2

6. Move decimal point to dam^2 and fill in 0's.
   1.4 m^2 = 0.014 dam^2

7. Move decimal point to m^2 and fill in 0's.
   8 km^2 = 8 000 000 m^2

Make your own chart on a piece of paper and see if you agree with the results in Examples 8–10.

EXAMPLES

8. 8.52 m^2 = 652 dm^2 = 65 200 cm^2

9. 147 cm^2 = 14 700 mm^2

10. 3.8 cm^2 = 0.038 dm^2 = 0.00038 m^2

A square with each side 10 meters long encloses an area of 1 **are** (a). A **hectare** (ha) is 100 ares. The are and hectare are used in measuring land area. (See Table 9.4.)

Table 9.4 Measures of Land Area

1 a = 100 m^2
1 ha = 100 a = 10 000 m^2

EXAMPLES

11. A square 10 m on each side encloses 100 m^2 or 1 are.

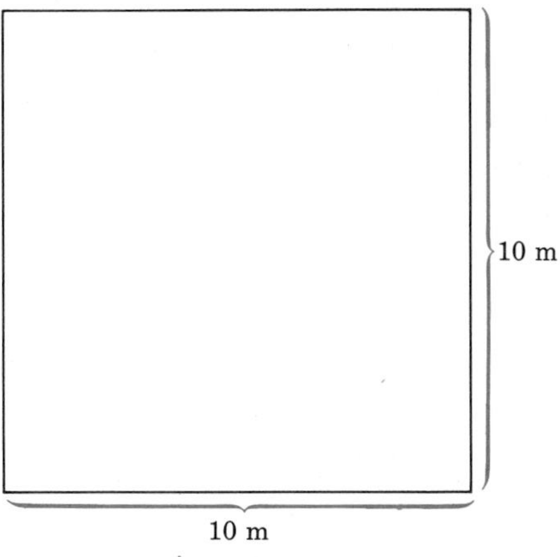

1 a = 100 m^2

12. (a) 3.2 a = 320 m^2

    (b) 65 m^2 = 0.65 a

13. How many ares are in 1 km^2? [NOTE: One km is about 0.6 mile, so 1 km^2 is about 0.6 × 0.6 = 0.36 square mile.]

    Remember that 1 km = 1000 m, so

$$\begin{aligned} 1 \text{ km}^2 &= (1000 \text{ m}) \times (1000 \text{ m}) \\ &= 1\,000\,000 \text{ m}^2 \\ &= 10\,000 \text{ a} \end{aligned}$$

Divide m^2 by 100 to get ares because every 100 m^2 is equal to 1 are.

14. A farmer plants corn and beans as shown in the figure. How many ares and how many hectares are planted in corn? In beans? (From Example 13 we know 1 km^2 = 10 000 a.)

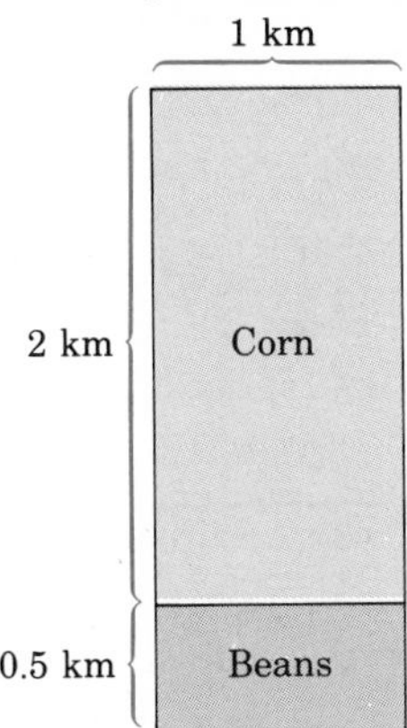

Corn: $(2 \text{ km}) \cdot (1 \text{ km}) = 2 \text{ km}^2 = 20\,000 \text{ a} = 200 \text{ ha}$

Beans: $(0.5 \text{ km}) \cdot (1 \text{ km}) = 0.5 \text{ km}^2 = 5000 \text{ a} = 50 \text{ ha}$

Geometric figures and the related formulas for area provide useful applications of measurements of area.

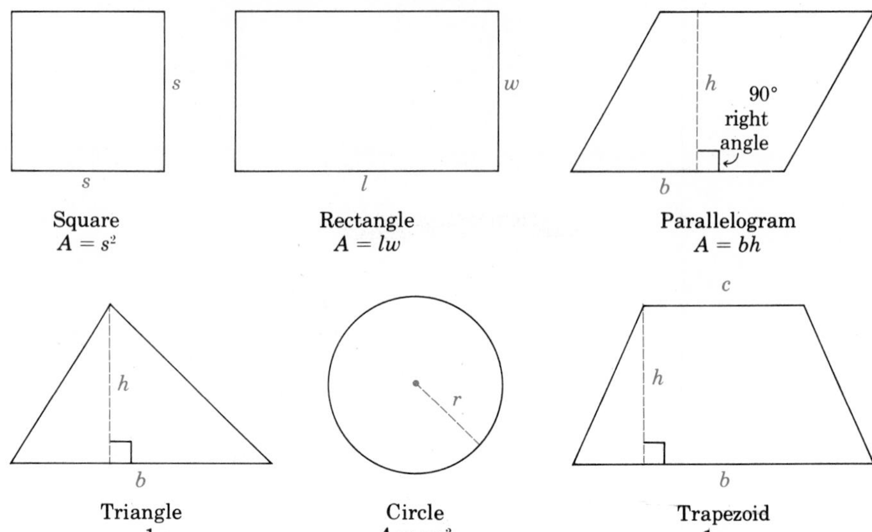

Square $A = s^2$

Rectangle $A = lw$

Parallelogram $A = bh$

Triangle $A = \frac{1}{2}bh$

Circle $A = \pi r^2$

Trapezoid $A = \frac{1}{2}h(b + c)$

EXAMPLES

15. Find the area of the figure shown here with the indicated dimensions.

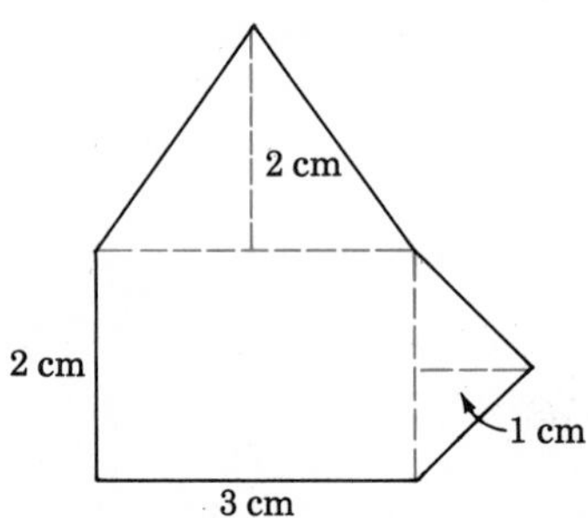

Solution

There are two triangles and one rectangle.

RECTANGLE	*LARGER TRIANGLE*	*SMALLER TRIANGLE*
$A = lw$	$A = \frac{1}{2}bh$	$A = \frac{1}{2}bh$
$A = 2 \cdot 3 = 6 \text{ cm}^2$	$A = \frac{1}{2} \cdot 3 \cdot 2 = 3 \text{ cm}^2$	$A = \frac{1}{2} \cdot 2 \cdot 1 = 1 \text{ cm}^2$

$$\begin{aligned}\text{Total area} &= 6 \text{ cm}^2 + 3 \text{ cm}^2 + 1 \text{ cm}^2 \\ &= 10 \text{ cm}^2\end{aligned}$$

16. Find the area of the washer (shaded portion) with dimensions as shown. (Use $\pi \approx 3.14$.)

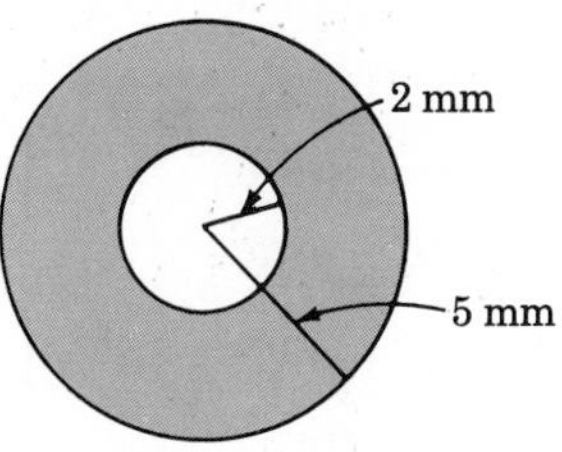

Solution

Subtract the area of the inside (smaller) circle from the area of the outside (larger) circle.

LARGER CIRCLE	*SMALLER CIRCLE*
$A = \pi r^2$	$A = \pi r^2$
$A = 3.14(5^2)$	$A = 3.14(2^2)$
$= 3.14(25)$	$= 3.14(4)$
$= 78.50 \text{ mm}^2$	$= 12.56 \text{ mm}^2$

WASHER

$$\begin{array}{r} 78.50 \text{ mm}^2 \\ -\,12.56 \text{ mm}^2 \\ \hline 65.94 \text{ mm}^2 \end{array}$$ area of washer

---

PRACTICE QUIZ

Change the following units as indicated.

1. $22 \text{ cm}^2 =$ ______ mm^2
2. $500 \text{ mm}^2 =$ ______ cm^2
3. $3.7 \text{ dm}^2 =$ ______ $\text{cm}^2 =$ ______ mm^2
4. $3.6 \text{ a} =$ ______ m^2
5. $0.73 \text{ ha} =$ ______ $\text{a} =$ ______ m^2

ANSWERS

1. 2200 mm^2
2. 5 cm^2
3. $370 \text{ cm}^2 = 37\,000 \text{ mm}^2$
4. 360 m^2
5. $73 \text{ a} = 7300 \text{ m}^2$

Exercises 9.2

Change the following units as indicated.

1. $3 \text{ cm}^2 =$ ______ mm^2
2. $5.6 \text{ cm}^2 =$ ______ mm^2
3. $8.7 \text{ cm}^2 =$ ______ mm^2
4. $3.61 \text{ cm}^2 =$ ______ mm^2

5. $600 \text{ mm}^2 = \_\_\_\_\_ \text{ cm}^2$ **6.** $28 \text{ mm}^2 = \_\_\_\_\_ \text{ cm}^2$

7. $1400 \text{ mm}^2 = \_\_\_\_\_ \text{ cm}^2$ **8.** $20\,000 \text{ mm}^2 = \_\_\_\_\_ \text{ cm}^2$

9. $4 \text{ dm}^2 = \_\_\_\_\_ \text{ cm}^2 = \_\_\_\_\_ \text{ mm}^2$

10. $7.3 \text{ dm}^2 = \_\_\_\_\_ \text{ cm}^2 = \_\_\_\_\_ \text{ mm}^2$

11. $57 \text{ dm}^2 = \_\_\_\_\_ \text{ cm}^2 = \_\_\_\_\_ \text{ mm}^2$

12. $0.6 \text{ dm}^2 = \_\_\_\_\_ \text{ cm}^2 = \_\_\_\_\_ \text{ mm}^2$

13. $17 \text{ m}^2 = \_\_\_\_\_ \text{ dm}^2 = \_\_\_\_\_ \text{ cm}^2 = \_\_\_\_\_ \text{ mm}^2$

14. $2.9 \text{ m}^2 = \_\_\_\_\_ \text{ dm}^2 = \_\_\_\_\_ \text{ cm}^2 = \_\_\_\_\_ \text{ mm}^2$

15. $0.03 \text{ m}^2 = \_\_\_\_\_ \text{ dm}^2 = \_\_\_\_\_ \text{ cm}^2 = \_\_\_\_\_ \text{ mm}^2$

16. $0.5 \text{ m}^2 = \_\_\_\_\_ \text{ dm}^2 = \_\_\_\_\_ \text{ cm}^2 = \_\_\_\_\_ \text{ mm}^2$

17. $142 \text{ mm}^2 = \_\_\_\_\_ \text{ cm}^2$ **18.** $5800 \text{ mm}^2 = \_\_\_\_\_ \text{ cm}^2$

19. $200 \text{ dm}^2 = \_\_\_\_\_ \text{ m}^2$ **20.** $35 \text{ dm}^2 = \_\_\_\_\_ \text{ m}^2$

21. $7.8 \text{ a} = \_\_\_\_\_ \text{ m}^2$ **22.** $300 \text{ a} = \_\_\_\_\_ \text{ m}^2$

23. $0.04 \text{ a} = \_\_\_\_\_ \text{ m}^2$ **24.** $0.53 \text{ a} = \_\_\_\_\_ \text{ m}^2$

25. $8.69 \text{ ha} = \_\_\_\_\_ \text{ a} = \_\_\_\_\_ \text{ m}^2$

26. $7.81 \text{ ha} = \_\_\_\_\_ \text{ a} = \_\_\_\_\_ \text{ m}^2$

27. $0.16 \text{ ha} = \_\_\_\_\_ \text{ a} = \_\_\_\_\_ \text{ m}^2$

28. $0.02 \text{ ha} = \_\_\_\_\_ \text{ a} = \_\_\_\_\_ \text{ m}^2$

29. $1 \text{ a} = \_\_\_\_\_ \text{ ha}$ **30.** $15 \text{ a} = \_\_\_\_\_ \text{ ha}$

31. $5 \text{ km}^2 = \_\_\_\_\_ \text{ a} = \_\_\_\_\_ \text{ ha}$

32. $4.76 \text{ km}^2 = \_\_\_\_\_ \text{ a} = \_\_\_\_\_ \text{ ha}$

33. $0.3 \text{ km}^2 = \_\_\_\_\_ \text{ a} = \_\_\_\_\_ \text{ ha}$

34. Match each formula for area to its corresponding geometric figure.

_____ (a) square (1) $A = lw$

_____ (b) parallelogram (2) $A = bh$

_____ (c) circle (3) $A = s^2$

_____ (d) rectangle (4) $A = \pi r^2$

_____ (e) trapezoid (5) $A = \frac{1}{2}bh$

_____ (f) triangle (6) $A = \frac{1}{2}h(b + c)$

Find the area of each of the following figures.

35. a rectangle 35 cm long and 25 cm wide

36. a triangle with base 2 cm and altitude 6 cm

37. a triangle with base 5 mm and altitude 8 mm

38. a circle of radius 5 m (use $\pi \approx 3.14$)

39. a circle of radius 1.5 cm (use $\pi \approx 3.14$)

40. a trapezoid with parallel sides of 8 cm and 10 cm and altitude of 35 cm

41. a trapezoid with parallel sides of 3.5 mm and 4.2 mm and altitude of 1 cm

42. a parallelogram of altitude 10 cm to a base of 5 mm

Find the area of each of the following figures with the indicated dimensions. (Use $\pi \approx 3.14$.)

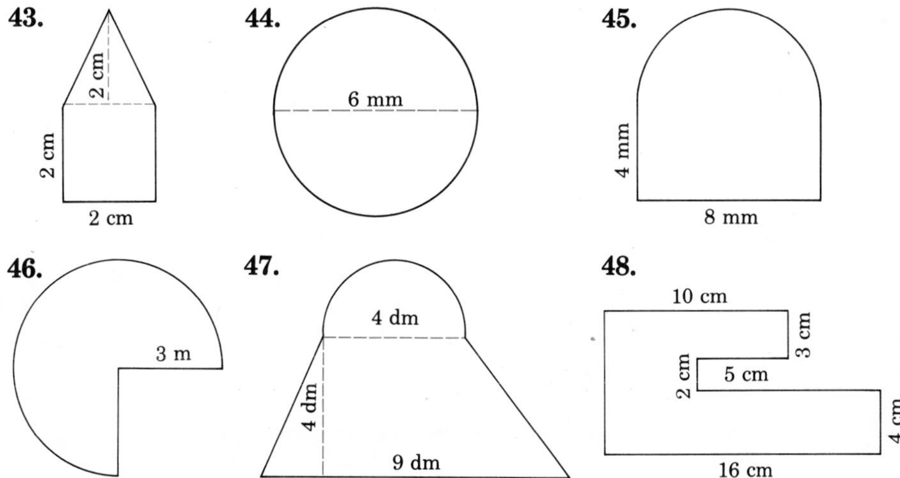

Find the areas of the shaded portions in ares. (Use $\pi \approx 3.14$.)

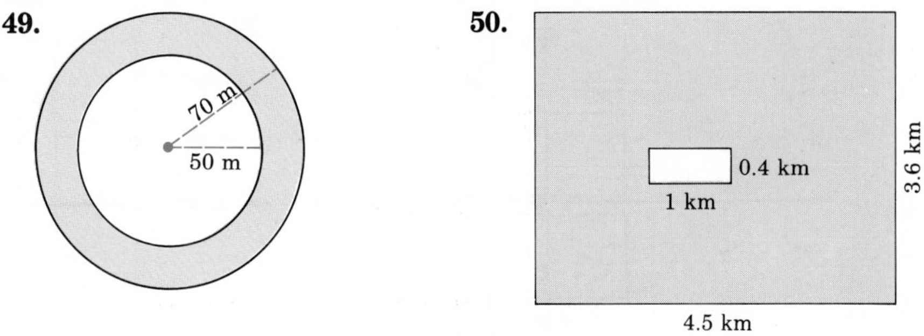

51. Find the area of a circle of radius 1 m.

52. Find the area of a circle with diameter 1 m.

53. Find the area of a rectangle 2 m long and 60 cm wide. Write your answer in both square meters and square centimeters.

54. Find the area of a square with sides of 50 cm. Write your answer in both cm^2 and m^2.

55. Find the area of a rectangle 0.5 m long and 35 cm wide. Write your answer in both cm^2 and m^2.

9.3 VOLUME

Volume is a measure of the space enclosed by a three-dimensional figure. The volume or space contained within a cube that is 1 cm on each edge is **one cubic centimeter,** or 1 cm^3, as shown in Figure 9.4. A cubic centimeter is about the size of a sugar cube.

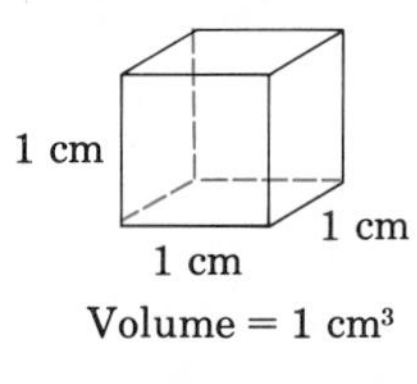

Figure 9.4

A rectangular solid that has edges of 3 cm and 2 cm and 5 cm has a volume of 3 cm $\times$ 2 cm $\times$ 5 cm = 30 cm^3. We can think of the rectangular solid as being three layers of ten cubic centimeters, as shown in Figure 9.5.

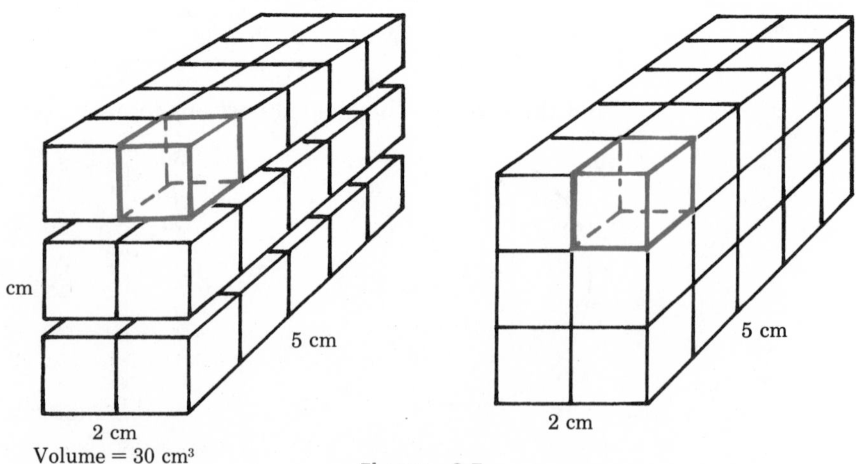

Figure 9.5

If a cube is 1 decimeter along each edge, then the volume of the cube is 1 cubic decimeter (or 1 dm^3). In terms of centimeters, this same cube has volume

$$10 \text{ cm} \times 10 \text{ cm} \times 10 \text{ cm} = 1000 \text{ cm}^3$$

That is, as shown in Figure 9.6,

$$1 \text{ dm}^3 = 1000 \text{ cm}^3$$

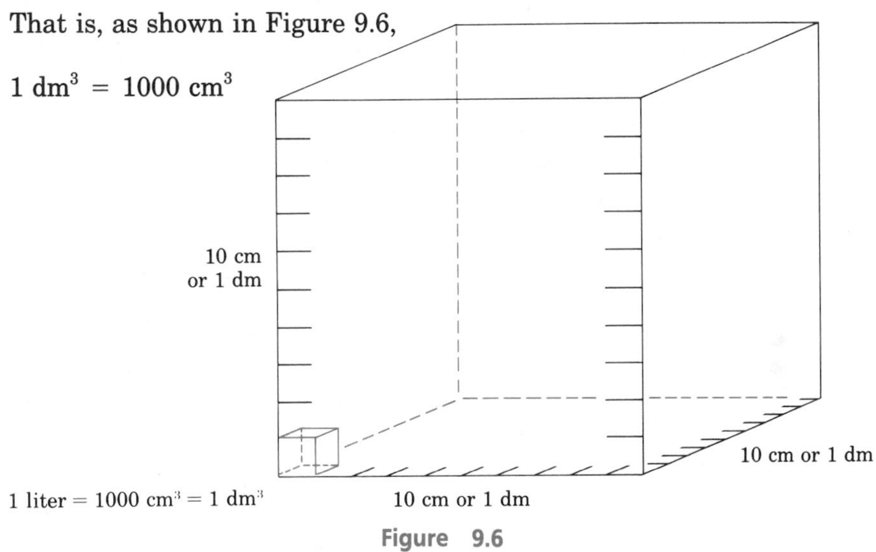

Figure 9.6

This relationship is true of cubic units in the metric system: equivalent cubic units can be found by multiplying the previous unit by 1000. Again, we can use a chart; however, this time **there must be three digits in each column.**

A chart can be used to change from one unit of volume to another.

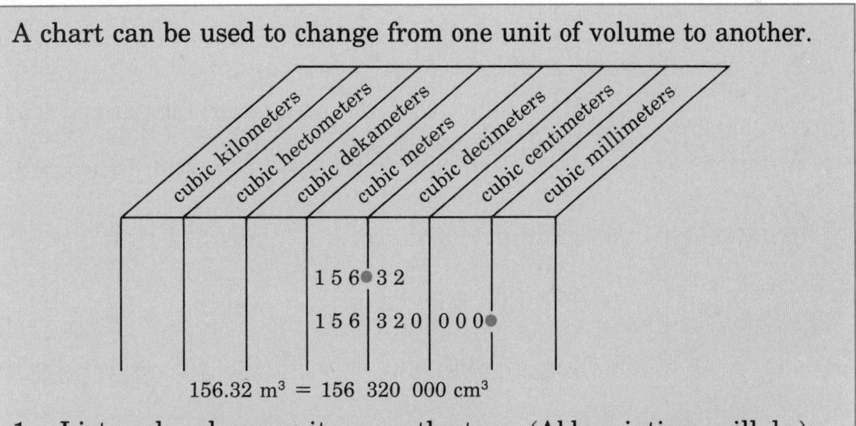

1. List each volume unit across the top. (Abbreviations will do.)
2. Enter the given number so that there are **three** digits in each column with the decimal point on the given unit line.
3. Move the decimal point to the desired unit line.
4. Fill in the spaces with 0's using three digits per column.

The chart shows how the following equivalent measures can be found.

EXAMPLES

1. $15\ \text{cm}^3 = 15\ 000\ \text{mm}^3$

2. $4.1\ \text{dm}^3 = 4100\ \text{cm}^3$

3. $8\ \text{dm}^3 = 0.008\ \text{m}^3$

4. $22.6\ \text{m}^3 = 22\ 600\ 000\ \text{m}^3$

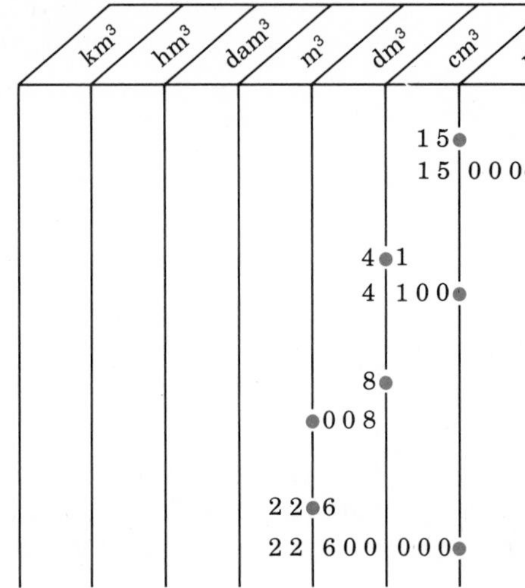

1. Move decimal point to mm³ and fill in 0's.
   $15\ \text{cm}^3 = 15\ 000\ \text{mm}^3$

2. Move decimal point to cm³ and fill in 0's.
   $4.1\ \text{dm}^3 = 4100\ \text{cm}^3$

3. Move decimal point to m³ and fill in 0's.
   $8\ \text{dm}^3 = 0.008\ \text{m}^3$

4. Move decimal point to cm³ and fill in 0's.
   $22.6\ \text{m}^3 = 22\ 600\ 000\ \text{cm}^3$

Mark your own chart on a piece of paper and see if you agree with the results in Examples 5–7.

EXAMPLES

5. $3.7\ \text{dm}^3 = 3700\ \text{cm}^3$

6. $0.8\ \text{m}^3 = 0.008\ \text{dam}^3$

7. $4\ \text{m}^3 = 4000\ \text{dm}^3 = 4\ 000\ 000\ \text{cm}^3 = 4\ 000\ 000\ 000\ \text{mm}^3$

Volumes measured in cubic kilometers are so large that they are not used in everyday situations. Possibly, some scientists work with these large volumes. More practically, we are interested in m^3, dm^3, cm^3, and mm^3.

Liquid Volume

Liquid volume is measured in **liters** (abbreviated L). You are probably familiar with 1 L and 2 L bottles of soda on your grocer's shelf. A **liter** is the volume enclosed in a cube that is 10 cm on each edge. So, 1 liter is equal to

$$10 \text{ cm} \times 10 \text{ cm} \times 10 \text{ cm} = 1000 \text{ cm}^3 \quad \text{or} \quad 1 \text{ liter} = 1000 \text{ cm}^3$$

That is, the cubic box shown in Figure 9.6 would hold 1 liter of liquid.

The prefixes kilo-, hecto-, deka-, deci-, centi-, and milli- all indicate the same parts of a liter as they do of the meter. The same type of chart used in Section 9.1 **with one digit per column** will be helpful for changing units. The centiliter (cL), deciliter (dL), and dekaliter (daL) are not commonly used and are not included in the tables or exercises.

Table 9.5 Measures of Liquid Volume

1 **milli**liter	(mL)	=	0.001 liter
1 liter	(L)	=	1.0 liter
1 **hecto**liter	(hL)	=	100 liters
1 **kilo**liter	(kL)	=	1000 liters

Table 9.6 Equivalent Measures of Volume

1000 mL = 1 L	1 mL = 1 cm^3
1000 L = 1 kL	1 L = 1 dm^3
10 hL = 1 kL	1 kL = 1 m^3

The use of the chart shows how the following equivalent measures can be found.

EXAMPLES

8. 6 L = 6000 mL = 0.06 hL

9. 500 mL = 0.5 L

10. 3 kL = 3000 L

11. 72 hL = 7.2 kL

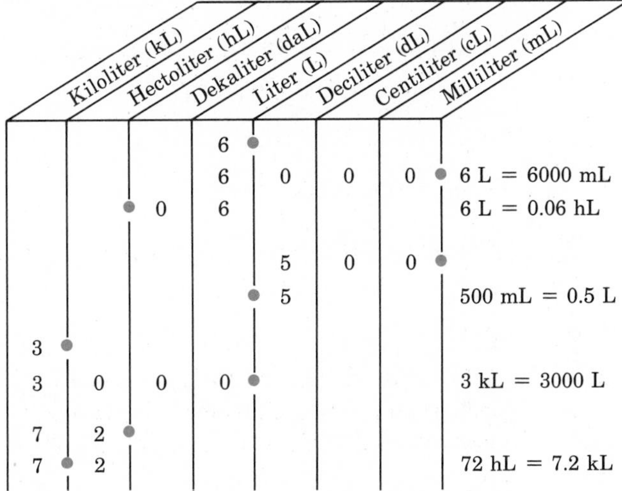

Kiloliter (kL)
Hectoliter (hL)
Dekaliter (daL)
Liter (L)
Deciliter (dL)
Centiliter (cL)
Milliliter (mL)
6
6 0 0 0
0 6
5 0 0
5
3
3 0 0 0
7 2
7 2
6 L = 6000 mL
6 L = 0.06 hL
500 mL = 0.5 L
3 kL = 3000 L
72 hL = 7.2 kL

There is an interesting "crossover" relationship between liquid volume measures and cubic volume measures. Since

$$1 \text{ L} = 1000 \text{ mL} \qquad \text{and} \qquad 1 \text{ L} = 1000 \text{ cm}^3$$

we have

$$\mathbf{1\ mL = 1\ cm^3}$$

Also,

$$1 \text{ kL} = 1000 \text{ L} = 1\,000\,000 \text{ cm}^3 \qquad \text{and} \qquad 1\,000\,000 \text{ cm}^3 = 1 \text{ m}^3$$

This gives

$$\mathbf{1\ kL = 1000\ L = 1\ m^3}$$

EXAMPLES

12. 6000 mL = 6 L
13. 3.2 L = 3200 mL
14. 60 hL = 6 kL
15. 637 mL = 0.637 L
16. 70 mL = 70 cm^3
17. 3.8 kL = 3.8 m^3

Five geometric solids and the formulas for their volumes are shown here.

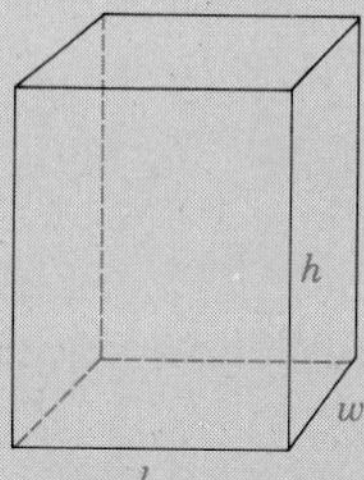

Rectangular solid
$V = lwh$

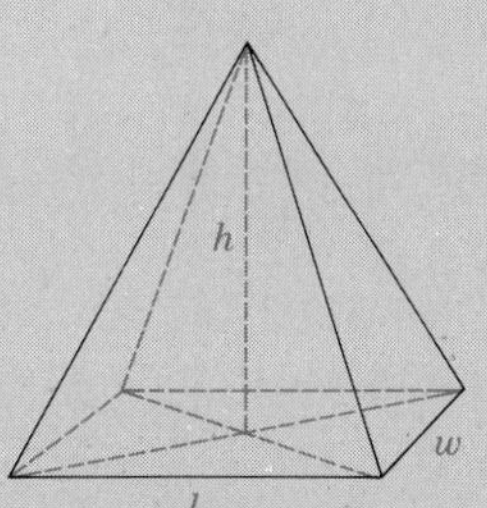

Rectangular pyramid
$V = \frac{1}{3}lwh$

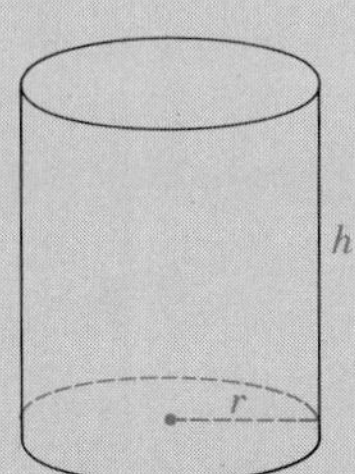

Right circular cylinder
$V = \pi r^2 h$

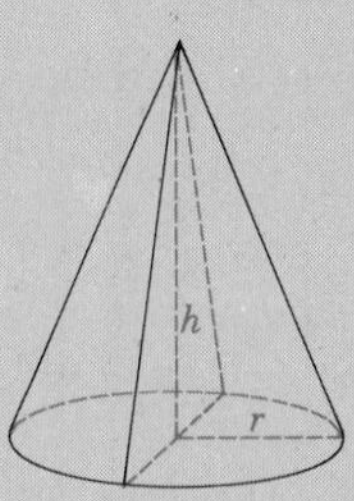

Right circular cone
$V = \frac{1}{3}\pi r^2 h$

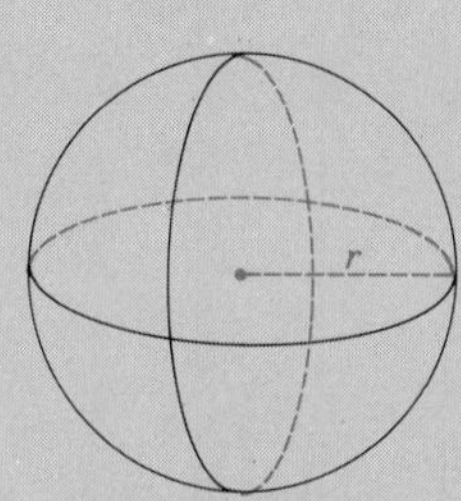

Sphere
$V = \frac{4}{3}\pi r^3$

EXAMPLE

18. Find the volume of the solid with the dimensions indicated. (Use $\pi \approx 3.14$.)

Solution

On top of the cylinder is a hemisphere (half a sphere). Find the volume of the cylinder and the hemisphere and add the results.

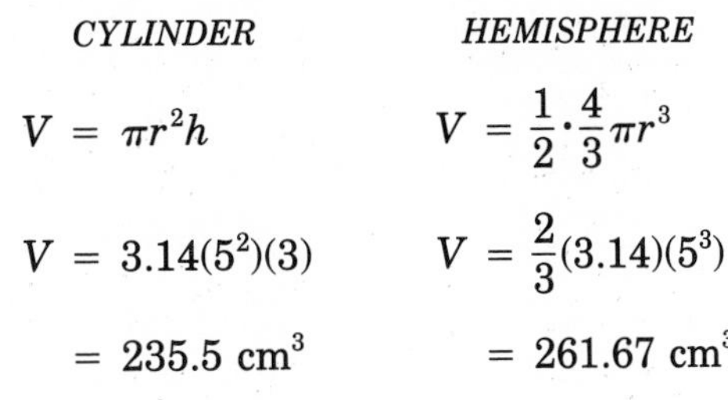

CYLINDER	*HEMISPHERE*
$V = \pi r^2 h$	$V = \frac{1}{2}\cdot\frac{4}{3}\pi r^3$
$V = 3.14(5^2)(3)$	$V = \frac{2}{3}(3.14)(5^3)$
$= 235.5 \text{ cm}^3$	$= 261.67 \text{ cm}^3$

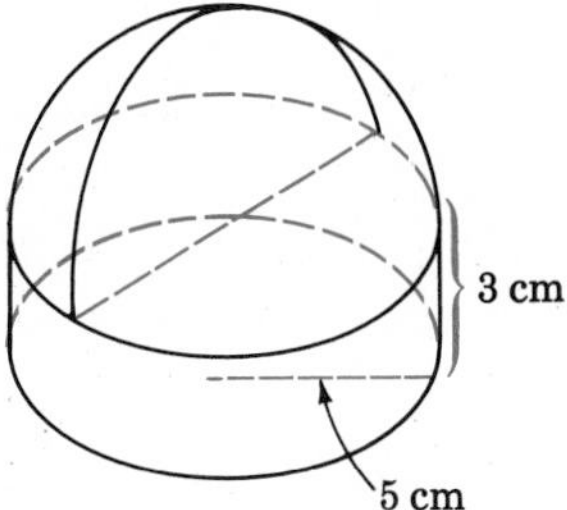

TOTAL VOLUME

235.50 cm^3
261.67 cm^3
497.17 cm^3 (or 497.17 mL) total volume

PRACTICE QUIZ

Change the following units as indicated.

1. 2 mL = ______ L
2. 3.6 kL = ______ L
3. 500 mL = ______ L
4. 500 mL = ______ cm^3
5. 42 hL = ______ kL

ANSWERS

1. 0.002 L
2. 3600 L
3. 0.5 L
4. 500 cm^3
5. 4.2 kL

Exercises 9.3

Copy and complete the following tables.

1.

1 cm^3 = ______ mm^3
1 dm^3 = ______ cm^3
1 m^3 = ______ dm^3
1 km^3 = ______ m^3

2.

1 dm = ______ cm
1 dm = ______ mm
1 dm^2 = ______ cm^2
1 dm^2 = ______ mm^2
1 dm^3 = ______ cm^3
1 dm^3 = ______ mm^3

3. $1 \text{ m} =$ ______ dm
$1 \text{ m} =$ ______ cm
$1 \text{ m}^2 =$ ______ dm^2
$1 \text{ m}^2 =$ ______ cm^2
$1 \text{ m}^3 =$ ______ dm^3
$1 \text{ m}^3 =$ ______ cm^3

4. $1 \text{ km} =$ ______ m
$1 \text{ km}^2 =$ ______ m^2
$1 \text{ km}^3 =$ ______ m^3
$1 \text{ km} =$ ______ dm
$1 \text{ km}^2 =$ ______ ha
$1 \text{ km}^3 =$ ______ kL

Change the following units as indicated.

5. $73 \text{ m}^3 =$ ______ dm^3
6. $0.9 \text{ m}^3 =$ ______ dm^3
7. $400 \text{ m}^3 =$ ______ cm^3
8. $525 \text{ cm}^3 =$ ______ m^3
9. $63 \text{ dm}^3 =$ ______ m^3
10. $8.7 \text{ m}^3 =$ ______ cm^3
11. $5 \text{ mm}^3 =$ ______ cm^3
12. $3.1 \text{ cm}^3 =$ ______ mm^3
13. $19 \text{ mm}^3 =$ ______ dm^3
14. $5 \text{ cm}^3 =$ ______ mm^3
15. $2 \text{ dm}^3 =$ ______ cm^3
16. 76.4 mL = ______ L
17. 5.3 L = ______ mL
18. $30 \text{ cm}^3 =$ ______ mL
19. $30 \text{ cm}^3 =$ ______ L
20. 5.3 mL = ______ L
21. 48 kL = ______ L
22. 72 000 L = ______ kL
23. 32 L = ______ hL
24. 80 L = ______ mL
25. 290 L = ______ kL
26. 569 mL = ______ L
27. 72 hL = ______ mL
28. 7 L = ______ mL
29. 95 hL = ______ L
30. 72 L = ______ hL

Find the volume of each of the following solids. (Use $\pi \approx 3.14$.)

31. a rectangular solid with a length of 5 dm, a width of 2 dm, and a height of 7 dm
32. a right circular cylinder 15 cm high and 1 dm in diameter (A diameter of a circle is a segment through the center with endpoints on the circle.)
33. a sphere with radius 4.5 cm
34. a sphere with diameter 12 dm
35. a right circular cone 3 dm high with a 2-dm radius
36. a rectangular pyramid with a length of 8 cm, a width of 10 mm, and a height of 3 dm

Find the volume of each of the solids with the dimensions indicated.

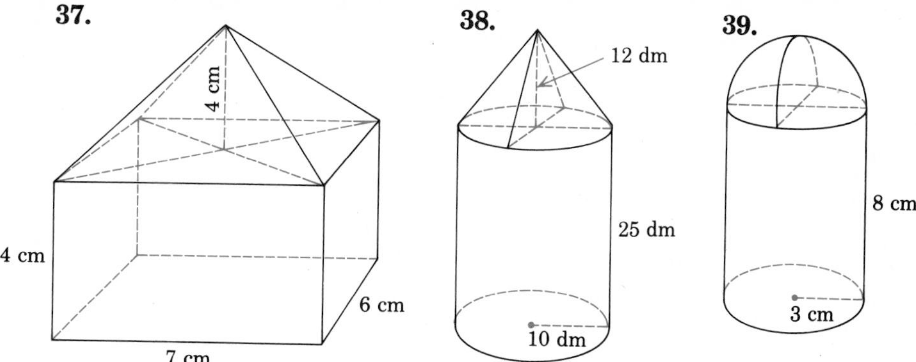

9.4 MASS (WEIGHT)

Mass is the amount of material in an object. Regardless of where the object is in space, its mass remains the same. (See Figure 9.7.) **Weight** is the force of the Earth's gravitational pull on an object. The farther an object is from Earth, the less the gravitational pull of the Earth. Thus, astronauts experience weightlessness in space, but their mass is unchanged.

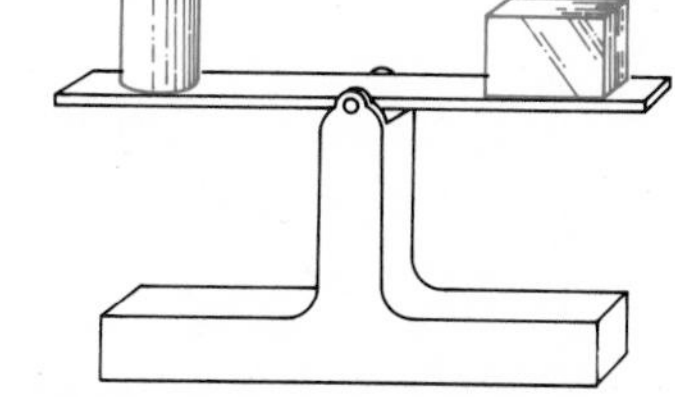

The two objects have the same *mass* and balance on an equal arm balance, regardless of their location in space.

Figure 9.7

Because most of us do not stray far from the Earth's surface, in this text weight and mass will be used interchangeably. Thus, a **mass** of 20 kilograms will be said to **weigh** 20 kilograms.

The basic unit of mass in the metric system is the **kilogram,*** about 2.2 pounds. In some fields, such as medicine, the **gram** (about the mass

---

*Technically, a kilogram is the mass of a certain cylinder of platinum-iridium alloy kept by the International Bureau of Weights and Measures in Paris.

Originally, the basic unit was a gram, defined to be the mass of 1 cm^3 of distilled water at 4° Celsius. This mass is still considered accurate for many purposes, so that

1 cm^3 of water has a mass of 1 g.
1 dm^3 of water has a mass of 1 kg.
1 m^3 of water has a mass of 1000 kg, or 1 metric ton.

of a paper clip) is more convenient as a basic unit than the kilogram.

Large masses, such as loaded trucks and railroad cars, are measured by the **metric ton** (1000 kilograms or about 2200 pounds). (See Tables 9.7 and 9.8.)

Table 9.7 Measures of Mass

1 **milli**gram	(mg)	=	0.001 gram
1 **centi**gram	(cg)	=	0.01 gram
1 **deci**gram	(dg)	=	0.1 gram
1 gram	(g)	=	1.0 gram
1 **deka**gram	(dag)	=	10 grams
1 **hecto**gram	(hg)	=	100 grams
1 **kilo**gram	(kg)	=	1000 grams
1 metric ton	(t)	=	1000 kilograms

Table 9.8 Equivalent Measures of Mass

1000 mg = 1g	0.001 g = 1 mg
1000 g = 1 kg	0.001 kg = 1 g
1000 kg = 1 t	0.001 t = 1 kg
1 t = 1000 kg = 1 000 000 g = 1 000 000 000 mg	

The centigram, decigram, dekagram, and hectogram have little practical use and are not included in the exercises. For completeness, they are all included in the headings of the chart used to change units.

A chart can be used to change from one unit of mass to another.

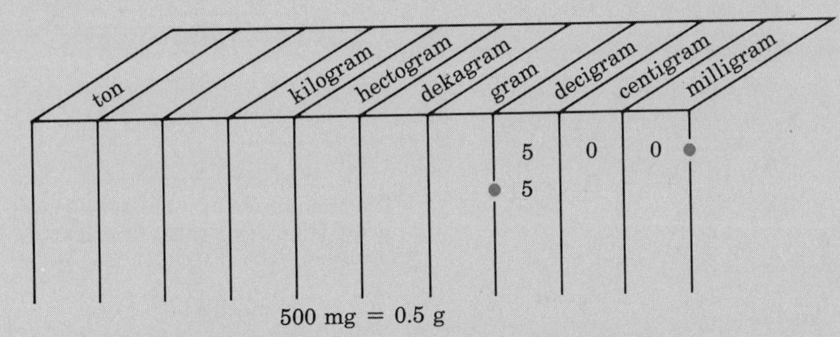

1. List each unit across the top.
2. Enter the given number so that there is **one** digit in each column with the decimal point on the given unit line.
3. Move the decimal point to the given unit line.
4. Fill in the spaces with 0's using one digit per column.

The use of the chart shows how the following equivalent measures can be found.

EXAMPLES

1. 23 mg = 0.023 g
2. 6 g = 6000 mg
3. 49 kg = 49 000 g
4. 5 t = 5000 kg
5. 70 kg = 0.07 t

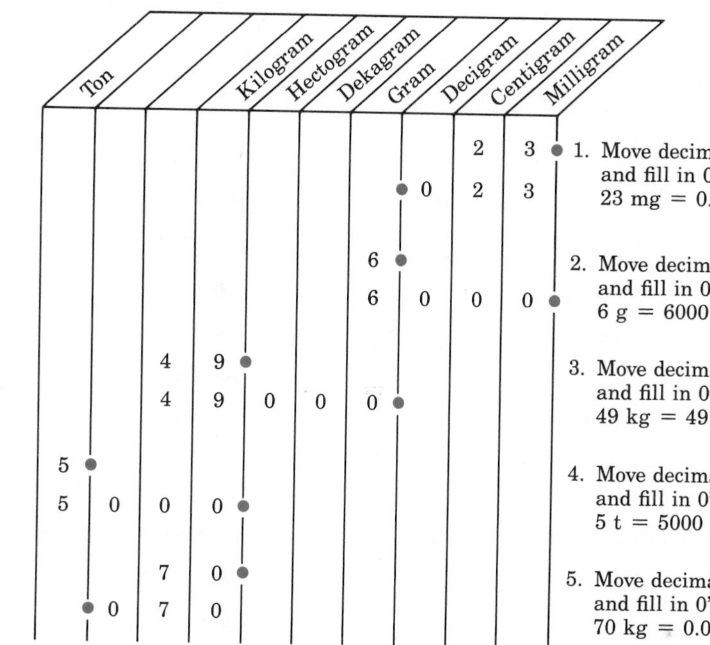
Ton
Kilogram
Hectogram
Dekagram
Gram
Decigram
Centigram
Milligram
1. Move decimal point to g and fill in 0's.
23 mg = 0.023 g
2. Move decimal point to mg and fill in 0's.
6 g = 6000 mg
3. Move decimal point to g and fill in 0's.
49 kg = 49 000 g
4. Move decimal point to kg and fill in 0's.
5 t = 5000 kg
5. Move decimal point to t and fill in 0's.
70 kg = 0.07 t

Make your own chart on a piece of paper and see if you agree with the results in Examples 6–10.

EXAMPLES

6. 60 mg = 0.06 g
7. 135 mg = 0.135 g
8. 5700 kg = 5.7 t
9. 100 g = 0.1 kg
10. 78 g = 78 000 kg

PRACTICE QUIZ

Change the following units as indicated.

1. 500 mg = ______ g
2. 500 kg = ______ t
3. 43 g = ______ mg
4. 62 g = ______ kg

ANSWERS

1. 0.5 g
2. 0.5 t
3. 43 000 mg
4. 0.062 kg

Exercises 9.4

Change the following units as indicated.

1. 7 g = ______ mg
2. 2 kg = ______ g
3. 34.5 mg = ______ g
4. 3700 kg = ______ t
5. 4000 kg = ______ t
6. 5600 g = ______ kg
7. 73 kg = ______ mg
8. 91 kg = ______ t
9. 0.54 g = ______ mg
10. 0.7 g = ______ mg
11. 5 t = ______ kg
12. 17 t = ______ kg
13. 2 t = ______ kg
14. 896 mg = ______ g
15. 896 g = ______ mg
16. 342 kg = ______ g
17. 75 000 g = ______ kg
18. 3000 mg = ______ g
19. 7 t = ______ g
20. 0.4 t = ______ g
21. 0.34 g = ______ kg
22. 0.78 g = ______ mg
23. 16 mg = ______ g
24. 2.5 g = ______ mg
25. 3.94 g = ______ mg
26. 92.3 g = ______ kg
27. 5.6 t = ______ kg
28. 7.58 t = ______ kg
29. 3547 kg = ______ t
30. 2963 kg = ______ t

SUMMARY: CHAPTER 9

The metric equivalents are given in tables throughout Chapter 9 and are not reproduced here. You should be familiar with all of them.

The prefixes used, from largest to smallest unit size, are

kilo, hecto, deka, deci, centi, milli

Abbreviations such as mm, cm, and dm do not have periods. Zero is written to the left of the decimal point if there is no whole number part. No commas are used in writing numbers.

The basic units are meter (length), gram (mass), are (area), liter (volume).

Some useful formulas related to geometric figures:

FIGURE	PERIMETER	AREA
square	$P = 4s$	$A = s^2$
rectangle	$P = 2l + 2w$	$A = lw$
parallelogram	$P = 2b + 2a$	$A = bh$
triangle	$P = a + b + c$	$A = \frac{1}{2}bh$
circle	$C = 2\pi r$ $C = \pi d$	$A = \pi r^2$
trapezoid	$P = a + b + c + d$	$A = \frac{1}{2}h(b + c)$

FIGURE	VOLUME
rectangular solid	$V = lwh$
rectangular pyramid	$V = \frac{1}{3}lwh$
right circular cylinder	$V = \pi r^2 h$
right circular cone	$V = \frac{1}{3}\pi r^2 h$
sphere	$V = \frac{4}{3}\pi r^3$

REVIEW QUESTIONS: CHAPTER 9

Change the following units as indicated.

1. 15 m = ______ cm

2. 35 mm = ______ dm

3. 37 cm^2 = ______ mm^2

4. 17 mm^2 = ______ cm^2

5. 3 ha = ______ a

6. 3 ha = ______ m^2

7. 5 L = ______ cm^3

8. 36 L = ______ mL

9. 13 dm^3 = ______ cm^3

10. 68 cm^3 = ______ mm^3

11. 5 kg = ______ g

12. 3.4 g = ______ mg

13. 6.71 t = ______ kg

14. 19 mg = ______ g

15. 8 kg = ______ g

16. 4290 g = ______ kg

Find the perimeter and area of each of the following figures with the dimensions indicated. (Use $\pi \approx 3.14$.)

17.

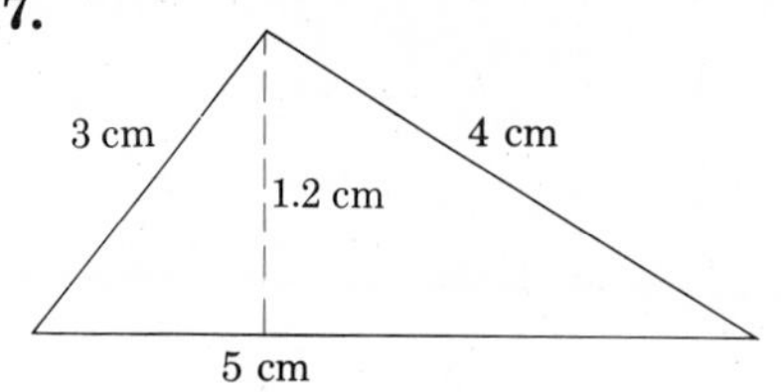

18.

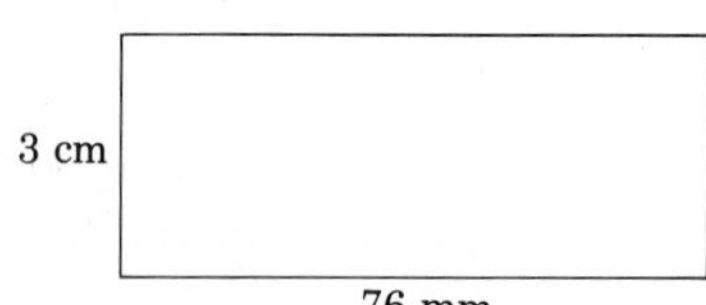

19.

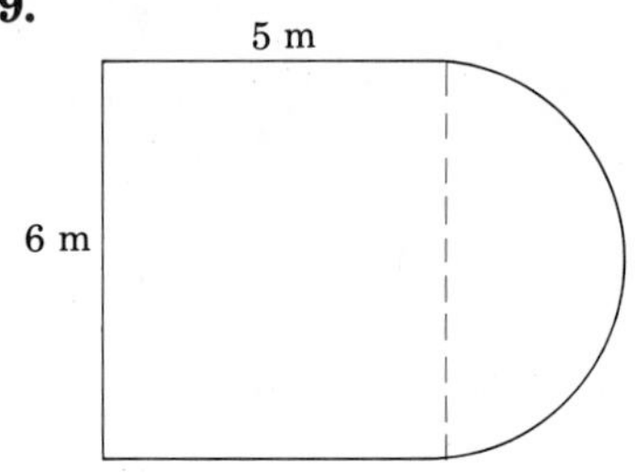

20.

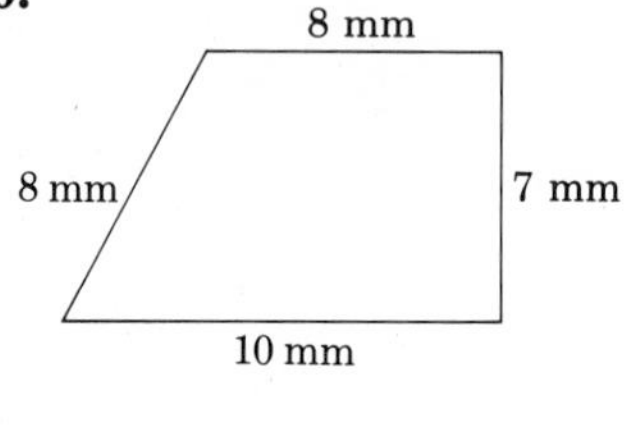

Find the volume in liters of each of the following solids with dimensions indicated. (Use $\pi \approx 3.14$.)

21.

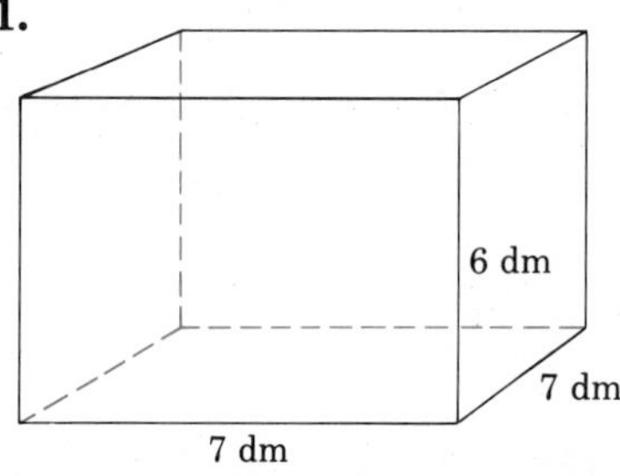

22.

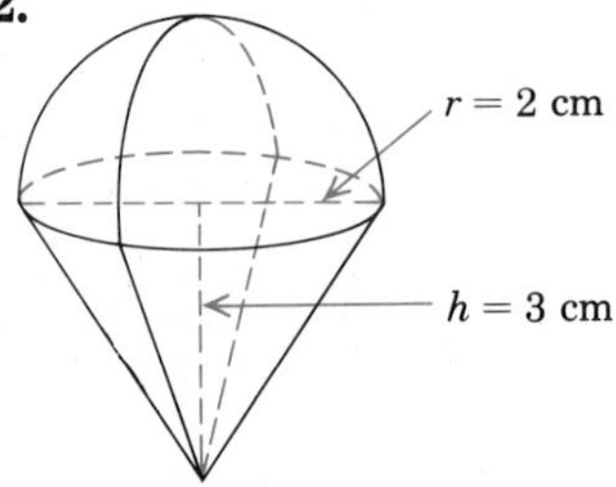

23. Find the area of a circle in square centimeters if the diameter is 0.4 m. (Use $\pi \approx 3.14$.)

24. Find the volume of a right circular cylinder in mm^3 if the radius is 2 cm and the height is 7 cm. (Use $\pi \approx 3.14$.)

25. Find the shaded area in square meters.

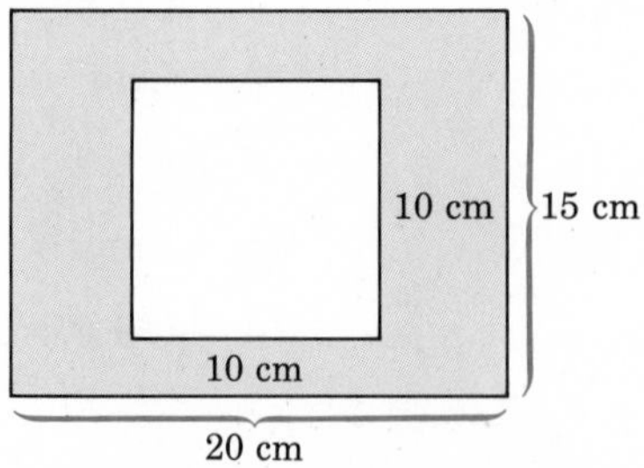

TEST: CHAPTER 9

1. Which is longer, 20 mm or 20 cm? How much longer?
2. Which is heavier, 10 g or 10 kg? How much heavier?
3. Which has the greater volume, 15 mL or 15 cm^3? How much greater?

Change the following units as indicated.

4. 37 cm = ______ m
5. 23 m = ______ cm
6. 2 L = ______ cm^3
7. 1200 g = ______ kg
8. 5.6 t = ______ kg
9. 75 a = ______ m^2
10. 11 000 mm = ______ m
11. 4 cm^3 = ______ mm^3
12. 960 mm^2 = ______ cm^2
13. 83.5 mg = ______ g

Find the perimeter and area of each figure.

14.

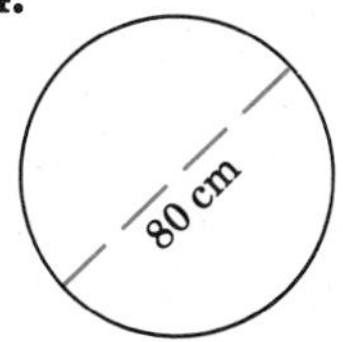

15.

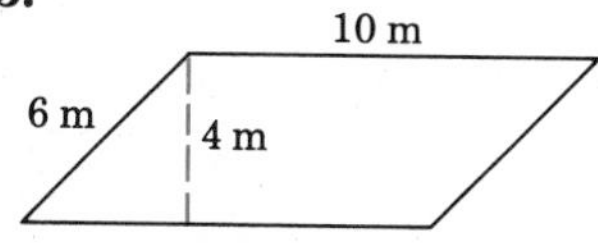

16. Find the volume in liters of a right circular cylinder with radius 3 cm and height 12 cm.
17. Find the volume in mm^3 of a rectangular box 0.2 m long, 0.1 m wide, and 0.05 m high.
18. Find the area in square centimeters of the triangle with a base of 175 mm and a height of 120 mm.
19. Find the volume in cubic meters of a sphere with diameter 60 cm. (Use $\pi \approx 3.14$.)
20. What is the volume of the sphere in Exercise 19 in dm^3?

10 Measurement: The U.S. Customary System

10.1 U.S. CUSTOMARY EQUIVALENTS

In the U.S. customary system (formerly the English system), the units are not systematically related as are the units in the metric system. Historically some of the units were associated with parts of the body, which would vary from person to person. For example, a foot was the length of a person's foot, and a yard was the distance from the tip of one's nose to the tip of one's fingers with arm outstretched. A king might dictate his own foot to be the official "foot," but, of course, the next king might have a different sized foot.

There is considerably more stability now because the official weights and measures are monitored by the government.

In this section we will discuss the common units for length: area, liquid volume, weight, and time, and how to find equivalent measures. The basic relationships are listed in Table 10.1. The measures of time are universal.

To change units, you must either have a table of equivalent values with you or memorize the basic equivalent values. Most people know some of these values but not all.

There are several methods used to change from one unit to another. One is to use proportions and solve these proportions; another is to substitute ratios (commonly used in science courses); and another is to substitute equivalent values for just one unit and multiply. We will illustrate the third technique because it is probably the simplest.

Table 10.1 U.S. Customary Units of Measure

Length		*Liquid Volume*	
1 foot (ft)	= 12 inches (in.)	1 pint (pt)	= 16 fluid ounces (fl oz)
1 yard (yd)	= 3 ft	1 quart (qt)	= 2 pt = 32 fl oz
1 mile (mi)	= 5280 ft	1 gallon (gal)	= 4 qt
Weight		*Time*	
1 pound (lb)	= 16 ounces (oz)	1 minute (min)	= 60 seconds (sec)
1 ton (t)	= 2000 lb	1 hour (hr)	= 60 min
		1 day	= 24 hr

EXAMPLES

Convert the following measures as indicated.

1. 4 ft = ______ in.

Solution

Think of 4 ft as 4(1 ft) and replace 1 ft with 12 in.

$$4 \text{ ft} = 4(1 \text{ ft}) = 4(12 \text{ in.}) = 48 \text{ in.}$$

2. 6 ft = ______ in.

Solution

$$6 \text{ ft} = 6(1 \text{ ft}) = 6(12 \text{ in.}) = 72 \text{ in.}$$

3. 36 in. = ______ ft

Solution

In this case, changing from a smaller to a larger unit, a fraction is necessary. We know

$$1 \text{ ft} = 12 \text{ in.}$$

so, $$\frac{1 \text{ ft}}{12} = \frac{12 \text{ in.}}{12} \quad \text{or} \quad \frac{1}{12} \text{ ft} = 1 \text{ in.}$$

Now, $$36 \text{ in.} = 36(1 \text{ in.}) = 36\left(\frac{1}{12} \text{ ft}\right) = 3 \text{ ft}$$

4. 12 ft = ______ yd

Solution

Since 3 ft = 1 yd, we have 1 ft $= \frac{1}{3}$ yd.

So, $$12 \text{ ft} = 12(1 \text{ ft}) = 12\left(\frac{1}{3} \text{ yd}\right) = 4 \text{ yd}$$

5. 6 qt = ______ pt

Solution

$$6 \text{ qt} = 6(1 \text{ qt}) = 6(2 \text{ pt}) = 12 \text{ pt}$$

6. 2.5 lb = ______ oz

Solution

$$2.5 \text{ lb} = 2.5(1 \text{ lb}) = 2.5(16 \text{ oz}) = 40 \text{ oz}$$

7. 36 hr = ______ days

Solution

Since 24 hr = 1 day, we know 1 hr $= \frac{1}{24}$ day.

Substituting,

$$36 \text{ hr} = 36(1 \text{ hr}) = 36\left(\frac{1}{24} \text{ day}\right) = 1\frac{1}{2} \text{ days or } 1.5 \text{ days}$$

8. 3 hr = ______ sec

Solution

In this case, two substitutions are made. First, change hours to minutes, then change minutes to seconds.

$$3 \text{ hr} = 3(1 \text{ hr}) = 3(60 \text{ min}) = 180 \text{ min}$$
$$180 \text{ min} = 180(1 \text{ min}) = 180(60 \text{ sec}) = 10{,}800 \text{ sec}$$

So, 3 hr = 10,800 sec

PRACTICE QUIZ	Convert the following measures as indicated.	ANSWERS
	1. 2 ft = ______ in.	**1.** 24 in.
	2. 8 in. = ______ ft	**2.** $\frac{2}{3}$ ft
	3. 2 gal = ______ qt	**3.** 8 qt
	4. 3 t = ______ lb	**4.** 6000 lb
	5. 45 min = ______ hr	**5.** $\frac{3}{4}$ hr

Exercises 10.1

Convert the following measures as indicated. Use Table 10.1 as a reference.

1. 5 ft = ______ in.

2. 3 ft = ______ in.

3. 1.5 ft = ______ in.

4. 2.5 ft = ______ in.

5. 48 in. = ______ ft

6. 120 in. = ______ ft

7. 30 in. = ______ ft

8. 18 in. = ______ ft

9. 3 yd = ______ ft

10. 4 yd = ______ ft

11. $2\frac{1}{3}$ yd = ______ ft

12. $1\frac{2}{3}$ yd = ______ ft

13. 3 mi = ______ ft

14. 4 mi = ______ ft

15. 6 ft = ______ yd

16. 9 ft = ______ yd

17. 10,560 ft = ______ mi

18. 15,840 ft = ______ mi

19. 2 pt = ______ fl oz

20. 3 pt = ______ fl oz

21. 3 qt = ______ pt

22. $1\frac{1}{2}$ qt = ______ pt

23. 5 gal = ______ qt

24. 3 gal = ______ qt

25. 12 qt = ______ gal

26. 20 qt = ______ gal

27. 15 qt = ______ gal

28. 13 qt = ______ gal

29. 5 lb = ______ oz

30. 3 lb = ______ oz

31. 3.5 lb = ______ oz

32. $1\frac{3}{4}$ lb = ______ oz

33. $2\frac{1}{2}$ t = ______ lb

34. 3 t = ______ lb

35. 5 hr = ______ min

36. 4 hr = ______ min

37. $1\frac{1}{2}$ hr = ______ min

38. $3\frac{1}{2}$ hr = ______ min

39. 30 min = ______ hr

40. 15 min = ______ hr

41. 3 days = ______ hr

42. 4 days = ______ hr

43. $\frac{1}{2}$ hr = ______ sec

44. $\frac{2}{3}$ hr = ______ sec

45. 5 min = ______ sec

46. 3 min = ______ sec

47. 240 sec = ______ min

48. 300 sec = ______ min

49. 48 hr = ______ days

50. 72 hr = ______ days

10.2 DENOMINATE NUMBERS

Numbers with no units of measure attached are called **abstract numbers.** Numbers with units of measure attached are called **denominate numbers.** The numbers discussed in Section 10.1 were all denominate numbers. In this section, we will do more than just convert from one unit to another. Here we will discuss:

(a) Mixed denominate numbers (denominate numbers with two or more units);

(b) Adding denominate numbers;

(c) Subtracting denominate numbers.

Examples of mixed denominate numbers that we commonly use are

5 ft 8 in.
3 lb 4 oz
and 1 hr 45 min

With mixed numbers, we make sure that the fraction part is less than 1. Thus, we write $5\frac{1}{2}$, **not** $4\frac{3}{2}$. Similarly, in **simplified mixed denominate numbers,** the number of the smaller unit is less than 1 of the larger unit. The following examples illustrate the technique for simplifying mixed denominate numbers.

EXAMPLES

Simplify the following mixed denominate numbers so that the number of smaller units is less than 1 of the larger unit.

1. 3 ft 14 in.

Solution

Since 14 in. is more than 1 ft, we write

$$\begin{aligned} 3 \text{ ft } 14 \text{ in.} &= 3 \text{ ft} + 12 \text{ in.} + 2 \text{ in.} \\ &= \underbrace{3 \text{ ft} + 1 \text{ ft}} + 2 \text{ in.} \\ &= 4 \text{ ft} + 2 \text{ in.} \\ &= 4 \text{ ft } 2 \text{ in.} \end{aligned}$$

2. 5 lb 30 oz

Solution

Since 30 oz is more than 1 lb, we write

$$
\begin{aligned}
5\text{ lb }30\text{ oz} &= 5\text{ lb} + \underset{\downarrow}{16\text{ oz}} + 14\text{ oz} \\
&= \underbrace{5\text{ lb} + 1\text{ lb}} + 14\text{ oz} \\
&= 6\text{ lb} + 14\text{ oz} \\
&= 6\text{ lb }14\text{ oz}
\end{aligned}
$$

3. 2 hr 70 min

Solution

Since 70 min is more than 1 hr, we write

$$
\begin{aligned}
2\text{ hr }70\text{ min} &= 2\text{ hr} + \underset{\downarrow}{60\text{ min}} + 10\text{ min} \\
&= \underbrace{2\text{ hr} + 1\text{ hr}} + 10\text{ min} \\
&= 3\text{ hr} + 10\text{ min} \\
&= 3\text{ hr }10\text{ min}
\end{aligned}
$$

Understanding how to simplify mixed denominate numbers helps in both adding and subtracting such numbers. **Like denominate numbers** are denominate numbers with the same units. For example, 5 ft 10 in. and 2 ft 3 in. are like denominate numbers. Also, 3 hr 5 min and 4 hr 15 min are like denominate numbers.

To add like denominate numbers:

1. Write the numbers in column form so that like units are aligned.
2. Add the numbers in each column.
3. Simplify the resulting sum if necessary.

EXAMPLES

Add the following like denominate numbers and simplify the sum if necessary.

4.
```
    3 ft  2 in.
    2 ft  8 in.
+   5 ft  5 in.
---------------
   10 ft 15 in. = 10 ft + 12 in. + 3 in.
                = 11 ft 3 in.
```

5.
```
   2 hr 15 min
 + 4 hr 50 min
--------------
   6 hr 65 min = 6 hr + 60 min + 5 min
               = 7 hr 5 min
```

6.
```
   3 gal 2 qt
   1 gal 3 qt
 + 5 gal 2 qt
-------------
   9 gal 7 qt = 9 gal + 4 qt + 3 qt
              = 10 gal 3 qt
```

To subtract like denominate numbers:

1. Write the numbers in column form so that like units are aligned.
2. If necessary for subtraction, borrow 1 of the larger units and rewrite the top number.
3. Subtract the like units.

EXAMPLES

Subtract the following like denominate numbers.

7.
```
   8 lb 14 oz
 - 3 lb 10 oz
-------------
   5 lb  4 oz
```

8. $\begin{array}{r} 6 \text{ ft } 5 \text{ in.} \\ -2 \text{ ft } 8 \text{ in.} \\ \hline \end{array}$ Here 5 in. is smaller than 8 in., and we cannot subtract. So, borrow 1 ft = 12 in. from 6 ft.

$$\begin{array}{rcrl} 6 \text{ ft } 5 \text{ in.} & = & 5 \text{ ft } 17 \text{ in.} & (12 + 5 \text{ in.} = 17 \text{ in.}) \\ -2 \text{ ft } 8 \text{ in.} & = & 2 \text{ ft } \ \ 8 \text{ in.} & \\ \hline & & 3 \text{ ft } \ \ 9 \text{ in.} & \end{array}$$

9. $\begin{array}{r} 13 \text{ hr } 20 \text{ min} \\ -10 \text{ hr } 50 \text{ min} \\ \hline \end{array}$ Here 20 min is smaller than 50 min, and we cannot subtract. So, borrow 1 hr = 60 min from 13 hr.

$$\begin{array}{rcrl} 13 \text{ hr } 20 \text{ min} & = & 12 \text{ hr } 80 \text{ min} & (60 \text{ min} + 20 \text{ min} = 80 \text{ min}) \\ -10 \text{ hr } 50 \text{ min} & = & 10 \text{ hr } 50 \text{ min} & \\ \hline & & 2 \text{ hr } 30 \text{ min} & \end{array}$$

PRACTICE QUIZ

Perform the indicated operations and simplify.

1. $\begin{array}{r} 2 \text{ ft } 3 \text{ in.} \\ 6 \text{ ft } 8 \text{ in.} \\ +1 \text{ ft } 4 \text{ in.} \\ \hline \end{array}$

2. $\begin{array}{r} 5 \text{ hr } 30 \text{ min} \\ -1 \text{ hr } 45 \text{ min} \\ \hline \end{array}$

ANSWERS

1. 10 ft 3 in.

2. 3 hr 45 min

Exercises 10.2

Simplify the following mixed denominate numbers.

1. 3 ft 20 in.

2. 4 ft 18 in.

3. 6 lb 20 oz

4. 3 lb 24 oz

5. 5 min 80 sec

6. 14 min 90 sec

7. 2 days 30 hr

8. 5 days 36 hr

9. 8 gal 5 qt

10. 2 gal 6 qt

11. 4 pt 20 fl oz

12. 3 pt 24 fl oz

Add and simplify if necessary.

13. 2 ft 8 in.
5 ft 4 in.
+1 ft 7 in.

14. 3 ft 5 in.
6 ft 5 in.
+2 ft 3 in.

15. 10 lb 10 oz
+ 7 lb 8 oz

16. 4 lb 5 oz
+4 lb 7 oz

17. 8 min 35 sec
+9 min 35 sec

18. 5 min 10 sec
+14 min 35 sec

19. 2 hr 15 min 45 sec
+1 hr 55 min 30 sec

20. 5 hr 20 min 30 sec
+2 hr 35 min 40 sec

21. 2 days 20 hr 50 min
+3 days 5 hr 45 min

22. 1 day 15 hr 40 min
+2 days 10 hr 20 min

23. 5 gal 2 qt
3 gal 2 qt
+4 gal 3 qt

24. 4 gal 3 qt
1 gal 2 qt
+3 gal 1 qt

25. 4 gal 3 qt 10 fl oz
+2 gal 3 qt 10 fl oz

26. 5 gal 3 qt 15 fl oz
+1 gal 2 qt 8 fl oz

27. 5 yd 2 ft 8 in.
+6 yd 2 ft 10 in.

28. 3 yd 1 ft 7 in.
+2 yd 2 ft 3 in.

Subtract.

29. 5 yd 1 ft 7 in.
−2 yd 2 ft 5 in.

30. 8 yd 2 ft 3 in.
−7 yd 1 ft 8 in.

31. 9 gal 2 qt 4 fl oz
−5 gal 3 qt 6 fl oz

32. 20 gal 1 qt 13 fl oz
−14 gal 2 qt 10 fl oz

33. 15 hr 30 min
−12 hr 45 min

34. 6 hr 20 min
−4 hr 40 min

35. 15 min 20 sec
−10 min 30 sec

36. 30 min 15 sec
−20 min 25 sec

37. 8 lb 4 oz
−3 lb 12 oz

38. 20 lb 10 oz
−10 lb 14 oz

39. 6 ft 5 in.
−2 ft 9 in.

40. 3 ft 8 in.
−1 ft 10 in.

10.3 U.S. CUSTOMARY AND METRIC EQUIVALENTS

In this section, we will discuss equivalent measures between the metric system and the U.S. customary system.

Temperature: U.S. Customary measure is in **degrees** Fahrenheit (°F)
Metric measure is in **degrees** Celsius (°C)

The two scales are shown here on thermometers. Approximate conversions can be found by reading along a ruler or the edge of a piece of paper held horizontally across the page.

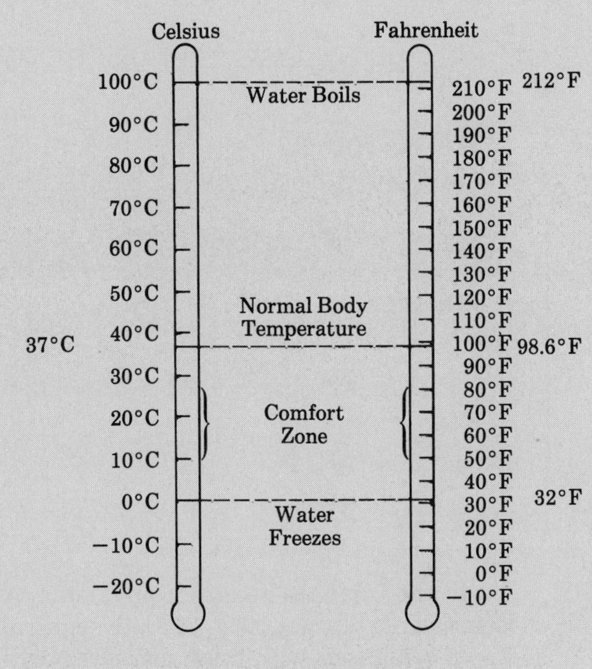

EXAMPLES

1. Hold a straight edge horizontally across the two thermometers and you will read:

100° C = 212° F	Water boils at sea level.
40° C = 105° F	hot day in the desert
20° C = 68° F	comfortable room temperature

Two formulas that give exact conversions are given here.

F = Fahrenheit temperature and C = Celsius temperature.

$$C = \frac{5(F - 32)}{9} \qquad F = \frac{9 \cdot C}{5} + 32$$

A calculator will give answers accurate to 8 digits. Answers that are not exact may be rounded off to whatever place of accuracy you choose.

2. Let F = 86° and find the equivalent measure in Celsius.

Solution

$$C = \frac{5(86 - 32)}{9} = \frac{5(54)}{9} = 30$$

86° F = 30° C.

3. Let C = 40° and convert this to degrees Fahrenheit.

Solution

$$F = \frac{9 \cdot 40}{5} + 32 = 72 + 32 = 104$$

40° C = 104° F.

In the tables of Length Equivalents, Area Equivalents, and Volume Equivalents (Tables 10.2–10.5), the equivalent measures are rounded off. Any calculations with these measures (with or without a calculator) cannot be any more accurate than the measure in the table. Figure 10.1 shows some length equivalents.

Table 10.2 Length Equivalents

U.S. TO METRIC	METRIC TO U.S.
1 in. = 2.54 cm (exact)	1 cm = 0.394 in.
1 ft = 0.305 m	1 m = 3.28 ft
1 yd = 0.915 m	1 m = 1.09 yd
1 mi = 1.61 km	1 km = 0.62 mi

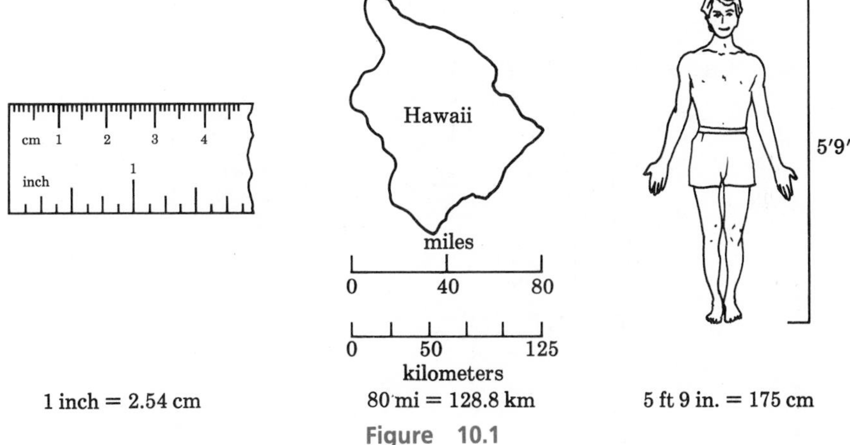

Figure 10.1

In Examples 4–7, use Table 10.2 to convert measurements as indicated.

EXAMPLES

4. 6 ft = ______ cm

$$6 \text{ ft} = 72 \text{ in.} = 72(2.54 \text{ cm}) = 183 \text{ cm (rounded off)}$$

or $$6 \text{ ft} = 6(0.305 \text{ m}) = 1.83 \text{ m} = 183 \text{ cm}$$

5. 25 mi = ______ km

$$25 \text{ mi} = 25(1.61 \text{ km}) = 40.25 \text{ km}$$

6. 30 m = ______ ft

$$30 \text{ m} = 30(3.28 \text{ ft}) = 98.4 \text{ ft}$$

7. 10 km = ______ mi

$$10 \text{ km} = 10(0.62 \text{ mi}) = 6.2 \text{ mi}$$

Table 10.3 Area Equivalents

U.S. TO METRIC	METRIC TO U.S.
1 in^2 = 6.45 cm^2	1 cm^2 = 0.155 in^2
1 ft^2 = 0.093 m^2	1 m^2 = 10.764 ft^2
1 yd^2 = 0.836 m^2	1 m^2 = 1.196 yd^2
1 acre = 0.405 ha	1 ha = 2.47 acres

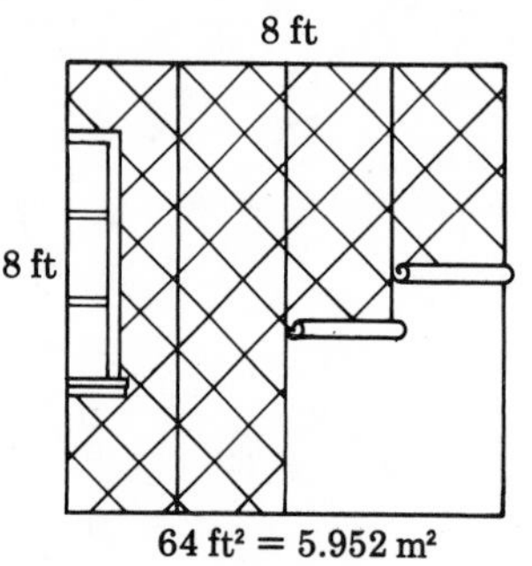

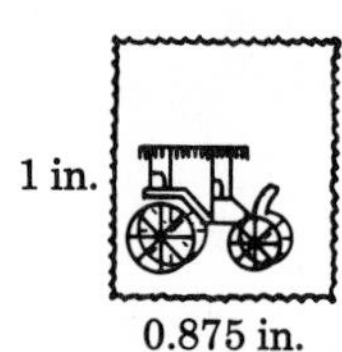

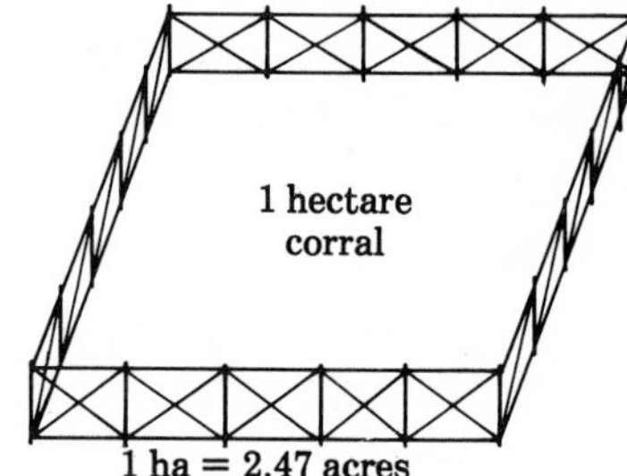

Figure 10.2

In Examples 8–11, use Table 10.3 to convert the measures as indicated. (Also see Figure 10.2.)

EXAMPLES

8. 40 yd^2 = ______ m^2

$$40 \text{ yd}^2 = 40(0.836 \text{ m}^2) = 33.44 \text{ m}^2$$

9. 5 acres = ______ ha

$$5 \text{ acres} = 5(0.405 \text{ ha}) = 2.025 \text{ ha}$$

10. 5 ha = ______ acres

$$5 \text{ ha} = 5(2.47 \text{ acres}) = 12.35 \text{ acres}$$

11. 100 cm^2 = ______ in^2

$$100 \text{ cm}^2 = 100(0.155 \text{ in}^2) = 15.5 \text{ in}^2$$

Table 10.4 Volume Equivalents

U.S. TO METRIC	METRIC TO U.S.
$1 \text{ in}^3 = 16.387 \text{ cm}^3$	$1 \text{ cm}^3 = 0.06 \text{ in}^3$
$1 \text{ ft}^3 = 0.028 \text{ m}^3$	$1 \text{ m}^3 = 35.315 \text{ ft}^3$
$1 \text{ qt} = 0.946 \text{ L}$	$1 \text{ L} = 1.06 \text{ qt}$
$1 \text{ gal} = 3.785 \text{ L}$	$1 \text{ L} = 0.264 \text{ gal}$

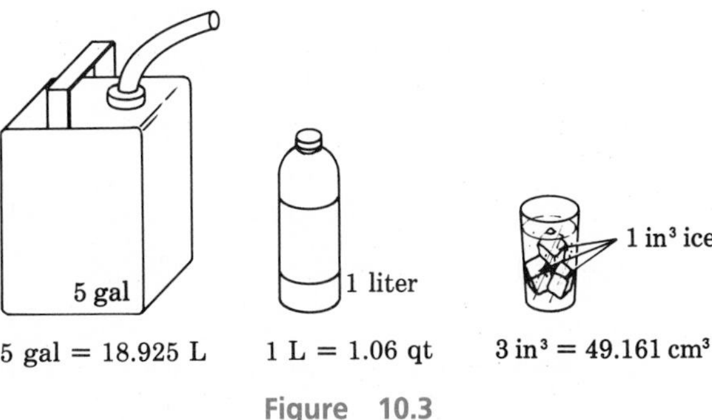

Figure 10.3

In Examples 12–15, use Table 10.4 to convert the measures as indicated. (Also see Figure 10.3.)

EXAMPLES

12. 20 gal = ______ L

$$20 \text{ gal} = 20(3.785 \text{ L}) = 75.7 \text{ L}$$

13. 42 L = ______ gal

$$42 \text{ L} = 42(0.264 \text{ gal}) = 11.088 \text{ gal}$$

or

$$42 \text{ L} = 11.1 \text{ gal (rounded off)}$$

14. 6 qt = ______ L

$$6 \text{ qt} = 6(0.946 \text{ L}) = 5.676$$

or

$$6 \text{ qt} = 5.7 \text{ L (rounded off)}$$

15. 10 cm^3 = ______ in^3

$$10 \text{ cm}^3 = 10(0.06 \text{ in}^3) = 0.6 \text{ in}^3$$

Table 10.5 Mass Equivalents

U.S. TO METRIC	METRIC TO U.S.
1 oz = 28.35 g	1 g = 0.035 oz
1 lb = 0.454 kg	1 kg = 2.205 lb

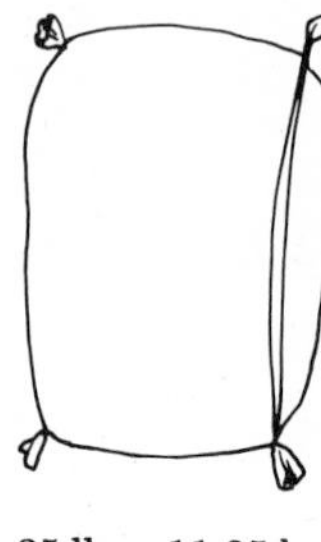

25 lb = 11.35 kg

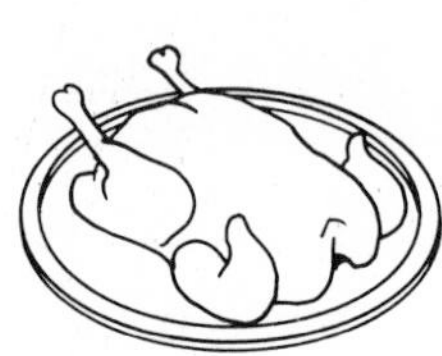

9 kg = 19.85 lb

Figure 10.4

In Examples 16 and 17, use Table 10.5 to convert the measures as indicated. (Also see Figure 10.4.)

EXAMPLES

16. 5 lb = ______ kg

$$5 \text{ lb} = 5(0.454 \text{ kg}) = 2.27 \text{ kg}$$

17. 15 kg = ______ lb

$$15 \text{ kg} = 15(2.205 \text{ lb}) = 33.075 \text{ lb}$$

or

$$15 \text{ kg} = 33.1 \text{ lb (rounded off)}$$

Exercises 10.3

Use the appropriate formula to convert the degrees as indicated.

1. 25°C = ______°F **2.** 80°C = ______°F **3.** 10°C = ______°F

4. 35°C = ______°F **5.** 50°F = ______°C **6.** 100°F = ______°C

7. 32°F = ______°C **8.** 41°F = ______°C

Use the appropriate table to convert the following measures as indicated.

9. 5 ft 2 in. = ______ cm **10.** 6 ft 3 in. = ______ cm

11. 3 yd = ______ m **12.** 5 yd = ______ m

13. 60 mi = ______ km

14. 100 mi = ______ km

15. 400 mi = ______ km

16. 350 mi = ______ km

17. 200 km = ______ mi

18. 65 km = ______ mi

19. 35 km = ______ mi

20. 450 km = ______ mi

21. 50 cm = ______ in.

22. 100 cm = ______ in.

23. 3 in^2 = ______ cm^2

24. 16 in^2 = ______ cm^2

25. 600 ft^2 = ______ m^2

26. 300 ft^2 = ______ m^2

27. 100 yd^2 = ______ m^2

28. 250 yd^2 = ______ m^2

29. 1000 acres = ______ ha

30. 250 acres = ______ ha

31. 300 ha = ______ acres

32. 400 ha = ______ acres

33. 5 m^2 = ______ ft^2

34. 10 m^2 = ______ yd^2

35. 30 cm^2 = ______ in^2

36. 50 cm^2 = ______ in^2

37. 10 qt = ______ L

38. 20 qt = ______ L

39. 25 gal = ______ L

40. 18 gal = ______ L

41. 10 L = ______ qt

42. 25 L = ______ qt

43. 42 L = ______ gal

44. 50 L = ______ gal

45. 200 in^3 = ______ cm^3

46. 10 m^3 = ______ ft^3

47. 10 lb = ______ kg

48. 16 oz = ______ g

49. 500 kg = ______ lb

50. 100 g = ______ oz

10.4 APPLICATIONS IN MEDICINE

This section is designed to provide an introduction into measurements used in medicine, techniques (proportions) for changing units, and formulas related to dosages of medicine.

Household fluid measures are based on the uses of household cooking and eating utensils. This method is **not** accurate, but it is useful. The **apothecaries' system of fluid measure** is used by pharmacists. (See Tables 10.6–10.7 and Figure 10.5.)

Table 10.6 Abbreviations

HOUSEHOLD FLUID MEASURE	APOTHECARIES' FLUID MEASURE
gtt = 1 drop	♏ = 1 minim (about 1 drop of water)
tsp = 1 teaspoonful	*f*ʒ = 1 fluidram (60 minims)
tbsp = 1 tablespoonful	*f*℥ = 1 fluidounce (480 minims)

Table 10.7 Fluid Measure Equivalents

HOUSEHOLD		METRIC		APOTHECARIES'
1 gtt	=	0.06 mL	=	1 ♏
15 gtt	=	1.0 mL	=	15 ♏
75 gtt	=	5.0 mL	=	75 ♏
1 tsp	=	5.0 mL	=	$1\frac{1}{3}$ *f*ʒ
3 tsp	=	15.0 mL	=	4 *f*ʒ
1 tbsp	=	15.0 mL	=	4 *f*ʒ = $\frac{1}{2}$ *f*℥
1 glassful	=	240.0 mL	=	8 *f*℥

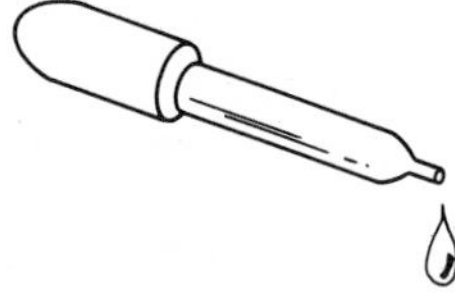

1 drop (gtt) = 1 minim (♏)
= 0.06 mL

1 teaspoonful (tsp) = 75 drops (gtt)
= 5.0 mL

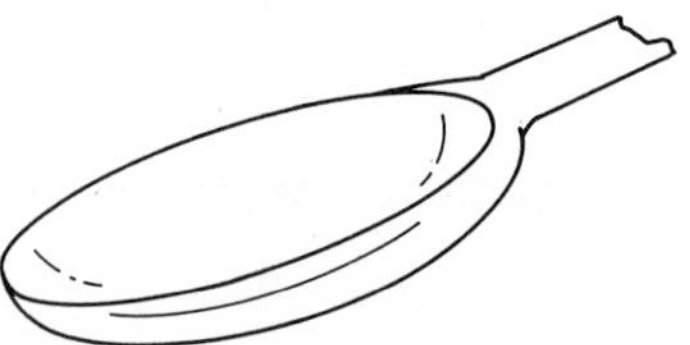

1 tablespoonful (tbsp)
= 3 teaspoonfulls (tsp)
= 15.0 mL

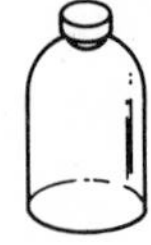

1 fluidram (*f*ʒ) = 60 minims (♏)
(about $\frac{4}{5}$) teaspoonful

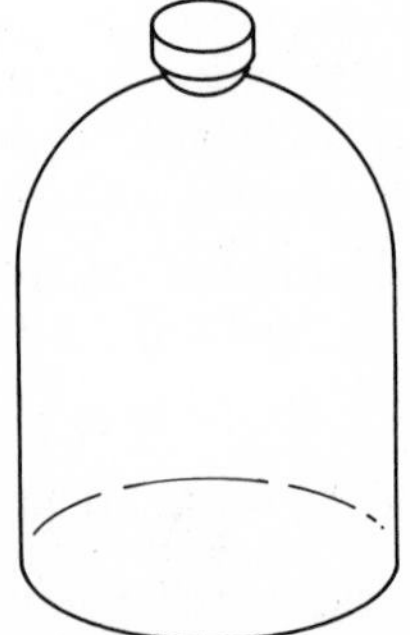

1 fluidounce (*f*℥) = 8 fluidrams (*f*ʒ)

Figure 10.5

To use proportions to change from one system to another:

1. Find the corresponding equivalent units in the table of Fluid Measure Equivalents (Table 10.7)
2. Write x for the unknown quantity.
3. Write a proportion using the ratio of equivalent measures and the ratio of x to the known measure being changed.
4. Solve the proportion.

EXAMPLES

1. How many milliliters (mL) are in 6 drops (gtt)?

Solution

From Table 10.7, 1 gtt = 0.06 mL.

Let x = unknown number of milliliters.

Set up the proportion:

$$\frac{x \text{ mL}}{6 \text{ gtt}} = \frac{0.06 \text{ mL}}{1 \text{ gtt}}$$

$$1 \cdot x = 6(0.06)$$

$$x = 0.36 \text{ mL}$$

So, 6 drops (gtt) = 0.36 milliliters (mL) or there are 0.36 milliliters in 6 drops.

2. How many tablespoonfuls (tbsp) are in 60 milliliters (mL)?

Solution

From Table 10.7, 1 tbsp = 15.0 mL.

Let x = unknown number of tablespoonfuls.

Set up the proportion:

$$\frac{x \text{ tbsp}}{60 \text{ mL}} = \frac{1 \text{ tbsp}}{15.0 \text{ mL}}$$

$$15 \cdot x = 60 \cdot 1$$

$$\frac{\cancel{15}x}{\cancel{15}} = \frac{60}{15}$$

$$x = 4 \text{ tbsp}$$

So, 60 mL = 4 tbsp or 4 tablespoonfuls are 60 milliliters.

3. How many teaspoonfuls are in 8 fluidrams?

Solution

From Table 10.7, 1 tsp $= 1\frac{1}{3} fʒ$.

Let $x =$ unknown number of teaspoonfuls.

Set up the proportion:

$$\frac{x \text{ tsp}}{8 f ʒ} = \frac{1 \text{ tsp}}{1\frac{1}{3} f ʒ}$$

$$\frac{4}{3} \cdot x = 8 \cdot 1$$

$$\frac{\cancel{3}}{\cancel{4}} \cdot \frac{\cancel{4}}{\cancel{3}} \cdot x = \frac{3}{4} \cdot 8$$

$$x = 6 \text{ tsp}$$

There are 6 teaspoonfuls in 8 fluidrams or 6 tsp $= 8 f ʒ$.

4. How many glassfuls are in 16 fluidounces?

Solution

From Table 10.7, 1 glassful $= 8\ f℥$.

Let $x =$ unknown number of glassfuls.

Set up the proportion:

$$\frac{x \text{ glassfuls}}{16\ f℥} = \frac{1 \text{ glassful}}{8\ f℥}$$

$$8 \cdot x = 16 \cdot 1$$

$$\frac{\cancel{8} \cdot x}{\cancel{8}} = \frac{16}{8}$$

$$x = 2 \text{ glassfuls}$$

There are 2 glassfuls in 16 fluidounces or 2 glassfuls $= 16\ f℥$.

5. How many glassfuls are in 480 milliliters?

Solution

From Table 10.7, 1 glassful = 240.0 mL.

Let $x =$ unknown number of glassfuls.

Set up the proportion:

$$\frac{x \text{ glassfuls}}{480 \text{ mL}} = \frac{1 \text{ glassful}}{240 \text{ mL}}$$

$$240 \cdot x = 480 \cdot 1$$

$$\frac{\cancel{240} \cdot x}{\cancel{240}} = \frac{480}{240}$$

$$x = 2 \text{ glassfuls}$$

Thus, 2 glassfuls = 480 mL or there are 2 glassfuls in 480 milliliters. From Example 4, we can also tell that 2 glassfuls = 480 mL = 16 *f*℥.

THREE FORMULAS USED TO CALCULATE DOSAGES OF MEDICINE FOR CHILDREN

1. **Fried's Rule** (time of birth to age 2)

$$\text{Child's dose} = \frac{\text{age in months} \times \text{adult dose}}{150}$$

2. **Young's Rule** (ages 1 to 12)

$$\text{Child's dose} = \frac{\text{age in years} \times \text{adult dose}}{\text{age in years} + 12}$$

3. **Clark's Rule** (ages 2 and over)

$$\text{Child's dose} = \frac{\text{weight in pounds} \times \text{adult dose}}{150}$$

EXAMPLES

6. Use Fried's Rule to find a child's dose, given:

Child's age = 12 months
Adult dose = 5 mL of paregoric

Solution

Substituting into the formula gives:

$$\text{Child's dose} = \frac{12 \text{ months} \times 5 \text{ mL}}{150}$$

$$= \frac{60}{150}$$

$$= 0.4 \text{ mL}$$

The child's dose would be 0.4 mL of paregoric.

7. Use Young's Rule to find a child's dose, given:

$$\text{Child's age} = 30 \text{ months } (2.5 \text{ years})$$
$$\text{Adult dose} = 15 \text{ mL of castor oil}$$

Solution

Substituting into the formula gives

$$\text{Child's dose} = \frac{2.5 \text{ yrs} \times 15 \text{ mL}}{2.5 + 12}$$
$$= \frac{37.5}{14.5}$$
$$= 2.6 \text{ mL (rounded off)}$$

According to Young's Rule, the child's dose would be 2.6 mL of castor oil.

Exercises 10.4

Use Table 10.7 and proportions to find the following equivalent measures.

1. 5 gtt = ______ mL

2. 7 mL = ______ gtt

3. 4 tbsp = ______ mL

4. 3 tbsp = ______ *f*ʒ

5. 6 *f*ʒ = ______ tsp

6. 10 mL = ______ tsp

7. 0.18 mL = ______ gtt

8. $2\frac{1}{2}$ glassfuls = ______ mL

9. 5 tbsp = ______ *f*ʒ

10. 4 tbsp = ______ *f*℥

11. 360 mL = ______ glassfuls

12. 4 gtt = ______ ♏

13. 5 ♏ = ______ gtt

14. 30 gtt = ______ mL

15. 30 mL = ______ tsp

16. 6 *f*℥ = ______ glassfuls

17. 45 mL = ______ tbsp

18. 5 tsp = ______ *f*ʒ

19. 150 ♏ = ______ gtt

20. 6 glassfuls = ______ mL

Use Fried's Rule to find the child's dose from the given information.

21. Child's age: 9 months
Adult dose: 250 mL liquid Ampicillin

22. Child's age: 5 months
Adult dose: 30 mL milk of magnesia

23. Child's age: 20 months
Adult dose: 15 mL castor oil

Use Young's Rule to find the child's dose from the given information.

24. Child's age: 5 years
Adult dose: 250 mL liquid Ampicillin

25. Child's age: 8 years
Adult dose: 10 mL Robitussin®

26. Child's age: 6 years
Adult dose: 15 mL castor oil

Use Clark's Rule to find the child's dose from the given information.

27. Child's weight: 20 lb
Adult dose: 5 mL Polaramine® syrup

28. Child's weight: 60 lb
Adult dose: 250 mL penicillin

29. Child's weight: 75 lb
Adult dose: 15 ml castor oil

SUMMARY: CHAPTER 10

Tables presented throughout the chapter are not reproduced here.

Denominate numbers are numbers with units of measure attached.

Abstract numbers are numbers with no related measurement units.

To add like denominate numbers:

1. Write the numbers in column form so that like units are aligned.
2. Add the numbers in each column.
3. Simplify the resulting sum if necessary.

To subtract like denominate numbers:

1. Write the numbers in column form so that like units are aligned.
2. If necessary for subtraction, borrow 1 of the larger units and rewrite the top number.
3. Subtract the like units.

For exact conversions between Fahrenheit and Celsius,

$$C = \frac{5(F - 32)}{9} \qquad F = \frac{9 \cdot C}{5} + 32$$

Household fluid measures are based on the uses of household cooking and eating utensils. The **apothecaries' system of fluid measure** is used by pharmacists.

To use proportions to change from one system to another:

1. Find the corresponding equivalent units in the table of Fluid Measure Equivalents (Table 10.7).
2. Write x for the unknown quantity.
3. Write a proportion using the ratio of equivalent measures and the ratio of x to the known measure being changed.
4. Solve the proportion.

THREE FORMULAS USED TO CALCULATE DOSAGES OF MEDICINE FOR CHILDREN

1. **Fried's Rule** (time of birth to age 2)

$$\text{Child's dose} = \frac{\text{age in months} \times \text{adult dose}}{150}$$

2. **Young's Rule** (ages 1 to 12)

$$\text{Child's dose} = \frac{\text{age in years} \times \text{adult dose}}{\text{age in years} + 12}$$

3. **Clark's Rule** (ages 2 and over)

$$\text{Child's dose} = \frac{\text{weight in pounds} \times \text{adult dose}}{150}$$

REVIEW QUESTIONS: CHAPTER 10

Convert the following measures as indicated.

1. 2.75 ft = ______ in.

2. $6\frac{1}{2}$ yd = ______ ft

3. 74 in. = ______ ft

4. 2 ft = ______ yd

5. $4\frac{3}{4}$ lb = ______ oz

6. 1.3 t = ______ lb

7. 54 oz = ______ lb

8. 800 lb = ______ t

9. 3.2 hr = ______ min

10. $2\frac{5}{12}$ days = ______ hr

11. 168 sec = ______ min

12. 84 hr = ______ days

13. 1.6 pt = ______ fl oz

14. $2\frac{3}{8}$ gal = ______ qt

15. $3\frac{1}{4}$ qt = ______ gal

16. 22 fl oz ______ pt

Simplify the following mixed denominate numbers.

17. 5 ft 19 in.

18. 2 yds 5 ft 28 in.

19. 3 lb 22 oz

20. 2 t 3120 lb

21. 5 days 28 hr

22. 62 min 73 sec

23. 1 pt 40 fl oz

24. 2 gal 11 qt 3 pt

Add and simplify if necessary.

25.
```
   2 lb 4 oz
 + 3 lb 7 oz
```

26.
```
   12 hr 23 min 12 sec
 +  6 hr 21 min  4 sec
```

27.
```
   1 ft 5 in.
   2 ft 3 in.
 + 1 ft 9 in.
```

28.
```
   1 gal 2 qt 1 pt
   2 gal      1 pt
 +       3 qt 1 pt
```

Find each difference.

29.
```
   5 lb 12 oz
 − 3 lb  5 oz
```

30.
```
   3 gal 3 qt 1 pt
 − 1 gal 2 qt 1 pt
```

31.
```
   6 hr 12 min
 − 5 hr 23 min
```

32.
```
   4 yd 2 ft 5 in.
 − 1 yd 2 ft 9 in.
```

Use the appropriate formula to convert the degrees as indicated.

33. 72° F = ______ ° C

34. 3° C = ______ ° F

Use the appropriate table to convert the following measures as indicated.

35. 5 in. = ______ cm

36. 2 ft 3 in. = ______ cm (Use in. to cm equivalents.)

37. 25.4 cm = ______ in.

38. 20 cm = ______ in.

39. $1\frac{1}{2}$ yd = ______ m

40. 2 ft 3 in. = ______ m (Use ft to m equivalents.)

41. 7 m = ______ ft

42. 3.2 m = ______ yd

43. 200 mi = ______ km

44. $12\frac{3}{4}$ mi = ______ km

45. 88 km = ______ mi

46. 0.4 km = ______ mi

47. 6 acres = ______ ha

48. 5 yd^2 = ______ m^2

49. 2.1 m^2 = ______ ft^2

50. $\frac{3}{5}$ ha = ______ acres

51. 40 lb = ______ kg

52. 1 lb 2 oz = ______ g

53. 454 g = ______ oz

54. 0.08 kg = ______ lb

55. $5\frac{1}{3}$ tbsp = ______ mL

56. 55 mL = ______ tsp

57. How many fluidounces are in 7 tablespoons?

58. How many teaspoons are in 10 fluidrams?

Use either Fried's Rule, Young's Rule, or Clark's Rule to find the child's dose from the given information.

59. Child's age: 9 months
Adult dose: 250 mL penicillin

60. Child's age: 4 years
Adult dose: 10 mL Robitussin®

61. Child's weight: 65 lb
Adult dose: 30 mL milk of magnesia

62. Child's age: 12 years
Adult dose: 15 mL castor oil

TEST: CHAPTER 10

Convert the following measures as indicated. Use the appropriate table where necessary.

1. $5\frac{3}{4}$ lb = ______ oz
2. 3 ft 8 in. = ______ yd
3. 1 day 4 hr = ______ min
4. 4 yd 2 ft = ______ m (Use ft to m equivalents.)
5. 50 cm = ______ in.
6. 1 lb 4 oz = ______ g (Use oz to g equivalents.)
7. 7.8 kg = ______ lb
8. 5 qt 1 pt = ______ L (Use qt to L equivalents.)
9. 25 L = ______ gal
10. $16\frac{3}{5}$ acres = ______ ha
11. 200 km = ______ mi
12. 10.5 gtt = ______ mL
13. 30 tsp = ______ *f*ʒ
14. Use the formula $C = \frac{5(F - 32)}{9}$ to convert 68° F to degrees Celsius.
15. Simplify the following mixed denominate number:

    1 gal 4 qt 3 pt 18 fl oz

Add the following and simplify if necessary.

16. 2 days 5 hr 7 min
    +5 days 7 hr 41 min

17. 2 ft 9 in.
    +1 ft 4 in.

Find the following differences.

18. 3 qt 1 pt 12 fl oz
    −1 qt 1 pt 5 fl oz

19. 10 lb 4 oz
    − 8 lb 12 oz

20. Use the appropriate rule for calculating the child's dose from the following information:

    Child's weight: 45 lb
    Adult dose: 250 mL liquid Ampicillin

11 Signed Numbers

11.1 NUMBER LINES AND ABSOLUTE VALUE

Draw a horizontal line, choose any point on the line, and label it with the number 0. (See Figure 11.1.)

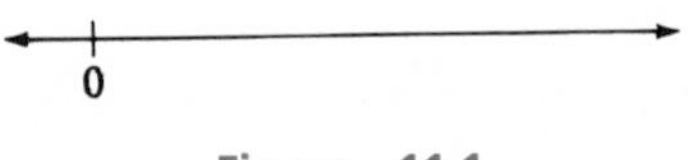

Figure 11.1

Now choose another point on the same line to the right of 0 and label it with the number 1 (Figure 11.2).

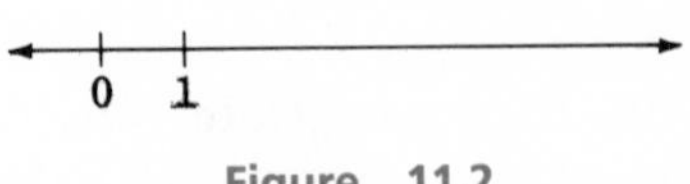

Figure 11.2

The line in Figure 11.2 is called a **number line.** The points corresponding to the whole numbers are determined by the distance between 0 and 1. The point corresponding to 2 is the same distance to the right of 1 as 1 is from 0, 3 is the same distance to the right of 2, and so on (Figure 11.3).

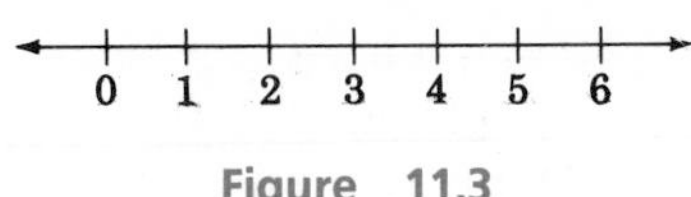

Figure 11.3

The **graph** of a number is indicated with a shaded dot at the corresponding point on the number line. The point or dot is the graph of the number, and the number is the **coordinate** of the point. We will use the terms **number** and **point** interchangeably. Thus, "number 3" and "point 3" are considered to have the same meaning.

The numbers 0, 1, and 5 form a set. Sets are indicated with braces and are named with capital letters. Thus, we can say $A = \{0, 1, 5\}$. The graph of A is shown in Figure 11.4.

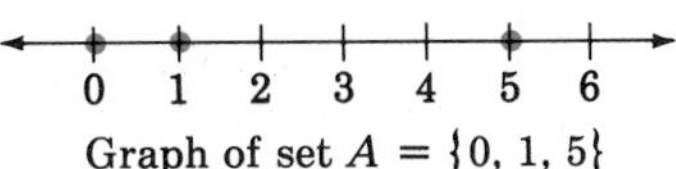

Graph of set $A = \{0, 1, 5\}$

Figure 11.4

On a horizontal number line, the point one unit to the left of 0 is the **opposite of 1.** It is also called **negative 1** and is symbolized -1. (The $-$ sign is called a negative sign.) The point two units to the left of 0 is the **opposite of 2** or **negative 2,** symbolized -2; and so on. (See Figure 11.5.)

The opposite of 1 is -1;
the opposite of 2 is -2;
the opposite of 3 is -3;
and so on.

The opposite of -1 is $-(-1) = +1$;
the opposite of -2 is $-(-2) = +2$;
the opposite of -3 is $-(-3) = +3$;
and so on.

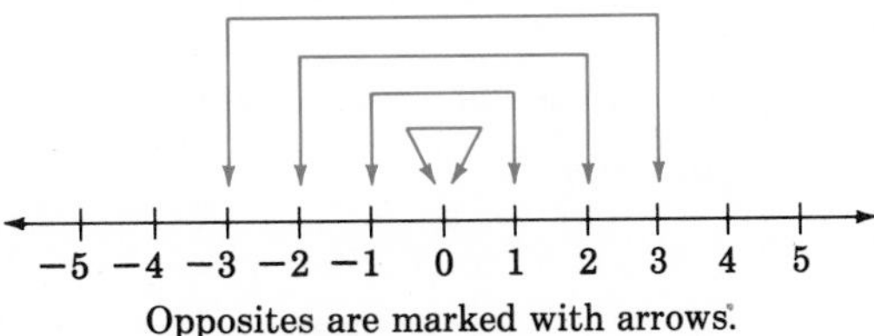

Opposites are marked with arrows.

Figure 11.5

The set of numbers consisting of the whole numbers and their opposites is called the set of **integers.** The counting numbers are called **positive integers.** (With positive integers the plus sign, $+$, is optional.

Thus, $+3$ and 3 both represent positive 3.) The opposites of the positive integers are called **negative integers.** **The number 0 is neither positive nor negative, and 0 is its own opposite.** $(0 = -0)$ (See Figure 11.6.) Note that the opposite of a positive integer is a negative integer, and the opposite of a negative integer is a positive integer.

Integers	$\ldots, -3, -2, -1, 0, 1, 2, 3, \ldots$
Positive Integers	$1, 2, 3, 4, 5, \ldots$
Negative Integers	$\ldots, -4, -3, -2, -1$

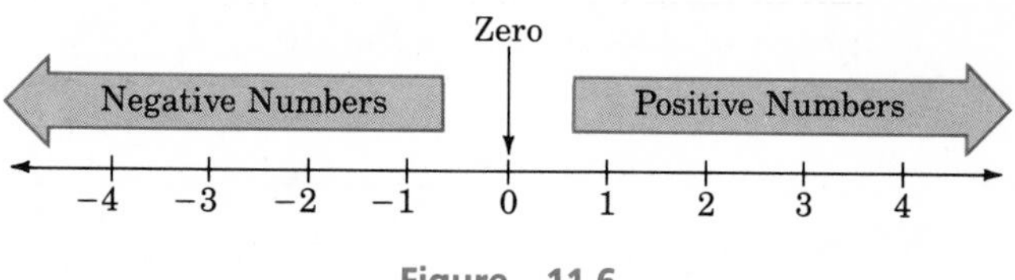

Figure 11.6

EXAMPLES

1. Find the opposite of 8.

   Solution -8

2. Find the opposite of -2.

   Solution

$-(-2)$	read "the opposite of negative 2"
or $+2$	read "positive 2"
or 2	A number is understood to be positive if there is no sign in front of it.

3. Graph the set $B = \{-2, -1, 3\}$.

   Solution

   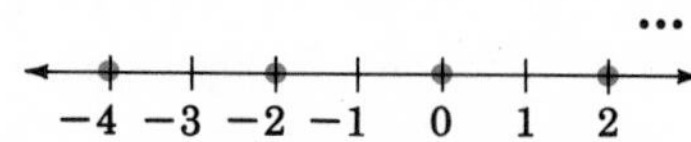

4. Graph the set $\{-4, -2, 0, 2, 4, \ldots\}$.

   Solution

   −4 −3 −2 −1 0 1 2 ...

The integers are not the only numbers that can be represented on a number line. Fractions and decimal numbers such as $\frac{1}{2}, \frac{2}{3}, -\frac{5}{3}, -2.2,$ and 1.7 can also be represented. (See Figure 11.7.)

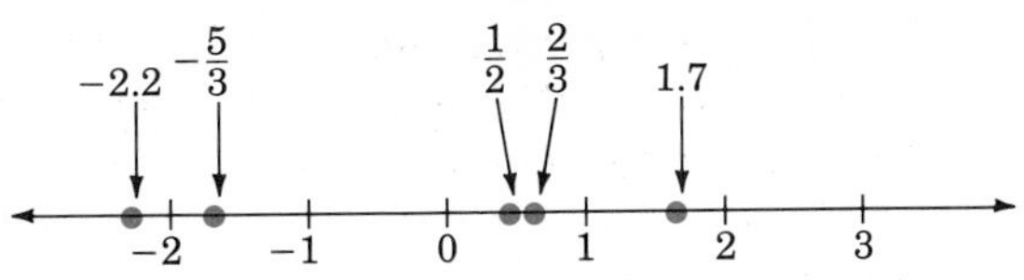

Figure 11.7

All the numbers that can be represented on a number line are called **signed numbers.** Signed numbers include all the integers, fractions, and decimals, as well as a type of number called **irrational numbers.** These numbers will be discussed in Chapter 13.

On a horizontal number line, **smaller numbers are always to the left of larger numbers.** Each number is smaller than any number to its right and larger than any number to its left. Two symbols used to indicate order are

	$<$	read "is less than"
and	$>$	read "is greater than"

USING $<$		*or*	*USING* $>$	
$\frac{1}{4} < \frac{3}{4}$	$\frac{1}{4}$ is less than $\frac{3}{4}$		$\frac{3}{4} > \frac{1}{4}$	$\frac{3}{4}$ is greater than $\frac{1}{4}$
$0 < 5$	0 is less than 5		$5 > 0$	5 is greater than 0
$-3 < 1$	-3 is less than 1		$1 > -3$	1 is greater than -3
$-6.1 < -2.4$	-6.1 is less than -2.4		$-2.4 > -6.1$	-2.4 is greater than -6.1

SYMBOLS FOR ORDER

$=$	is equal to	$\neq$	is not equal to
$<$	is less than	$>$	is greater than
$\leq$	is less than or equal to	$\geq$	is greater than or equal to

EXAMPLES

5. Determine whether each of the following statements is true or false.

(a) $6 < 10$	True, since 6 is less than 10.
(b) $2 > -1$	True, since 2 is greater than -1.
(c) $5 \geq 5$	True, since 5 is equal to 5.
(d) $-2.3 \leq 0$	True, since -2.3 is less than 0.
(e) $-3 < -10$	False, since -3 is greater than -10.

6. Graph the set of numbers $\left\{-\frac{1}{2}, 0, 1.5, 2\right\}$ on a number line.

Solution

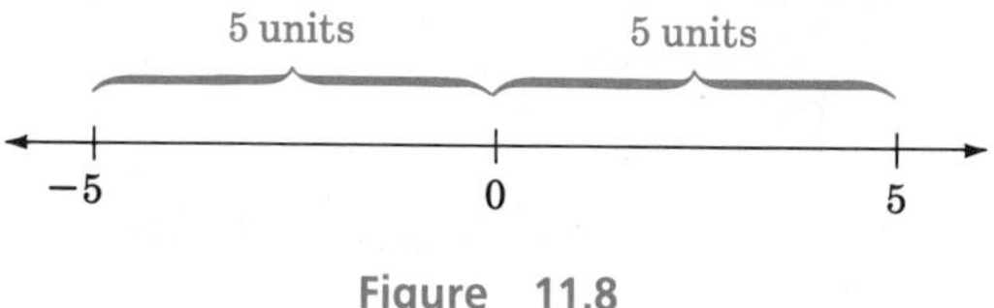

On a number line, any number and its opposite lie the same number of units from 0. For example, both $+5$ and -5 are five units from 0. (See Figure 11.8.) The $+$ and $-$ signs indicate direction and the 5 indicates distance.

5 units 5 units

−5 0 5

Figure 11.8

The distance a number is from 0 on a number line is called its **absolute value** and is symbolized by two vertical bars, $|\ |$. Thus $|+5| = 5$ and $|-5| = 5$. Figure 11.9 illustrates this concept.

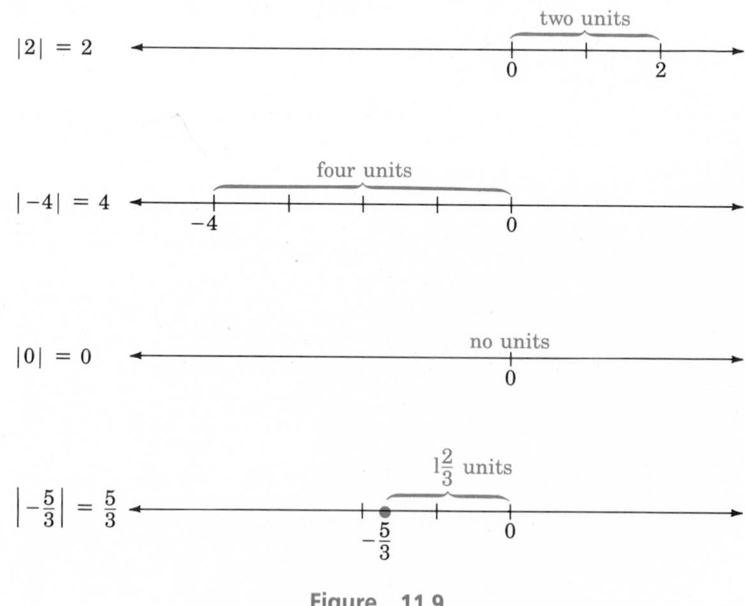

Figure 11.9

The absolute value of a number is never negative.

EXAMPLES

7. $|5.6| = 5.6$

8. $|-3.2| = 3.2$

9. True or False: $|-5| \leq 5$

Solution

True, since $|-5| = 5$ and $5 \leq 5$.

10. True or False: $\left|-3\frac{1}{2}\right| > 3\frac{1}{2}$.

Solution

False, since $\left|-3\frac{1}{2}\right| = 3\frac{1}{2}$ and $3\frac{1}{2}$ is not greater than $3\frac{1}{2}$.

11. If $|x| = 3$, what are the possible values for x?

Solution

$$x = 3 \quad \text{or} \quad x = -3 \qquad \text{since } |3| = 3 \text{ and } |-3| = 3$$

12. If $|x| = -2.1$, what are the possible values for x?

Solution

There are no values for x for which $|x| = -2.1$. The absolute value can never be negative.

13. (a) True or False: $-6 > -13$.

Solution

True, since -6 is to the right of -13 on the number line.

(b) True or False: $|-6| > |-13|$.

Solution

False, since $|-6| = 6$ and $|-13| = 13$ and 6 is less than 13.

Exercises 11.1

Find the opposite of each number.

1. 14 **2.** 12 **3.** -10 **4.** -9 **5.** $1\frac{1}{3}$

6. $2\frac{1}{4}$ **7.** -5.3 **8.** -7.1 **9.** 30 **10.** 40

In each of the following exercises, draw a number line and graph the given set of numbers.

11. $\{0, 1, 2\}$

12. $\{0, 2, 4\}$

13. $\{-3, -1, 1\}$

14. $\{-3, -2, 0\}$

15. $\{1, 2, 3, \ldots, 10\}$

16. $\{-2, -1, 0, \ldots, 5\}$

17. $\{|-5|, |-2|, 0\}$

18. $\{|-3|, 0, 1\}$

19. $\left\{-2\frac{1}{2}, -1.3, -\frac{1}{8}, 0.75\right\}$

20. $\left\{-1\frac{3}{4}, -0.25, 0.125, 1\right\}$

Fill in the blank with the appropriate symbol, $<$, $>$, or $=$.

21. 5 ______ 3

22. -4 ______ 1

23. -3 ______ $-(-6)$

24. 7 ______ $-(-7)$

25. $\frac{2}{3}$ ______ $\frac{1}{2}$

26. $\frac{3}{4}$ ______ $\frac{1}{3}$

27. $-\frac{2}{3}$ ______ $-\frac{1}{2}$

28. $-\frac{3}{4}$ ______ $-\frac{1}{3}$

29. $|-2.3|$ ______ 2.3

30. $|4.6|$ ______ -4.6

Determine whether each statement is true or false.

31. $0 = -0$

32. $-10 < -8$

33. $-8 > -7.4$

34. $-6 \leq -6$

35. $-5.2 \leq -5.2$

36. $4 \leq -13$

37. $\left|3\frac{1}{2}\right| = -3\frac{1}{2}$

38. $|-4| \leq 4$

39. $-2.6 \leq |-2.6|$

40. $|-7| \geq 7$

41. $-7.1 < -8.1$

42. $-10.3 \leq -12$

Find the possible values for x in each of the following equations.

43. $|x| = 6$

44. $|x| = 9$

45. $|x| = 4.3$

46. $|x| = \frac{3}{2}$

47. $|x| = -2$

48. $|x| = -5$

49. $|x| = \frac{2}{3}$

50. $|x| = \frac{1}{4}$

11.2 ADDITION WITH SIGNED NUMBERS

The sum of two signed numbers can be indicated with a plus (+) sign between the two numbers. For example,

$$\underset{\text{positive 2}}{\underset{\uparrow}{(+2)}} \quad \underset{\text{plus}}{\underset{\uparrow}{+}} \quad \underset{\text{positive 7}}{\underset{\uparrow}{(+7)}}$$

The sum in this example is obviously +9 since we already know how to add 2 and 7:

$$(+2) + (+7) = +9$$

But the sums

$$(+2) + (-7) = ?$$
$$(-2) + (+7) = ?$$
and
$$(-2) + (-7) = ?$$

are not so obvious.

By using a number line, we can develop an intuitive idea of how to add signed numbers. Start at the first number to be added, then

1. Move right if the second number is positive, or
2. Move left if the second number is negative.

The distance moved is the absolute value of the second number.

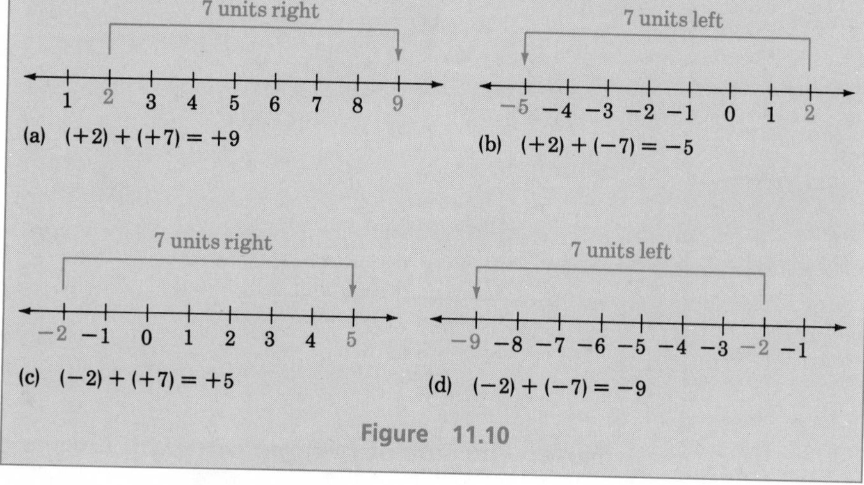

(a) $(+2) + (+7) = +9$ (b) $(+2) + (-7) = -5$

(c) $(-2) + (+7) = +5$ (d) $(-2) + (-7) = -9$

Figure 11.10

EXAMPLES

Find the following sums, using a number line as illustrated in Figure 11.10.

1. $(-5) + (+4) = ?$

Solution

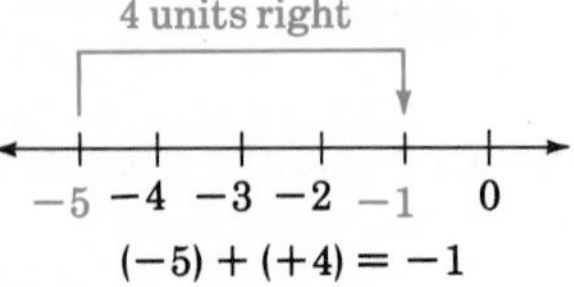

2. $(-3) + (-8) = ?$

Solution

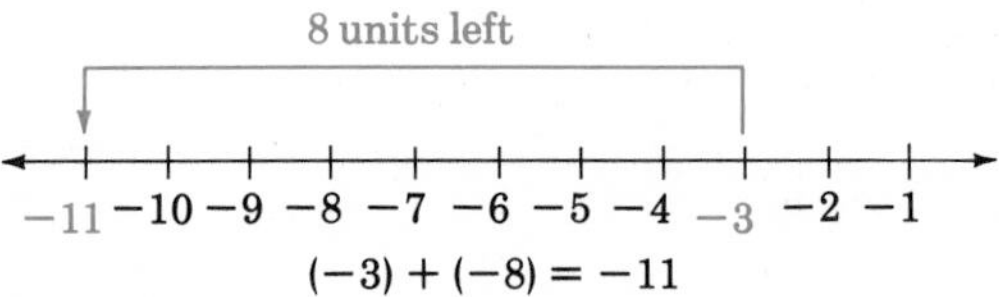

3. $(+6) + (-10) = ?$

Solution

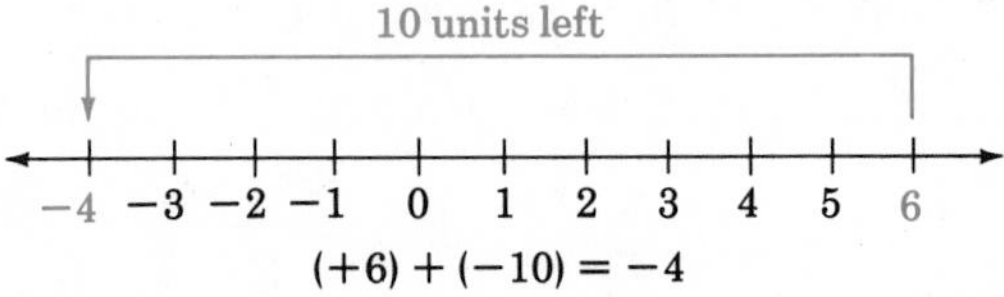

4. $(-7) + (+7) = ?$

Solution

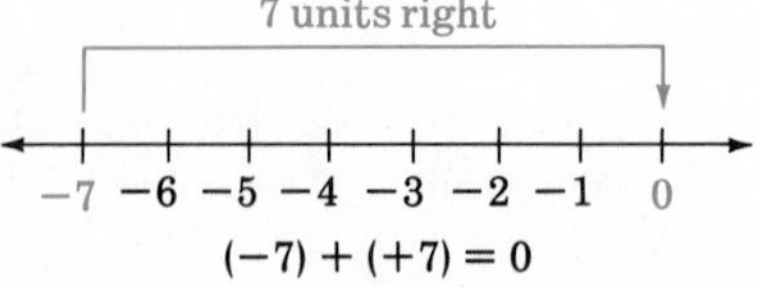

NOTE: The sum of two opposites will always be 0.

5. $(-8.6) + (+5.2) = ?$

Solution

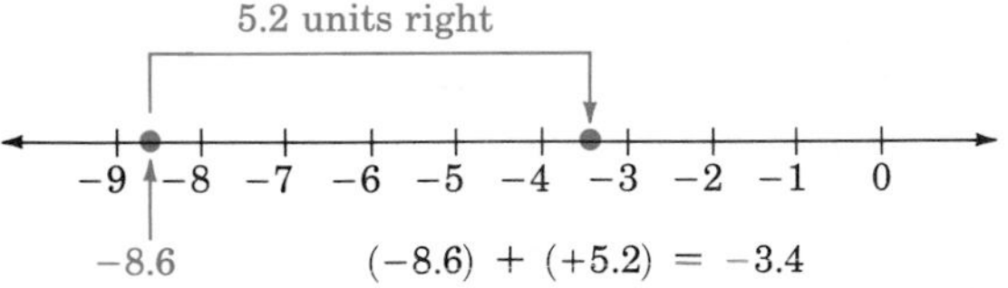

Signed numbers can be added mentally, without the use of number lines, once you know the following rules.

RULES FOR ADDING SIGNED NUMBERS

1. **To add two signed numbers with like signs,** add their absolute values and use the common sign.

$$(+6) + (+5) = +[|+6| + |+5|] = +[6 + 5] = +11$$
$$(-2) + (-8) = -[|-2| + |-8|] = -[2 + 8] = -10$$

2. **To add two signed numbers with unlike signs,** subtract their absolute values (smaller from larger) and use the sign of the number with the larger absolute value.

$$(+10) + (-15) = -[|-15| - |+10|] = -[15 - 10] = -5$$
$$(+12) + (-9) = +[|+12| - |-9|] = +[12 - 9] = +3$$

EXAMPLES

6. $(+4) + (-1) = +(4 - 1) = +3$ (unlike signs)

7. $(-5) + (-3) = -(5 + 3) = -8$ (like signs)

8. $(+7) + (+10) = +(7 + 10) = +17$ (like signs)

9. $(+6) + (-19) = -(19 - 6) = -13$ (unlike signs)

10. $(+30) + (-3) = (30 - 30) = 0$ (opposites)

11. $(+5.8) + (-7.2) = -(7.2 - 5.8) = -1.4$ (unlike signs)

12. $\left(-3\frac{1}{2}\right) + \left(-4\frac{1}{4}\right) = -\left(3\frac{1}{2} + 4\frac{1}{4}\right) = -7\frac{3}{4}$ (like signs)

If more than two numbers are to be added, add any two, then add their sum to one of the other numbers until all numbers have been added.

In algebra, equations are written horizontally, so the ability to work with sums written horizontally is very important. However, there are some situations (as in long division) where sums are written vertically. The rules for adding are the same whether the numbers are written horizontally or vertically.

EXAMPLES

13. $(+9) + (+2) + (-6) = +|9 + 2| + (-6)$
$= +|11| + (-6)$
$= +5$

14.
$$\begin{array}{r} -12 \\ +8 \\ \hline -4 \end{array}$$

15.
$$\begin{array}{r} -5 \\ 6 \\ -14 \\ \hline -13 \end{array}$$

16.
$$\begin{array}{r} -1.4 \\ -2.3 \\ -4.9 \\ \hline -8.6 \end{array}$$

PRACTICE QUIZ

Find each sum.

	ANSWERS
1. $(-10) + (+3) =$	**1.** -7
2. $(-5) + (+9) =$	**2.** $+4$
3. $(-2) + (-3) =$	**3.** -5
4. $(+1.6) + (-6.3) =$	**4.** -4.7

Exercises 11.2

Find each sum.

1. $(+6) + (-4)$ **2.** $(+8) + (-7)$ **3.** $(4) + (+6)$

4. $(5) + (-8)$ **5.** $(16) + (+3)$ **6.** $(-8) + (-2)$

7. $(-3) + (-6)$ **8.** $(-2) + (+2)$ **9.** $(+4) + (-4)$

10. $(13) + (12)$ **11.** $(6) + (-10)$ **12.** $(14) + (-17)$

13. $(+5) + (-3)$ **14.** $(+15) + (-18)$ **15.** $(-4) + (-12)$

16. $(-8) + (+8)$ **17.** $(+2) + (-6)$ **18.** $(-9) + (+5)$

19. $(-16) + (+3) + (+13)$ **20.** $(-5) + (+5) + (14)$

21. $(-1) + (-2) + (+7)$ **22.** $(+3) + (-4) + (-5)$

23. $(+6) + (+3) + (+5)$ **24.** $(-18) + (-5) + (-7)$

25. $(-1) + (+2) + (-4) + (+2)$

26. $\begin{array}{r} -4 \\ \underline{+8} \end{array}$ **27.** $\begin{array}{r} -5 \\ \underline{-10} \end{array}$ **28.** $\begin{array}{r} -13 \\ \underline{-6} \end{array}$ **29.** $\begin{array}{r} +16 \\ \underline{+25} \end{array}$ **30.** $\begin{array}{r} +14 \\ \underline{-8} \end{array}$

31. $\begin{array}{r} +20 \\ \underline{-7} \end{array}$ **32.** $\begin{array}{r} +2 \\ -5 \\ \underline{-3} \end{array}$ **33.** $\begin{array}{r} +8 \\ +3 \\ \underline{-1} \end{array}$ **34.** $\begin{array}{r} +10 \\ -4 \\ \underline{+2} \end{array}$ **35.** $\begin{array}{r} -16 \\ -8 \\ \underline{+12} \end{array}$

36. $\begin{array}{r} -15 \\ -20 \\ \underline{-6} \end{array}$ **37.** $\begin{array}{r} -4 \\ -17 \\ \underline{+11} \end{array}$ **38.** $\begin{array}{r} +13 \\ -5 \\ +17 \\ \underline{-25} \end{array}$ **39.** $\begin{array}{r} +14 \\ -14 \\ +37 \\ \underline{-37} \end{array}$ **40.** $\begin{array}{r} -8 \\ -5 \\ -13 \\ \underline{-22} \end{array}$

41. $\begin{array}{r} -100 \\ -50 \\ \underline{-85} \end{array}$ **42.** $\begin{array}{r} -96 \\ +14 \\ \underline{-83} \end{array}$ **43.** $\begin{array}{r} +750 \\ -632 \\ -198 \\ \underline{+200} \end{array}$ **44.** $\begin{array}{r} -300 \\ +450 \\ +325 \\ \underline{+500} \end{array}$ **45.** $\begin{array}{r} -25 \\ -95 \\ -48 \\ -20 \\ \underline{+67} \end{array}$

46. $\left(-3\frac{1}{6}\right) + \left(-2\frac{1}{3}\right)$ **47.** $\left(-1\frac{3}{5}\right) + \left(-2\frac{1}{5}\right)$

48. $\left(-10\frac{3}{4}\right) + \left(-9\frac{3}{8}\right)$ **49.** $\left(-4\frac{1}{5}\right) + \left(+6\frac{3}{10}\right)$

50. $\left(+2\frac{1}{2}\right) + \left(-7\frac{3}{4}\right)$ **51.** $(-10.3) + (-8.51)$

52. $(-25.6) + (+30)$ **53.** $(-18.4) + (+20)$

54. $\begin{array}{r} 10.4 \\ \underline{-16.5} \end{array}$ **55.** $\begin{array}{r} -33.82 \\ \underline{-35.46} \end{array}$ **56.** $\begin{array}{r} -16.9 \\ \underline{14.3} \end{array}$ **57.** $\begin{array}{r} +25.4 \\ \underline{-19.5} \end{array}$

58. $\begin{array}{r} -10.1 \\ -3.4 \\ \underline{5.5} \end{array}$ **59.** $\begin{array}{r} 8.6 \\ -2.5 \\ \underline{-7.2} \end{array}$ **60.** $\begin{array}{r} +24.3 \\ -18.5 \\ \underline{+16.1} \end{array}$

11.3 SUBTRACTION WITH SIGNED NUMBERS

In the intuitive discussion of addition using number lines, we

1. Moved right if the second number was positive, or
2. Moved left if the second number was negative.

With **subtraction,** we simply reverse the moves. That is, we

1. Move left if the second number is positive, or
2. Move right if the second number is negative.

In other words, in subtraction, move in the opposite direction from the direction indicated by the number being subtracted. (See Figure 11.11.)

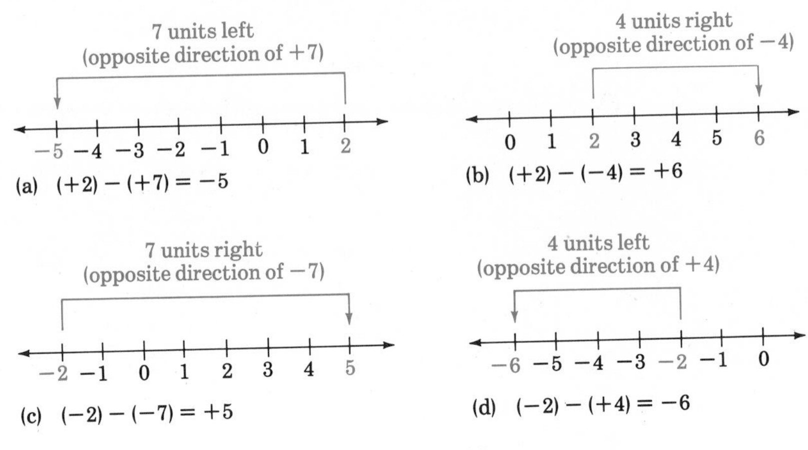

Figure 11.11

We can restate the previous discussion and the moves outlined in Figure 11.10 in terms of addition. **In subtraction, add the opposite of the number being subtracted.** For example,

$(+4)$	$-$	$(+6)$	$=$	$(+4)$	$+$	(-6)	$= -2$
↑	↑	↑		↑	↑	↑	
positive 4	minus	positive 6		positive 4	plus	opposite of +6	

$(+5)$	$-$	(-8)	$=$	$(+5)$	$+$	$(+8)$	$= +13$
↑	↑	↑		↑	↑	↑	
positive 5	minus	negative 8		positive 5	plus	opposite of −8	

DEFINITION

The **difference** of two integers a and b is the sum of a and the opposite of b. Symbolically, $(a) - (b) = (a) + (-b)$.

EXAMPLES

1. $(+2) - (-6) = (+2) + (+6) = +8$

   ↑ minus (under −); ↑ plus (under +); ↑ opposite of −6 (under (+6))

2. $(-3) - (-7) = (-3) + (+7) = +4$

   ↑ opposite of −7 (under (+7))

3. $(-5) - (+2) = (-5) + (-2) = -7$
   ↑ opposite of $+2$

4. $(-9) - (-9) = (-9) + (+9) = 0$
   ↑ opposite of -9

The numbers may also be written vertically, one underneath the other. In this case, change the sign of the number being subtracted (the bottom number), then add.

EXAMPLES

	SUBTRACT	*(ADD)*
5.	$\begin{array}{r} -10 \\ -3 \\ \hline \end{array}$	$\begin{array}{r} -10 \\ +3 \\ \hline -7 \end{array}$
6.	$\begin{array}{r} +14 \\ +9 \\ \hline \end{array}$	$\begin{array}{r} +14 \\ -9 \\ \hline 5 \end{array}$
7.	$\begin{array}{r} -8.6 \\ +4.3 \\ \hline \end{array}$	$\begin{array}{r} -8.6 \\ -4.3 \\ \hline -12.9 \end{array}$

We have been using a plus (+) sign for addition, a minus (−) sign for subtraction, and parentheses around each number being added or subtracted. In another notation more commonly used in algebra, the parentheses and the plus (+) and minus (−) signs are dropped. **The numbers are written horizontally, and the problem is considered as adding positive and negative numbers.** All positive and negative signs between numbers must be included.

If the first number to the left is positive, the + sign may be omitted with the understanding that the number is positive. For example,

$$9 - 12 \text{ is the same as } (+9) + (-12)$$

So,

$$9 - 12 = (+9) + (-12) = -3$$

After some practice, you will be able to do the second step mentally and will simply write

$$9 - 12 = -3$$

EXAMPLES

8. $7 - 3 = (+7) + (-3) = 4$

or

$7 - 3 = 4$

9. $-13 + 6 = (-13) + (+6) = -7$

or

$-13 + 6 = -7$

10. $-4 - 8 = (-4) + (-8) = -12$

or

$-4 - 8 = -12$

11. $6 - 9 = -3$

12. $-20 + 12 = -8$

13. $-3 - 14 = -17$

14. $16.3 - 4.1 - 5.7 = 12.2 - 5.7 = 6.5$

15. $$\begin{aligned} -14.3 - 20 + 5 + 1.8 &= -34.3 + 5 + 1.8 \\ &= -29.3 + 1.8 \\ &= -27.5 \end{aligned}$$

PRACTICE QUIZ

Evaluate each of the following expressions.

	ANSWERS
1. $(+8) - (-3)$	**1.** 11
2. $(-4.1) - (-5.7)$	**2.** 1.6
3. $-10 + 4$	**3.** -6
4. $13 - 5 + 6.1$	**4.** 14.1

Exercises 11.3

Find each difference.

1. $(+5) - (+2)$ **2.** $(+16) - (+3)$ **3.** $(+8) - (-3)$

4. $(+12) - (-4)$ **5.** $(-5) - (+2)$ **6.** $(-10) - (+3)$

7. $(-10) - (-1)$ **8.** $(-15) - (-1)$ **9.** $(-3) - (-7)$

10. $(-2) - (-12)$ **11.** $(-4) - (+6)$ **12.** $(-9) - (+13)$

13. $(-13) - (-14)$ **14.** $(-12) - (-15)$ **15.** $(+9) - (-9)$

16. $(+11) - (-11)$ **17.** $(+15) - (-2)$ **18.** $(+20) - (-3)$

19. $(-17) - (+14)$ **20.** $(-16) - (+10)$ **21.** $(+3) - (+8)$

22. $(+1) - (+5)$ **23.** $(-5) - (-5)$ **24.** $(-7) - (-7)$

25. $(+7) - (+12)$ **26.** $(+8.4) - (+10.5)$ **27.** $(-7.31) - (5.2)$

28. $(-6.3) - (-6.3)$

29. $\left(-11\frac{7}{10}\right) - \left(-11\frac{7}{10}\right)$ **30.** $\left(4\frac{1}{2}\right) - \left(-2\frac{3}{4}\right)$

Subtract the bottom number from the top number.

31. $\begin{array}{r} 18 \\ -12 \\ \hline \end{array}$ **32.** $\begin{array}{r} 24 \\ 16 \\ \hline \end{array}$ **33.** $\begin{array}{r} -8 \\ -12 \\ \hline \end{array}$ **34.** $\begin{array}{r} -13 \\ -18 \\ \hline \end{array}$ **35.** $\begin{array}{r} -4 \\ +5 \\ \hline \end{array}$

36. $\begin{array}{r} 32 \\ -48 \\ \hline \end{array}$ **37.** $\begin{array}{r} -6.2 \\ -30.1 \\ \hline \end{array}$ **38.** $\begin{array}{r} -25.6 \\ -13.7 \\ \hline \end{array}$ **39.** $\begin{array}{r} -45.18 \\ -16.39 \\ \hline \end{array}$ **40.** $\begin{array}{r} 28.76 \\ -15.81 \\ \hline \end{array}$

Evaluate each of the following expressions.

41. $6 + 2$ **42.** $4 + 8$ **43.** $7 - 1$ **44.** $9 - 4$

45. $4 + 6$ **46.** $8 + 9$ **47.** $-3 - 1$ **48.** $-2 - 6$

49. $12 - 6$ **50.** $9 - 3$ **51.** $-13 + 4$ **52.** $-20 + 14$

53. $-10 + 9$ **54.** $-18 + 3$ **55.** $24 - 32$ **56.** $14 - 17$

57. $-12 - 6$ **58.** $-2 - 8$ **59.** $-15 + 18$ **60.** $-25 + 30$

61. $-20 + 21$ **62.** $-30 + 32$ **63.** $-7 + 7$ **64.** $-6 + 6$

65. $18 - 3$ **66.** $-4 + 16 - 8$ **67.** $-5 + 12 - 3$

68. $-20 - 2 + 6$ **69.** $14 - 5 - 12$ **70.** $13.1 + 15.2 - 6$

71. $-6.4 - 8.5 - 13.7$ **72.** $-4.1 - 7.3$ **73.** $30\frac{1}{2} + 12 - 18\frac{1}{4}$

74. $16\frac{1}{5} + 4\frac{3}{5} - 20\frac{3}{10}$ **75.** $15.63 + 6.14 - 21.77$

76. $13 - 4 + 6 - 5$ **77.** $16 - 3 - 7 - 1$

78. $19 - 5 - 8 - 6$ **79.** $-4 + 10 - 12 + 1$

80. $-8 + 14 - 10 + 3$

11.4 MULTIPLICATION AND DIVISION WITH SIGNED NUMBERS

Multiplication with whole numbers is shorthand for repeated addition. For example,

$$7 + 7 + 7 + 7 = 4 \cdot 7 = 28$$
$$3 + 3 + 3 + 3 + 3 + 3 = 6 \cdot 3 = 18$$

Multiplication with signed numbers can also be considered shorthand for repeated addition.

$$(-7) + (-7) + (-7) = 3(-7) = -21$$
$$(-1.5) + (-1.5) + (-1.5) + (-1.5) = 4(-1.5) = -6.0$$

Repeated addition of a negative number results in the product of a positive number and a negative number. Since the sum of negative numbers is negative, the product of a positive number with a negative number will be negative. In fact, **the product of any positive number with a negative number will be negative.**

EXAMPLES

1. $4(-3) = (-3) + (-3) + (-3) + (-3) = -12$
2. $5(-2) = -10$
3. $3.1(-5) = -15.5$
4. $-\frac{1}{3}\left(\frac{1}{2}\right) = -\frac{1}{6}$

The product of two negative numbers is not explained in terms of repeated addition. The following patterns lead, intuitively, to the rule for multiplying two negative numbers. Try to supply the missing products.

$$\begin{aligned}(+3)(-5) &= -15\\ (+2)(-5) &= -10\\ (+1)(-5) &= -5\\ (0)(-5) &= 0\\ (-1)(-5) &= ?\\ (-2)(-5) &= ?\\ (-3)(-5) &= ?\end{aligned}$$

Did you notice that the products are increasing 5 at a time? If you did, your intuition should lead to

$$\begin{aligned}(-1)(-5) &= +5\\ (-2)(-5) &= +10\\ (-3)(-5) &= +15\end{aligned}$$

Although one example does not prove a rule, this time intuition is correct. **The product of two negative numbers is positive.**

EXAMPLES

5. $(-6)(-4) = +24$

6. $-7(-9) = +63$

7. $-2.3(-5.1) = +11.73$

8. $\left(-\frac{2}{3}\right)\left(-\frac{1}{5}\right) = +\frac{2}{15}$

9. $(0)(-12) = 0$

10. $(-2.7)(0) = 0$

As Examples 9 and 10 illustrate, **the product of 0 with any number is 0.**

RULES FOR MULTIPLYING SIGNED NUMBERS

1. The product of two positive numbers is positive.
2. The product of two negative numbers is positive.
3. The product of a positive number and a negative number is negative.
4. The product of 0 with any number is 0.

The same rules can be stated using symbols.

If a and b are positive numbers,

1. $a \cdot b = ab$
2. $(-a)(-b) = ab$
3. $a(-b) = -ab$
4. $a \cdot 0 = 0$

Now, consider a division problem, such as $42 \div 6$. We know that

$$\frac{42}{6} = 7 \qquad \text{because} \qquad 42 = 6 \cdot 7$$

That is, division is defined in terms of multiplication.

DEFINITION For any numbers a and b where $b \neq 0$,

$$\frac{a}{b} = x \quad \text{means} \quad a = b \cdot x$$

If a is any number, then $\frac{a}{0}$ is undefined. (We cannot divide by 0.)

Using the rules for multiplication with signed numbers, we can develop the rules for division with signed numbers.

EXAMPLES

11. $\frac{+28}{+4} = +7$ because $+28 = +4(+7)$.

12. $\frac{-28}{+4} = -7$ because $-28 = (+4)(-7)$.

13. $\frac{+28}{-4} = -7$ because $+28 = (-4)(-7)$.

14. $\frac{-28}{-4} = +7$ because $-28 = (-4)(+7)$.

The rules for division with signed numbers can be stated as follows.

RULES FOR DIVIDING SIGNED NUMBERS

1. The quotient of two positive numbers is positive.
2. The quotient of two negative numbers is positive.
3. The quotient of a positive number and a negative number is negative.

The rules can be stated using symbols.

If a and b are positive numbers,

1. $\frac{a}{b} = \frac{a}{b}$
2. $\frac{-a}{-b} = \frac{a}{b}$
3. $\frac{-a}{b} = -\frac{a}{b}$ and $\frac{a}{-b} = -\frac{a}{b}$

EXAMPLES

15. $\dfrac{21.7}{-7} = -3.1$

16. $\dfrac{-36.28}{-2} = +18.14$

17. $\dfrac{-\frac{3}{4}}{+\frac{1}{8}} = -\dfrac{3}{4}\left(\dfrac{8}{1}\right) = -\dfrac{6}{1} = -6$

If a product involves more than two signed numbers, multiply any two, then continue to multiply until all the numbers have been multiplied.

EXAMPLES

18. $(-3)(4)(-2) = [(-3)(4)](-2)$
$= [-12](-2)$
$= +24$

19. $(-5)(-3)(-10) = +15(-10)$
$= -150$

20. $(-2)^3(-6.1) = (-2)(-2)(-2)(-6.1)$
$= +4(-2)(-6.1)$
$= -8(-6.1)$
$= +48.8$

PRACTICE QUIZ	Find the value of each expression.	ANSWERS
	1. $(-5)(5)$	**1.** -25
	2. $(-3)(-3)(0)$	**2.** 0
	3. $\dfrac{-4}{0}$	**3.** undefined
	4. $\dfrac{-15.5}{-5}$	**4.** 3.1
	5. $(-4)^3$	**5.** -64

Exercises 11.4

Find the following products.

1. $5(-3)$ **2.** $4(-6)$ **3.** $-6(-4)$

4. $-2(-7)$ **5.** $-5(4)$ **6.** $-8(3)$

7. $14(2)$ **8.** $13(3)$ **9.** $-10(5)$

10. $-11(3)$
11. $(-7)3$
12. $(-2)9$
13. $6(-8)$
14. $9(-4)$
15. $-7(-9)$
16. $-8(-9)$
17. $0(-6)$
18. $0(-4)$
19. $(-6)(-5)(3)$
20. $(-2)(-1)(7)$
21. $4(-2)(-3)$
22. $5(-6)(-1)$
23. $(-5)(3)(-4)$
24. $(-3)(7)(-5)$
25. $(-7)(-2)(-3)$
26. $(-4)(-4)(-4)$
27. $(-3)(-3)(-5)$
28. $(-2)(-2)(-8)$
29. $(-3)(-4)(-5)$
30. $(-2)(-5)(-7)$
31. $(-5)(0)(-6)$
32. $(-6)(0)(-2)$
33. $(-1)^3$
34. $(-3)^3$
35. $(-2)^4$
36. $(-4)^3$

Find the following quotients.

37. $\frac{-12}{4}$
38. $\frac{-18}{2}$
39. $\frac{-14}{7}$
40. $\frac{-28}{7}$
41. $\frac{-20}{-5}$
42. $\frac{-30}{-3}$
43. $\frac{-50}{-10}$
44. $\frac{-30}{-5}$
45. $\frac{30}{-6}$
46. $\frac{40}{-8}$
47. $\frac{75}{-25}$
48. $\frac{80}{-4}$
49. $\frac{12}{6}$
50. $\frac{24}{8}$
51. $\frac{36}{9}$
52. $\frac{22}{11}$
53. $\frac{-39}{13}$
54. $\frac{27}{-9}$
55. $\frac{32}{-4}$
56. $\frac{23}{-23}$
57. $\frac{-34}{-17}$
58. $\frac{-60}{-15}$
59. $\frac{-8}{-8}$
60. $\frac{26}{-13}$
61. $\frac{-31}{0}$
62. $\frac{17}{0}$
63. $\frac{0}{-20}$
64. $\frac{0}{-16}$
65. $\frac{35}{0}$
66. $\frac{0}{25}$

Evaluate each of the following expressions.

67. $7(-2.8)$
68. $15(-3.1)$
69. $(-0.017)(-5.3)$
70. $(-7.3)(-0.26)$
71. $\frac{3.366}{-0.612}$
72. $\frac{-79.53}{-24.1}$
73. $\left(-\frac{2}{3}\right) \div \left(-\frac{9}{7}\right)$
74. $\left(-\frac{5}{8}\right) \div \left(+\frac{1}{2}\right)$
75. $\left(+\frac{3}{4}\right) \div \left(-\frac{3}{5}\right)$

11.5 ORDER OF OPERATIONS WITH SIGNED NUMBERS

The rules for order of operations are the same for expressions with signed numbers as they were stated in Section 2.2 for expressions with whole numbers. The rules are restated here for easy reference.

RULES FOR ORDER OF OPERATIONS

1. First, simplify within grouping symbols, such as parentheses ( ), brackets [ ], or braces { }. Start with the innermost grouping.
2. Second, find any powers indicated by exponents.
3. Third, moving from **left to right,** perform any multiplications or divisions in the order they appear.
4. Fourth, moving from **left to right,** perform any additions or subtractions in the order they appear.

EXAMPLES

Evaluate the following expressions using the rules for order of operations and your knowledge of positive and negative numbers.

1. $14.6 + 3(-2.5)$

Solution

$$14.6 + 3(-2.5) = 14.6 + (-7.5) \quad \text{Multiply first.}$$
$$= 7.1 \quad \text{Add.}$$

2. $-3(5 - 6) - 4.2$

Solution

$$-3(5 - 6) - 4.2 = -3(-1) - 4.2 \quad \text{Simplify within parentheses.}$$
$$= +3 - 8 \quad \text{Multiply, from left to right.}$$
$$= -5 \quad \text{Add (or subtract).}$$

3. $(-8)^2 \cdot 5 + 3.1(-2) \div \frac{1}{2}$

Solution

$$(-8)^2 \cdot 5 + 3.1(-2) \div \frac{1}{2} = 64 \cdot 5 + 3.1(-2) \div \frac{1}{2} \quad \text{exponents}$$
$$= 320 + (-6.2) \div \frac{1}{2} \quad \text{Multiply, from left to right.}$$
$$= 320 + (-6.2) \cdot \frac{2}{1} \quad \text{Divide.}$$
$$= 320 - 12.4$$
$$= 307.6 \quad \text{Subtract.}$$

4. $[5 + 3(-2 - 4)] \div 13$

Solution

$$\begin{aligned}[5 + 3(-2 - 4)] \div 13 &= [5 + 3(-6)] \div 13 \\ &= [5 - 18] \div 13 \\ &= [-13] \div 13 \\ &= -1\end{aligned}$$

Exercises 11.5

Evaluate the following expressions using the rules for order of operations.

1. $4 \cdot 3 - 5 \cdot 7$

2. $5 \cdot 3 - 4 \cdot 8$

3. $18 \cdot 2 + 6 \div (-2)$

4. $7 \cdot 5 + 12 \div (-3)$

5. $16 \div (-4) \cdot 2$

6. $24 \div (-8) \div 3$

7. $-18 \cdot 2 \div 3$

8. $-20 \cdot 15 \div 3$

9. $\frac{15}{-3} + 4(-8)$

10. $\frac{16}{-2} + 3(-5)$

11. $(+12)(-6) \div 3 \cdot 2$

12. $(+14)(-2) \div 7 \cdot 5$

13. $(-5)^2 + (-5)^3$

14. $(-4)^2 + (-4)^3$

15. $(16 - 25)(32 - 21)$

16. $(8 - 10)(15 - 12)$

17. $(8.3 - 4.1)(-3.2)$

18. $(6.4 - 5.2)(-1.6)$

19. $[4 + 3(-1 - 5)] \div 2$

20. $[10 + 2(-3 - 8)] \div 3$

21. $(6 \cdot 10 - 5) \div 11 \cdot 5 - 3^2$

22. $(4 \cdot 11 - 4) \div 10 \cdot 4 - 5^2$

did to here → **23.** $6^2 - 4 \cdot 2 + 5(7 + 3^2)$

24. $5^2 - 6 \cdot 3 + 2(5 + 2^3)$

25. $(1.3)^2 - (1.4)^2$

26. $(1.1)^2 - (1.2)^2$

27. $(-8.4) \div \left(-\frac{1}{2}\right) \cdot \frac{3}{4}$

28. $(-7.2) \div \left(-\frac{1}{3}\right) \cdot \frac{5}{8}$

29. $\left(-\frac{2}{3}\right) \div \left(-\frac{7}{8}\right) \div (-2)$

30. $\left(-\frac{3}{4}\right) \div \left(-\frac{5}{16}\right) \div (-3)$

Fill in the blanks with the correct term (positive, negative, 0, undefined).

31. The product of two negative numbers is ______.

32. The quotient of two negative numbers is ______.

33. The quotient of two positive numbers is ______.

34. The product of two positive numbers is ______.

35. The quotient of a positive number and a negative number is ______.

36. The product of three negative numbers is ______.

37. If x is any signed number, then $0 \cdot x$ is ______.

38. If x is any nonzero number, then $\frac{0}{x}$ is ______.

39. If x is any signed number, then $\frac{x}{0}$ is ______.

40. If x is a negative signed number, then x^2 is ______.

SUMMARY: CHAPTER 11

A shaded dot on a number line is the **graph** of the number corresponding to that point, and the number is the **coordinate** of the point.

The whole numbers and their opposites form the set of **integers.**

Integers	$\ldots, -3, -2, -1, 0, 1, 2, 3, \ldots$
Positive integers	$1, 2, 3, 4, 5, \ldots$
Negative integers	$\ldots, -3, -2, -1$

The number 0 is neither positive nor negative. (See Figure 11.12.)

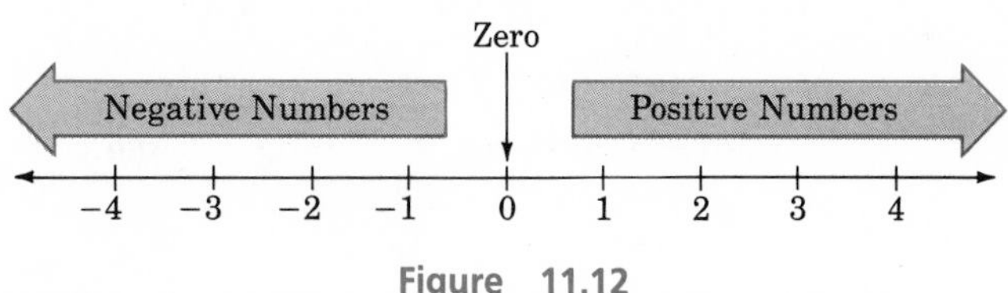

Figure 11.12

Signed numbers are numbers that can be represented on a number line.

SYMBOLS FOR ORDER

$=$	is equal to	$\neq$	is not equal to
$<$	is less than	$>$	is greater than
$\leq$	is less than or equal to	$\geq$	is greater than or equal to

The **absolute value** of a number is its distance from 0 on a number line. **The absolute value of a number is never negative.**

RULES FOR ADDING SIGNED NUMBERS

1. **To add two signed numbers with like signs,** add their absolute values and use the common sign.

$$(+6) + (+5) = +[|+6| + |+5|] = +[6 + 5] = +11$$
$$(-2) + (-8) = -[|-2| + |-8|] = -[2 + 8] = -10$$

2. **To add two signed numbers with unlike signs,** subtract their absolute values (smaller from larger) and use the sign of the number with the larger absolute value.

$$(+10) + (-15) = -[|-15| - |+10|] = -[15 - 10] = -5$$
$$(+12) + (-9) = +[|+12| - |-9|] = +[12 - 9] = +3$$

DEFINITION The **difference** of two integers a and b is the sum of a and the opposite of b. Symbolically,

$$(a) - (b) = (a) + (-b)$$

If a and b are positive numbers,

1. $a \cdot b = ab$
2. $(-a)(-b) = ab$
3. $a(-b) = -ab$
4. $a \cdot 0 = 0$

DEFINITION For any numbers a and b where $b \neq 0$,

$$\frac{a}{b} = x \quad \text{means} \quad a = b \cdot x$$

If a and b are positive numbers,

1. $\frac{a}{b} = \frac{a}{b}$
2. $\frac{-a}{-b} = \frac{a}{b}$
3. $\frac{-\mathrm{a}}{\mathrm{b}} = -\frac{\mathrm{a}}{\mathrm{b}}$ and $\frac{a}{-b} = -\frac{a}{b}$

REVIEW QUESTIONS: CHAPTER 11

Find the opposite of each number.

1. -4 **2.** $\frac{3}{8}$ **3.** 0

For each of the following, draw a number line and graph the given set of numbers.

4. $\{-1, 0, |-1|\}$ **5.** $\left\{-1\frac{1}{2}, -0.3, \frac{3}{4}, 1.8\right\}$ **6.** $\{-4, -3, \ldots, 2\}$

Fill in the blank with the appropriate symbol, <, >, or =.

7. -4 ______ $|-4|$ **8.** $\frac{2}{3}$ ______ $\frac{5}{8}$ **9.** $-(-6)$ ______ $|-6|$

Determine whether each statement is true or false.

10. $-\left(-\frac{1}{2}\right) \leq \frac{1}{2}$ **11.** $-18 < -12$ **12.** $3.2 > |-3.2|$

Name the possible values for x.

13. $|x| = 0.4$ **14.** $|x| = -7$ **15.** $|x| = \frac{9}{10}$

Find each sum.

16. $(-14) + (+18)$ **17.** $(-17) + (+12)$

18. $(-15) + (+4) + (-9)$ **19.** $(+14) + (-23) + (+9)$

20. $\left(-2\frac{3}{5}\right) + \left(-1\frac{7}{10}\right)$ **21.** $(-17.3) + (+14)$

22. $\begin{array}{r} -312 \\ -422 \\ \underline{+610} \end{array}$ **23.** $\begin{array}{r} -7.63 \\ 11.01 \\ \underline{2.24} \end{array}$

Find each difference.

24. $(+12) - (+7)$ **25.** $(+8) - (-8)$ **26.** $(-16) - (+5)$

27. $(-9) - (-9)$

Subtract the bottom number from the top number.

28. $\begin{array}{r} 16 \\ \underline{-10} \end{array}$ **29.** $\begin{array}{r} 16 \\ \underline{10} \end{array}$ **30.** $\begin{array}{r} -62 \\ \underline{-12} \end{array}$ **31.** $\begin{array}{r} -62 \\ \underline{12} \end{array}$

Evaluate each of the following expressions.

32. $-12.4 + 5.11 - 3$

33. $5\frac{3}{8} - 2\frac{1}{4} - 1\frac{2}{3}$

34. $-8 - 14 + 29 - 7$

Find the following products.

35. $12(-6)$

36. $-4(-16)$

37. $\left(-\frac{2}{3}\right)\left(-\frac{6}{7}\right)\left(\frac{7}{8}\right)$

38. $(-4.732)(0.75104)(0)(-5.2)$

39. $(3)(0.4)(-0.02)$

40. $(-5)^3$

Find the following quotients.

41. $\frac{-51}{3}$

42. $\frac{-24}{-72}$

43. $\frac{0}{-73}$

44. $\frac{-52}{0}$

45. $(-14) \div \left(-\frac{7}{8}\right)$

46. $\frac{435}{-1.5}$

Evaluate the following expressions using the rules for order of operations.

47. $3(-4) + 12 \div 2$

48. $(-4)^2 \div (8 - 2 \cdot 3)$

49. $\frac{1}{4} + \frac{2}{3} \div \frac{5}{9} + \frac{3}{10}$

50. $[4 + 3(1 - 2^2)] \div 4$

Fill in the blanks with the correct term (positive, negative, 0, undefined).

51. The quotient when any signed number is divided by zero is ______.

52. The product of a negative number and zero is ______.

TEST: CHAPTER 11

1. Find the opposite of 0.34.
2. Draw a number line and graph the following:

$$\left\{-3,\ -0.5,\ \frac{3}{4},\ |-1|,\ 2.4\right\}$$

Complete the following with the appropriate symbol, <, >, or =.

3. -12 _____ -15
4. $|-12|$ _____ $-(-15)$
5. Name the possible values for x if $|x| = 17$.
6. Find the sum of 1.4 and -6.
7. Add: $(-3) + (+7) + (-4) + (-9)$.
8. Subtract $3\frac{1}{2}$ from $-5\frac{3}{4}$.
9. Evaluate the following: $17 - 4 - 12 + 15$.
10. Find the product of -0.4 and -0.07.

Multiply each of the following.

11. $(-3)(-5)(2)(-3)$
12. $\left(-\frac{3}{8}\right)\left(\frac{4}{5}\right)\left(-\frac{5}{6}\right)$
13. Evaluate $(-2.1)^2$.

Find each quotient.

14. $\frac{-5.6}{-7}$
15. $\frac{-3}{0}$
16. $\frac{65}{-13}$
17. $3\frac{1}{4} \div \left(-1\frac{1}{2}\right)$

Evaluate each of the following using the rules for order of operations.

18. $\left(-6\frac{1}{2}\right) \div \left(1\frac{1}{6}\right) - 3\frac{1}{14}$
19. $(-7)^2 - 6 \cdot 8 \div 2 + 3(6 - 5^2)$
20. Complete the following with positive, negative, 0, or undefined: If x is any integer, then

$x(-x)$ is either _____ or _____

12 Solving Equations

12.1 SIMPLIFYING AND EVALUATING EXPRESSIONS

Any one number is called a **constant.** A **variable** is a symbol (usually a letter) that can represent more than one number. A number written next to a letter (as in $3x$), or two variables written next to each other (as in xy), indicates multiplication. In $3x$, the constant 3 is called the **coefficient** of x.

A **term** is an expression that involves only multiplication and/or division with constants and/or variables. Examples of terms are:

$$5x, \quad -6y, \quad 13x^2, \quad \frac{-15x}{3a}, \quad \text{and} \quad -42.6$$

Expressions, such as

$$5 + x, \quad 2y - 7x, \quad \text{and} \quad 2x^2 - 4x + 6$$

that involve addition or subtraction are **not** terms. They are the algebraic sums of terms.

Like terms (or **similar terms**) are terms that are constants or terms that contain the same variables that are of the same power.

LIKE TERMS	*UNLIKE TERMS*	
$-2x$ and $5x$	$2x$ and $4x^2$	(x is not of the same power in both terms.)
6 and -10		
	-7 and $5x$	(The term -7 does not have the variable x in it.)
$3a^2$ and $-8a^2$		
$5x^2y^3$ and $3.1x^2y^3$	$-4a^2b$ and $3a^2$	(The term $3a^2$ does not have b in it.)

EXAMPLE

1. From the following list of terms, pick out the like terms.

$$7, \quad 3x, \quad 32, \quad -7x, \quad 8yz^2, \quad 0, \quad 13yz^2$$

Solution

(a) 7, 32, and 0 are like terms. All are constants.

(b) $3x$ and $-7x$ are like terms.

(c) $8yz^2$ and $13yz^2$ are like terms.

If no coefficient is written next to a variable, the coefficient is understood to be 1. For example,

$$x = 1 \cdot x, \quad x^3 = 1 \cdot x^3, \quad \text{and} \quad -a = -1 \cdot a$$

To simplify expressions that contain like terms, we want to **combine like terms.** For example,

$$5x + 3x = 8x$$
$$-10x - x = -11x$$
and
$$4a - a + 7a = 10a$$

An explanation of how to combine like terms involves the distributive property as applied to signed numbers.

DISTRIBUTIVE PROPERTY OF MULTIPLICATION OVER ADDITION

If a, b, and c are signed numbers, then

$$a(b + c) = ab + ac$$

Another form of the distributive property is

$$ba + ca = (b + c)a$$

This form is particularly useful when b and c are numerical coefficients because it leads directly to the explanation of combining like terms. Using the previous examples to illustrate simplifying by combining like terms,

$$5x + 3x = (5 + 3)x = 8x$$

$$-10x - x = (-10 - 1)x = -11x$$ (-1 is the coefficient of $-x$.)

$$4a - a + 7a = (4 - 1 + 7)a = 10a$$ (-1 is the coefficient of $-a$.)

EXAMPLES

Simplify by combining like terms whenever possible.

2. $7x + 10x = (7 + 10)x = 17x$ With practice, you will learn to do this step mentally.

3. $-4x + 5x + 3x = (-4 + 5 + 3)x = 4x$

4. $-6a + 4b - 8$ The expression has no like terms, so it is already simplified.

5. $3a^2 + 2x + a^2 - x = 3a^2 + a^2 + 2x - x$
$= (3 + 1)a^2 + (2 - 1)x$ $a^2 = 1a^2$; $-x = -1x$.
$= 4a^2 + x$

After simplifying an expression, we may want to evaluate that expression for one or more values of the variables. Only one value may be used for each variable during a single evaluation. Such an evaluation involves applying the rules for order of operations.

RULES FOR ORDER OF OPERATIONS

1. First, simplify within grouping symbols, such as parentheses ( ), brackets [ ], or braces { }. Start with the innermost grouping.
2. Second, find any powers indicated by exponents.
3. Third, moving from **left to right,** perform any multiplications or divisions in the order they appear.
4. Fourth, moving from **left to right,** perform any additions or subtractions in the order they appear.

To evaluate an algebraic expression:

1. Combine like terms.
2. Substitute the given value(s) for the variable(s).
3. Follow the rules for order of operations.

EXAMPLES

Evaluate each expression for the given values of the variables. Use the rules for order of operations.

6. $2y - 6y - 1; \quad y = -2$

Solution

Simplify first: $2y - 6y - 1 = -4y - 1$

Substitute $y = -2$:

$$\begin{aligned} -4y - 1 &= -4(-2) - 1 \\ &= 8 - 1 \\ &= 7 \end{aligned}$$

7. $5ab + ab + 8a - a; \quad a = 2, b = -3$

Solution

$$\begin{aligned} 5ab + ab + 8a - a &= 6ab + 7a \\ 6ab + 7a &= 6(2)(-3) + 7(2) \\ &= -36 + 14 \\ &= -22 \end{aligned}$$

Substitute $a = 2$, $b = -3$.

Multiply before adding.

8. $6x^2 - 7x^2 + 3x - 10 + 2x + 1; \quad x = -5$

Solution

$$\begin{aligned} 6x^2 - 7x^2 + 3x - 10 + 2x + 1 &= 6x^2 - 7x^2 + 3x + 2x - 10 + 1 \\ &= -x^2 + 5x - 9 \end{aligned}$$

$$\begin{aligned} -x^2 + 5x - 9 &= -(-5)^2 + 5(-5) - 9 \\ &= -(25) + 5(-5) - 9 \\ &= -25 - 25 - 9 \\ &= -50 - 9 \\ &= -59 \end{aligned}$$

Substitute $x = -5$.

Use the exponent first.

PRACTICE QUIZ	Simplify each expression.	ANSWERS
	1. $5x - 6x$	1. $-1x$ or $-x$
	2. $3y + 4y - 2y + 6 - 8$	2. $5y - 2$
	3. $2x^2 + 3x^2 + ab - 4ab$	3. $5x^2 - 3ab$
	Evaluate each expression for $x = -3$.	
	4. $-7x - 14$	4. 7
	5. $2x^2 - x + 1$	5. 22

Exercises 12.1

Simplify the following expressions by combining like terms whenever possible.

1. $6x + 2x$
2. $4x - 3x$
3. $5x + x$
4. $7x - 3x$
5. $-10a + 3a$
6. $-11y + 4y$
7. $-18y + 6y$
8. $-2x - 5x$
9. $-5x - 4x$
10. $-x - 2x$
11. $-7x - x$
12. $2x - 2x$
13. $5x - 5x$
14. $16p - 17p$
15. $9c - 10c$
16. $3x - 5x + 12x$
17. $2a + 14a - 25a$
18. $6c - 13c + 5c$
19. $40p - 30p - 10p$
20. $16x - 15x - 3x$
21. $2x + 3x - 7$
22. $5x - 6x + 2$
23. $7x - 8x + 5$
24. $-5x - 7x - 4$
25. $-8a - 3a - 2$
26. $-4x + x + 1 - 3$
27. $-2x + 5x + 6 - 5$
28. $4x + 7 - 8 + 3x$
29. $-5x - 1 + 8 + 9x$
30. $10y + 3 - 4 - 6y$
31. $6y^2 - y^2$
32. $15x^2 - 5x^2$
33. $2x^2 + 3x + 1$
34. $5x^2 - 2x + 3$
35. $x + y - x - 2y$
36. $a + b - 2a - b$
37. $3ab + a^2 - ab + 2a^2$
38. $xy + y^2 + 4xy + y^2$
39. $3(x + 2y) + 2(x - y)$ [HINT: Use the distributive property.]
40. $5(x + 4y) + 3(2x - y)$ [HINT: Use the distributive property.]

Evaluate each of the following expressions for $x = -3, y = 2, z = 3, a = -1$, and $c = -2$.

41. $x - 2$
42. $y - 2$
43. $z - 3$
44. $2x + z$
45. $3y - x$
46. $x - 4z$
47. $20 - 2a$
48. $10 + 2c$
49. $3c - 5$
50. $2x + 3x - 7$
51. $7a - a + 3$
52. $-3y - 4y + 6 - 2$
53. $-2x - 3x + 1 - 4$
54. $5y - 2y - 3y + 4$
55. $2x - 3x + x - 8$
56. $5y^2 - y^2 + 2y$
57. $4x^2 + 3x - 1$
58. $a^2 + 2c - 3ac$
59. $2a^2 + 3c - 4ac$
60. $x^2 + y^2 - z^2$

12.2 SOLVING EQUATIONS ($ax + b = c$)

If an equation contains a variable, such as $2x + 4 = 10$, we want to find the value (or values) for the variable that will give a true statement when substituted for the variable. This procedure is called **solving the equation,** and the value (or values) found is called the **solution** (or **solutions**) of the equation. **To solve an equation means to find all the solutions.**

EXAMPLES

1. $2x = 14$

Show that 7 is a solution and that 5 is **not** a solution.

Solution

Since $2 \cdot 7 = 14$ is true, 7 is a solution.
But, $2 \cdot 5 = 14$ is false, so 5 is **not** a solution.

2. $y + 11 = 3$

Show that -8 is a solution and that 8 is **not** a solution.

Solution

Since $-8 + 11 = 3$ is true, -8 is a solution.
But, $8 + 11 = 3$ is false, so 8 is **not** a solution.

DEFINITION

A **first-degree equation in x** is any equation that can be written in the form

$$ax + b = c \quad \text{where } a, b, \text{ and } c \text{ are constants and } a \neq 0$$

(The variable may be something other than x.)

HOW TO SOLVE AN EQUATION

There are two basic principles to use.

1. The **addition principle**

   If $A = B$ is true, then
   $A + C = B + C$ is also true for any number C.

2. The **multiplication principle**

   If $A = B$ is true, then
   $A \cdot C = B \cdot C$ is also true for any number C.

   $\frac{A}{C} = \frac{B}{C}$ is also true for $C \neq 0$.

NOTE: Dividing by C is the same as multiplying by the reciprocal $\frac{1}{C}$.

EXAMPLES

3.
$$x + 14 = 10$$

$$x + 14 - 14 = 10 - 14$$ Using the **addition principle**, add -14 to both sides.

$$x + 0 = -4$$ Simplify.

$$x = -4$$ The solution.

4.
$$8y = 72$$

$$\frac{8y}{8} = \frac{72}{8}$$ Using the **multiplication principle**, divide both sides by 8.

$$y = 9$$ Simplify to find the solution.

To understand solving equations:

1. If a constant is added to a variable, add its opposite to both sides of the equation.
2. If a constant is multiplied by a variable, divide both sides by that constant.
3. Remember that the object is to isolate the variable on one side of the equation, right side or left side.

More difficult problems involving several steps are illustrated in Examples 5–8. Study each step carefully. Write the equations one under the other. Remember, the object of the procedures is to get the variable terms on one side and the constant terms on the other side.

EXAMPLES

5.

$4x + 3 = 11$	Write the equation.
$4x + 3 - 3 = 11 - 3$	Add -3 to both sides.
$4x = 8$	Simplify.
$\frac{4x}{4} = \frac{8}{4}$	Divide both sides by 4, the coefficient of x.
$x = 2$	Simplify.

6.

$2x - 4 + 3x = 26$	Write the equation.
$5x - 4 = 26$	Combine like terms on the left side.
$5x - 4 + 4 = 26 + 4$	Add -4 to both sides.
$5x = 30$	Simplify.
$\frac{5x}{5} = \frac{30}{5}$	Divide both sides by 5.
$x = 6$	Simplify.

7.

$5x + 2 = 3x - 8$	Write the equation.
$5x + 2 - 2 = 3x - 8 - 2$	Add -2 to both sides.
$5x = 3x - 10$	Simplify.
$5x - 3x = 3x - 10 - 3x$	Add $-3x$ to both sides. Simplify; now one side has the term with variable, and the other side has the constant term.
$2x = -10$	
$\frac{2x}{2} = \frac{-10}{2}$	Divide both sides by 2.
$x = -5$	Simplify.

8.

$3(x + 2) = 18 - x$	Write the equation.
$3x + 6 = 18 - x$	Use the distributive property.
$3x + 6 - 6 = 18 - x - 6$	Add -6 to both sides.
$3x = 12 - x$	Simplify.
$3x + x = 12 - x + x$	Add $+x$ to both sides.
$4x = 12$	Simplify.
$\frac{4x}{4} = \frac{12}{4}$	Divide both sides by 4.
$x = 3$	Simplify.

CHECKING SOLUTIONS TO EQUATIONS

Checking can be done by substituting the solution into the original equation. However, checking can be time-consuming and need not be done for every problem. Particularly on an exam, check only after you have finished the entire exam.

Checking for Example 6

$$\begin{aligned} 2x - 4 + 3x &= 26 \\ 2(6) - 4 + 3(6) &\stackrel{?}{=} 26 \\ 12 - 4 + 18 &\stackrel{?}{=} 26 \\ 26 &= 26 \end{aligned}$$

Checking for Example 7

$$\begin{aligned} 5x + 2 &= 3x - 8 \\ 5(-5) + 2 &\stackrel{?}{=} 3(-5) - 8 \\ -25 + 2 &\stackrel{?}{=} -15 - 8 \\ -23 &= -23 \end{aligned}$$

If there are fractions in an equation, multiply each term on both sides of the equation by the LCM (least common multiple) of all the denominators. Then solve the equation just as before. The following example illustrates how to proceed with fractions.

EXAMPLE

9. $\frac{1}{2}x + \frac{3}{4}x = \frac{1}{6}x - 26$

$$12\left(\frac{1}{2}x\right) + 12\left(\frac{3}{4}x\right) = 12\left(\frac{1}{6}x\right) - 12(26) \qquad \text{(12 is the LCM of 2, 4, 6)}$$

$$\begin{aligned} 6x + 9x &= 2x - 312 \\ 15x - 2x &= 2x - 312 - 2x \\ 13x &= -312 \\ \frac{13x}{13} &= \frac{-312}{13} \\ x &= -24 \end{aligned}$$

PRACTICE QUIZ	Solve the following equations.	ANSWERS
	1. $10x + 4 = 14$	**1.** $x = 1$
	2. $x + 5 - 2x = 7 + x$	**2.** $x = -1$
	3. $2(x - 7) = 5(x + 2)$	**3.** $x = -8$
	4. $\frac{2}{3}x = \frac{1}{2}x + 1$	**4.** $x = 6$

Exercises 12.2

Solve the following equations.

1. $x + 4 = 10$ **2.** $x + 13 = 20$ **3.** $y - 5 = 17$

4. $y - 12 = 4$ **5.** $y + 8 = 3$ **6.** $x + 10 = 7$

7. $x - 5 = -7$ **8.** $x - 14 = -10$ **9.** $y - 8 = -6$

10. $x - 12 = -5$ **11.** $5x = 30$ **12.** $3y = 15$

13. $10y = -40$ **14.** $6x = -48$ **15.** $-2x = 12$

16. $-4x = 24$ **17.** $-8y = -40$ **18.** $-12y = -36$

19. $16 = x + 3$ **20.** $25 = x + 14$ **21.** $2x + 3 = 5$

22. $3x - 4 = 8$ **23.** $4y + 1 = 9$ **24.** $3x - 10 = 11$

25. $6x + 4 = -14$ **26.** $7y - 8 = -1$ **27.** $3 + 6y = 15$

28. $6 + 5y = 21$ **29.** $2x + 3 = -9$ **30.** $3x - 1 = -4$

31. $5y + 12 = -3$ **32.** $10y + 3 = -17$ **33.** $15 = 2x - 3$

34. $20 = 3x - 1$ **35.** $-17 = 5y - 2$ **36.** $30 = 4y + 6$

37. $4 = 5x + 9$ **38.** $28 = 10x - 2$ **39.** $-24 = 7x - 3$

40. $96 = 25y - 4$ **41.** $3x = x - 10$ **42.** $5y = 2y + 12$

43. $7y = 6y + 5$ **44.** $6x = 2x + 20$ **45.** $5x = 2x$

46. $4x = 3x$ **47.** $4x + 3 = 2x + 9$

48. $5y - 2 = 4y - 6$ **49.** $7x + 14 = 10x + 5$

50. $5x + 20 = 8x - 4$ **51.** $5(x - 2) = 3(x - 8)$

52. $2(y + 1) = 3y + 3$ **53.** $4(x - 1) = 2x + 6$

54. $6y - 3 = 3(y + 2)$ **55.** $7y - 6y + 12 = 4y$

56. $6x + 5 + 3x = 3x - 13$ **57.** $5x - 2x + 4 = 3x + x - 1$

58. $7x + x - 6 = 2(x + 9)$

59. $x - 5 + 4x = 4(x - 3)$

60. $3(-x + 6) = 3x + 2(x + 1)$

61. $\frac{1}{2}x + \frac{3}{4}x = -15$

62. $\frac{2}{3}y - 5 = \frac{1}{3}y + 20$

63. $\frac{3}{5}x - 4 = \frac{1}{5}x - 5$

64. $\frac{5}{8}y - \frac{1}{4} = \frac{2}{5}y + \frac{1}{3}$

65. $\frac{x}{8} + \frac{1}{6} = \frac{x}{10} + 2$

12.3 SOLVING WORD PROBLEMS

Look for key words in word problems:

Key Words

ADDITION	SUBTRACTION	MULTIPLICATION	DIVISION
add	subtract	multiply	divide
sum	difference	product	quotient
plus	minus	times	
more than	less than	twice	
increased by	decreased by	of	

First we will translate some English phrases into algebraic expressions using the meanings of key words.

EXAMPLES

ENGLISH PHRASE	*ALGEBRAIC EXPRESSION*
1. 7 **multiplied by** the variable x the **product** of 7 and x 7 **times** x	$7x$
2. 5 **added to** the unknown y the **sum** of 5 and y 5 **plus** y	$5 + y$
3. 8 **subtracted from** a number **times** 6 the **difference** between $6x$ and 8 $6x$ **minus** 8	$6x - 8$

4. **twice** a number **plus** 3
   3 **added to** 2 **times** a number
   $2x$ **increased by** 3
   } $2x + 3$

5. three **times** the **quantity** found by **adding** 1 to a number
   the **product** of 3 with the **quantity** $x + 1$
   three **times** the **sum** of a number and 1
   } $3(x + 1)$

STEPS IN SOLVING WORD PROBLEMS

1. Read the problem carefully. Read the problem a second time.
2. Decide what is unknown and represent it with a letter. Draw a picture, if possible, to illustrate the problem.
3. Translate the English phrases into mathematical phrases and form an equation indicated by the problem.
4. Solve the equation.
5. Check to see that the solution of the equation makes sense in the problem.

EXAMPLES

6. Five times a number is increased by 3, and the result is 38. What is the number?

Solution

Let n = the number
Translate "five times a number is increased by 3" to "$5n + 3$."
Translate "the result is" to "=."
The equation to be solved is

$$5n + 3 = 38$$
$$5n + 3 - 3 = 38 - 3$$
$$5n = 35$$
$$\frac{5n}{5} = \frac{35}{5}$$
$$n = 7$$

Check

5 times 7 increased by 3 is $5 \cdot 7 + 3$, and $5 \cdot 7 + 3 = 35 + 3 = 38$.

7. Seven less than four times a number is equal to twice the number increased by five. Find the number.

Solution

Let n = the number
Translate "seven less than four times a number" to $4n - 7$.
Translate "twice the number increased by five" to $2n + 5$.
The equation to be solved is

$$4n - 7 = 2n + 5$$
$$4n - 7 - 2n = 2n + 5 - 2n$$
$$2n - 7 = +5$$
$$2n - 7 + 7 = 5 + 7$$
$$2n = 12$$
$$\frac{\cancel{2}n}{\cancel{2}} = \frac{12}{2}$$
$$n = 6$$

Check

$$4(6) - 7 \stackrel{?}{=} 2(6) + 5$$
$$24 - 7 \stackrel{?}{=} 12 + 5$$
$$17 = 17$$

8. A rectangular swimming pool has a perimeter of 160 meters and a length of 50 meters. How wide is the pool?

Solution

Use the formula for the perimeter of a rectangle.

$$P = 2l + 2w$$

Draw a picture.
Let w = width.
Substituting $P = 160$
and $l = 50$
we get

50 m

w

$$160 = 2 \cdot 50 + 2w$$
$$160 = 100 + 2w$$
$$160 - 100 = 100 + 2w - 100$$
$$60 = 2w$$
$$\frac{60}{2} = \frac{\cancel{2}w}{\cancel{2}}$$
$$30 = w$$

The width of the pool is 30 meters.

Consecutive integers: Each integer is one more than the previous integer

18, 19, 20 are three consecutive integers and can be represented by n, $n + 1$, $n + 2$.

Consecutive odd integers: Each integer is odd and two more than the previous integer

31, 33, 35 are three consecutive odd integers and can be represented by n, $n + 2$, $n + 4$.

Consecutive even integers: Each integer is even and two more than the previous integer

-8, -6, -4 are three consecutive even integers and can be represented by n, $n + 2$, $n + 4$.

EXAMPLE

9. The sum of three consecutive odd integers is -57. What are the integers?

Solution

Let n = first odd integer
$n + 2$ = second consecutive odd integer
$n + 4$ = third consecutive odd integer
The equation to be solved is:

$$\begin{aligned} n + (n + 2) + (n + 4) &= -57 \\ n + n + 2 + n + 4 &= -57 \\ 3n + 6 &= -57 \\ 3n + 6 - 6 &= -57 - 6 \\ 3n &= -63 \\ \frac{\not{3}n}{\not{3}} &= \frac{-63}{3} \\ n &= -21 \\ n + 2 &= -19 \\ n + 4 &= -17 \end{aligned}$$

Check

$(-21) + (-19) + (-17) = -57$

Exercises 12.3

Write the following English phrases as algebraic expressions. Use any letter as the unknown number.

1. 5 more than a number
2. 6 added to a number
3. a number plus 10
4. a number increased by 1
5. 8 less than a number
6. a number decreased by 4
7. 14 subtracted from a number
8. a number minus 3
9. the sum of a number and 11
10. the difference between 6 and a number
11. the product of 2 and a number
12. 3 multiplied by a number
13. the quotient of a number and -7
14. -18 divided by a number
15. 4 times a number minus 3
16. -2 times a number plus 17
17. 4 times the difference between a number and 3
18. 5 times the quantity $x - 7$
19. -2 times the quantity $x + 5$
20. 10 minus twice a certain number
21. 13 plus twice a certain number
22. a certain number increased by 3 times that number
23. 6 times the sum of a number and 1
24. the difference of 9 and twice a number
25. the sum of -3 and 3 times a number
26. 5 more than 8 times a number
27. the product of 9 with the quantity $x - 3$
28. -4 times the quantity found by decreasing a number by 2
29. 3 less than the product of a number and 7
30. 16 increased by twice a number

Solve each of the following problems.

31. Find a number whose product with 3 is 57.
32. Find a number that when multiplied by 7 gives 84.

33. The sum of a number and 32 is 86. What is the number?

34. The difference of a number and 16 is -48. What is the number?

35. If the product of a number and 4 is decreased by 10, the result is 50. Find the number.

36. If the product of a number and 5 is added to 12, the result is 7. Find the number.

37. If the product of a number and 8 is increased by 24, the result is twice the number. What is the number?

38. The sum of a number and 2 is equal to three times the number. What is the number?

39. If twice a number is decreased by 4, the result is the number. Find the number.

40. Three times the sum of a number and 4 is -60. What is the number?

41. Twice a number plus 5 is equal to 20 more than the number. What is the number?

42. If 7 is subtracted from a number, the result is 8 times the number. Find the number.

43. Twenty plus a number is equal to twice the number plus three times the same number. What is the number?

44. The sum of two consecutive integers is 37. Find the two integers.

45. The sum of three consecutive integers is -42. Find the three integers.

46. The sum of three consecutive odd integers is 27. Find the three integers.

47. Find three consecutive odd integers whose sum is 81.

48. Find three consecutive even integers whose sum is 30.

49. Find four consecutive even integers whose sum is 54 more than the smallest one.

50. If the product of 4 and the sum of a number and 3 is diminished by 6, the result is 26. Find the number.

51. What is the number whose sum with 18 is ten times the number?

52. If 7 times a number decreased by four times the number is equal to the sum of the number and 10, what is the number?

53. What is the number whose product with 6 is equal to twice the number decreased by 12?

54. The difference of a number and 3 is equal to the difference of five times the number and 15. What is the number?

55. If the sum of two consecutive integers is multiplied by 3, the result is -15. What are the two integers?

56. The perimeter of a rectangular garden is 240 feet. If the width is 40 feet, what is the length of the garden?

57. The perimeter of a triangle is 18 centimeters. If two sides are equal and the third side is 4 centimeters long, what is the length of each of the other two sides?

58. A length of wire is bent to form a triangle with two sides equal. If the wire is 30 cm long and the two equal sides are each 8 cm long, how long is the third side?

59. The perimeter of a rectangular parking lot is 400 yards. If the width is 75 yards, what is the length of the parking lot?

60. The circumference of a circular swimming pool is measured to be 78.5 feet. What is the diameter of the pool? [HINT: Use $C = \pi d$ with $\pi = 3.14$.]

12.4 WORKING WITH FORMULAS

Formulas play a major role in mathematics and all the related sciences as well as in economics, business, medicine, and so on. A formula is a general statement of the relationship between various quantities. We have used formulas in this text to discuss perimeter, area, and volume of geometric figures (Chapter 9), to discuss simple interest, and to relate Fahrenheit and Celsius temperatures.

Some formulas and their meanings are stated here for your interest.

FORMULA	*MEANING*
1. $I = PRT$	The simple interest (I) earned investing money is equal to the product of the principal (P), the rate of interest (R), and the time (T) in years.
2. $d = rt$	The distance (d) traveled is equal to the product of the rate of speed (r) and the time (t).

3. $C = \pi d$ — The circumference (C) of a circle is equal to the product of pi (π) and the diameter (d).

4. $C = \frac{5}{9}(F - 32)$ — Temperature in degrees Celsius (C) equals $\frac{5}{9}$ of the difference between the Fahrenheit temperature (F) and 32.

5. $\alpha + \beta + \gamma = 180$ — The sum of the angles (α, β, and γ) of a triangle is 180°. [α, β, and γ are the Greek letters alpha, beta, and gamma, respectively.]

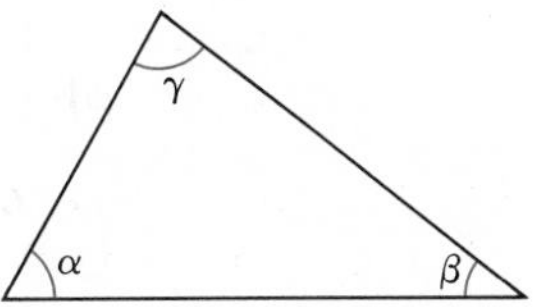

If you know the values for all but one variable in a formula, you can substitute those values; then solve the equation for the unknown variable. If the equation is first-degree, then the techniques for solving equations discussed in Section 12.2 can be used.

EXAMPLES

1. Given the formula $C = \frac{5}{9}(F - 32)$, find the value of F if C = 30°.

Solution

C = 30°, so substitute 30 for C in the formula:

$$30 = \frac{5}{9}(F - 32)$$

$$9 \cdot 30 = \cancel{9} \cdot \frac{5}{\cancel{9}}(F - 32) \qquad \text{Multiply both sides by 9.}$$

$$270 = 5F - 160 \qquad \text{Use the distributive property.}$$

$$420 = 5F \qquad \text{Add 160 to both sides.}$$

$$\frac{420}{5} = \frac{\cancel{5}F}{\cancel{5}} \qquad \text{Divide both sides by 5.}$$

$$84 = F \qquad \text{Simplify.}$$

Thus, 30° C is a warm summer day of 84° F.

2. Two angles of a triangle are measured as 65° and 87°. What is the measure of the third angle?

Solution

We know that $\alpha + \beta + \gamma = 180$ for any triangle (see Formula 5). In this case, $\alpha = 65°$ and $\beta = 87°$.

$$\alpha + \beta + \gamma = 180$$

$$65 + 87 + \gamma = 180 \quad \text{Substitute for } \alpha \text{ and } \beta.$$

$$152 + \gamma = 180$$

$$\gamma = 38 \quad \text{Subtract 152 from both sides.}$$

The third angle measures 38°.

Many times we want to use a formula but not in the form given. That is, we want the same relationship between the variables, but we want to **solve** the formula **for one of the other variables.** In this case, we solve just as in Section 12.2, but we **temporarily treat the variables other than the chosen one as if they were constants.** Study the following examples carefully.

EXAMPLES

3. Given the formula $d = rt$, solve for t in terms of d and r.

Solution

$$d = rt \quad \text{Treat } d \text{ and } r \text{ as if they were constants.}$$

$$\frac{d}{r} = \frac{rt}{r} \quad \text{Divide both sides by } r.$$

$$\frac{d}{r} = t \quad \text{Simplify.}$$

4. Given $\mathrm{C} = \frac{5}{9}(\mathrm{F} - 32)$, solve for F in terms of C.

Solution

$$\mathrm{C} = \frac{5}{9}(\mathrm{F} - 32) \quad \text{Treat C as a constant.}$$

$$\frac{9}{5}\cdot\mathrm{C} = \frac{9}{5}\cdot\frac{5}{9}(\mathrm{F} - 32) \quad \text{Multiply both sides by } \frac{9}{5}.$$

$$\frac{9}{5}\mathrm{C} = \mathrm{F} - 32 \quad \text{Simplify.}$$

$$\frac{9}{5}\mathrm{C} + 32 = \mathrm{F} \quad \text{Add 32 to both sides.}$$

Thus, two forms of the same formula are

$$C = \frac{5}{9}(F - 32) \qquad \text{(solved for C)}$$

$$F = \frac{9}{5}C + 32 \qquad \text{(solved for F)}$$

5. Given $2x + 3y = 12$, (a) solve for x in terms of y; and (b) solve for y in terms of x. Such an equation can represent the graph of a straight line, as we will see in Chapter 13.

Solution

(a) Solving for x:

$$\begin{aligned} 2x + 3y &= 12 \\ 2x + 3y - 3y &= 12 - 3y && \text{Subtract } 3y \text{ from both sides.} \\ 2x &= 12 - 3y && \text{Simplify.} \\ \frac{\not{2}x}{\not{2}} &= \frac{12 - 3y}{2} && \text{Divide both sides by 2.} \\ x &= \frac{12 - 3y}{2} \end{aligned}$$

(b) Solving for y:

$$\begin{aligned} 2x + 3y &= 12 \\ 2x + 3y - 2x &= 12 - 2x && \text{Subtract } 2x \text{ from both sides.} \\ 3y &= 12 - 2x && \text{Simplify.} \\ \frac{\not{3}y}{\not{3}} &= \frac{12 - 2x}{3} && \text{Divide both sides by 3.} \\ y &= \frac{12 - 2x}{3} \end{aligned}$$

Or we can write

$$y = 4 - \frac{2x}{3} \qquad \text{Both forms are correct.}$$

Exercises 12.4

In each formula, substitute the given values and find the value of the unknown variable.

1. $d = rt$; $r = 50$ mph, $t = 2.5$ hr

2. $A = lw$; $l = 10$ ft, $w = 40$ ft

3. $A = \pi r^2$; $r = 3$ cm, use $\pi = 3.14$

4. $C = \pi d$; $d = 5$ m, use $\pi = 3.14$

5. $2x + 3y = 10$; $x = -1$

6. $I = PRT$; $I = \$150$, $R = 10\%$, $T = \frac{1}{2}$ yr

[HINT: Change 10% to a decimal.]

7. $P = 2l + 2w$; $P = 50$ cm, $l = 15$ cm

8. $3x - y = 15$; $y = 3$

9. $\alpha + \beta + \gamma = 180$; $\beta = 10°$, $\gamma = 80°$

10. $V = lwh$; $V = 30\text{ mm}^3$, $l = 5$ mm, $w = 1.2$ mm

Solve each formula for the indicated variable in terms of the other variables.

11. $f = ma$; solve for m

12. $C = 2\pi r$; solve for r

13. $L = 2\pi rh$; solve for h

14. $I = PRT$; solve for P

15. $y = mx + b$; solve for x

16. $p = b + 2s$; solve for s

17. $p = a + b + c$; solve for a

18. $\alpha + \beta + \gamma = 180$; solve for β

19. $P = 2l + 2w$; solve for l

20. $A = P + PRT$; solve for T

21. $V = lwh$; solve for w

22. $V = \pi r^2 h$; solve for h

23. $2x + y = 4$; solve for y

24. $2x + y = 4$; solve for x

25. $x - y = 7$; solve for y

26. $3x - y = 5$; solve for y

27. $3x + 2y = 6$; solve for y

28. $2x - 4y = 5$; solve for x

29. $6x - 2y = -3$; solve for x

30. $5x - 3y = -1$; solve for y

SUMMARY: CHAPTER 12

A **variable** is a symbol or letter that can represent more than one number. An expression that involves only multiplications and/or divisions with constants and/or variables is called a **term.**

Like terms (or **similar terms**) can be constants, or they can be terms that contain variables that are of the same power in each term.

DISTRIBUTIVE PROPERTY OF MULTIPLICATION OVER ADDITION

If a, b, and c are signed numbers, then

$$a(b + c) = ab + ac$$

To evaluate an algebraic expression:

1. Combine like terms.
2. Substitute the given value(s) for the variable(s).
3. Follow the rules for order of operations.

DEFINITION

A **first-degree equation in x** is any equation that can be written in the form

$$\mathbf{ax + b = c} \qquad \text{where } a, b, \text{ and } c \text{ are constants and } a \neq 0$$

(The variable may be something other than x.)

HOW TO SOLVE AN EQUATION

There are two basic principles to use:

1. The **addition principle**

   If $A = B$ is true, then
   $A + C = B + C$ is also true for any number C.

2. The **multiplication principle**

   If $A = B$ is true, then
   $A \cdot C = B \cdot C$ is also true for any number C.

   $\frac{A}{C} = \frac{B}{C}$ is also true for $C \neq 0$.

NOTE: Dividing by C is the same as multiplying by the reciprocal $\frac{1}{C}$.

Look for key words in word problems:

Key Words

ADDITION	SUBTRACTION	MULTIPLICATION	DIVISION
add	subtract	multiply	divide
sum	difference	product	quotient
plus	minus	times	
more than	less than	twice	
increased by	decreased by	of	

$n, n + 1, n + 2$	represent **consecutive integers.**
$n, n + 2, n + 4$	represent **consecutive odd integers** if n is odd.
$n, n + 2, n + 4$	represent **consecutive even integers** if n is even.

REVIEW QUESTIONS: CHAPTER 12

Simplify the following expressions by combining like terms whenever possible.

1. $8x + 7x$

2. $9y + 3y$

3. $4x - x$

4. $-5x - x$

5. $13w + w$

6. $10y - 10y$

7. $8y - 11y + y$

8. $-30p + 2p - 6p$

9. $-7a - 2a + a + 6$

10. $5x + 13 - 7x + 1$

11. $18y - 10 + 3 - 4y$

12. $-2x - 8x - 2 - 8$

13. $3x^2 - x^2$

14. $y^2 - 4y^2$

15. $5x^2 - 3x + 1$

16. $4y^2 + 2y - 1$

17. $3x^2 - 4 - x^2 + 1$

18. $4y^2 - 5 + y^2 + 9$

19. $5(x + y) + 4(x - y)$
[HINT: Use the distributive property.]

20. $3(x - 2y) + 2(2x - 3)$
[HINT: Use the distributive property.]

Evaluate each expression for $x = 3$ and $y = -2$.

21. $20 - 2y + x$

22. $-10 + 6x + 2y$

23. $7x - 3y + x - 5$

24. $4y^2 - 2y + 1$

25. $x^2 + y^2 - 4x^2 + 4y$

Solve the following equations.

26. $x + 7 = 11$

27. $y - 8 = 8$

28. $x + 9 = 5$

29. $y - 3 = 0$

30. $x + 4 = -3$

31. $y - 2 = -4$

32. $7x = 28$

33. $4x = -20$

34. $-2x = 12$

35. $-3y = -15$

36. $3x + 4 = 13$

37. $4x - 1 = 15$

38. $2y + 5 = -5$

39. $3x - 7 = -7$

40. $5x = 4x + 9$

41. $5y = 2y - 9$

42. $7x = 4x$

43. $5y = -3y$

44. $4x + 3 = x + 6$

45. $4y - 3 = 11y - 17$

46. $2(x - 3) = 3(x + 2)$

47. $5(y - 2) + y = 4(y - 1)$

48. $\frac{3}{4}x - 1 = \frac{2}{3}x$

49. $\frac{2}{5}y - \frac{1}{4} = \frac{3}{10}y + \frac{1}{2}$

Write the following English phrases as algebraic expressions. Use x for the unknown number.

50. 8 more than a number

51. 3 less than 5 times a number

52. the quotient of a number and 9

53. -3 times the sum of a number and 2

54. the sum of 10 and 4 times a number

55. 18 decreased by twice a number

For each of the following, write an equation and use it to solve the problem.

56. If the product of a number and 10 is decreased by 15, the result is -35. Find the number.

57. If 14 is added to twice a number, the result is that number minus 13. Find the number.

58. The sum of three consecutive integers is 78. Find the three integers.

59. Find three consecutive odd integers with the property that $\frac{1}{3}$ of the first number is equal to the sum of the other two minus 41.

60. The product of 7 and a number is equal to the product of 5 with the sum of the number and 2. What is the number?

61. What is the number whose sum with -8 is equal to 6 less than 4 times the number?

62. The perimeter of a rectangle is 680 meters. Find the width if the length of the rectangle is 200 m.

In each formula, substitute the given values and find the value of the unknown variable.

63. $\alpha + \beta + \gamma = 180$; $\alpha = 23°$, $\gamma = 57°$

64. $d = rt$; $d = 300$ mi, $t = 12$ hr

Solve each formula for the indicated variable in terms of the other variables.

65. $C = \pi d$; solve for d

66. $3x - 4y = 2$; solve for y

TEST: CHAPTER 12

Simplify the following expressions by combining like terms.

1. $-5 - 6x - 2 + 3x$ **2.** $x^2 + 1 - 4x + 3x^2 - 5$

Evaluate the following expressions for $x = -1$ and $y = 4$.

3. $-3y - 2x + 18 + 5x - 4$ **4.** $5x^2 - 3xy - y^2$

Solve the following equations.

5. $x - 20 = 20$ **6.** $y + 4 = 4$

7. $-7x = -42$ **8.** $4y = -2y$

9. $81 = 21 - 5y$ **10.** $5x + 6 = 4x - 4$

11. $2(2x + 3) = 3(-7 - x)$ **12.** $\frac{1}{2}y + 2 = \frac{3}{4}y - 3$

Write the following English phrases as algebraic expressions. Use x for the unknown number.

13. 3 less than the quotient of a number and 4

14. 1 more than twice the quantity of a number and 3

15. the difference of $\frac{3}{4}$ of a number and 5

Solve each of the following word problems.

16. The sum of three consecutive even integers is equal to 252. Find the integers.

17. Twice the sum of a number and 3 is equal to the difference of the number and 10. What is the number?

18. Find the width of a rectangle if the perimeter is 30 centimeters and the length is $8\frac{1}{4}$ centimeters.

19. Substitute the given values and solve for the unknown variable:

$$I = PRT; \quad I = \$480, \quad P = \$2400, \quad R = 10\%.$$

20. Solve for y in terms of the other variables:

$$4x - 3y = -2$$

13 Real Numbers and Graphing

13.1 SIMPLIFYING RADICAL EXPRESSIONS

A number is squared when it is multiplied by itself. If a whole number is squared, the result is called a **perfect square.** For example, squaring 7 gives $7^2 = 49$, and 49 is a perfect square. Table 13.1 shows the perfect square numbers found by squaring the whole numbers from 1 to 20. A more complete table is located on the inside back cover of this text.

Table 13.1 Squares of Whole Numbers From 1 to 20

$1^2 = 1$	$6^2 = 36$	$11^2 = 121$	$16^2 = 256$
$2^2 = 4$	$7^2 = 49$	$12^2 = 144$	$17^2 = 289$
$3^2 = 9$	$8^2 = 64$	$13^2 = 169$	$18^2 = 324$
$4^2 = 16$	$9^2 = 81$	$14^2 = 196$	$19^2 = 361$
$5^2 = 25$	$10^2 = 100$	$15^2 = 225$	$20^2 = 400$

Since $5^2 = 25$, the **square root** of 25 is 5. The symbol for square root is $\sqrt{}$, called a **square root sign** or **radical sign.** Thus,

$$\sqrt{25} = 5 \quad \text{since} \quad 5^2 = 25$$
$$\sqrt{49} = 7 \quad \text{since} \quad 7^2 = 49$$
$$\sqrt{64} = 8 \quad \text{since} \quad 8^2 = 64$$

Table 13.2 contains the square roots of the perfect square numbers from 1 to 400. Both Table 13.1 and Table 13.2 should be memorized.

Table 13.2 Square Roots of Perfect Squares From From 1 to 400

$\sqrt{1} = 1$	$\sqrt{36} = 6$	$\sqrt{121} = 11$	$\sqrt{256} = 16$
$\sqrt{4} = 2$	$\sqrt{49} = 7$	$\sqrt{144} = 12$	$\sqrt{289} = 17$
$\sqrt{9} = 3$	$\sqrt{64} = 8$	$\sqrt{169} = 13$	$\sqrt{324} = 18$
$\sqrt{16} = 4$	$\sqrt{81} = 9$	$\sqrt{196} = 14$	$\sqrt{361} = 19$
$\sqrt{25} = 5$	$\sqrt{100} = 10$	$\sqrt{225} = 15$	$\sqrt{400} = 20$

The square roots of some numbers are not as easily found as those in the tables. In fact, most square roots are **irrational numbers (nonrepeating infinite decimals);** that is, most square roots can only be approximated with decimals.

Decimal approximations to $\sqrt{2}$ are shown here so that you will understand that there is no finite decimal number whose square is 2.

```
 1.4        1.414         1.41421        1.41422
 1.4        1.414         1.41421        1.41422
  56         5656          141421         282844
 1 4        1414          282842         282844
1.96       5656          565684         565688
          1 414         141421         141422
          1.999396     565684         565688
                     1 41421        1 41422
                     1.9999899241   2.0000182084
```

So, $\sqrt{2}$ is between 1.41421 and 1.41422.

Many numbers that are square roots ($\sqrt{}$), cube roots ($\sqrt[3]{}$), fourth roots ($\sqrt[4]{}$), and so on are infinite nonrepeating decimals and are classified as **irrational numbers.** All the numbers we have studied—rational numbers (including whole numbers, integers, and positive and negative fractions) and irrational numbers—come under the classification of **real numbers.**

Real numbers are the basis for the discussions in the remainder of Chapter 13. Real numbers and their properties, such as the commutative properties of addition and multiplication, are an important part of the courses in algebra that follow this course and will not be discussed here.

The symbol $\sqrt{}$ is called a radical sign, and the number under the radical sign is called the **radicand.** In $\sqrt{18}$, 18 is the radicand. Also, $\sqrt{18}$ is an irrational number, since 18 is not a perfect square. Expressions with radical signs, such as $\sqrt{18}$, are called **radicals.**

To simplify radicals in general, we need the following property.

If a and b are positive real numbers, then

$$\sqrt{ab} = \sqrt{a}\sqrt{b}$$

With this property, we can look for a perfect square factor and write

$$\sqrt{18} = \sqrt{9 \cdot 2} = \sqrt{9} \cdot \sqrt{2} = 3\sqrt{2} \quad \text{(9 is a square factor.)}$$

The expression $3\sqrt{2}$ is simplified because the radicand, 2, has no perfect square factor. That is, **a radical is considered to be in simplest form when the radicand has no square number factor.**

EXAMPLES

Simplify the following radicals.

1. $\sqrt{50}$

$$\sqrt{50} = \sqrt{25 \cdot 2} = \sqrt{25} \cdot \sqrt{2} = 5\sqrt{2} \quad \text{(25 is a square factor.)}$$

2. $\sqrt{75}$

$$\sqrt{75} = \sqrt{25 \cdot 3} = \sqrt{25} \cdot \sqrt{3} = 5\sqrt{3}$$

3. $\sqrt{450}$

$$\sqrt{450} = \sqrt{9 \cdot 50} = \sqrt{9 \cdot 25 \cdot 2} = \sqrt{9} \cdot \sqrt{25} \cdot \sqrt{2} = 3 \cdot 5\sqrt{2} = 15\sqrt{2}$$

or

$$\sqrt{450} = \sqrt{225 \cdot 2} = \sqrt{225} \cdot \sqrt{2} = 15\sqrt{2}$$

Just as $8x$ and $3x$ are like terms, the radicals $8\sqrt{2}$ and $3\sqrt{2}$ are **like radicals. Like radicals** are terms that either have the same radicals or can be simplified so that the radicals are the same. For example,

(a) $9\sqrt{2}$ and $4\sqrt{2}$ are like radicals.
(b) $3\sqrt{5}$ and $2\sqrt{3}$ are **not** like radicals.
(c) $\sqrt{75}$ and $2\sqrt{3}$ are like radicals because $\sqrt{75}$ simplifies to $5\sqrt{3}$, and $5\sqrt{3}$ and $2\sqrt{3}$ are like radicals.

To find a sum of like radicals, we proceed to use the distributive property, just as we do when adding like terms. For example,

$$8x + 3x = (8 + 3)x = 11x$$

and

$$8\sqrt{2} + 3\sqrt{2} = (8 + 3)\sqrt{2} = 11\sqrt{2}$$

The procedure is the same if the radical contains a variable.

To find the sum of radicals:

1. Simplify each radical expression.
2. Use the distributive property to combine any like radicals.

EXAMPLES

Simplify the following radical expressions.

4. $5\sqrt{3} - \sqrt{3} = (5 - 1)\sqrt{3} = 4\sqrt{3}$

5. $\sqrt{18} + \sqrt{8} = \sqrt{9}\sqrt{2} + \sqrt{4}\sqrt{2} = 3\sqrt{2} + 2\sqrt{2} = (3 + 2)\sqrt{2} = 5\sqrt{2}$

6. $\sqrt{7} + \sqrt{12} - \sqrt{3} = \sqrt{7} + \sqrt{4}\sqrt{3} - \sqrt{3}$
$= \sqrt{7} + 2\sqrt{3} - \sqrt{3} = \sqrt{7} + \sqrt{3}$

7. $5\sqrt{x} + 7\sqrt{x} + 2\sqrt{y} - 3\sqrt{y} = (5 + 7)\sqrt{x} + (2 - 3)\sqrt{y}$
$= 12\sqrt{x} - \sqrt{y}$

8. $5\sqrt{3a} + 3\sqrt{12a} = 5\sqrt{3a} + 3\sqrt{4}\sqrt{3a} = 5\sqrt{3a} + 3\cdot 2\sqrt{3a}$
$= 5\sqrt{3a} + 6\sqrt{3a} = 11\sqrt{3a}$

PRACTICE QUIZ

Simplify the following radical expressions.

	ANSWERS
1. $\sqrt{8}$	**1.** $2\sqrt{2}$
2. $\sqrt{20}$	**2.** $2\sqrt{5}$
3. $\sqrt{135}$	**3.** $3\sqrt{15}$
4. $3\sqrt{2} + 6\sqrt{2} + \sqrt{2}$	**4.** $10\sqrt{2}$
5. $\sqrt{75} + 3\sqrt{5} + \sqrt{3}$	**5.** $6\sqrt{3} + 3\sqrt{5}$

Exercises 13.1

In the following exercises, you may use Tables 13.1 and 13.2 and the table located on the inside back cover as aids.

State whether or not each number is a perfect square.

1. 144	**2.** 169	**3.** 81	**4.** 16	**5.** 400
6. 225	**7.** 242	**8.** 48	**9.** 45	**10.** 40

11. Show by squaring that $\sqrt{3}$ is between 1.732 and 1.733.

12. Show by squaring that $\sqrt{5}$ is between 2.236 and 2.237.

Answer the following questions using the set $A = \left\{-10, \frac{3}{4}, 7.2, \sqrt{3}\right\}$.

13. Which of the numbers in set A is an integer?

14. Which of the numbers in set A is an irrational number?

15. Which of the numbers in set A is a real number?

Simplify the following radical expressions.

16. $\sqrt{12}$ **17.** $\sqrt{28}$ **18.** $\sqrt{24}$ **19.** $\sqrt{32}$ **20.** $\sqrt{48}$

21. $\sqrt{288}$ **22.** $\sqrt{363}$ **23.** $\sqrt{242}$ **24.** $\sqrt{500}$ **25.** $\sqrt{300}$

26. $\sqrt{128}$ **27.** $\sqrt{125}$ **28.** $\sqrt{72}$ **29.** $\sqrt{98}$ **30.** $\sqrt{605}$

31. $\sqrt{150}$ **32.** $\sqrt{169}$ **33.** $\sqrt{196}$ **34.** $\sqrt{800}$ **35.** $\sqrt{80}$

36. $\sqrt{90}$ **37.** $\sqrt{40}$ **38.** $\sqrt{256}$ **39.** $\sqrt{361}$ **40.** $\sqrt{108}$

41. $3\sqrt{2} + 4\sqrt{2}$ **42.** $5\sqrt{2} + 8\sqrt{2}$ **43.** $5\sqrt{3} - 4\sqrt{3}$

44. $3\sqrt{5} - 2\sqrt{5}$ **45.** $7\sqrt{10} - 9\sqrt{10}$ **46.** $5\sqrt{11} - 8\sqrt{11}$

47. $\sqrt{24} + \sqrt{6}$ **48.** $\sqrt{28} + 2\sqrt{7}$ **49.** $\sqrt{72} + \sqrt{108}$

50. $2\sqrt{90} - \sqrt{40}$ **51.** $2\sqrt{a} + 5\sqrt{b} - 5\sqrt{a} + \sqrt{b}$

52. $\sqrt{x} + \sqrt{y} + 3\sqrt{x}$ **53.** $5\sqrt{2x} - \sqrt{8x} + \sqrt{2x}$

54. $4\sqrt{3x} + 2\sqrt{12x} - \sqrt{3x}$ **55.** $\sqrt{20a} + 2\sqrt{45a} - \sqrt{80a}$

13.2 INTERVALS AND FIRST-DEGREE INEQUALITIES ($ax + b < c$)

In Section 11.2, we discussed number lines and graphing sets of integers on number lines. The graph of the set $A = \{-2, -1, 1, 3\}$ is shown in Figure 13.1.

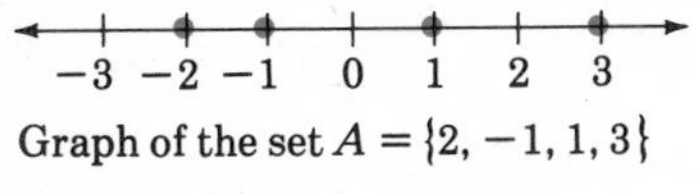

Graph of the set $A = \{2, -1, 1, 3\}$

Figure 13.1

On each number line there are many points (an infinite number) between the integers. **Each point on a number line corresponds to one real number, and each real number corresponds to one point on a number line.** In fact, number lines are called **real number lines.**

The locations of some real numbers, other than integers, are shown in Figure 13.2.

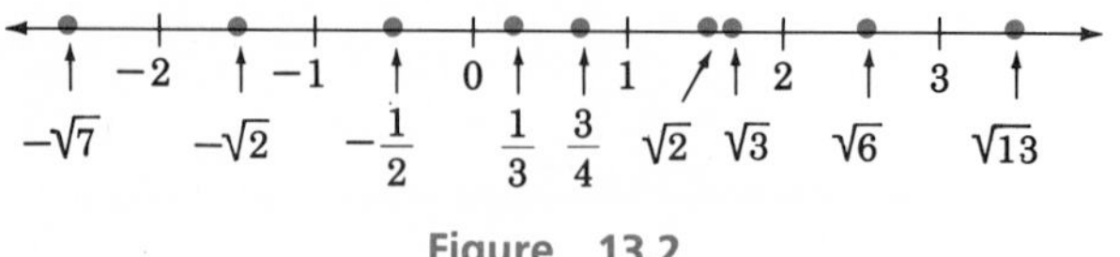

Figure 13.2

If we want to compare two numbers, we can use the following symbols of equality and inequality (reading from left to right as shown):

$a = b$ a is equal to b
$a < b$ a is less than b
$a \leq b$ a is less than or equal to b
$a > b$ a is greater than b
$a \geq b$ a is greater than or equal to b

(A slash, /, through a symbol negates that symbol. For example, $\neq$ is read "is not equal to" and $\not<$ is read "is not less than.") We can also read the symbols from right to left. For example, we can read $a < b$ as "b is greater than a."

On a number line, smaller numbers are to the left of larger numbers. Thus, referring to Figure 13.3, we have

$$-1 < 2 \qquad -2 < -1 \qquad c > a \qquad b > 0$$

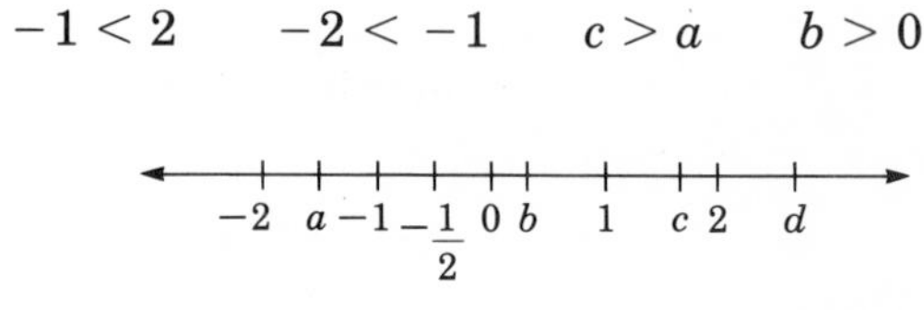

Figure 13.3

Suppose that a and b are two real numbers and that $a < b$. We may be interested not in a or b but in the numbers between a and b. The real numbers between two real numbers are in an **interval of real numbers.** Intervals are classified and graphed in the following manner. **(x is understood to represent real numbers.)**

Open Interval $a < x < b$

x is any real number greater than a **and** less than b.

Closed Interval $a \leq x \leq b$

x is any real number greater than or equal to a **and** less than or equal to b.

Half-Open Interval $a \le x < b$

$a < x \le b$

One of the endpoints is not included.

Open Interval $x > a$

$x < a$

Half-Open Interval $x \ge b$

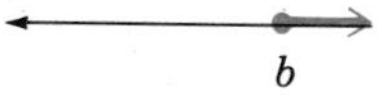

$x \le b$

EXAMPLES

1. Graph the closed interval $4 \le x \le 6$.

Solution

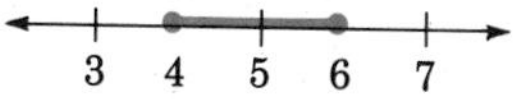

Note that all real numbers between 4 and 6 and including 4 and 6 are in this interval. For example, $4\frac{1}{2}$ and 5.99 are both in this interval because

$$4 \le 4\frac{1}{2} \le 6 \quad \text{and} \quad 4 \le 5.99 \le 6$$

2. Represent the following graph using interval notation and tell what kind of interval it is.

Solution

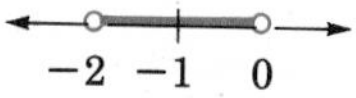

$-2 < x < 0$ is an open interval.

Equations of the form $ax + b = c$ are called **linear equations** or **first-degree equations** since the variable x is first degree. Similarly, inequalities of the form

$$ax + b < c, \qquad ax + b > c$$
$$ax + b \le c, \qquad ax + b \ge c$$

and

$$c < ax + b < d, \qquad c \le ax + b \le d$$

are called **linear inequalities** or **first-degree inequalities.**

Intervals of real numbers are the solutions to linear inequalities, and we can solve linear inequalities just as we can solve linear equations. The rules are the same with one important exception: multiplying both sides of an inequality by a negative number "reverses the sense" of the inequality. Consider the following examples.

We know that $4 < 10$:

ADD 3	*ADD* -5
$4 < 10$	$4 < 10$
$4 + 3 < 10 + 3$	$4 - 5 < 10 - 5$
$7 < 13$	$-1 < 5$

MULTIPLY BY 2	*MULTIPLY BY* -2	
$4 < 10$	$4 < 10$	
$2 \cdot 4 < 2 \cdot 10$	$-2(4) > -2(10)$	(The sense is reversed
$8 < 20$	$-8 > -20$	from $<$ to $>$.)

Figure 13.4 illustrates the effect of multiplying both sides of the inequality $-1 < 2$ by -3.

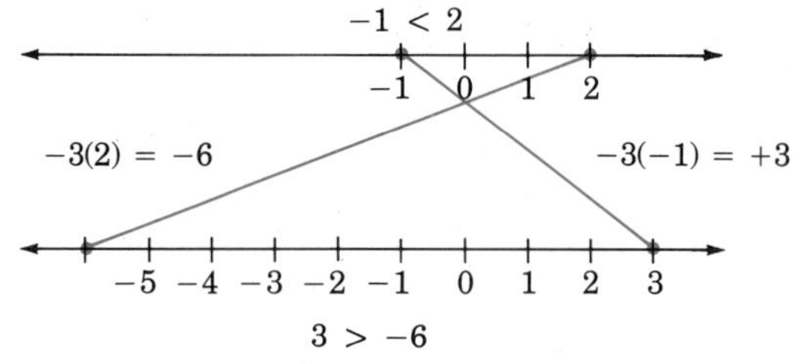

Figure 13.4

RULES FOR SOLVING FIRST-DEGREE INEQUALITIES

1. The same number may be added to both sides, and the sense of the inequality will remain the same.
2. Both sides may be multiplied by (or divided by) the same positive number, and the sense of the inequality will remain the same.
3. Both sides may be multiplied by (or divided by) the same negative number, but the sense of the inequality must be reversed.

As with solving equations, the object of solving inequalities is to find equivalent inequalities that are simpler than the original and to isolate the variable on one side of the inequality. The following examples illustrate the techniques.

EXAMPLES

Solve the inequalities in Examples 3–5 and graph each solution on a number line.

3. $x - 3 < 2$

Solution

$$x - 3 < 2$$

$$x - 3 + 3 < 2 + 3 \qquad \text{Add 3 to both sides.}$$

$$x < 5$$

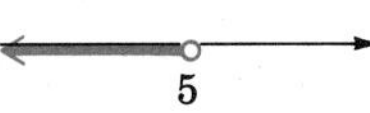

4. $-2x + 6 \geq 4$

Solution

$$-2x + 6 \geq 4$$

$$-2x + 6 - 6 \geq 4 - 6 \qquad \text{Add } -6 \text{ to both sides.}$$

$$-2x \geq -2$$

$$\frac{-2x}{-2} \leq \frac{-2}{-2}$$

Divide both sides by -2 and **reverse the sense.**

$$x \leq 1$$

5. $7y - 8 > y + 10$

Solution

$$7y - 8 > y + 10$$

$$7y - 8 - y > y + 10 - y \qquad \text{Add } -y \text{ to both sides.}$$

$$6y - 8 > 10$$

$$6y - 8 + 8 > 10 + 8 \qquad \text{Add 8 to both sides.}$$

$$6y > 18$$

$$\frac{6y}{6} > \frac{18}{6}$$

Divide both sides by 6. The sense of the inequality is unchanged.

$$y > 3$$

3

6. Find the values of x that satisfy both $5 < 2x + 3$ **and** $2x + 3 < 10$ and graph the solution.

Solution

We can write these two inequalities in one expression and solve both at the same time since both are to be satisfied by the same values of x.

$$5 < 2x + 3 < 10$$

$$5 - 3 < 2x + 3 - 3 < 10 - 3 \quad \text{Add } -3 \text{ to each member.}$$

$$2 < 2x < 7$$

$$\frac{2}{2} < \frac{2x}{2} < \frac{7}{2} \quad \text{Divide each member by 2.}$$

$$1 < x < \frac{7}{2}$$

Can be read: "x is greater than 1 **and** x is less than $\frac{7}{2}$."

1 $\quad \frac{7}{2}$

PRACTICE QUIZ

Solve each inequality and graph the solution.

1. $2x - 1 > 7$
2. $3x + 2 \leq 5x + 1$
3. $-4 \leq 5y + 1 < 11$

ANSWERS

1. $x > 4$

4

2. $\frac{1}{2} \leq x$

$\frac{1}{2}$

3. $-1 \leq y < 2$

$-1 \quad 2$

Exercises 13.2

State whether each of the following inequalities is true or false.

1. $3 \neq -3$
2. $-5 < -2$
3. $-13 > -1$
4. $|-7| \neq |+7|$
5. $|-7| < |+7|$
6. $|-4| > |+3|$
7. $-\frac{1}{2} < -\frac{3}{4}$
8. $\sqrt{2} > \sqrt{3}$
9. $-4 < -6$
10. $|-6| > 0$

Represent each of the following graphs with interval notation and tell what kind of interval it is.

11. 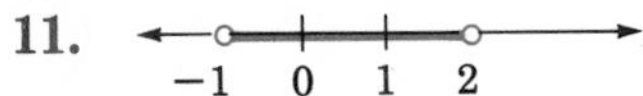

12. 0 1 2 3

13. −1 0

14. −1 0 1 2 3

15. −1 0 1

Graph each of the following intervals and tell what kind of interval it is.

16. $7 < x < 10$ **17.** $-2 < x < 0$ **18.** $1 \le y \le 4$

19. $-3 \le y \le 5$ **20.** $-1 < y \le 2$ **21.** $4 \le x < 7$

22. $68 \le x \le 72$ **23.** $-12 < z < -7$ **24.** $x > -3$

25. $x \ge 0$ **26.** $x < -\sqrt{3}$ **27.** $x \le \frac{2}{3}$

Solve and graph the solution for each of the following inequalities.

28. $x + 4 < 7$ **29.** $x - 6 > -2$ **30.** $y - 3 \ge -1$

31. $y + 5 \le 2$ **32.** $2y \le 3$ **33.** $5y > -6$

34. $6y + 1 > 5$ **35.** $7x - 2 < 9$ **36.** $x + 2 < 3x + 2$

37. $x - 4 > 2x + 1$ **38.** $2x - 5 \ge x + 2$ **39.** $3x - 8 \le x + 2$

40. $\frac{1}{2}x - 4 < 2$ **41.** $\frac{1}{3}x + 1 > -1$

42. $-x + 3 < -2$ **43.** $-x - 5 \ge -4$

44. $7y - 1 \ge 5y + 1$ **45.** $6x + 3 > x - 2$

46. $-4x - 4 \ge -4 + x$ **47.** $3x + 15 < x + 5$

48. $4 \le x + 7 \le 5$ **49.** $-2 \le x - 3 \le 1$

50. $-1 \le 2y + 1 \le 0$ **51.** $0 \le 3x - 1 \le 5$

52. $-3 < 4x + 1 \le 5$ **53.** $-5 \le y - 2 \le -1$

54. $7 \le -2x - 3 < 9$ **55.** $14 < 5x - 1 \le 24$

13.3 GRAPHING ORDERED PAIRS OF REAL NUMBERS

Equations such as $d = 40t$, $I = .18P$, and $y = 2x - 5$ represent relationships between pairs of variables. The first equation, $d = 40t$, can be interpreted as follows: The distance (d) traveled in time (t) at a rate of 40 miles per hour is found by multiplying 40 by t (assuming t is measured in hours). Thus, if $t = 3$ hours, then $d = 40 \cdot 3 = 120$ miles. The pair (3, 120) is called an **ordered pair** and is in the form (t, d). Similarly, (5, 200) represents $t = 5$ and $d = 200$. The ordered pairs (5, 200) and (200, 5) are not the same because the numbers are not in the same order.

We say that the ordered pair (3, 120) is **a solution of** or **satisfies** the equation $d = 40t$. In the same way, (100, 18) satisfies $I = .18P$ where $P = 100$ and $I = .18(100) = 18$. Also, (3, 1) satisfies $y = 2x - 5$ where $x = 3$ and $y = 2 \cdot 3 - 5 = 1$.

In an ordered pair such as (x, y), x is called the **first component** (or **first coordinate**), and y is called the **second component** (or **second coordinate**). To find ordered pairs that satisfy an equation in two variables, we can **choose any value** for one variable and find the corresponding value for the other variable by substituting into the equation. For example, for the equation $y = 2x - 5$,

choose $x = 2$, then $y = 2 \cdot 2 - 5 = -1$
choose $x = -1$, then $y = 2(-1) - 5 = -7$
choose $x = 0$, then $y = 3 \cdot 0 - 5 = -5$
choose $x = 5$, then $y = 3 \cdot 5 - 5 = 10$

All the ordered pairs, (2, −1), (−1, −7), (0, −5), and (5, 10), satisfy the equation $y = 2x - 5$.

Since the value of y "depends" on the choice of x, the first component (x) is called the **independent variable** and the second component (y) is called the **dependent variable.** The ordered pairs can be written in table form. Remember that the choices for the independent variable are arbitrary; other values could have been chosen.

$d = 40t$

t	d
1	40
2	80
5	200
7	280

$I = .18P$

P	I
100	18
200	36
1000	180
5000	900

$y = 2x - 5$

x	y
−2	−9
0	−5
1	−3
4	3
5	5

We can graph ordered pairs of real numbers as points on a plane using the **Cartesian* coordinate system.** In this system, the plane is separated into four **quadrants** by two number lines that are perpendicular to

*The system is named after the famous mathematician René Descartes (1596–1650).

each other. The lines intersect at a point called the **origin,** represented by the ordered pair (0, 0). The horizontal number line is called the **horizontal axis** or **x-axis.** The vertical number line is called the **vertical axis** or **y-axis.** (See Figure 13.5.)

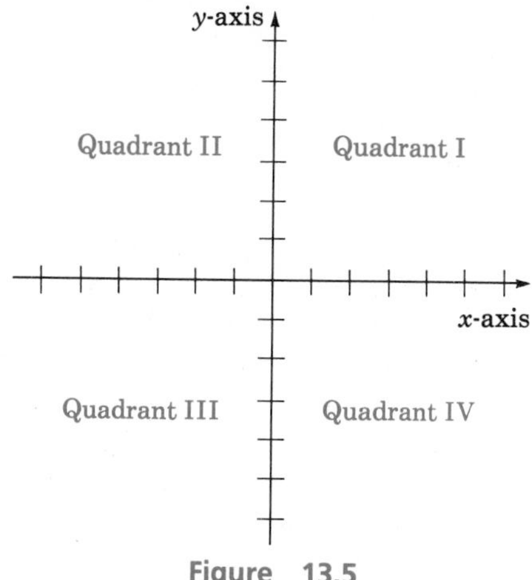

Figure 13.5

On the x-axis, positive numbers are marked to the right and negative numbers to the left. On the y-axis, positive numbers are marked up and negative numbers down. **Each point in a plane corresponds to one ordered pair of real numbers, and each ordered pair of real numbers corresponds to one point in a plane.** (See Figure 13.6.)

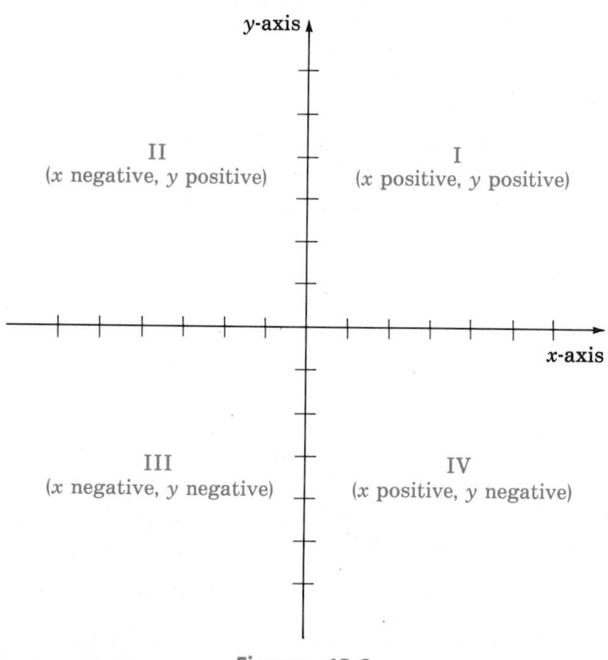

Figure 13.6

The graphs of the points $A(3, 1)$, $B(-2, 3)$, $C(-3, -1)$, $D(1, -2)$, and $E(2, 0)$ are shown in Figure 13.7. The point $E(2, 0)$ is on an axis and not in any quadrant. Each ordered pair is called the **coordinates** of the corresponding point.

POINT	QUADRANT
$A(3, 1)$	I
$B(-2, 3)$	II
$C(-3, -1)$	III
$D(1, -2)$	IV
$E(2, 0)$	x-axis

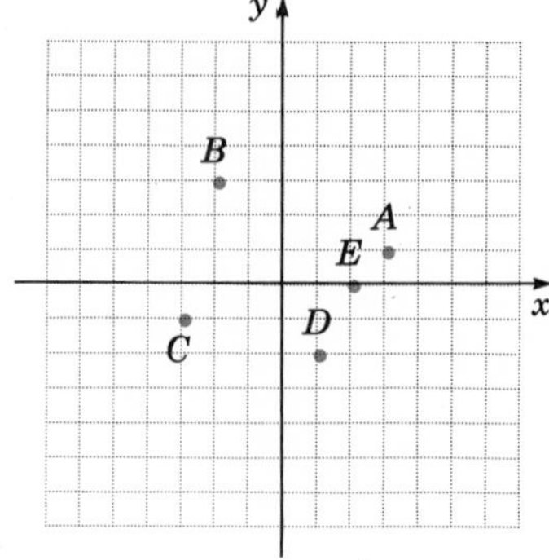

Figure 13.7

EXAMPLES

1. Graph the set of ordered pairs:

$$\{(-2, 1), (0, 3), (1, 2), (2, -2)\}$$

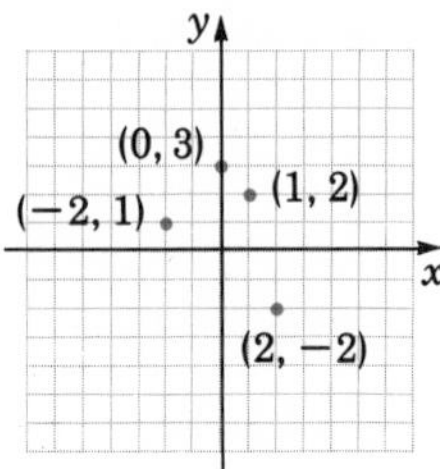

2. Graph the set of ordered pairs:

$$\{(-3, -5), (-2, -3), (-1, -1), (0, 1), (1, 3)\}$$

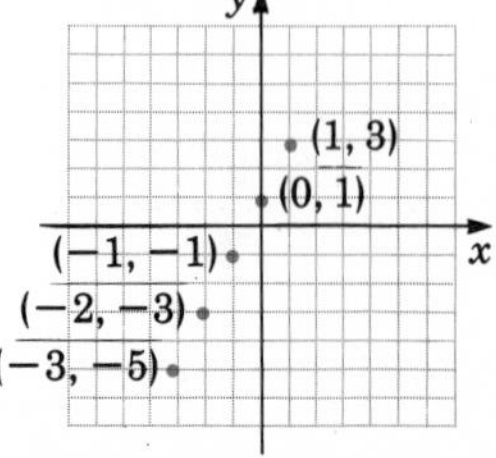

3. The graph of a set of ordered pairs is given. List the set of ordered pairs in the graph.

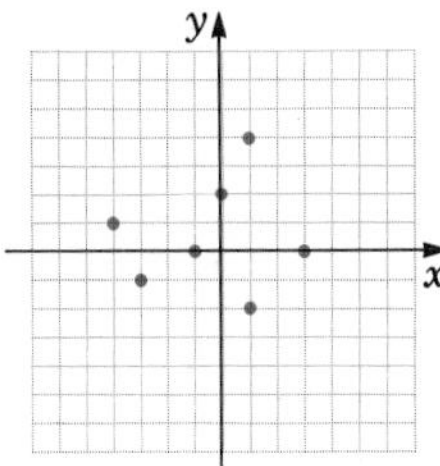

Solution

$\{(-4, 1), (-3, -1), (-1, 0), (0, 2), (1, 4), (1, -2), (3, 0)\}$

Exercises 13.3

List the set of ordered pairs that corresponds to the points in the graph.

1.

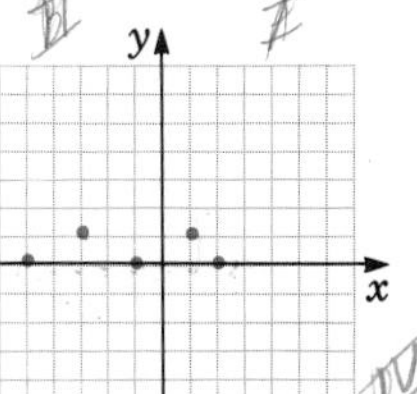

2.

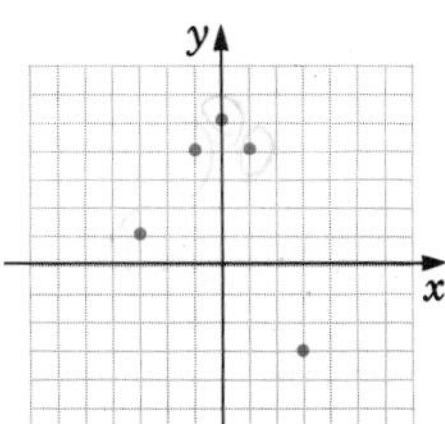

3.

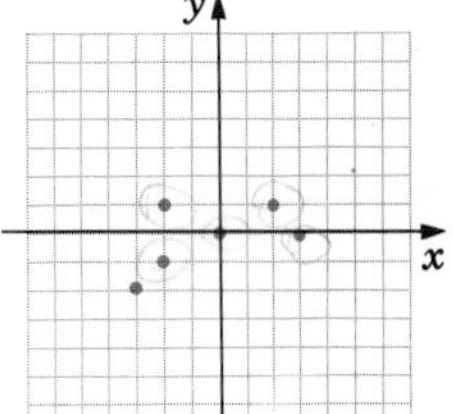

4.

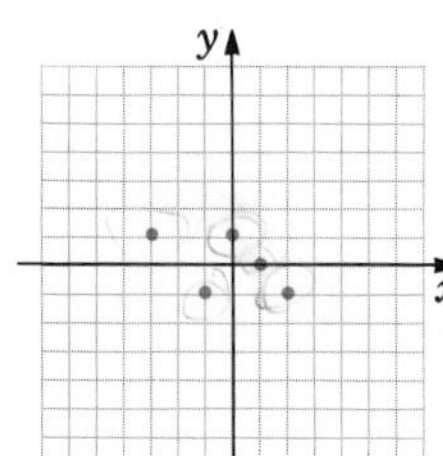

5.

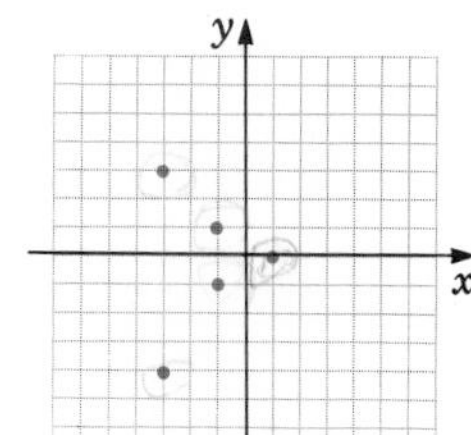

6.

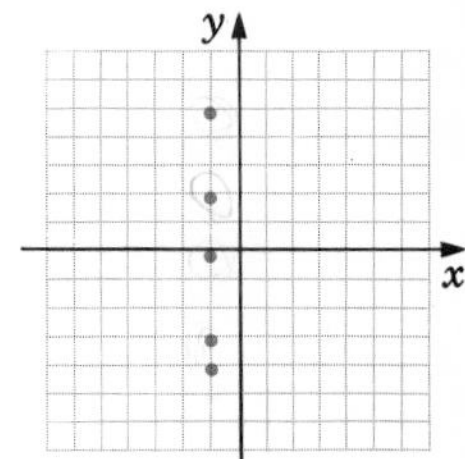

7.

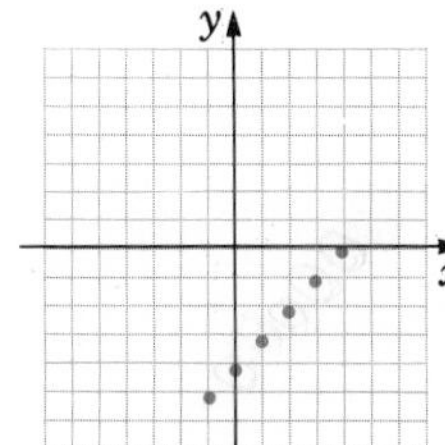

8.

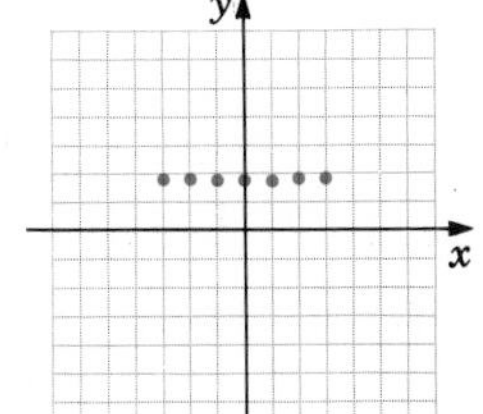

9.

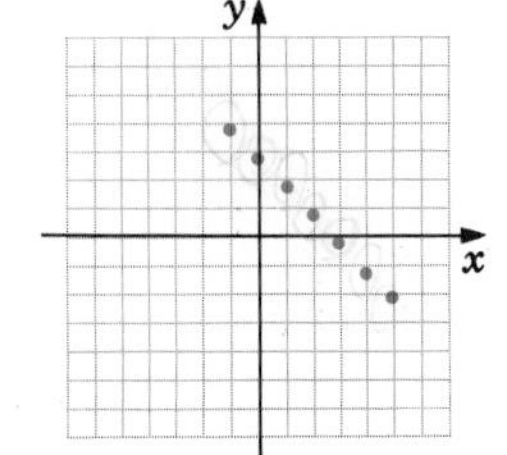

10. 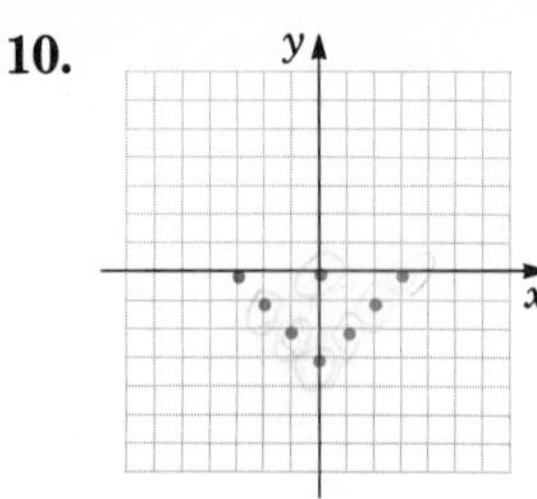

Graph each of the following sets of ordered pairs.

11. $\{(-2, 4), (-1, 3), (0, 1), (1, -2), (1, 3)\}$
12. $\{(-5, 1), (-3, 2), (-2, -1), (0, 2), (2, -1)\}$
13. $\{(-1, 2), (1, 3), (2, -2), (3, 4), (4, -2)\}$
14. $\{(-2, 3), (-1, 0), (0, -3), (2, 3), (4, -1)\}$
15. $\{(0, -3), (1, -1), (2, 1), (3, 3), (4, 5)\}$
16. $\{(-3, 3), (-2, 2), (-1, 1), (0, 0), (1, -1)\}$
17. $\{(-2, -1), (0, -1), (2, -1), (4, -1), (6, -1)\}$
18. $\{(-3, 1), (-2, 1), (-1, 1), (0, 1), (1, 1)\}$
19. $\{(-3, 3), (-1, 1), (0, 0), (1, 1), (3, 3)\}$
20. $\{(-2, -2), (-1, -1), (0, 0), (1, -1), (2, -2)\}$
21. $\{(-3, 9), (-2, 4), (-1, 1), (1, 1), (2, 4), (3, 9)\}$
22. $\{(-3, -9), (-2, -4), (-1, -1), (1, -1), (2, -4), (3, -9)\}$
23. $\{(-4, 0), (-2, 0), (0, 0), (2, 0), (4, 0)\}$
24. $\{(0, -3), (0, -1), (0, 0), (0, 1), (0, 3)\}$
25. $\{(-2, 1), (1, 4), (2, 5), (3, 6), (4, 7)\}$
26. $\{(-1, -5), (0, -2), (1, 1), (2, 4), (3, 7)\}$
27. $\{(-2, -7), (-1, -5), (2, 1), (3, 3), (4, 5)\}$
28. $\{(0, 1), (1, -1), (2, -3), (3, -5), (5, -9)\}$
29. $\{(-3, 11), (-2, 8), (0, 2), (2, -4), (3, -7)\}$
30. $\{(-2, -1), (-1, 1), (1, 5), (2, 7), (3, 9)\}$

13.4 GRAPHING LINEAR EQUATIONS ($Ax + By = C$)

There are an infinite number of ordered pairs of real numbers that satisfy the equation $y = 3x + 1$. In Section 13.3, we substituted only integers for x and calculated integer values for y. We can also substitute fractions and radicals.

For example,

$$\text{if } x = \frac{1}{3}, \quad \text{then } y = 3\cdot\frac{1}{3} + 1 = 1 + 1 = 2$$

$$\text{if } x = -\frac{3}{4}, \text{ then } y = 3\left(-\frac{3}{4}\right) + 1 = -\frac{9}{4} + \frac{4}{4} = -\frac{5}{4}$$

$$\text{If } x = \sqrt{2}, \text{ then } y = 3\sqrt{2} + 1$$

Further studies of these kinds of substitutions will be reserved for later courses. The important idea here is that we do not have enough time or paper to substitute all real values for x. But, graphing a few points will show the trend that is of interest here. (See Figure 13.8.)

x	$y = 3x + 1$
-2	$y = 3(-2) + 1 = -5$
-1	$y = 3(-1) + 1 = -2$
0	$y = 3(0) + 1 = 1$
$\frac{2}{3}$	$y = 3\left(\frac{2}{3}\right) + 1 = 3$
2	$y = 3(2) + 1 = 7$

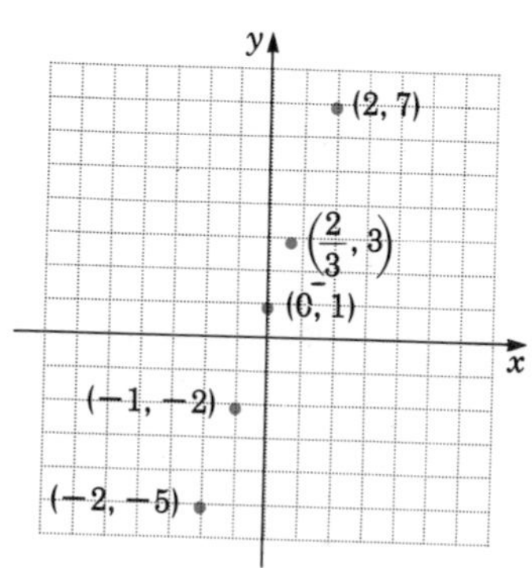

Figure 13.8

It appears that the points in Figure 13.8 lie on a straight line. In fact, they do lie on a straight line. We can draw a straight line through all the points, as shown in Figure 13.9, and **any point that lies on the line will satisfy the equation.**

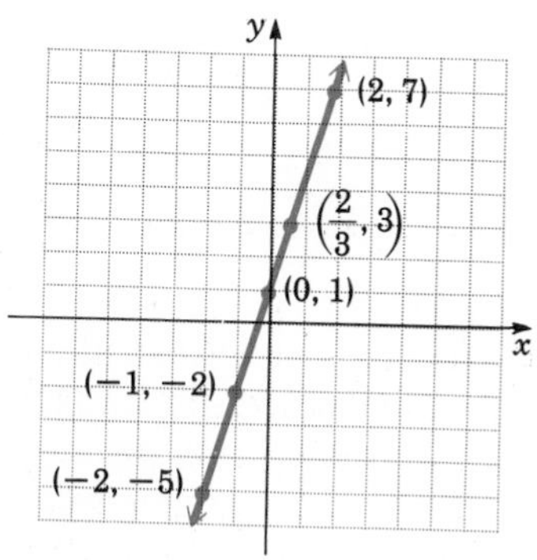

Figure 13.9

> The points (ordered pairs) that satisfy any equation of the form
>
> $$\mathbf{Ax + By = C} \quad (A \text{ and } B \text{ not both } 0)$$
>
> will lie on a straight line. The equation is called a **linear equation** and is in the **standard form** for the equation of a line.

The linear equation $y = 3x + 1$ can be written in the standard form $-3x + y = 1$. Both forms are acceptable and correct.

Since we know that the graph of a linear equation is a straight line, we need only graph two points (two points determine a line) and draw the line through these two points. Choose any two values of x or any two values of y. (As a check against possible error, it is a good idea to locate three points instead of just two.)

EXAMPLE

1. Draw the graph of the linear equation $x + 2y = 6$.

Solution

$$\begin{aligned} x &= -2 \\ -2 + 2y &= 6 \\ 2y &= 8 \\ y &= 4 \end{aligned} \qquad \begin{aligned} x &= 0 \\ 0 + 2y &= 6 \\ 2y &= 6 \\ y &= 3 \end{aligned} \qquad \begin{aligned} x &= 2 \\ 2 + 2y &= 6 \\ 2y &= 4 \\ y &= 2 \end{aligned}$$

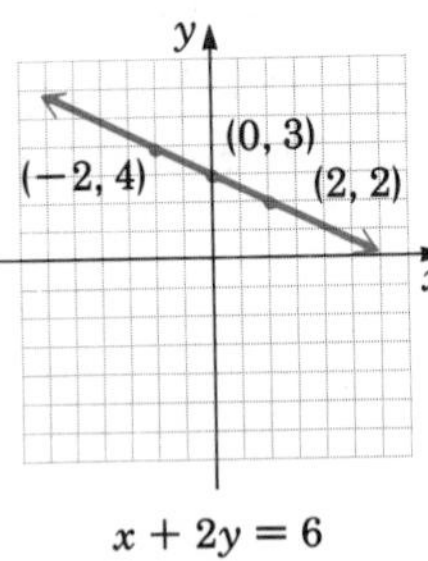

$x + 2y = 6$

(Locating three points helps in avoiding errors. Avoid choosing points close together.)

Letting $x = 0$ will locate the point where the line crosses the y-axis. This point is called the **y-intercept.** Letting $y = 0$ will locate the point where the line crosses the x-axis. This point is called the **x-intercept.** These two points are generally easy to locate and are frequently used for drawing the graph of a linear equation.

EXAMPLES

2. Draw the graph of the linear equation $x - 2y = 8$ by locating the y-intercept and the x-intercept.

Solution

$$\begin{aligned} x &= 0 \\ 0 - 2y &= 8 \\ y &= -4 \end{aligned} \qquad \begin{aligned} y &= 0 \\ x - 2 \cdot 0 &= 8 \\ x &= 8 \end{aligned}$$

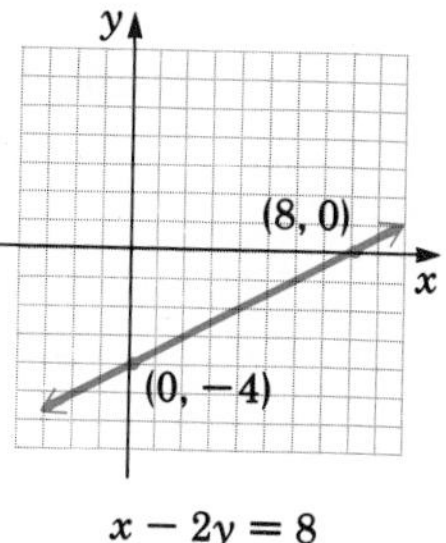

$x - 2y = 8$

3. Locate the y-intercept and the x-intercept and draw the graph of the linear equation $3x - y = 3$.

Solution

$$\begin{aligned} x &= 0 \\ 3 \cdot 0 - y &= 3 \\ y &= -3 \end{aligned} \qquad \begin{aligned} y &= 0 \\ 3x - 0 &= 3 \\ x &= 1 \end{aligned}$$

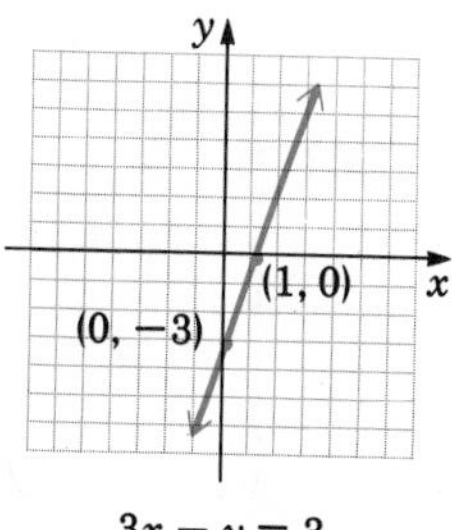

$3x - y = 3$

If $A = 0$ in the standard form $Ax + By = C$, the equation takes the form $By = C$ or $y = \frac{C}{B}$. For example, we can write $0x + 3y = 6$ as $y = 2$. Thus, no matter what value x has, the value of y is 2. The graph of the equation $y = 2$ is a horizontal line, as shown in Figure 13.10.

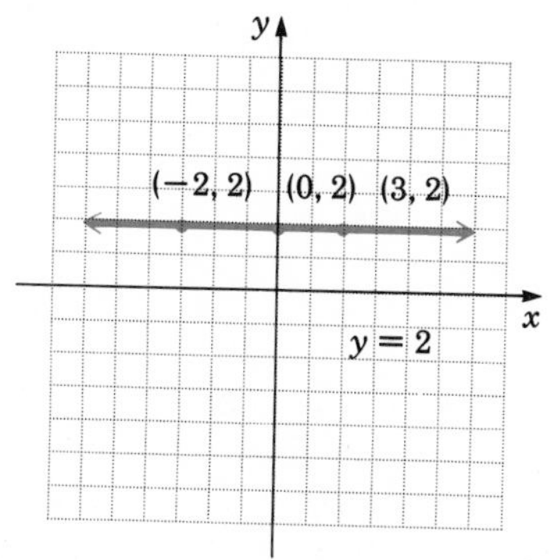

The y-coordinate is 2 for all points on the line $y = 2$.

Figure 13.10

If $B = 0$ in the standard form $Ax + By = C$, the equation takes the form $Ax = C$ or $x = \frac{C}{A}$. For example, we can write $5x + 0y = -5$ as $x = -1$. Thus, no matter what value y has, the value of x is -1. The graph of the equation $x = -1$ is a vertical line, as shown in Figure 13.11.

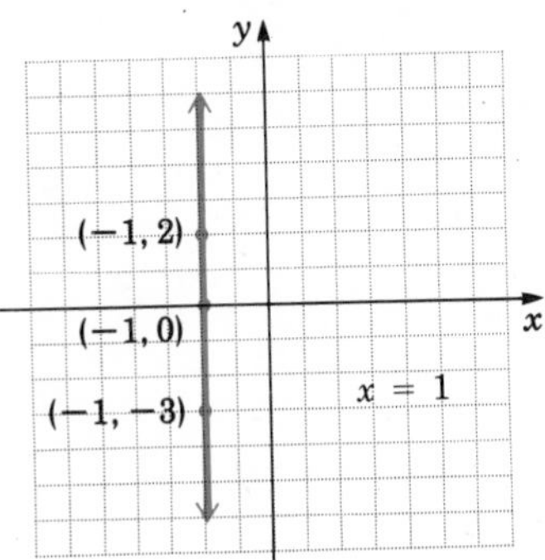

The x-coordinate is -1 for all points on the line $x = -1$.

Figure 13.11

EXAMPLE

4. Graph the horizontal line $y = -1$ and the vertical line $x = 3$ using the same axes.

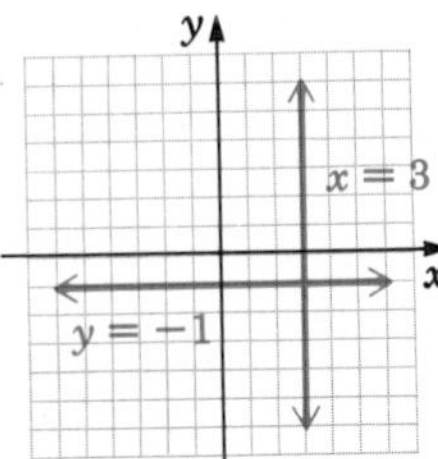

Exercises 13.4

Graph the following linear equations.

1. $y = x + 1$	**2.** $y = x + 2$	**3.** $y = x - 4$
4. $y = x - 6$	**5.** $y = 2x$	**6.** $y = 3x$
7. $y = -x$	**8.** $y = -4x$	**9.** $y = x$
10. $y = 2 - x$	**11.** $y = 3 - x$	**12.** $y = 5 - x$
13. $y = 2x + 1$	**14.** $y = 2x - 1$	**15.** $y = 2x - 3$
16. $y = 2x + 5$	**17.** $y = -2x + 1$	**18.** $y = -2x - 2$
19. $y = -3x + 2$	**20.** $y = -3x - 4$	**21.** $x - 2y = 4$
22. $x - 3y = 6$	**23.** $-2x + 3y = 6$	**24.** $2x - 5y = 10$

25. $-2x + y = 4$ **26.** $2x + 3y = 6$ **27.** $3x + 5y = 15$

28. $4x + y = 8$ **29.** $x + 4y = 8$ **30.** $3x - 4y = 12$

31. $x = 4$ **32.** $y = 5$ **33.** $y = -5$

34. $x = -2$ **35.** $x - 5 = 0$ **36.** $y + 3 = 0$

37. $2y - 3 = 0$ **38.** $3x = -1$ **39.** $4x - 5 = 0$

40. $3y + 7 = 0$

13.5 GRAPHING PARABOLAS ($y = ax^2 + c$)

This section will show that not all graphs are straight lines. The student will see that the relationship between x and y is what determines the nature of the graph. This section is only an introduction to this topic, and the student should understand that more detailed discussions are found in later courses in algebra.

> The graph of any equation of the form
>
> $$\mathbf{y = ax^2 + c}$$
>
> where a and c are real constants and x is a real variable is a **parabola.**

We will start by discussing a specific example: $y = x^2$. The nature of the graph can be seen by plotting several points and noting that x^2 is never negative. So, the graph of $y = x^2$ will not be below the x-axis. (See Figure 13.12.)

$\mathbf{x}$	$\mathbf{y = x^2}$
-3	$(-3)^2 = 9$
-2	$(-2)^2 = 4$
-1	$(-1)^2 = 1$
$-\frac{1}{2}$	$\left(-\frac{1}{2}\right)^2 = \frac{1}{4}$
0	$(0)^2 = 0$
$\frac{1}{2}$	$\left(\frac{1}{2}\right)^2 = \frac{1}{4}$
1	$(1)^2 = 1$
2	$(2)^2 = 4$
3	$(3)^2 = 9$

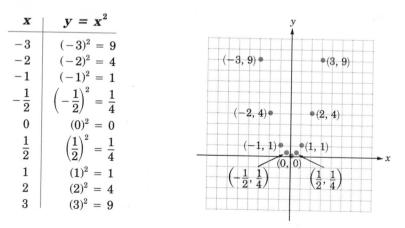

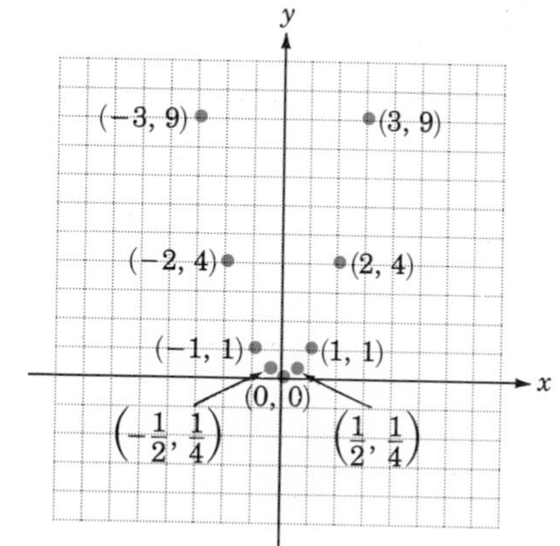

Figure 13.12

Now we assume (correctly) that the graph has points corresponding to all real values of x and that the graph is a smooth curve. The "turning point" of the parabola is called the **vertex.** (See Figure 13.13.)

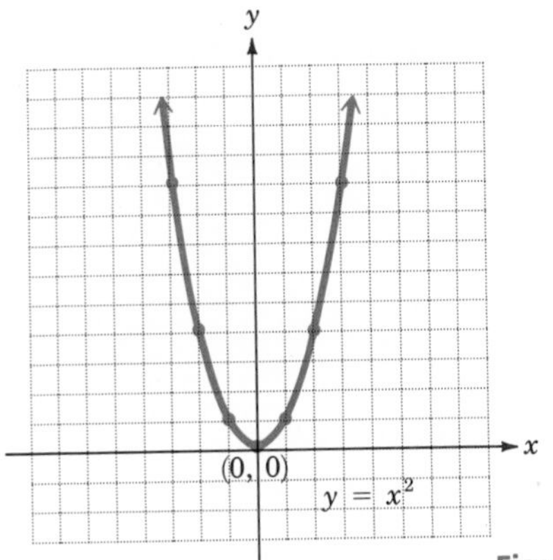

The graph of $y = x^2$ is a parabola. The origin (0, 0) is the **vertex.** We say the parabola **"opens upward."**

Figure 13.13

Whether a parabola "opens upward" or "opens downward" depends on the coefficient of x^2. We know that x^2 is positive. So, if a is negative, then $ax^2 \leq 0$, and the graph of $y = ax^2$ will be on or below the x-axis because y must be 0 or negative. If a is positive, then $ax^2 \geq 0$, and the graph of $y = ax^2$ will be on or above the x-axis.

Several graphs are shown in Figure 13.14.

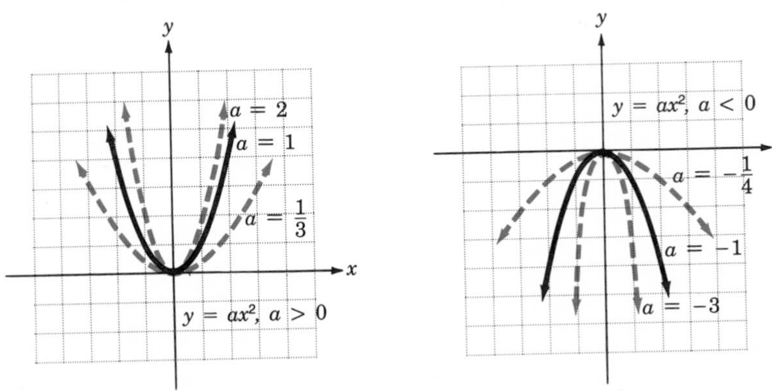

Figure 13.14

As the graphs in Figure 13.14 illustrate, the bigger $|a|$ is, the wider the opening. You can double check these graphs by setting up a table of values similar to that in Figure 13.12.

The more general case is $y = ax^2 + c$. Adding c to ax^2 simply increases (or decreases) each y-value by c units. Thus, the graph of $y = ax^2 + c$ is found by shifting the graph of $y = ax^2$. The shift is called a **vertical shift** and will be up $|c|$ units if $c > 0$ and down $|c|$ units if $c < 0$. (See Figure 13.15.) In any case, **the vertex is at the point (0, c).**

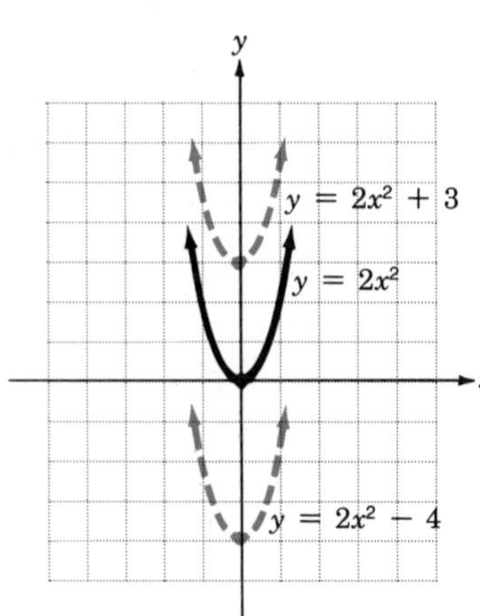

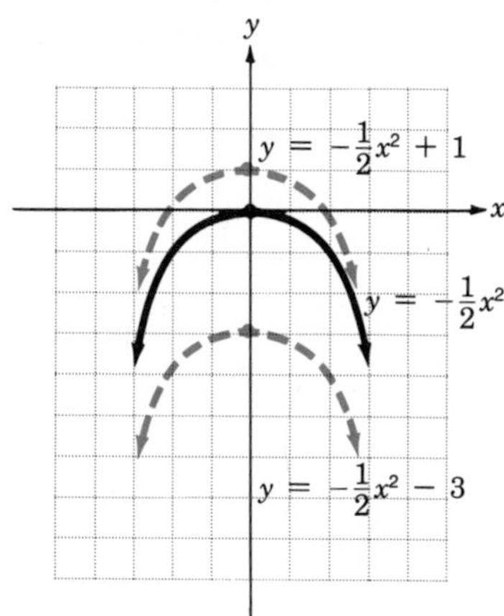

Figure 13.15

EXAMPLES

1. Make a table of values and draw the graph of $y = -2x^2$. Label the vertex.

Solution

x	$\mathbf{y = -2x^2}$
-2	$-2(-2)^2 = -8$
-1	$-2(-1)^2 = -2$
0	$-2(0)^2 = 0$
1	$-2(1)^2 = -2$
2	$-2(2)^2 = -8$

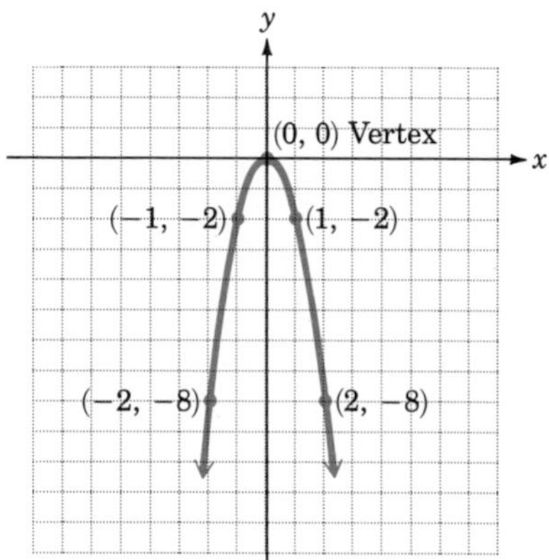

2. Make a table of values and draw the graph of $y = x^2 - 3$. Label the vertex.

Solution

x	$\mathbf{y = x^2 - 3}$
-3	$(-3)^2 - 3 = 6$
-2	$(-2)^2 - 3 = 1$
-1	$(1)^2 - 3 = -2$
0	$(0)^2 - 3 = -3$
1	$(1)^2 - 3 = -2$
2	$(2)^2 - 3 = 1$
3	$(3)^2 - 3 = 6$

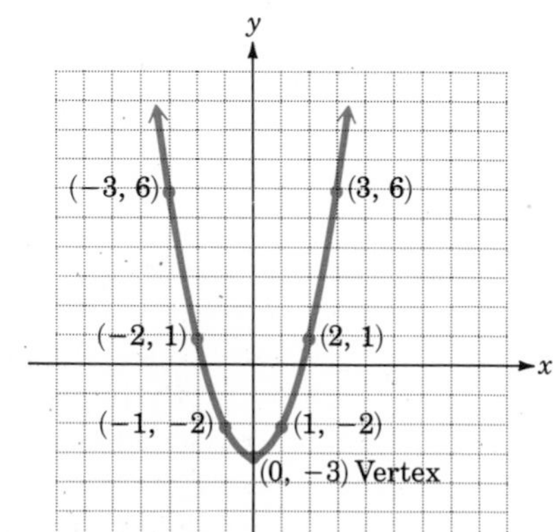

Exercises 13.5

Sketch the graph of each of the following equations and label the vertex.

1. $y = -x^2$
2. $y = -\frac{1}{2}x^2$
3. $y = 3x^2$
4. $y = 4x^2$
5. $y = \frac{1}{3}x^2$
6. $y = \frac{1}{4}x^2$
7. $y = -\frac{1}{2}x^2 + 1$
8. $y = \frac{1}{3}x^2 - 1$
9. $y = 3x^2 - 2$
10. $y = 2x^2 + 2$
11. $y = 3x^2 + 2$
12. $y = 2x^2 - 3$
13. $y = \frac{1}{4}x^2 + 1$
14. $y = \frac{1}{5}x^2 + 3$
15. $y = -\frac{1}{4}x^2 + 1$
16. $y = -\frac{1}{5}x^2 + 3$
17. $y = -5x^2 - 6$
18. $y = -4x^2 - 5$
19. $y = 5x^2 - 6$
20. $y = 4x^2 - 5$

13.6 DISTANCE BETWEEN TWO POINTS $[d = \sqrt{(x_2 - x_1)^2 + (y_2 - y_1)^2}]$ AND THE MIDPOINT FORMULA $\left(\frac{x_1 + x_2}{2}, \frac{y_1 + y_2}{2}\right)$

If two sides of a triangle are perpendicular to each other, the angle has a measure of 90° and is called a **right angle.** If a triangle has a right angle, it is called a **right triangle.** The longest side (opposite the 90° angle) is called the **hypotenuse.** (See Figure 13.16.)

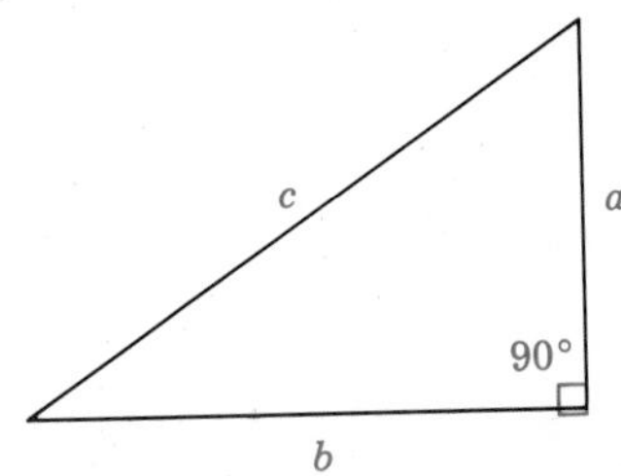

a and b are the shortest sides.
c is the hypotenuse.

Figure 13.16

Pythagoras, a famous Greek mathematician, is given credit for discovering the following theorem. He discovered that if he squared each side of a right triangle, the sum of the squares of the two sides was equal to the square of the hypotenuse.

PYTHAGOREAN THEOREM

In a right triangle the square of the hypotenuse is equal to the sum of the squares of the two sides:

$$c^2 = a^2 + b^2$$

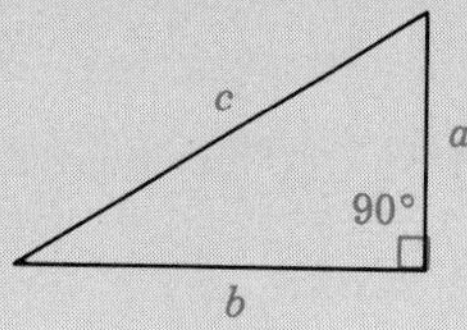

In effect, Pythagoras and his Society had discovered a new kind of number, irrational numbers. But they were living in an age when numbers were considered mystical, and the Society was so unpopular that its members had to repress their knowledge of the new numbers or risk punishment and ridicule.

The following examples illustrate the Pythagorean Theorem.

EXAMPLES

1.

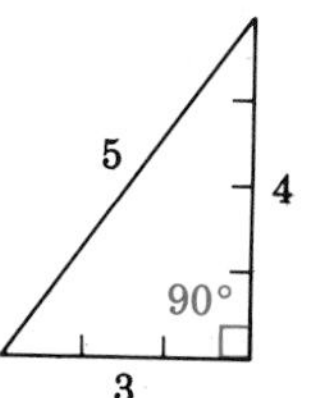

$$5^2 = 3^2 + 4^2$$
$$25 = 9 + 16$$

2.

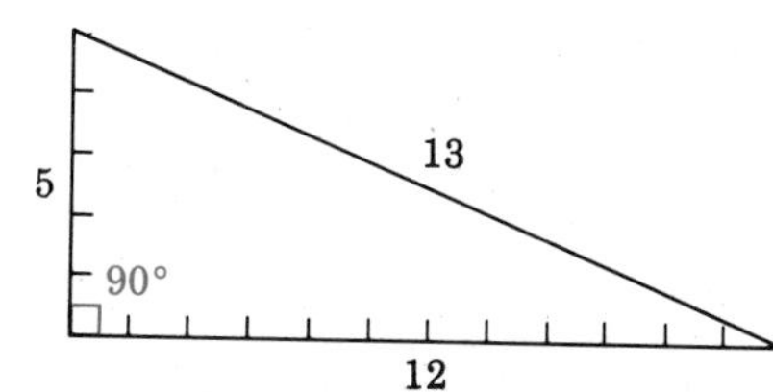

$$13^2 = 5^2 + 12^2$$
$$169 = 25 + 144$$

In most right triangles, all three sides do not have integer values. Examples 3 and 4 illustrate irrational numbers.

3.

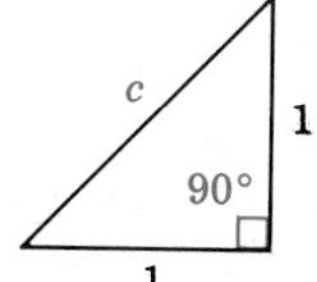

$$c^2 = 1^2 + 1^2$$
$$c^2 = 1 + 1 = 2$$
$$c = \sqrt{2}$$

That is, the length of the hypotenuse is $\sqrt{2}$.

4.

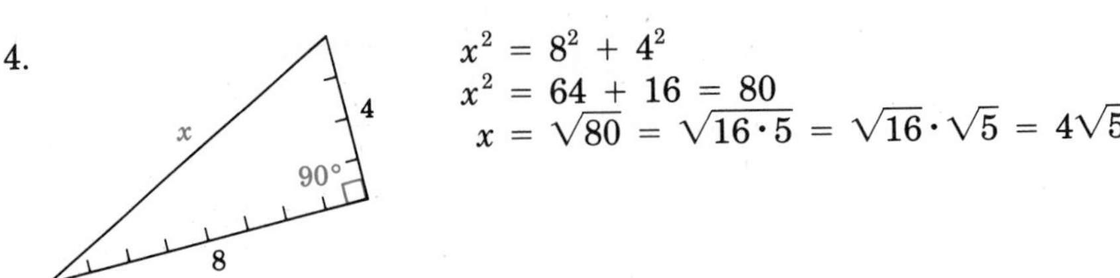

$$x^2 = 8^2 + 4^2$$
$$x^2 = 64 + 16 = 80$$
$$x = \sqrt{80} = \sqrt{16 \cdot 5} = \sqrt{16} \cdot \sqrt{5} = 4\sqrt{5}$$

The length of the hypotenuse is $4\sqrt{5}$.

Subscript notation is very useful in working with graphs and points. A **subscript** is a small number or letter written below and to the right of another letter. For example, P_1 (read "P sub 1") can represent one point, P_2 (read "P sub 2") can represent a second point, P_3 a third point, and so on. Similarly, the coordinates of P_1 might be (x_1, y_1), the coordinates of P_2 might be (x_2, y_2), and so on.

The Pythagorean Theorem can be used to find the distance between two points. Suppose P_1 has coordinates (2, 3) and P_2 has coordinates (6, 8), as shown in Figure 13.17.

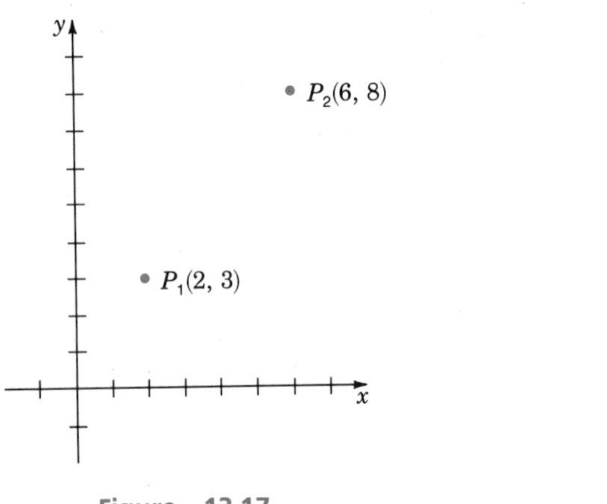

Figure 13.17

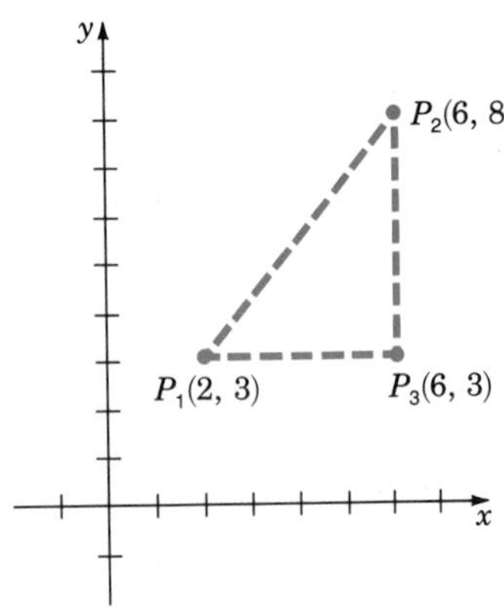

Figure 13.18

Form a right triangle by drawing a line through P_1 parallel to the x-axis and a line through P_2 parallel to the y-axis. The line segment joining P_1 and P_2 is the hypotenuse of the triangle. The point P_3 has coordinates (6, 3) since it is on a vertical line through P_2 and is on a horizontal line through P_1. (See Figure 13.18.)

The distance between P_1 and P_3 is the difference in the x-coordinates, $a = |6 - 2| = 4$. The distance between P_2 and P_3 is the difference in the y-coordinates, $b = |8 - 3| = 5$. By letting d be the distance between P_1 and P_2 (the hypotenuse) and applying the Pythagorean Theorem, we get

$$d^2 = a^2 + b^2 = 4^2 + 5^2 = 16 + 25 = 41$$

So, $d = \sqrt{41}$. (See Figure 13.19.)

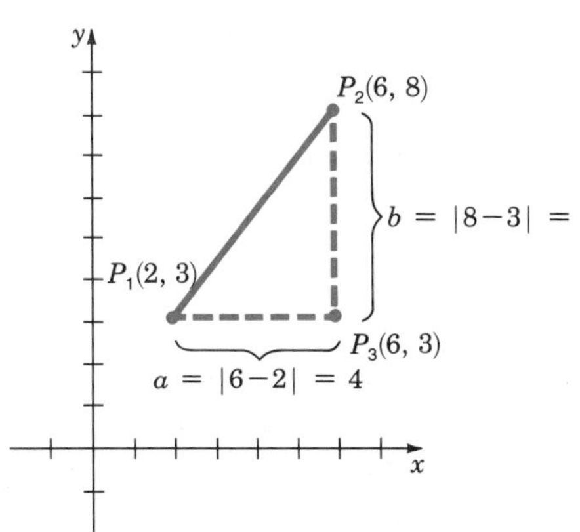

$$\begin{aligned} d^2 &= a^2 + b^2 \\ &= 4^2 + 5^2 \\ &= 16 + 25 = 41 \\ d &= \sqrt{41} \end{aligned}$$

Figure 13.19

This technique leads to the **distance formula** (see Figure 13.20).

$$\mathbf{d = \sqrt{(x_2 - x_1)^2 + (y_2 - y_1)^2}}$$

In the calculation, be sure to add the squares before taking the square root.

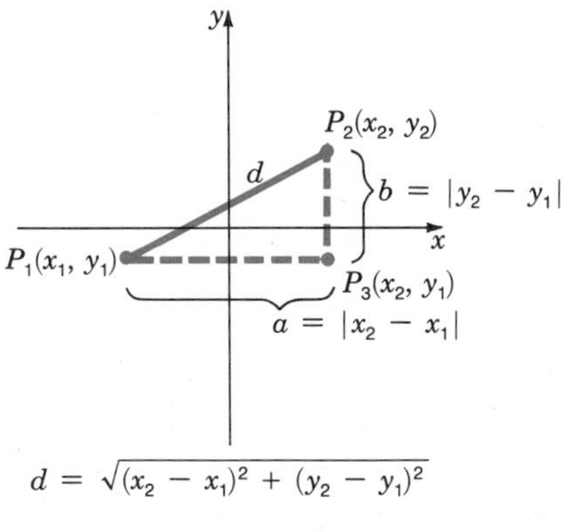

Figure 13.20

EXAMPLES

In Examples 5 and 6, draw a right triangle similar to the one in Figure 13.18 and label the points. Then use the distance formula to find the distance between the given points as illustrated in Figures 13.19 and 13.20.

5. $P_1(-2, -1)$ and $P_2(5, 3)$

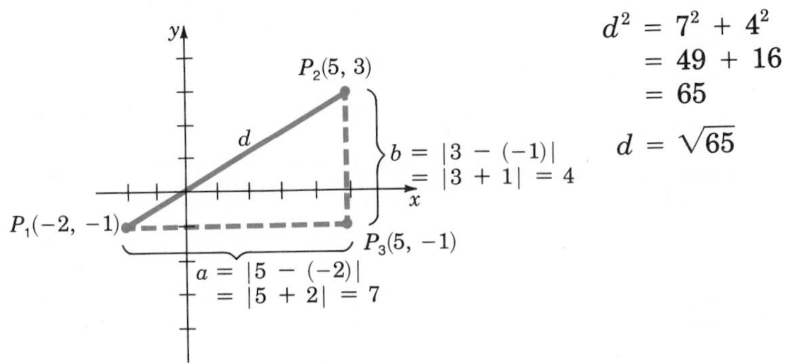

NOTE: We subtract negative numbers to find a and b.

6. $P_1(-2, 6)$ and $P_2(4, -2)$

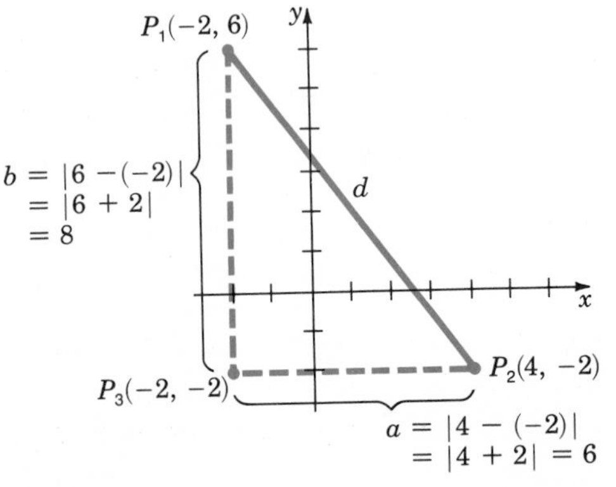

$$
\begin{aligned}
d^2 &= 6^2 + 8^2 \\
&= 36 + 64 \\
&= 100 \\
d &= \sqrt{100} = 10
\end{aligned}
$$

The distance formula gives the length of the line segment joining the two points $P_1(x_1, y_1)$ and $P_2(x_2, y_2)$. Another closely related topic is to find the coordinates of some point on the line segment. In particular, we are interested in finding **the midpoint of the line segment P_1P_2.** (See Figure 13.21.)

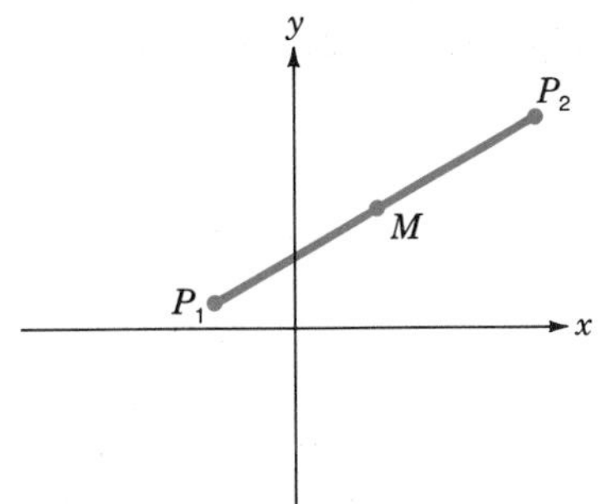

M is the midpoint of the line segment P_1P_2.

Figure 13.21

First, consider the coordinate of the point on a number line halfway between 3 and 7 as shown in Figure 13.22.

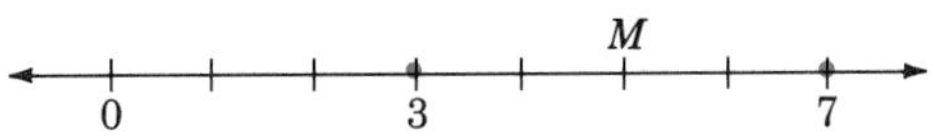

Figure 13.22

The distance between 3 and 7 is $7 - 3 = 4$. But this is **not** what we want. We don't want the distance. We want the coordinate. Obviously, from the graph, the coordinate is 5. We can get 5 by averaging 7 and 3; adding $7 + 3$ and dividing by 2.

$$\frac{7+3}{2} = \frac{10}{2} = 5$$

Thinking of the x-axis and y-axis as number lines (they are), we realize that the midpoint of a line segment is found by **averaging the x-coordinates and averaging the y-coordinates.** (See Figure 13.23.)

Thus, the **midpoint formula** is

$$\left(\frac{x_1 + x_2}{2}, \frac{y_1 + y_2}{2}\right)$$

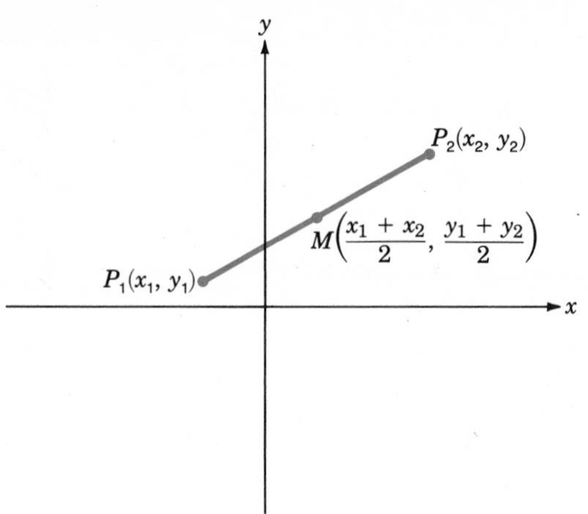

Figure 13.23

EXAMPLES

Plot each pair of points on a graph, and find the coordinates of the midpoint of the line segment joining these points.

7. $P_1(2, 0)$ and $P_2(5, 6)$

Solution

$$M = \left(\frac{2 + 5}{2}, \frac{0 + 6}{2}\right) = \left(\frac{7}{2}, 3\right)$$

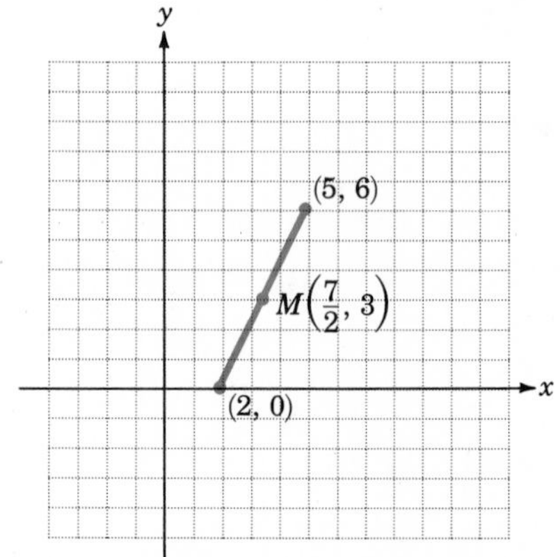

8. $P_1(-3, -1)$ and $P_2(4, -7)$

Solution

$$M = \left(\frac{-3 + 4}{2}, \frac{-1 + (-7)}{2}\right)$$

$$= \left(\frac{1}{2}, -4\right)$$

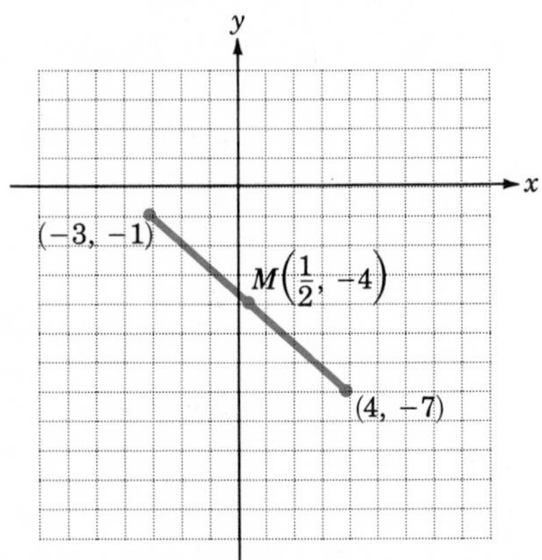

Exercises 13.6

For each problem, draw a right triangle similar to the one in Figure 13.18 and label the points. Then use the distance formula to find the distance between the given points.

1. $P_1(2, 0)$ and $P_2(3, 5)$

2. $P_1(3, 0)$ and $P_2(5, 2)$

3. $P_1(0, 3)$ and $P_2(5, 7)$

4. $P_1(0, 2)$ and $P_2(6, 7)$

5. $P_1(-4, 0)$ and $P_2(0, 3)$

6. $P_1(-3, 0)$ and $P_2(0, 4)$

7. $P_1(0, -12)$ and $P_2(5, 0)$

8. $P_1(-5, 0)$ and $P_2(0, 12)$

9. $P_1(-3, 2)$ and $P_2(3, -6)$

10. $P_1(1, -3)$ and $P_2(9, 3)$

11. $P_1(1, 5)$ and $P_2(4, 2)$

12. $P_1(1, 4)$ and $P_2(4, 1)$

13. $P_1(-6, 1)$ and $P_2(-1, -4)$

14. $P_1(-7, -2)$ and $P_2(0, -5)$

15. $P_1(-2, 5)$ and $P_2(2, -5)$

16. $P_1(-3, 4)$ and $P_2(4, -3)$

17. $P_1(-1, 3)$ and $P_2(6, -2)$

18. $P_1(1, -7)$ and $P_2(8, -1)$

19. $P_1(-3, 7)$ and $P_2(9, 2)$

20. $P_1(6, 4)$ and $P_2(8, -2)$

Use the distance formula to find the length of each side and use the Pythagorean Theorem to determine whether or not each triangle is a right triangle. Graph the points and draw the triangle.

21. $A(1, 1), B(7, 4), C(5, 8)$

22. $A(-5, -1), B(2, 1), C(-1, 6)$

Plot each pair of points on a graph, and find the coordinates of the midpoint of the line segment joining these points.

23. $P_1(1, 1)$ and $P_2(5, -1)$

24. $P_1(3, 1)$ and $P_2(5, -2)$

25. $P_1(-2, -1)$ and $P_2(4, 3)$

26. $P_1(-3, -2)$ and $P_2(4, -1)$

27. $P_1(-5, 5)$ and $P_2(3, 1)$

28. $P_1(-4, 3)$ and $P_2(3, -3)$

29. $P_1\left(\frac{1}{2}, \frac{2}{3}\right)$ and $P_2\left(\frac{17}{2}, \frac{16}{3}\right)$

30. $P_1\left(-\frac{1}{2}, -\frac{1}{3}\right)$ and $P_2\left(\frac{5}{8}, -\frac{2}{3}\right)$

31. $P_1\left(-\frac{5}{8}, \frac{1}{2}\right)$ and $P_2\left(\frac{3}{16}, -\frac{1}{2}\right)$

32. $P_1\left(-\frac{10}{3}, -\frac{2}{3}\right)$ and $P_2\left(-\frac{1}{3}, \frac{5}{6}\right)$

33. $P_1(-1.2, 1.6)$ and $P_2(5.4, 1.6)$

34. $P_1(2.3, -1.7)$ and $P_2(2.3, 5.5)$

35. $P_1(1.4, 3.6)$ and $P_2(4.2, -1.4)$

SUMMARY: CHAPTER 13

The symbol for **square root** is $\sqrt{}$, called a **square root sign** or **radical sign.** A **radicand** is the number under the radical sign.

Irrational numbers are infinite nonrepeating decimals. (There is no pattern to their decimal representation.) Most radicals are irrational numbers.

The **real numbers** include all rational numbers (whole numbers, integers, and positive and negative fractions) and all irrational numbers.

> If a and b are positive real numbers, then
>
> $$\sqrt{ab} = \sqrt{a}\sqrt{b}$$

> **To find the sum of radicals:**
>
> 1. Simplify each radical expression.
> 2. Use the distributive property to combine any like radicals.

Each point on a number line corresponds to one real number, and each real number corresponds to one point on a number line.

Intervals are classified as

Open $a < x < b$

$x > a$

$x < a$

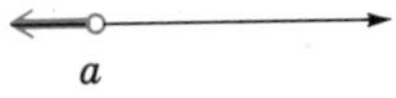

Closed $a \leq x \leq b$

Half-Open $a \leq x < b$

$a < x \leq b$

$x \geq b$

$x \leq b$

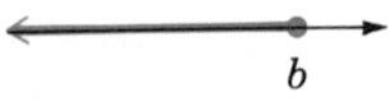

Linear inequalities are of the form $\mathbf{ax + b < c}$. ($>$, $\leq$, or $\geq$ might appear instead of $<$.)

RULES FOR SOLVING INEQUALITIES

1. The same number may be added to both sides, and the sense of the inequality will remain the same.
2. Both sides may be multiplied by (or divided by) the same positive number, and the sense of the inequality will remain the same.
3. Both sides may be multiplied by (or divided by) the same negative number, but the sense of the inequality must be reversed.

In an **ordered pair** such as (x, y), x is called the **first component** (or **first coordinate**), and y is called the **second component** (or **second coordinate**). x is also called the **independent variable** and y is called the **dependent variable.**

In the **Cartesian coordinate system,** two perpendicular number lines (called **axes**) intersect at a point (called the **origin**) and separate a plane into four **quadrants.** The horizontal number line is called the **horizontal axis** or **x-axis.** The vertical number line is called the **vertical axis** or **y-axis.**

Each point in a plane corresponds to one ordered pair of real numbers, and each ordered pair of real numbers corresponds to one point in a plane.

The points (ordered pairs) that satisfy any equation of the form

$$Ax + By = C \quad (A \text{ and } B \text{ not both } 0)$$

will lie on a straight line. The equation is called a **linear equation** and is in the **standard form** for the equation of a line.

The graph of any equation of the form

$$y = ax^2 + c$$

where a and c are real constants and x is a real variable is a **parabola.**

PYTHAGOREAN THEOREM

In a right triangle, the square of the hypotenuse is equal to the sum of the squares of the two sides:

$$c^2 = a^2 + b^2$$

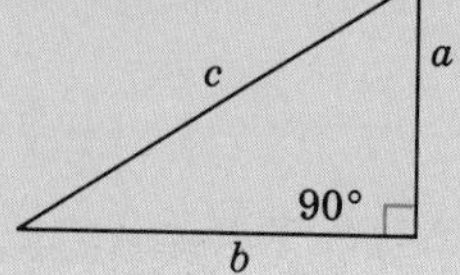

The **distance formula** is

$$d = \sqrt{(x_2 - x_1)^2 + (y_2 - y_1)^2}$$

The **midpoint formula** is

$$\left(\frac{x_1 + x_2}{2}, \frac{y_1 + y_2}{2}\right)$$

REVIEW QUESTIONS: CHAPTER 13

1. Show by squaring that $\sqrt{7}$ is between 2.645 and 2.646.

Answer the following questions using the set $B = \left\{-\sqrt{5}, -\frac{1}{2}, 0, 6.13\right\}$.

2. Which of the numbers in set B is an integer?

3. Which of the numbers in set B is a rational number?

4. Which of the numbers in set B is a real number?

Simplify the following radical expressions.

5. $\sqrt{169}$ **6.** $\sqrt{225}$ **7.** $\sqrt{128}$ **8.** $\sqrt{63}$

9. $\sqrt{54}$ **10.** $\sqrt{242}$ **11.** $\sqrt{300}$ **12.** $\sqrt{250}$

13. $4\sqrt{3} + \sqrt{3}$ **14.** $2\sqrt{5} + 3\sqrt{2} - 5\sqrt{5} - \sqrt{2}$ **15.** $\sqrt{75} + \sqrt{45}$

16. $2\sqrt{12a} + \sqrt{27a} - 4\sqrt{48a}$

Graph each of the following intervals and tell what kind of interval it is.

17. $-\frac{1}{2} < x < \frac{3}{4}$ **18.** $0 \le x < 5$ **19.** $x \ge \sqrt{2}$

20. $-3 \le x \le 3.1$ **21.** $y < \frac{1}{3}$ **22.** $14 < y \le 15$

Represent each of the following graphs using interval notation and tell what kind of interval it is.

23. 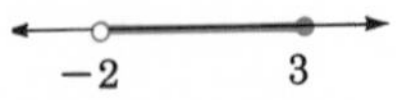

24. 

Solve and graph the solution for each of the following inequalities.

25. $x + 3 < -1$ **26.** $y - 5 \le 2$ **27.** $3y \ge 8$

28. $x - 1 > 3x + 5$ **29.** $-4x + 6 < 16 + x$ **30.** $9 \le 5x - 1 \le 14$

List the set of ordered pairs that correspond to the points in the graph.

31.

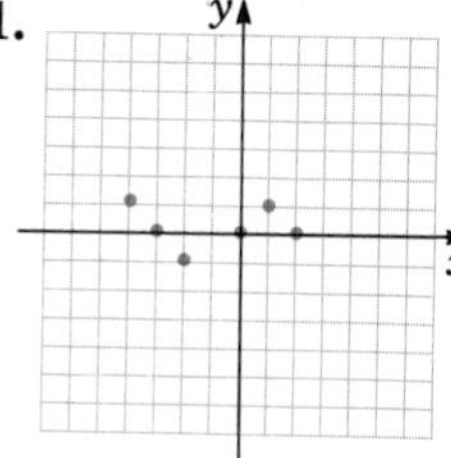

32.

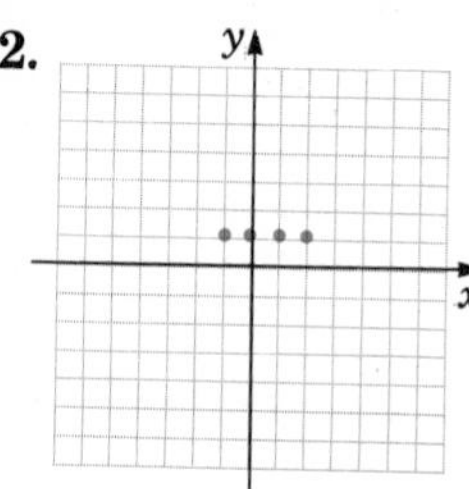

33.

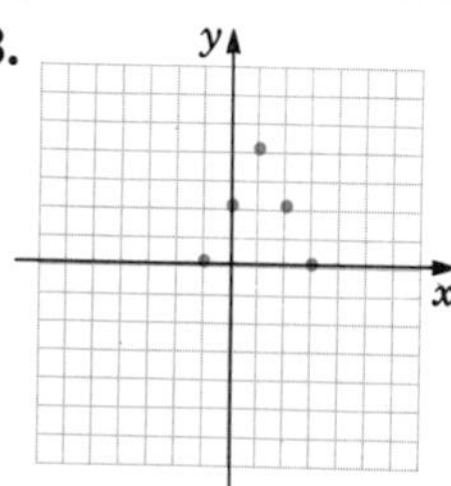

Graph each of the following sets of ordered pairs.

34. $\{(-3, 1), (-2, 1), (-1, 2), (0, 2), (1, 3)\}$

35. $\{(-4, 5), (-3, 2), (0, -4), (1, 1), (3, 1)\}$

36. $\{(1, 4), (1, 3), (1, 1), (1, 0)\}$

Graph the following linear equations.

37. $y = 2x - 1$ **38.** $y = x + 5$ **39.** $y = -x - 2$

40. $3x + y = 6$ **41.** $2x - 3y = 12$ **42.** $y = -3$

Graph the following parabolas and label each vertex.

43. $y = -x^2 + 4$ **44.** $y = 2x^2 - 3$

Use the distance formula to find the distance between the given points.

45. $P_1(0, 3)$ and $P_2(4, 6)$ **46.** $P_1(-2, 1)$ and $P_2(5, -1)$

47. Graph the three points $A(-1, 2)$, $B(3, 4)$, and $C(3, -4)$ and draw the triangle. Use the Pythagorean Theorem to determine whether or not the triangle is a right triangle.

Plot each pair of points on a graph, and find the coordinates of the midpoint of the line segment joining these points.

48. $P_1(3, -4)$ and $P_2(3, 5)$ **49.** $P_1(-1, 4)$ and $P_2(1, -4)$

TEST: CHAPTER 13

Answer Questions 1 and 2 using the set

$$A = \left\{-5, -\sqrt{9}, -1.2, 0, \frac{\sqrt{5}}{2}, 1\frac{3}{4}\right\}$$

1. Which of the numbers in set A is a rational number?

2. Which of the numbers in set A is an integer?

Simplify the following radical expressions.

3. $3\sqrt{5} - \sqrt{2} + 2\sqrt{5} + 2\sqrt{2}$ **4.** $\sqrt{12} - \sqrt{18}$

5. $\sqrt{50} - 3\sqrt{8} + \sqrt{72}$

6. Represent the following graph using interval notation and tell what kind of interval it is:

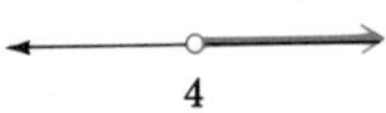

7. Graph the following interval and tell what kind of interval it is:

$$-\frac{1}{2} < x < \frac{3}{4}$$

Solve and graph the solutions for each of the following inequalities.

8. $2x - 5 \leq 5x - 2$ **9.** $-3 < 7x + 4 < 11$

10. List the set of ordered pairs that correspond to the points in the graph.

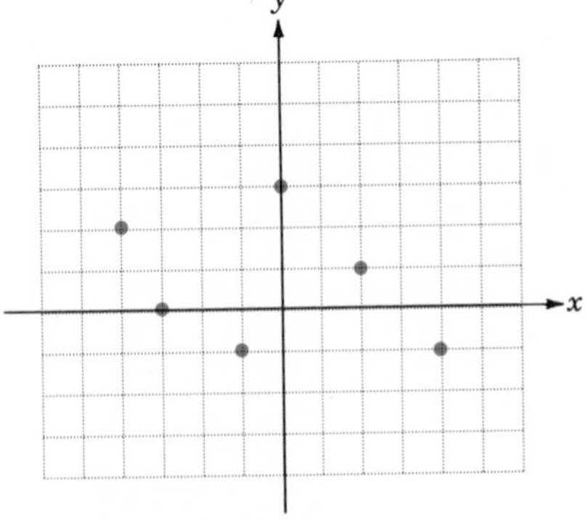

11. Graph the set of ordered pairs:

$$\{(-2, 5), (-1, 3), (0, 1), (1, -1), (2, 0)\}$$

Graph the following linear equations.

12. $y = -2x + 1$

13. $2x - 3y = 6$

14. $x = 2\frac{1}{2}$

Graph the following parabolas and label each vertex.

15. $y = x^2 + 2$

16. $y = -\frac{1}{2}x^2 + 5$

Use the distance formula to find the distance between the points for each of the following.

17. $P_1(-3, 4)$ and $P_2(1, 8)$

18. $P_1(7, -5)$ and $P_2(3, -2)$

19. Use the Pythagorean Theorem to show that the points $A(-3, -1)$, $B(2, -1)$, and $C(2, 5)$ do form a right triangle.

20. Plot the following pair of points on a graph, and find the coordinates of the midpoint of the line segment joining the points.

$$P_1(-5, -3) \text{ and } P_2(3, 5)$$

CUMULATIVE TEST III: CHAPTERS 1–13

All answers should be in simplest form. Reduce all fractions to lowest terms and express all improper fractions as mixed numbers.

Find each sum.

1. $\frac{2}{3} + \frac{5}{6} + \frac{2}{9}$

2. $17\frac{5}{8} + 12\frac{7}{10}$

3. $6.09 + 10.6 + 7$

4. (4 yd 2 ft 8 in.) + (1 yd 6 in.)

5. $(-2.03) + (16.7) + (-5.602)$

Find each difference.

6. $\frac{3}{4} - \frac{5}{9}$

7. $4\frac{1}{6} - 2\frac{2}{3}$

8. $142.01 - 67.135$

9. (3 days 9 hr 15 min) − (9 hr 30 min)

10. Subtract -4.5 from -0.03.

Find each product.

11. $2400 \cdot 30{,}000$

12. $\frac{5}{12} \cdot \frac{3}{5} \cdot \frac{4}{7}$

13. $\left(3\frac{2}{3}\right)\left(4\frac{4}{11}\right)$

14. $(7.03)(0.28)$

15. $(-9)(1.6)$

Find each quotient.

16. $\frac{4}{9} \div \frac{8}{15}$

17. $5\frac{1}{4} \div \frac{7}{10}$

18. $1.0336 \div 1.7$

19. $-3.7 \div 0$

20. Find the prime factorization of 792.

21. Which of the numbers 2, 3, 4, 5, 9, 10 will exactly divide 31,752?

22. True or false: $\frac{2}{3} : 72 = \frac{3}{4} : 56$.

Round off each of the following as indicated.

23. 5987.042 to the nearest ten

24. 723.042 to the nearest tenth

25. Find the mean of the following decimal numbers (correct to the nearest hundredth): 6.07, 9.72, 8.51, and 7.23.

Evaluate each of the following expressions.

26. $(2^2 \cdot 3 \div 4 + 2) \cdot 5 - 3$

27. $4(x - 2) + 2(3y + 4)$, if $x = -3$ and $y = 2$

28. Simplify: $3\sqrt{20} + 4\sqrt{75} - \sqrt{45}$.

29. Find the distance between the points $P_1(0, -2)$ and $P_2(6, 4)$.

30. Fill in the blank with the appropriate symbol: <, >, or =.

$$-|-7| \_\_\_\_\_ -(-7)$$

31. Write an algebraic expression for the following English phrase: 5 less than the sum of twice a number and 6.

Solve each of the following.

32. $-2(3 - x) = 3(x + 2)$

33. $\frac{5}{6}x - 1 = \frac{2}{3}x + 2$

34. Solve and graph the solution for $2 \leq 3 - 4x < 5$.

35. Solve $y = mx + b$ for x in terms of the other variables.

Graph each of the following.

36. $3x + 4y = 0$

37. $y = 5 - \frac{1}{3}x^2$

38. $16\frac{1}{2}\%$ of what number is $6\frac{3}{5}$?

39. What percent of 2.1 is 4.41?

40. Given that 1 pound = 0.454 kilograms, 1 lb 4 oz = _____ g?

41. Given that 1 inch = 2.54 centimeters, 1 yd 2 ft 10 in. = _____ m?

42. Use either $C = \frac{5(F - 32)}{9}$ or $F = \frac{9}{5}C + 32$ to convert 24.5° C to degrees Fahrenheit.

43. Find the volume in liters of a rectangular box 80 cm long, 40 cm wide, and 20 cm high.

44. Find the area in square centimeters of a rectangle 0.12 m long and 0.07 m wide.

45. Find the circumference of a circle that has a radius of 20 centimeters. (Use $\pi \approx 3.14$.)

46. Find the area in square meters of a circle that has a diameter of 50 cm. (Use $\pi \approx 3.14$.)

47. Two towns are shown as 13.5 centimeters apart on a map that has a scale of 3 centimeters to 50 miles. What is the actual distance between the towns (in miles)?

48. During the last week, Cara jogged the following distances: 2 miles on both Thursday and Saturday; 0 miles on Sunday; 3 miles per day on Monday, Tuesday and Wednesday, and 1 mile on Friday. What was her average number of miles per day for last week?

49. If $1550 is deposited in an account paying 8% simple interest, what is the total amount in the account at the end of the first year (to the nearest dollar)?

50. Find two consecutive integers such that 3 more than twice the smaller equals 13 less than the larger integer.

Appendix I
Ancient Numeration Systems

I.1 EGYPTIAN, MAYAN, ATTIC GREEK, AND ROMAN SYSTEMS

The number systems used by ancient peoples are interesting from a historical point of view, but from a mathematical point of view they are difficult to work with. One of the many things that determine the progress of any civilization is its system of numeration. Humankind has made its most rapid progress since the invention of the zero and the place value system (which we will discuss in the next section) by the Hindu-Arabic peoples about A.D. 800.

Egyptian Numerals (Hieroglyphics)

The ancient Egyptians used a set of symbols called hieroglyphics as early as 3500 B.C. (See Table I.1 on page A2.) To write the numeral for a number, the Egyptians wrote the symbols next to each other from left to right, and the number represented was the sum of the values of the symbols. The most times any symbol was used was nine. Instead of using a symbol ten times, they used the symbol for the next higher number. They also grouped the symbols in threes or fours.

Table I.1 Egyptian Hieroglyphic Numerals

SYMBOL	NAME	VALUE	
I	Staff (vertical stroke)	1	one
∩	Heel bone (arch)	10	ten
𓍢	Coil of rope (scroll)	100	one hundred
𓆼	Lotus flower	1000	one thousand
𓂭	Pointing finger	10,000	ten thousand
𓆐	Bourbot (tadpole)	100,000	one hundred thousand
𓁨	Astonished man	1,000,000	one million

EXAMPLE

1. 𓆼 𓍢𓍢𓍢𓍢𓍢𓍢 ∩∩ IIII III represents the number one thousand six hundred twenty-seven, or 1000 + 600 + 20 + 7 = 1627

Mayan System

The Mayans used a system of dots and bars (for numbers from 1 to 19) combined with a place value system. A dot represented one and a bar represented five. They had a symbol, 𝋠, for zero and based their system, with one exception, on twenty. (See Table I.2.) The symbols were arranged vertically, smaller values starting at the bottom. The value of the third place up was 360 (18 times the value of the second place), but all other places were 20 times the value of the previous place.

Table I.2 Mayan Numerals

SYMBOL	VALUE	
•	1	one
—	5	five
𝋠	0	zero

EXAMPLES

2. $\underline{\cdots}$ (3 + 5 = 8)

3. $\underline{\underline{\underline{\cdots\cdot}}}$ (3·5 + 4 = 19)

4. $\cdots$ 3 20's
   ◯ 0 units
   (3·20 + 0 = 60)

5. $\cdot\cdot$ 2 7200's
   ◯ 0 360's
   $\underline{\cdot}$ 6 20's
   $\underline{\cdot\cdot}$ 7 units
   (2·7200 + 0·360 + 6·20 + 7 = 14,527)

[**Note:** ◯ is used as a place holder.]

Attic Greek System

The Greeks used two numeration systems, the Attic (see Table I.3) and the Alexandrian (see Section I.2 for information on the Alexandrian system). In the Attic system, no numeral was used more than four times. When a symbol was needed five or more times, the symbol for five was used, as shown in the examples.

Table I.3 Attic Greek Numerals

SYMBOL	VALUE	
I	1	one
Γ	5	five
Δ	10	ten
H	100	one hundred
X	1000	one thousand
M	10,000	ten thousand

EXAMPLES

6. X X 𐅅 H H 𐅄 I I I I (2·1000 + 7·100 + 5·10 + 4 = 2754)

7. 𐅇 H H H H Δ Δ Γ (5·1000 + 4·100 + 2·10 + 5 = 5425)

Roman System

The Romans used a system (Table I.4) that we still see in evidence as hours on clocks and dates on buildings.

Table I.4 Roman Numerals

SYMBOL	VALUE	
I	1	one
V	5	five
X	10	ten
L	50	fifty
C	100	one hundred
D	500	five hundred
M	1000	one thousand

The symbols were written largest to smallest, from left to right. The value of the numeral was the sum of the values of the individual symbols. Each symbol was used as many times as necessary, with the following exceptions: When the Romans got to 4, 9, 40, 90, 400, or 900, they used a system of subtraction.

IV = 5 − 1 = 4
IX = 10 − 1 = 9
XL = 50 − 10 = 40

XC = 100 − 10 = 90
CD = 500 − 100 = 400
CM = 1000 − 100 = 900

EXAMPLES

8. VII represents 7

9. DXLIV represents 544

10. MCCCXXVIII represents 1328

Exercises I.1

Find the values of the following ancient numbers.

1. 𓍢𓍢 𓎆𓎆𓎆𓎆𓎆 𓏺𓏺𓏺

2. 𓆐 𓂭𓂭𓂭𓂭𓂭𓂭 𓆼𓆼𓆼 𓎆𓎆𓎆𓎆 𓏺

3. 𓂭𓂭 𓆼 𓍢𓍢𓍢 𓎆𓎆 𓏺𓏺𓏺𓏺𓏺𓏺𓏺

4. ••• over two bars

5.

6.

7. ΓΔ I I I

8. ΓH H H Δ Δ Δ ΓΔ I

9. X X H H H ΓΔ Δ I I

10. XCVII

11. DCCXLIV

12. MMMCDLXV

13. CMLXXVIII

14. Write 64
    (a) as an Egyptian numeral
    (b) as a Mayan numeral
    (c) as an Attic Greek numeral
    (d) as a Roman numeral

15. Follow the instructions for Problem 14, using 532 in place of 64.

16. Follow the same instructions, using 1969.

17. Follow the same instructions, using 846.

I.2 BABYLONIAN, ALEXANDRIAN GREEK, AND CHINESE-JAPANESE SYSTEMS

Babylonian System (Cuneiform Numerals)

The Babylonians (about 3500 B.C.) used a place value system based on the number sixty, called a sexagesimal system. They had only two symbols, V and <. (See Table I.5.) These wedge shapes are called cuneiform numerals, since **cuneus** means **wedge** in Latin.

Table I.5 Cuneiform Numerals

SYMBOL	VALUE	
V	1	one
<	10	ten

The symbol for one was used as many as nine times, and the symbol for ten as many as five times; however, since there was no symbol for zero, many Babylonian numbers could be read several ways. For our purposes, we will group the symbols to avoid some of the ambiguities inherent in the system.

EXAMPLES

1. VVV <<VVV VV <<<VV

$$(3 \cdot 60^2) + (25 \cdot 60^1) + (32 \cdot 1) = (3 \cdot 3600) + (25 \cdot 60) + 32$$
$$= 10{,}800 + 1500 + 32$$
$$= 12{,}332$$

2. V <<<<VVV VVV <VVVV VVV

$$(1 \cdot 60^2) + (46 \cdot 60^1) + (17 \cdot 1) = 3600 + 2760 + 17 = 6377$$

Alexandrian Greek System

The Greeks used two numeration systems, the Attic and the Alexandrian. We discussed the Attic Greek system in Section I.1.

In the Alexandrian system (Table I.6), the letters were written next to each other, largest to smallest, from left to right. Since the numerals were also part of the Greek alphabet, an accent mark or bar was sometimes used above a letter to indicate that it represented a number. Multiples of 1000 were indicated by strikes in front of the unit symbols, and multiples of 10,000 were indicated by placing the unit symbols above the symbol M.

Table I.6 Alexandrian Greek Symbols

SYMBOL	NAME	VALUE	SYMBOL	NAME	VALUE
Α	Alpha	1 one	Ξ	Xi	60 sixty
Β	Beta	2 two	Ο	Omicron	70 seventy
Γ	Gamma	3 three	Π	Pi	80 eighty
Δ	Delta	4 four	C	Koppa	90 ninety
Ε	Epsilon	5 five	Ρ	Rho	100 one hundred
F	Digamma (or Vau)	6 six	Σ	Sigma	200 two hundred
Ζ	Zeta	7 seven	Τ	Tau	300 three hundred
Η	Eta	8 eight	Υ	Upsilon	400 four hundred
Θ	Theta	9 nine	Φ	Phi	500 five hundred
Ι	Iota	10 ten	Χ	Chi	600 six hundred
Κ	Kappa	20 twenty	Ψ	Psi	700 seven hundred
Λ	Lambda	30 thirty	Ω	Omega	800 eight hundred
Μ	Mu	40 forty	ΤΤ	Sampi	900 nine hundred
Ν	Nu	50 fifty			

EXAMPLES

3. $\overline{\Phi\ \Xi\ \mathrm{Z}}$ (500 + 60 + 7 = 567)

4. $\overline{\mathrm{B}\qquad\qquad}$
   M T N Δ (20,000 + 300 + 50 + 4 = 20,354)

Chinese-Japanese System

The Chinese-Japanese system (Table I.7) uses a different numeral for each of the digits up to ten, then a symbol for each power of ten. A digit written above a power of ten is to be multiplied by that power, and all such results are to be added to find the value of the numeral.

Table I.7 Chinese-Japanese Numerals

SYMBOL	VALUE	SYMBOL	VALUE
一	1 one	七	7 seven
二	2 two	八	8 eight
三	3 three	九	9 nine
四	4 four	十	10 ten
五	5 five	百	100 one hundred
六	6 six	千	1000 one thousand

EXAMPLES

5.
三 } 30
十 }
九 9

(30 + 9 = 39)

6.
五 } 5000
千 }
四 } 400
百 }
八 } 80
十 }
二 } 2

(5000 + 400 + 80 + 2 = 5482)

Exercises I.2

Find the value of each of the following ancient numerals.

1. V <<VVV VVV 2. <<<VV 3. <<< VV

4. $\overline{\text{Y N E}}$ 5. $\overline{\Delta}$ M /Z π 6. $\overline{\Sigma \text{ K B}}$

7. 四
十
六

8. 五
千
一
百
八

9. 九
百
九
十
九

Write the following numbers as (a) Babylonian numerals, (b) Alexandrian Greek numerals, and (c) Chinese-Japanese numerals.

10. 472 11. 596 12. 5047 13. 3665 14. 7293 15. 10,852

Appendix II
Base Two and Base Five

II.1 THE BINARY SYSTEM (BASE TWO)

In the decimal system, ten is the base. You might ask if another number could be chosen as the base in a place value system. And, if so, would the system be any better or more useful than the decimal system? The fact is that computers do operate under a place value system whose base is two. In the **binary system** (or base two system), only two digits are needed, 0 and 1. These two digits correspond to the two possible conditions of an electric current, either **on** or **off.**

Any number can be represented in base two or in base ten. However, base ten has a definite advantage when large numbers are involved, as you will see. The advantage of base two over base ten is that for base two only two digits are needed, while ten digits are needed for base ten.

If the base of a place value system were not ten but two, then the beginning point would be not a decimal point but a **binary point.** The value of each place would be a power of two, as shown in Figure II.1.

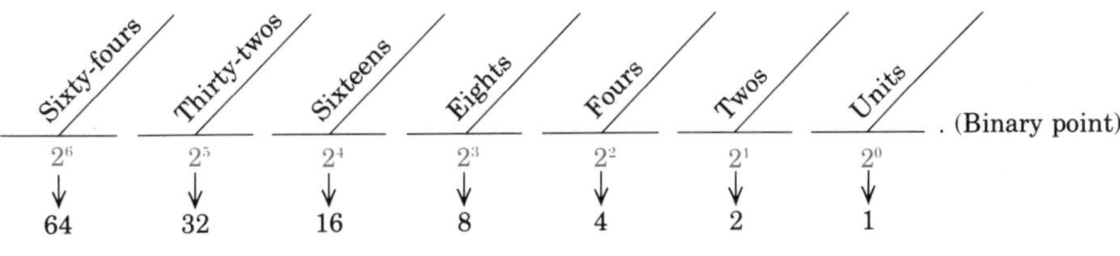

Figure II.1

TO WRITE NUMBERS IN THE BASE TWO SYSTEM, REMEMBER THREE THINGS

1. {0, 1} is the set of digits we can use.
2. The value of each place from the binary point is in powers of two.
3. The symbol 2 does not exist in the binary system, just as there is no digit for ten in the decimal system.

To avoid confusion with the base ten numerals, we will write $_{(2)}$ to the lower right of each base two numeral. We could write $_{(10)}$ to the lower right of each base ten numeral, but this would not be practical since most of the numerals we work with are in base ten. Therefore, **if no base is indicated, the numeral will be understood to be in base ten.**

EXAMPLES

1. Find the value of the numeral $1101._{(2)}$.
   Writing the value of each place under the digit gives

$$\frac{1}{2^3}\ \frac{1}{2^2}\ \frac{0}{2^1}\ \frac{1}{2^0}\ ._{(2)}$$

In expanded notation,

$$\begin{aligned} 1101._{(2)} &= 1(2^3) + 1(2^2) + 0(2^1) + 1(2^0) \\ &= 1(8) + 1(4) + 0(2) + 1(1) \\ &= 8 + 4 + 0 + 1 \\ &= 13 \end{aligned}$$

Thus, to a computer, the symbol $1101._{(2)}$ means "thirteen."

2. $\underline{1}._{(2)} = 1$

3. $\underline{1}\,\underline{0}._{(2)} = 1(2) + 0 = 2$

4. $\underline{1}\,\underline{1}._{(2)} = 1(2) + 1 = 2 + 1 = 3$

5. $\underline{1}\,\underline{0}\,\underline{0}._{(2)} = 1(2^2) + 0(2) + 0(1) = 4 + 0 + 0 = 4$

6. $\underline{1}\,\underline{0}\,\underline{1}._{(2)} = 1(2^2) + 0(2) + 1(1) = 4 + 0 + 1 = 5$

7. $\underline{1}\,\underline{1}\,\underline{0}._{(2)} = 1(2^2) + 1(2) + 0(1) = 4 + 2 + 0 = 6$

8. $\underline{1}\,\underline{1}\,\underline{1}._{(2)} = 1(2^2) + 1(2) + 1(1) = 4 + 2 + 1 = 7$

9. $\underline{1}\,\underline{0}\,\underline{0}\,\underline{0}._{(2)} = 1(2^3) + 0(2^2) + 0(2) + 0(1) = 8 + 0 + 0 + 0 = 8$

10. $\underline{1}\,\underline{0}\,\underline{0}\,\underline{1}._{(2)} = 1(2^3) + 0(2^2) + 0(2) + 1(1) = 8 + 0 + 0 + 1 = 9$

11. $\underline{1}\,\underline{0}\,\underline{1}\,\underline{0}._{(2)} = 1(2^3) + 0(2^2) + 1(2) + 0(1) = 8 + 0 + 2 + 0 = 10$

Do **not** read $100_{(2)}$ as "one hundred" because the 1 is not in the hundreds place. The 1 is in the fours place. So, $100_{(2)}$ is read "four" or "one, zero, zero–base two." Similarly, $111_{(2)}$ is read "seven" or "one, one, one—base two."

Exercises II.1

Write the following base ten numerals in expanded form using components.

Example: $273 = 2(10^2) + 7(10^1) + 3(10^0)$

1. 35 **2.** 761 **3.** 8469 **4.** 500 **5.** 62,322

Write the following base two numerals in expanded form and find the value of each numeral.

Example:
$$\begin{aligned} 110_{(2)} &= 1(2^2) + 1(2^1) + 0(2^0) \\ &= 1(4) + 1(2) + 0(1) \\ &= 4 + 2 + 0 \\ &= 6 \end{aligned}$$

6. $11_{(2)}$ **7.** $101_{(2)}$ **8.** $111_{(2)}$ **9.** $1011_{(2)}$

10. $1101_{(2)}$ **11.** $110111_{(2)}$ **12.** $11110_{(2)}$ **13.** $101011_{(2)}$

14. $11010_{(2)}$ **15.** $1000_{(2)}$ **16.** $1000010_{(2)}$ **17.** $11101_{(2)}$

18. $10110_{(2)}$ **19.** $111111_{(2)}$ **20.** $1111_{(2)}$

21. A computer is directed to place some infomation in memory space number $1101111_{(2)}$. What is the number of this memory space in base ten?

II.2 THE QUINARY SYSTEM (BASE FIVE)

Many numbers may be used as bases for place value systems. To illustrate this point and to emphasize the concept of place value, we will discuss one more base system, base five. Interested students may want to try writing numerals in base three or base eight or base eleven.

Again, the system relies on powers of the base and a set of digits. In the **quinary system** (base five system), the powers of 5 are 5^0, 5^1, 5^2, 5^3,

5^4, and so on, and the digits to be used are {0, 1, 2, 3, 4}. The **quinary point** is the beginning point, as shown in Figure II.2.

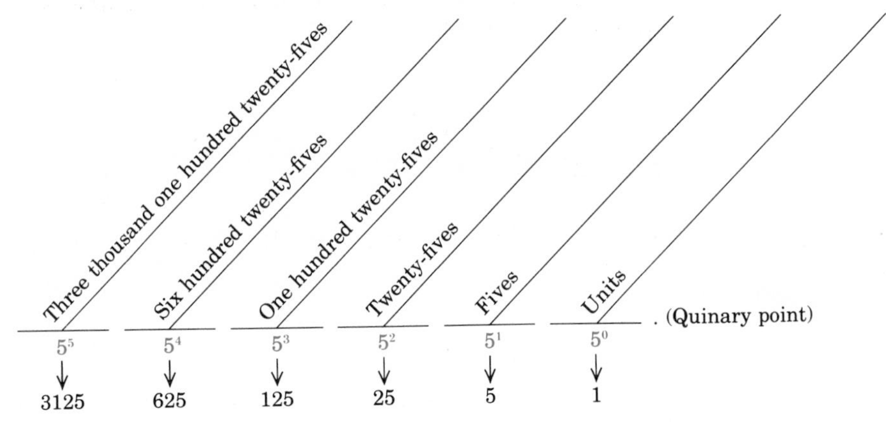

Figure II.2

EXAMPLES

1. $\underline{1}._{(5)} = 1$
2. $\underline{2}._{(5)} = 2$
3. $\underline{3}._{(5)} = 3$
4. $\underline{4}._{(5)} = 4$
5. $\underline{1}\,\underline{0}._{(5)} = 1(5) + 0(1) = 5 + 0 = 5$
6. $\underline{1}\,\underline{1}._{(5)} = 1(5) + 1(1) = 5 + 1 = 6$
7. $\underline{1}\,\underline{2}._{(5)} = 1(5) + 2(1) = 5 + 2 = 7$
8. $\underline{1}\,\underline{3}._{(5)} = 1(5) + 3(1) = 5 + 3 = 8$
9. $\underline{1}\,\underline{4}._{(5)} = 1(5) + 4(1) = 5 + 4 = 9$
10. $\underline{2}\,\underline{0}._{(5)} = 2(5) + 0(1) = 10 + 0 = 10$
11. $\underline{2}\,\underline{1}._{(5)} = 2(5) + 1(1) = 10 + 1 = 11$
12. $\underline{2}\,\underline{2}._{(5)} = 2(5) + 2(1) = 10 + 2 = 12$
13. $\underline{3}\,\underline{2}\,\underline{4}._{(5)} = 3(5^2) + 2(5^1) + 4(5^0)$
    $= 3(25) + 2(5) + 4(1)$
    $= 75 + 10 + 4$
    $= 89$

Exercises II.2

Write the following base five numerals in expanded form and find the value of each.

1. $24_{(5)}$ **2.** $13_{(5)}$ **3.** $10_{(5)}$ **4.** $43_{(5)}$ **5.** $104_{(5)}$

6. $312_{(5)}$ **7.** $32_{(5)}$ **8.** $230_{(5)}$ **9.** $423_{(5)}$ **10.** $444_{(5)}$

11. $1034_{(5)}$ **12.** $4124_{(5)}$ **13.** $244_{(5)}$ **14.** $3204_{(5)}$ **15.** $13042_{(5)}$

16. Do the numerals $101_{(2)}$ and $10_{(5)}$ represent the same number? If so, what is the number?

17. Answer the questions in Problem 16 about the numerals $1101_{(2)}$ and $23_{(5)}$.

18. Answer the questions in Problem 16 about the numerals $11{,}100_{(2)}$ and $103_{(5)}$.

19. What set of digits do you think would be used in a base eight system?

20. What set of digits do you think would be used in a base twelve system? [HINT: New symbols for some new digits must be introduced.]

II.3 ADDITION AND MULTIPLICATION IN BASE TWO AND BASE FIVE

Now that we have two new numeration systems, base two and base five, a natural question to ask is, How are addition and multiplication* performed in these systems? The basic techniques are the same as for base ten, since place value is involved. However, because different bases are involved, the numerals will be different. For example, to add five plus seven in base ten, we write $5 + 7 = 12$. In base two, this same sum is written $101_{(2)} + 111_{(2)} = 1100_{(2)}$.

Writing the numerals vertically (one under the other) gives

$$\begin{array}{r} 5 \\ 7 \\ \hline 12 \end{array} \qquad \begin{array}{r} 101_{(2)} \\ 111_{(2)} \\ \hline 1100_{(2)} \end{array}$$

Now a step-by-step analysis of the sum in base two will be provided.

*Subtraction and division may also be performed in base two and base five, but will not be discussed here for reasons of time. Some students may want to investigate these operations on their own.

EXAMPLE

1. (a) $\begin{array}{r} 101_{(2)} \\ \underline{111_{(2)}} \end{array}$

The numerals are written so that the digits of the same place value line up.

(b) $\begin{array}{r} {}^{1} \\ 101_{(2)} \\ \underline{111_{(2)}} \\ 0_{(2)} \end{array}$

Adding 1 + 1 in the units column gives "two," which is written $10_{(2)}$. 0 is written in the units column, and 1 is "carried" to the "twos" column.

(c) $\begin{array}{r} {}^{1\,1} \\ 101_{(2)} \\ \underline{111_{(2)}} \\ 00_{(2)} \end{array}$

Now, in the twos column, 1 + 0 + 1 is again "two," or $10_{(2)}$. Again 0 is written, and 1 is "carried" to the next column, the "fours" column.

(d) $\begin{array}{r} {}^{1\,1} \\ 101_{(2)} \\ \underline{111_{(2)}} \\ 1100_{(2)} \end{array}$

In the fours column (or third column), 1 + 1 + 1 is "three," or $11_{(2)}$. Since there are no digits in the "eights" column (or fourth column), 11 is written, and the sum is $1100_{(2)}$.

Checking in Base Ten

$$\begin{array}{rcrcrr} 101_{(2)} & = & 1(2^2) + 0(2^1) + 1(2^0) = & 4 + 0 + 1 = & 5 & 5 \\ 111_{(2)} & = & 1(2^2) + 1(2^1) + 1(2^0) = & 4 + 2 + 1 = & 7 & \underline{7} \\ 1100_{(2)} & & 1(2^3) + 1(2^2) + 0(2^1) + 0(2^0) = & 8 + 4 + 0 + 0 = & 12 & 12 \end{array}$$

Addition in base five is similar. Although the thinking is done in base ten, which is familiar to us, the numerals written must be in base five.

EXAMPLE

2. (a) $\begin{array}{r} 143_{(5)} \\ \underline{34_{(5)}} \end{array}$

The numerals are written so that the digits of the same place value line up.

(b) $\begin{array}{r} {}^{1} \\ 143_{(5)} \\ \underline{34_{(5)}} \\ 2_{(5)} \end{array}$

Adding 3 + 4 in the units column gives "seven," which is written $12_{(5)}$. 2 is written in the units column, and 1 is carried to the fives column.

(c) $\begin{array}{r} {}^{1\,1} \\ 143_{(5)} \\ \underline{34_{(5)}} \\ 32_{(5)} \end{array}$

In the fives column, 1 + 4 + 3 gives "eight," which is $13_{(5)}$. The 3 is written, and 1 is carried to the next column (the twenty-fives column).

$$\begin{array}{r} {}^{1\,1} \\ 143_{(5)} \\ 34_{(5)} \\ \hline 232_{(5)} \end{array}$$

(d) In the third column, 1 + 1 gives 2. The sum is $232_{(5)}$.

Checking in Base Ten

$$\begin{array}{lcl} 143_{(5)} = 1(5^2) + 4(5^1) + 3(5^0) = 25 + 20 + 3 = 48 & & 48 \\ 34_{(5)} = \phantom{1(5^2) +{}} 3(5^1) + 4(5^0) = \phantom{25 +{}} 15 + 4 = 19 & & 19 \\ \hline 232_{(5)} = 2(5^2) + 3(5^1) + 2(5^0) = 50 + 15 + 2 = 67 & & 67 \end{array}$$

Multiplication in each base is performed and checked in the same manner as addition. Of course, the difference is that you multiply instead of add. When you multiply, be sure to write the correct symbol for the number in the base being used. Also remember to add in the correct base.

EXAMPLES

3.
$$\begin{array}{r} 101_{(2)} \\ 111_{(2)} \\ \hline {}_{1}101 \\ {}_{1}101 \\ 101 \\ \hline 100011_{(2)} \end{array}$$

Multiplication in base two is easy, since we are multiplying by only 1's or 0's. The adding must be done in base two.

Checking gives

$$\begin{array}{rcr} 101_{(2)} & = & 5 \\ 111_{(2)} & = & 7 \\ \hline 101 & & 35 \\ 101 & & \\ 101 & & \\ \hline 100011_{(2)} & & \end{array}$$

Remember to **multiply** the checking numbers.

$100{,}011_{(2)} = 1(2^5) + 0(2^4) + 0(2^3) + 0(2^2) + 1(2) + 1(1)$
$= 32 + 2 + 1 = 35$

4.
$$\begin{array}{r} {}^{2} \\ {}^{1}34_{(5)} \\ 23_{(5)} \\ \hline 212 \\ 123 \\ \hline 1442_{(5)} \end{array}$$

Multiplying 3 × 4 gives "twelve," which is $22_{(5)}$. Write 2 and carry 2 just as in regular multiplication. Then, 3 × 3 is "nine," and "nine" plus 2 is "eleven"; but in base five, "eleven" is $21_{(5)}$. Similarly, 2 × 4 is "eight," or $13_{(5)}$. Write the 3, carry the 1. 2 × 3 is "six," and "six" plus 1 is "seven," or $12_{(5)}$.

Checking gives

$$\begin{array}{rcr} 34_{(5)} & = & 19 \\ 23_{(5)} & = & 13 \\ \hline 212 & & 57 \\ 123 & & 19 \\ \hline 1442_{(5)} & & 247 \end{array}$$

Remember to **multiply** the checking numbers.
$1442_{(5)} = 1(5^3) + 4(5^2) + 4(5) + 2(1)$
$= 125 + 100 + 20 + 2 = 247$

Exercises II.3

Add in the base indicated and check your work in base ten.

1. $101_{(2)}$ + $11_{(2)}$
2. $43_{(5)}$ + $213_{(5)}$
3. $1101_{(2)}$ + $1011_{(2)}$
4. $111_{(2)}$ + $1010_{(2)}$
5. $134_{(5)}$ + $243_{(5)}$
6. $11_{(2)}$ + $10_{(2)}$ + $11_{(2)}$
7. $11_{(2)}$ + $11_{(2)}$ + $101_{(2)}$
8. $214_{(5)}$ + $343_{(5)}$
9. $14_{(5)}$ + $321_{(5)}$ + $43_{(5)}$
10. $431_{(5)}$ + $214_{(5)}$ + $102_{(5)}$
11. $11_{(2)}$ + $101_{(2)}$ + $111_{(2)}$ + $101_{(2)}$
12. $111_{(2)}$ + $11_{(2)}$ + $110_{(2)}$ + $111_{(2)}$
13. $101_{(2)}$ + $101_{(2)}$ + $101_{(2)}$ + $101_{(2)}$
14. $23_{(5)}$ + $103_{(5)}$ + $214_{(5)}$ + $322_{(5)}$
15. $414_{(5)}$ + $211_{(5)}$ + $334_{(5)}$ + $222_{(5)}$

Multiply in the base indicated and check your work in base ten.

16. $1101_{(2)}$ × $111_{(2)}$
17. $1011_{(2)}$ × $101_{(2)}$
18. $423_{(5)}$ × $30_{(5)}$
19. $104_{(5)}$ × $23_{(5)}$
20. $223_{(5)}$ × $44_{(5)}$
21. $423_{(5)}$ × $32_{(5)}$
22. $1111_{(2)}$ × $111_{(2)}$
23. $111_{(2)}$ × $111_{(2)}$
24. $2212_{(5)}$ × $43_{(5)}$
25. $10111_{(2)}$ × $110_{(2)}$

Appendix III

Greatest Common Divisor (GCD)

Consider the two numbers 12 and 18. Is there a number (or numbers) that will divide into **both** 12 and 18? To help answer this question, the divisors for 12 and 18 are listed below.

Set of divisors for 12: {1, 2, 3, 4, 6, 12}
Set of divisors for 18: {1, 2, 3, 6, 9, 18}

The **common divisors** for 12 and 18 are 1, 2, 3, and 6. The **greatest common divisor (GCD)** for 12 and 18 is 6; that is, of all the common divisors of 12 and 18, 6 is the largest divisor.

EXAMPLE

1. List the divisors of each number in the set {36, 24, 48} and find the greatest common divisor (GCD).

Set of divisors for 36: {**1, 2, 3, 4, 6,** 9, **12,** 18, 36}
Set of divisors for 24: {**1, 2, 3, 4, 6,** 8, **12,** 24}
Set of divisors for 48: {**1, 2, 3, 4, 6,** 8, **12,** 16, 24, 48}

The common divisors are **1, 2, 3, 4, 6,** and **12. GCD = 12.**

DEFINITION The **Greatest Common Divisor (GCD)*** of a set of natural numbers is the largest natural number that will divide into all the numbers in the set.

As Example 1 illustrates, listing all the divisors of each number before finding the GCD can be tedious and difficult. **The use of prime factorizations leads to a simple technique for finding the GCD.**

TECHNIQUE FOR FINDING THE GCD OF A SET OF NATURAL NUMBERS

1. Find the prime factorization of each number.
2. Find the prime factors common to all factorizations.
3. Form the product of these primes, using each prime the number of times it is common to **all** factorizations.
4. This product is the GCD. If there are no primes common to all factorizations, the GCD is 1.

EXAMPLES

2. Find the GCD for {36, 24, 48}.

$$\left.\begin{aligned} 36 &= 2\cdot 2\cdot 3\cdot 3 \\ 24 &= 2\cdot 2\cdot 2\cdot 3 \\ 48 &= 2\cdot 2\cdot 2\cdot 2\cdot 3 \end{aligned}\right\} \quad \text{GCD} = 2\cdot 2\cdot 3 = 12$$

The factor 2 appears twice, and the factor 3 appears once in **all** the prime factorizations.

3. Find the GCD for {360, 75, 30}.

$$\left.\begin{aligned} 360 &= 36\cdot 10 = 4\cdot 9\cdot 2\cdot 5 = 2\cdot 2\cdot 2\cdot 3\cdot 3\cdot 5 \\ 75 &= 3\cdot 25 = 3\cdot 5\cdot 5 \\ 30 &= 6\cdot 5 = 2\cdot 3\cdot 5 \end{aligned}\right\} \quad \begin{aligned} \text{GCD} &= 3\cdot 5 \\ &= 15 \end{aligned}$$

Each of the factors 3 and 5 appears only once in **all** the prime factorizations.

---

*The largest common divisor is, of course, the largest common factor, and the GCD could be called the **greatest common factor** and be abbreviated **GCF.**

4. Find the GCD for {168, 420, 504}.

$$\left.\begin{aligned} 168 &= 8\cdot 21 = 2\cdot 2\cdot 2\cdot 3\cdot 7 \\ 420 &= 10\cdot 42 = 2\cdot 5\cdot 6\cdot 7 \\ &= 2\cdot 2\cdot 3\cdot 5\cdot 7 \\ 504 &= 4\cdot 126 = 2\cdot 2\cdot 6\cdot 21 \\ &= 2\cdot 2\cdot 2\cdot 3\cdot 3\cdot 7 \end{aligned}\right\} \quad \text{GCD} = 2\cdot 2\cdot 3\cdot 7 = 84$$

In **all** the prime factorizations, 2 appears twice, 3 once, and 7 once.

If the GCD of two numbers is 1 (that is, they have no common prime factors), then the two numbers are said to be **relatively prime.** The numbers themselves may be prime or they may be composite.

EXAMPLES

5. Find the GCD for {15, 8}.

$$\left.\begin{aligned} 15 &= 3\cdot 5 \\ 8 &= 2\cdot 2\cdot 2 \end{aligned}\right\} \quad \text{GCD} = 1 \qquad \text{8 and 15 are relatively prime.}$$

6. Find the GCD for {20, 21}.

$$\left.\begin{aligned} 20 &= 2\cdot 2\cdot 5 \\ 21 &= 3\cdot 7 \end{aligned}\right\} \quad \text{GCD} = 1 \qquad \text{20 and 21 are relatively prime.}$$

Exercises III

Find the GCD for each of the following sets of numbers.

1. {12, 8}
2. {16, 28}
3. {85, 51}
4. {20, 75}
5. {20, 30}
6. {42, 48}
7. {15, 21}
8. {27, 18}
9. {18, 24}
10. {77, 66}
11. {182, 184}
12. {110, 66}
13. {8, 16, 64}
14. {121, 44}
15. {28, 52, 56}
16. {98, 147}
17. {60, 24, 96}
18. {33, 55, 77}
19. {25, 50, 75}
20. {30, 78, 60}
21. {17, 15, 21}
22. {520, 220}
23. {14, 55}
24. {210, 231, 84}
25. {140, 245, 420}

Which of the following pairs of numbers are relatively prime?

26. $\{35, 24\}$ **27.** $\{11, 23\}$ **28.** $\{14, 36\}$ **29.** $\{72, 35\}$

30. $\{42, 77\}$ **31.** $\{16, 51\}$ **32.** $\{20, 21\}$ **33.** $\{8, 15\}$

34. $\{66, 22\}$ **35.** $\{10, 27\}$

Appendix IV
Topics from Plane Geometry

IV.1 ANGLES

Plane geometry is the study of the properties of figures in a plane. The most basic ideas or concepts are **point, line,** and **plane.** These terms name ideas that cannot be formally defined. They are called **undefined terms.**

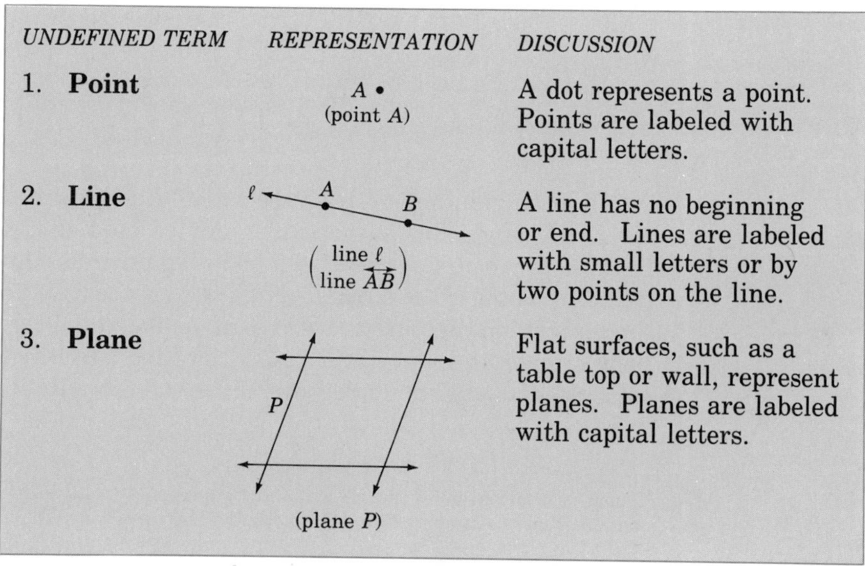

UNDEFINED TERM	*REPRESENTATION*	*DISCUSSION*
1. **Point**	A • (point A)	A dot represents a point. Points are labeled with capital letters.
2. **Line**	ℓ, A, B (line ℓ, line $\overleftrightarrow{AB}$)	A line has no beginning or end. Lines are labeled with small letters or by two points on the line.
3. **Plane**	P (plane P)	Flat surfaces, such as a table top or wall, represent planes. Planes are labeled with capital letters.

We make the assumption that **two points determine a line.** That is, there is only one line through the two points A and B (Figure IV.1).

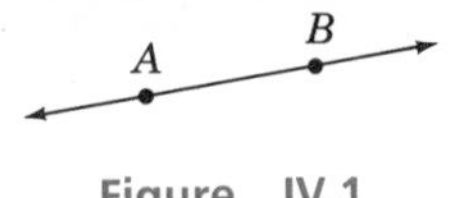

Figure IV.1

We say that the line $\overleftrightarrow{AB}$ **contains** the points A and B. A line contains an infinite number of points.

Using the undefined terms, we can define the following important terms.

TERM	*DEFINITION*	*ILLUSTRATION*
1. **Line segment** (or **segment**)	A **line segment** consists of two points on a line and all the points between* them.	A, B (segment $\overline{AB}$)
2. **Ray**	A **ray** consists of a point (called the **endpoint**) and all the points on a line on one side of that point.	P, Q (ray $\overrightarrow{PQ}$ with end point P)
3. **Angle**	An **angle** consists of two rays with a common endpoint (called a **vertex**).	O, A, B ($\angle$ AOB with vertex O)

In an angle, the two rays are called the **sides** of the angle.

Every angle has a **measurement** or **measure** associated with it. Suppose that a circle is divided into 360 equal arcs. If two rays are drawn from the center of the circle through two consecutive points of division on the circle, then that angle is said to **measure one degree** (symbolized 1°). For example, in Figure IV.2, a device called a protractor shows that the measure of $\angle AOB$ is 60 degrees. (We write $m\angle AOB = 60°$.)

---

*There is a formal definition of the word "between" as used in geometry. Here we will assume that the students' intuitive notion of "between" is sufficient.

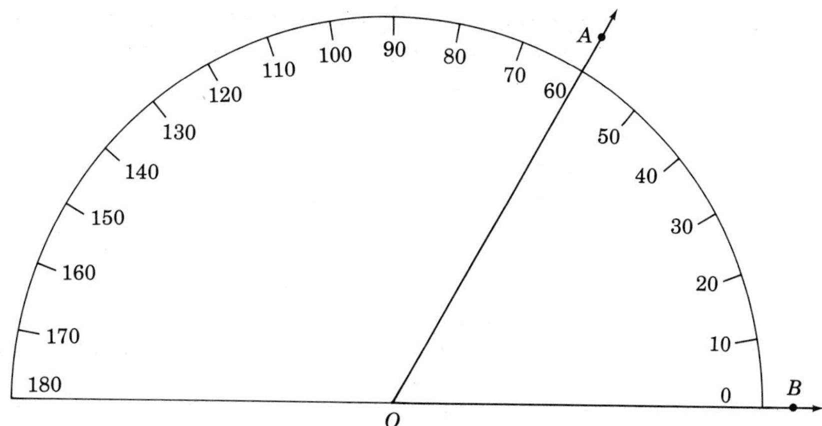

Figure IV.2 The protractor shows $m\angle AOB = 60°$

To measure an angle with a protractor, lay the bottom edge of the protractor along one side of the angle with the vertex at the marked center point. Then read the measure from the protractor where the other side of the angle crosses it. (See Figure IV.3.)

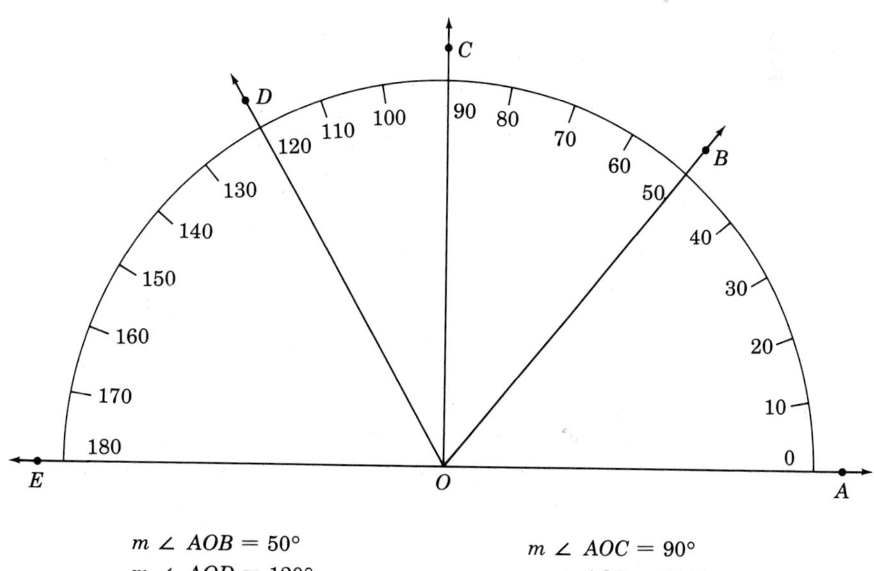

Figure IV.3

There are four common ways of labeling angles (Figure IV.4):

1. Using three capital letters with the vertex as the middle letter.
2. Using single small Greek letters such as α (alpha), β (beta), and γ (gamma).
3. Using single numbers such as 1, 2, 3.
4. Using the single capital letter at the vertex when the meaning is clear.

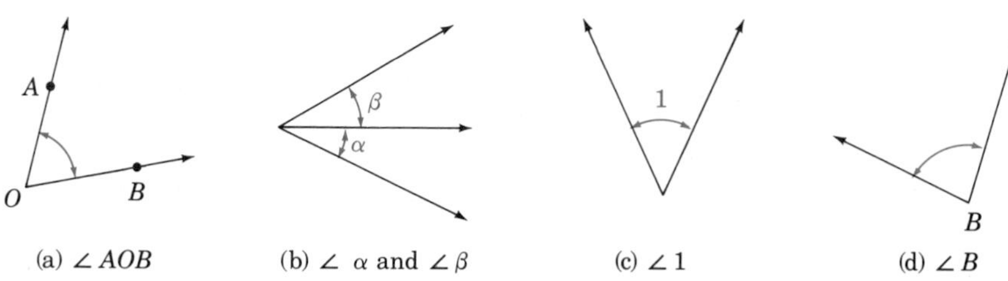

Figure IV.4 Four Ways of Labeling Angles

Angles can be classified (or named) according to their measures.

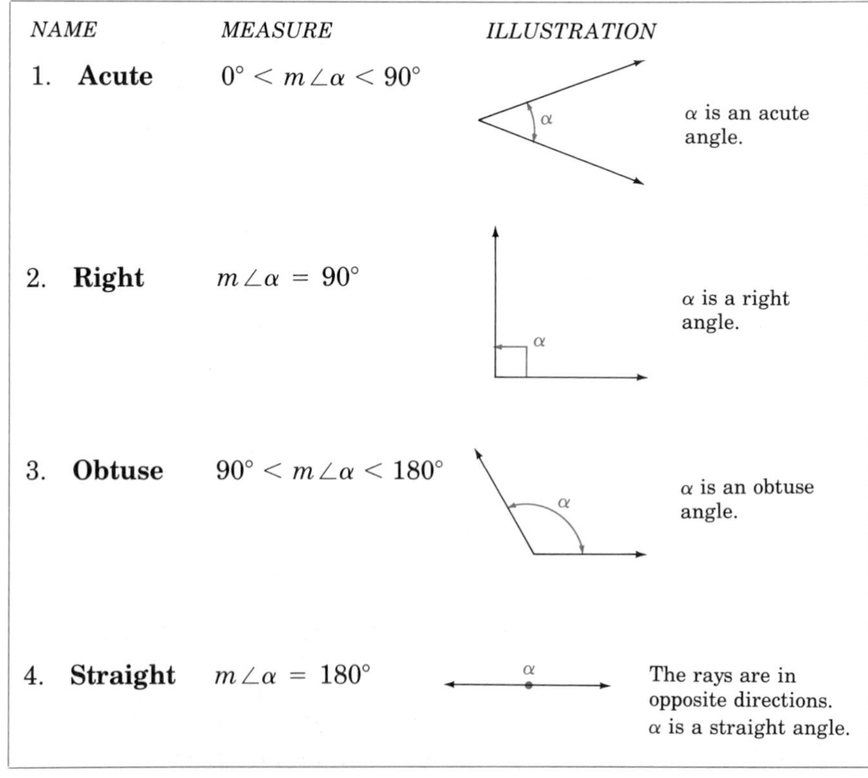

NAME	*MEASURE*	*ILLUSTRATION*	
1. **Acute**	$0^\circ < m\angle\alpha < 90^\circ$	α	α is an acute angle.
2. **Right**	$m\angle\alpha = 90^\circ$	α	α is a right angle.
3. **Obtuse**	$90^\circ < m\angle\alpha < 180^\circ$	α	α is an obtuse angle.
4. **Straight**	$m\angle\alpha = 180^\circ$	α	The rays are in opposite directions. α is a straight angle.

EXAMPLES

The following figure is used for Examples 1–4.

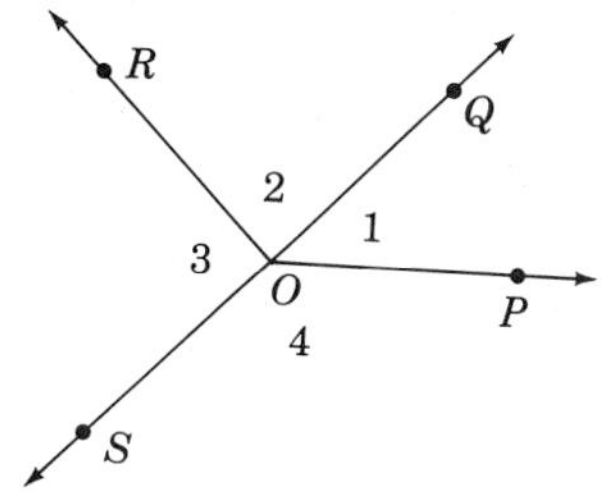

1. Use a protractor to check the measures of the following angles.

   (a) $m\angle 1 = 45°$

   (b) $m\angle 2 = 90°$

   (c) $m\angle 3 = 90°$

   (d) $m\angle 4 = 135°$

2. Tell whether each of the following angles is acute, right, obtuse, or straight.

   (a) $\angle 1$

   Solution

   $\angle 1$ is acute since $m\angle 1 < 90°$.

   (b) $\angle 2$

   Solution

   $\angle 2$ is a right angle since $m\angle 2 = 90°$.

   (c) $\angle POR$

   Solution

   $\angle POR$ is obtuse since $m\angle POR = 45° + 90° = 135° > 90°$.

3. In the figure, $\overline{OP}$ is a line segment. Name any other line segments shown.

   Solution

   $\overline{OQ}$, $\overline{OR}$, and $\overline{OS}$ are line segments.

4. Name any rays and lines shown in the figure.

   Solution

   $\overrightarrow{OP}$, $\overrightarrow{OQ}$, $\overrightarrow{OR}$, and $\overrightarrow{OS}$ are rays.
   $\overleftrightarrow{SQ}$ is the only line in the figure.

Two angles are:

1. **Complementary** if the sum of their measures is 90°.
2. **Supplementary** if the sum of their measures is 180°.
3. **Equal** if they have the same measure.

EXAMPLES

5. In the figure,

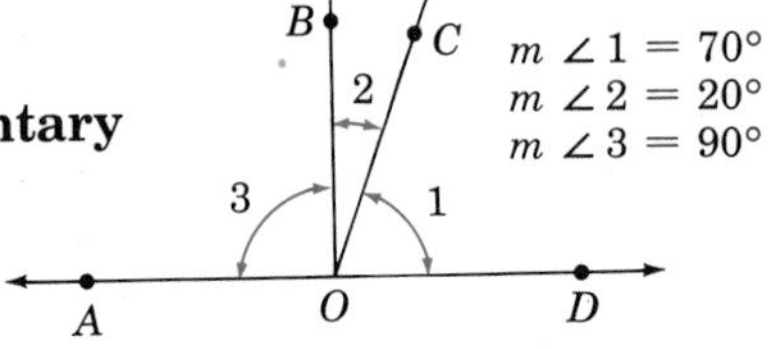

(a) $\angle 1$ and $\angle 2$ are **complementary** since $m\angle 1 + m\angle 2 = 90°$

(b) $\angle COD$ and $\angle COA$ are **supplementary** since $m\angle COD + m\angle COA = 70° + 100° = 180°$.

(c) $\angle AOD$ is a straight angle since $m\angle AOD = 180°$.

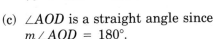

6. In the figure, $\overleftrightarrow{PS}$ is a straight line and $m\angle QOP = 30°$. Find the measures of (a) $\angle QOS$ and (b) $\angle SOP$. (c) Are any pairs supplementary?

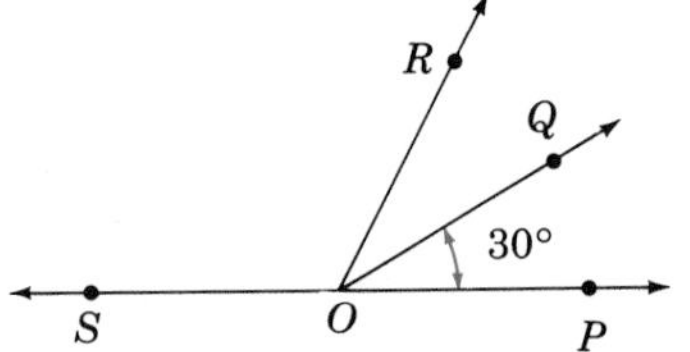

Solution

(a) $m\angle QOS = 150°$

(b) $m\angle SOP = 180°$

(c) Yes. $\angle QOP$ and $\angle QOS$ are supplementary and $\angle ROP$ and $\angle ROS$ are supplementary.

Exercises IV.1

1. Name all of the line segments in the figure.

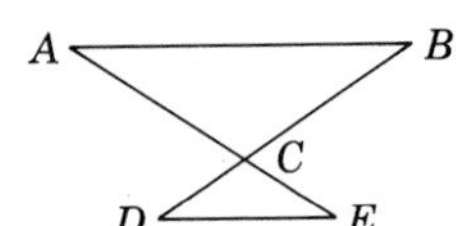

2. If $\angle 1$ and $\angle 2$ are complementary, and
(a) $m\angle 1 = 15°$, what is $m\angle 2$?
(b) $m\angle 1 = 3°$, what is $m\angle 2$?
(c) $m\angle 1 = 45°$, what is $m\angle 2$?
(d) $m\angle 1 = 75°$, what is $m\angle 2$?

3. If $\angle 3$ and $\angle 4$ are supplementary and
(a) $m\angle 3 = 45°$, what is $m\angle 4$?
(b) $m\angle 3 = 90°$, what is $m\angle 4$?

(c) $m\angle 3 = 110°$, what is $m\angle 4$?
(d) $m\angle 3 = 135°$, what is $m\angle 4$?

4. The supplement of an acute angle is an obtuse angle.
(a) What is the supplement of a right angle?
(b) What is the supplement of an obtuse angle?
(c) What is the complement of an acute angle?

5. Suppose that the line $\overleftrightarrow{AB}$ is in the plane P and the point X is on the line as shown. How many rays in plane P with endpoint X
(a) make an angle of 20° with $\overrightarrow{XB}$?
(b) make an angle of 160° with $\overrightarrow{XB}$?
(c) Are the answers to (a) and (b) the same rays?

6. In the figure shown, $\overleftrightarrow{DC}$ is a straight line and $m\angle BOA = 90°$.

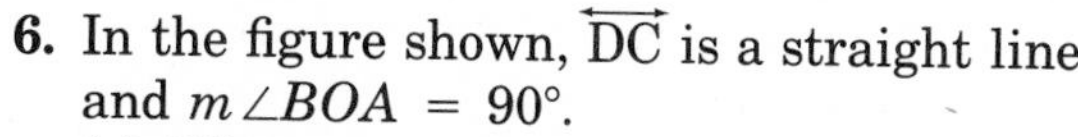

(a) What type of angle is $\angle AOC$?
(b) What type of angle is $\angle BOC$?
(c) What type of angle is $\angle BOA$?
(d) Name a pair of complementary angles.
(e) Name two pairs of supplementary angles.

7. An **angle bisector** is a ray that divides an angle into two angles with equal measures. If $\overrightarrow{OX}$ bisects $\angle COD$ and $m\angle COD = 50°$, what is the measure of each of the equal angles formed?

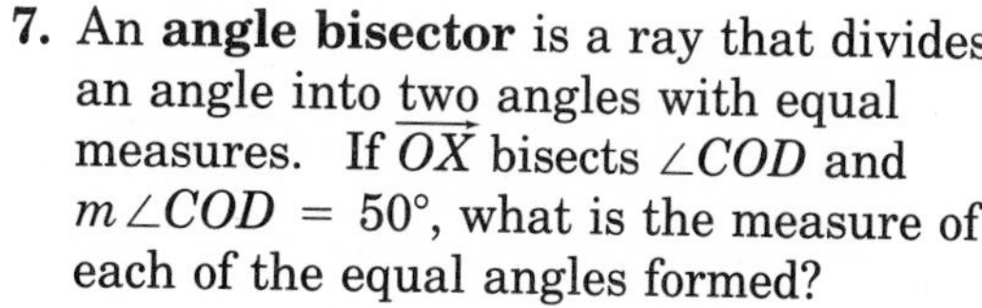

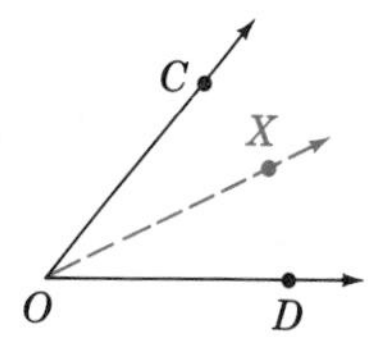

Given $m\angle AOB = 30°$ and $m\angle BOC = 80°$ and $\overrightarrow{OX}$ and $\overrightarrow{OY}$ are angle bisectors, find the measures of the following angles.

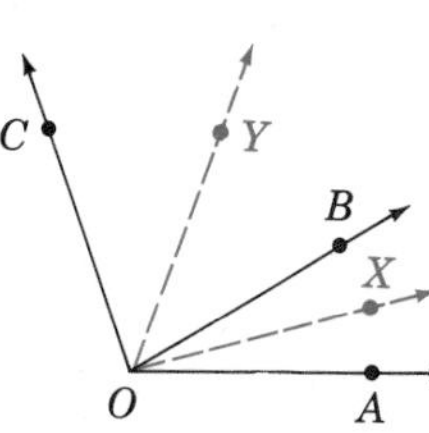

8. $\angle AOX$ **9.** $\angle BOY$

10. $\angle COX$ **11.** $\angle YOX$

12. $\angle AOY$ **13.** $\angle AOC$

14. In the figure shown,
(a) Name all the pairs of supplementary angles.
(b) Name all the pairs of complementary angles.

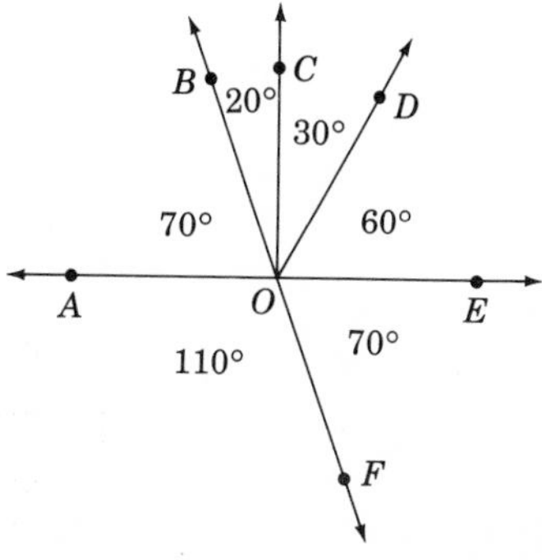

Use a protractor to measure all the angles in each figure. Each line segment may be extended as a ray to form the side of an angle.

15.

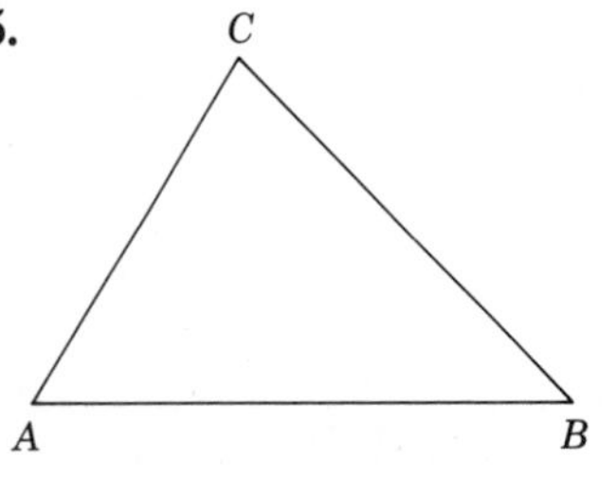

16.

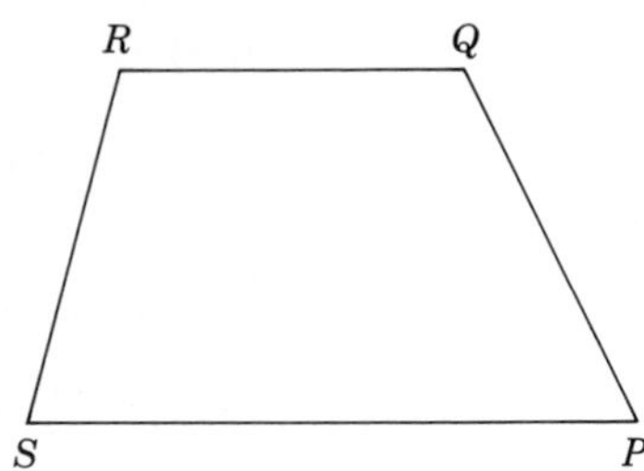

17.

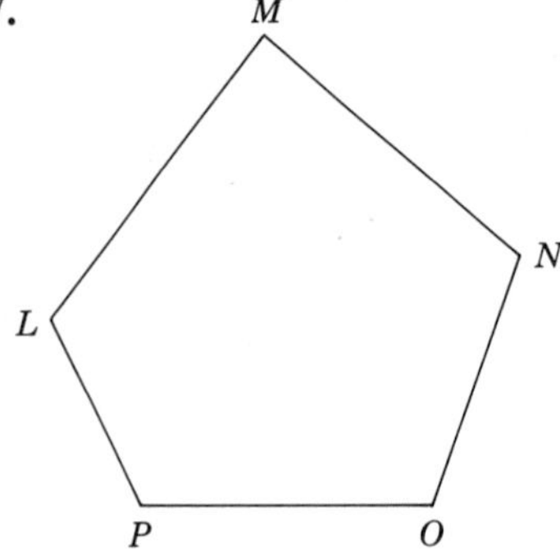

18.

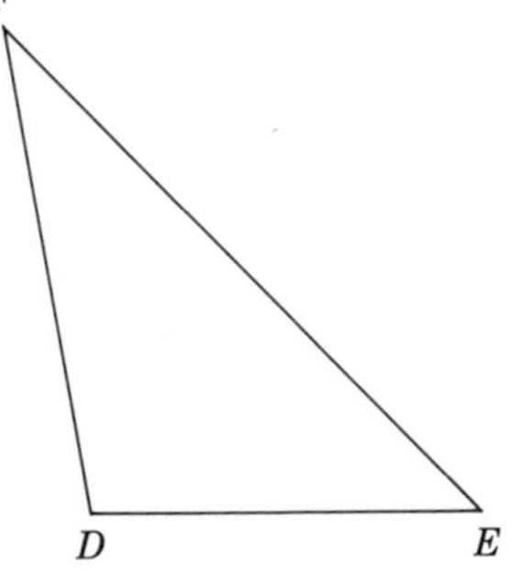

19. Name the type of angle formed by the hands on a clock:
(a) at six o'clock
(b) at three o'clock
(c) at one o'clock
(d) at five o'clock

20. What is the measure of each angle formed by the hands of the clock in Exercise 19?

IV.2 LINES IN A PLANE

We begin this section by defining and illustrating two terms.

TERMS	*DEFINITION*	*ILLUSTRATION*
1. **Collinear**	Three or more points are **collinear** if they lie on one line.	A B C • D *A*, *B*, and *C*, are collinear. *A*, *B*, and *D* are not collinear.

2. **Intersect**	Two lines **intersect** if there is one point on both lines.	Lines ℓ and m intersect at point A.

In Section IV.1, we stated that **two points determine a line.** This is a **postulate,** or statement accepted as true without proof. A similar postulate is that **three noncollinear points determine a plane.** In fact, a plane can be determined in any one of the following three ways (Figure IV.5):

1. Three noncollinear points (A, B, and C).
2. Two intersecting lines (ℓ and m).
3. A line and a point not on the line (line m and point C).

Figure IV.5

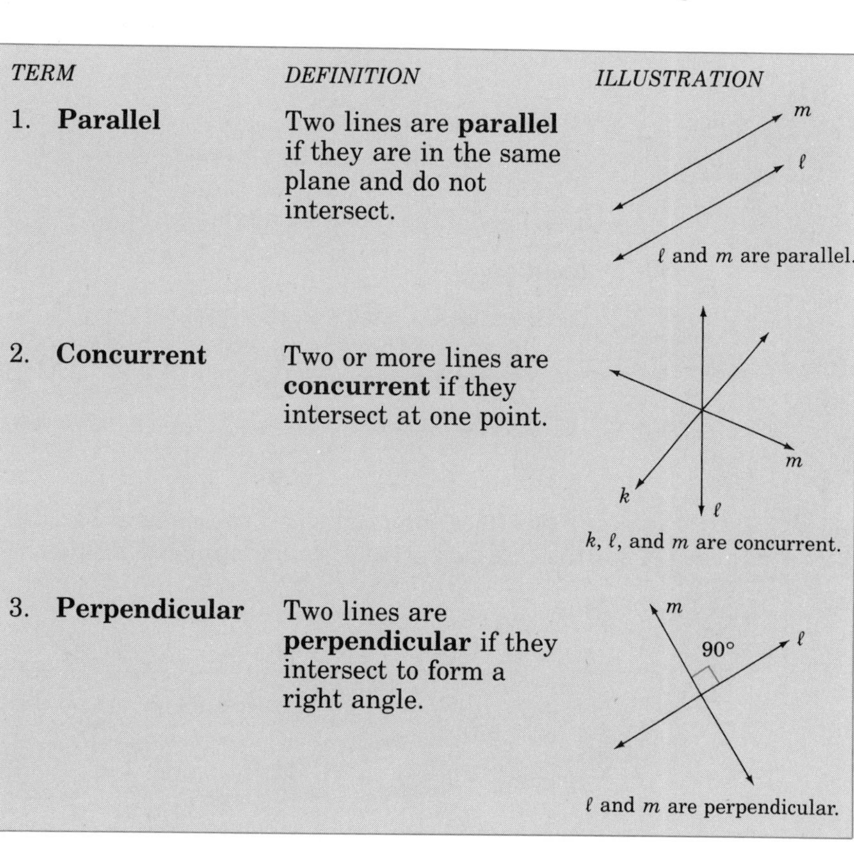

TERM	*DEFINITION*	*ILLUSTRATION*
1. **Parallel**	Two lines are **parallel** if they are in the same plane and do not intersect.	ℓ and m are parallel.
2. **Concurrent**	Two or more lines are **concurrent** if they intersect at one point.	k, ℓ, and m are concurrent.
3. **Perpendicular**	Two lines are **perpendicular** if they intersect to form a right angle.	ℓ and m are perpendicular.

We also define parallel and perpendicular segments and rays in terms of lines in a plane. **Rays and segments are parallel if the lines that contain them are parallel. Rays and segments are perpendicular if the lines that contain them are perpendicular.**

A **transversal** is a line in a plane that intersects two or more lines in that plane in different points. (See Figure IV.6.)

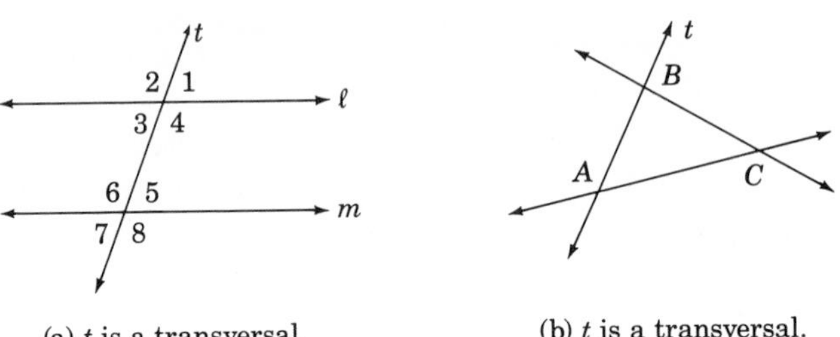

(a) t is a transversal. (b) t is a transversal.

Figure IV.6

As shown in Figure IV.6, a transversal that intersects two lines forms eight angles. There are several important properties related to these pairs of angles and parallel lines cut by a transversal. The following example illustrates some of these properties.

EXAMPLE

1. Suppose that lines ℓ and m are parallel and that t is a transversal. If $m\angle 1 = m\angle 5 = 50°$, find the measures of the other six angles.

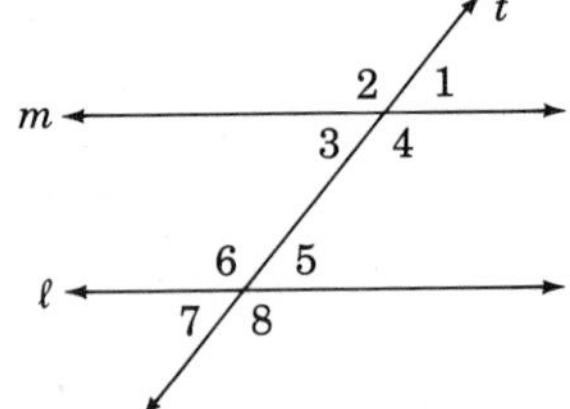

Solution

(a) Because $\angle 1$ and $\angle 4$ are supplementary, $m\angle 4 = 130°$. But this is also true of the pair $\angle 1$ and $\angle 2$, the pair $\angle 5$ and $\angle 8$, and the pair $\angle 5$ and $\angle 6$. So, $m\angle 2 = m\angle 8 = m\angle 6 = 130°$.

(b) Similar reasoning gives $m\angle 3 = m\angle 7 = 50°$.

If two lines intersect, then two pairs of vertical angles are formed. **Vertical angles** are also called **opposite angles.** (See Figure IV.7.)

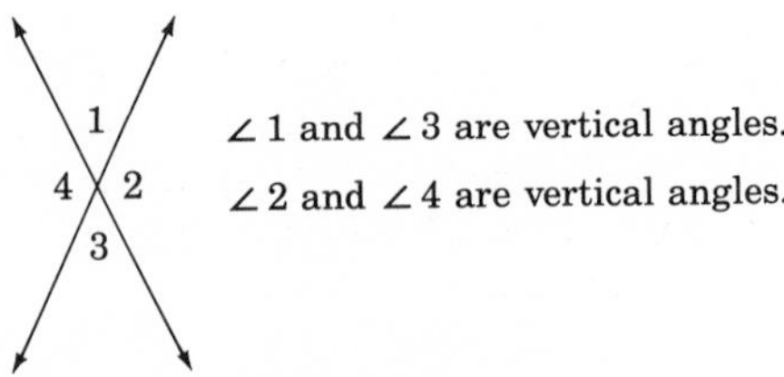

1
4
2
3
∠1 and ∠3 are vertical angles.
∠2 and ∠4 are vertical angles.

Figure IV.7

Vertical angles are equal. That is, **vertical angles have the same measure.** This property of vertical angles is also useful when working with parallel lines and transversals. We will develop this idea further in the exercises. (In Figure IV.7, $m\angle 1 = m\angle 3$ and $m\angle 2 = m\angle 4$.)

Two angles are **adjacent** if they have a common side. (See Figure IV.8.)

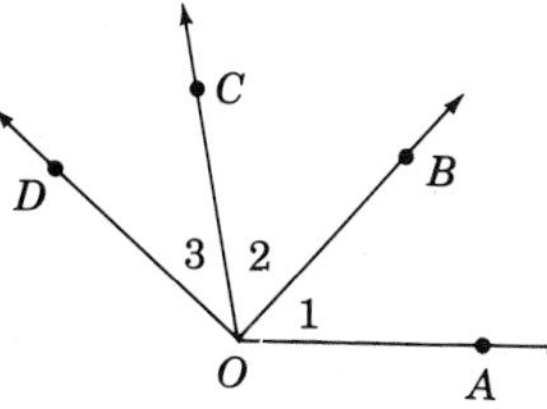

$\angle 1$ and $\angle 2$ are adjacent. They have the common side $\overrightarrow{OB}$.

$\angle 2$ and $\angle 3$ are adjacent. They have the common side $\overrightarrow{OC}$.

Figure IV.8

Exercises IV.2

1. Define each of the following terms.
  (a) parallel
  (b) perpendicular
  (c) intersect

2. Illustrate each of the following terms with a figure.
  (a) collinear points
  (b) concurrent lines
  (c) adjacent angles

3. **(a)** Draw two rays that are parallel.
  (b) Draw two segments that are perpendicular but do not intersect.

4. The figure shows two intersecting lines.
  (a) If $m\angle 1 = 30°$, what is $m\angle 2$?
  (b) Is $m\angle 3 = 30°$? Give a reason for your answer other than the fact that $\angle 1$ and $\angle 3$ are vertical angles.
  (c) Name four pairs of adjacent angles.

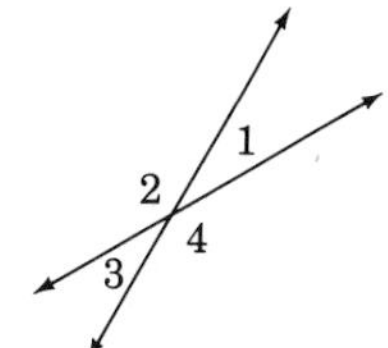

5. Why do you think a three-legged stool is so solid and not wobbly even though the three legs may not be the same length?

6. In the figure shown, $\overleftrightarrow{AB}$ is a straight line.
  (a) Name two pairs of adjacent angles.
  (b) Name two vertical angles if there are any.

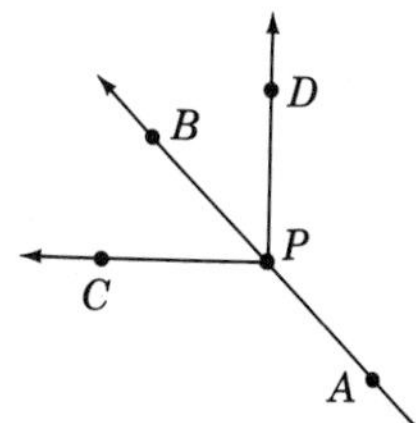

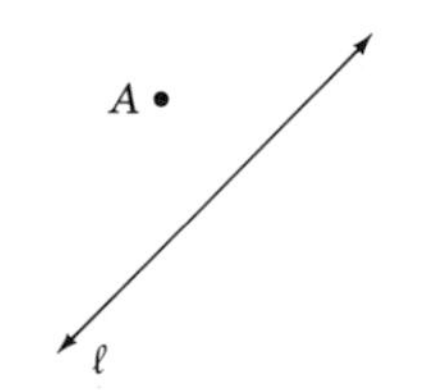

7. Line ℓ and point A are shown.

(a) How many planes contain both ℓ and A?

(b) Draw all the lines you can through A parallel to ℓ. How many lines did you draw?

(c) Draw all the lines you can through A perpendicular to ℓ. How many lines did you draw?

Use the figure at the right to answer Exercises 8–11.

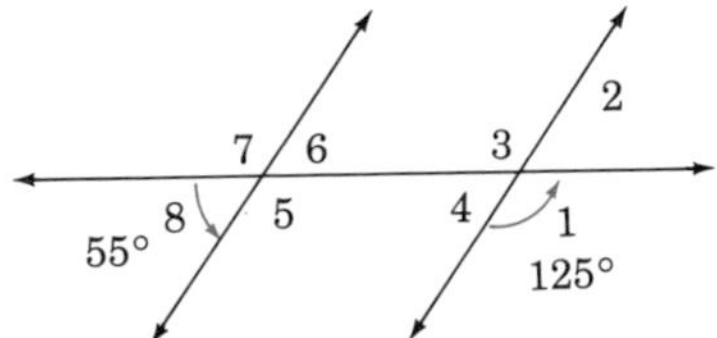

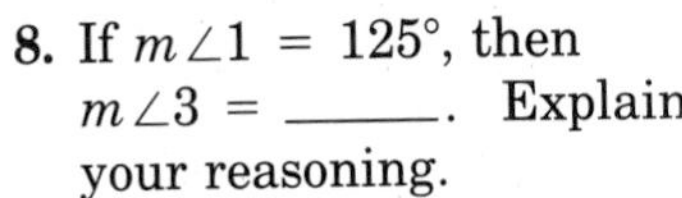

8. If $m\angle 1 = 125°$, then $m\angle 3 =$ ______. Explain your reasoning.

9. If $m\angle 8 = 55°$, then $m\angle 6 =$ ______. Explain your reasoning.

10. What is $m\angle 7$? Explain your reasoning.

11. Does $m\angle 2 = m\angle 6$? Explain your reasoning.

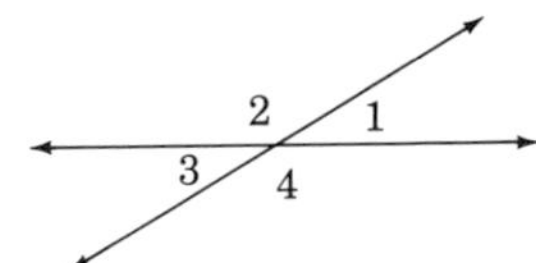

12. Given that $m\angle 1 = 30°$ in the figure at the right, find the measures of the other three angles.

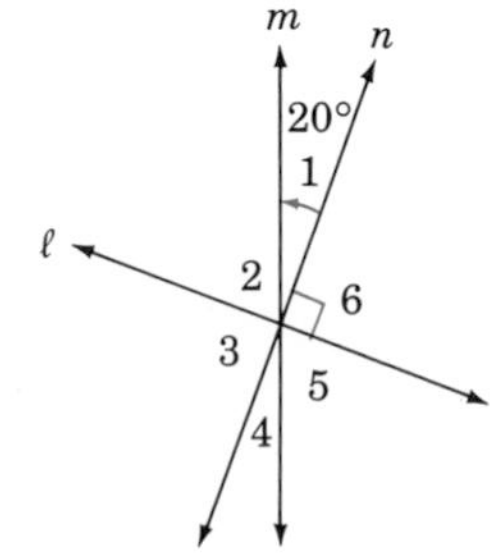

13. In the figure at the right, ℓ, m, and n are straight lines with $m\angle 1 = 20°$ and $m\angle 6 = 90°$.

(a) Find the measures of the other four angles.

(b) Which angle is supplementary to $\angle 6$?

(c) Which angles are complementary to $\angle 1$?

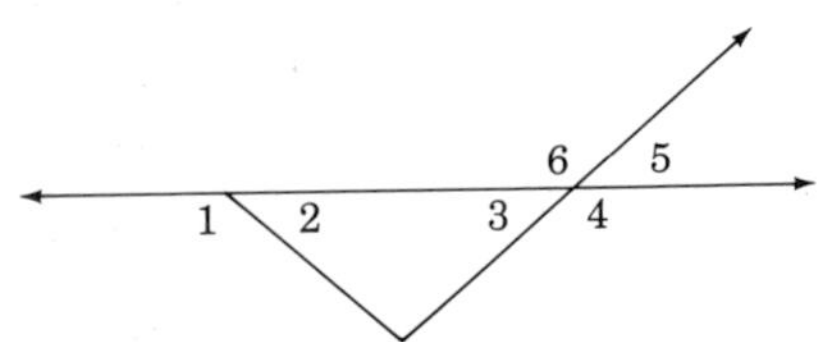

14. In the figure at the right, $m\angle 2 = m\angle 3 = 40°$. Find all other pairs of angles that have equal measures.

15. Suppose $\overleftrightarrow{AB}$ and $\overrightarrow{PQ}$ as shown in the figure are perpendicular.
(**a**) Which angles are acute?
(**b**) Which angles are obtuse?
(**c**) Which angles are right angles?
(**d**) Which pairs of angles are vertical angles?
(**e**) Which pairs of angles are complementary?
(**f**) Which pairs of angles are supplementary?
(**g**) Which pairs of angles are adjacent?

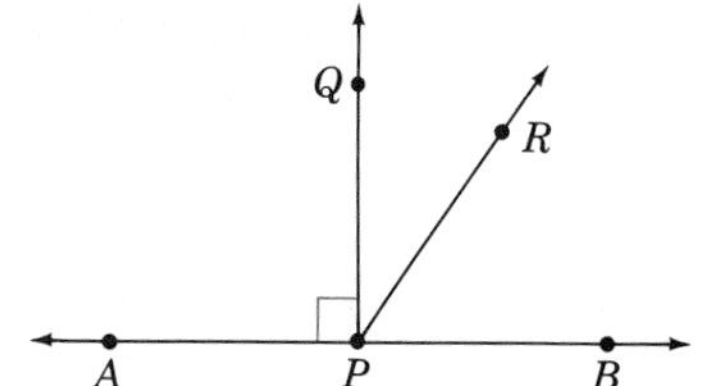

IV.3 TRIANGLES

A **triangle** consists of the three line segments that join three noncollinear points. The line segments are called the **sides** of the triangle, and the points are called the **vertices** of the triangle. If the points are labeled A, B, and C, the triangle is symbolized ΔABC. (See Figure IV.9.)

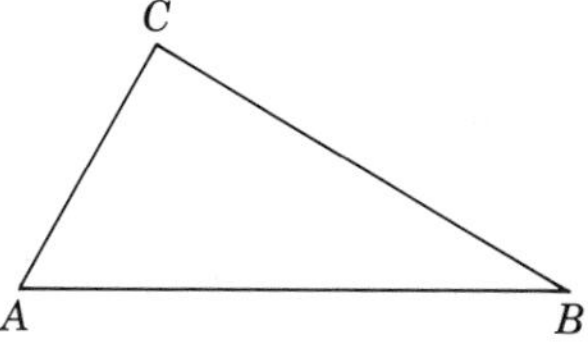

$\triangle ABC$ with vertices A, B, and C and sides $\overline{AB}$, $\overline{BC}$, and $\overline{AC}$.

Figure IV.9

The sides of a triangle are said to determine three angles, and these angles are labeled by the vertices. Thus, the angles of ΔABC are $\angle A$, $\angle B$, and $\angle C$. (Since the definition of angle involves rays, we can think of the sides of the triangle extended as rays to form these angles.)

Triangles are classified in two ways: (1) according to the lengths of their sides, and (2) according to the measures of their angles. The corresponding names and properties are listed in the following tables.

(**Note:** We will indicate the length of a line segment, such as $\overline{AB}$, by writing only the letters, such as AB.)

TRIANGLES CLASSIFIED BY SIDES

NAME	*PROPERTY*	*ILLUSTRATION*
1. **Scalene**	No two sides are equal.	$\triangle ABC$ is scalene since no sides are equal. 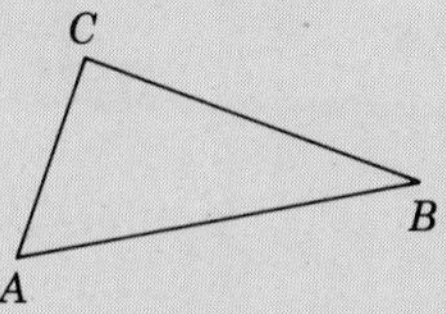
2. **Isosceles**	At least two sides are equal.	$\triangle PQR$ is isosceles since $PR = QR$. 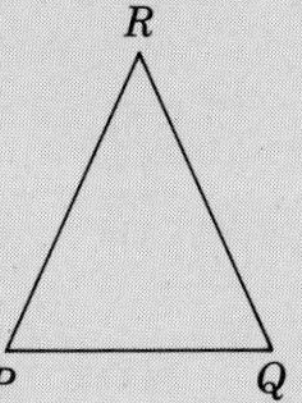
3. **Equilateral**	All three sides are equal.	$\triangle XYZ$ is equilateral since $XY = XZ = YZ$. 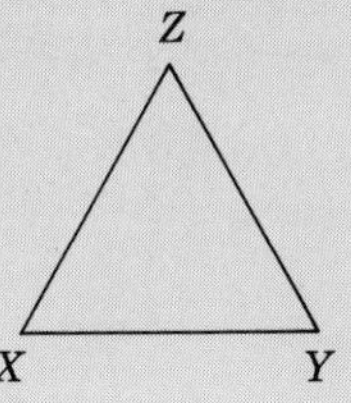

TRIANGLES CLASSIFIED BY ANGLES

NAME	*PROPERTY*	*ILLUSTRATION*
1. **Acute**	All three angles are acute.	$\angle A$, $\angle B$, $\angle C$ are all acute, so $\triangle ABC$ is acute. 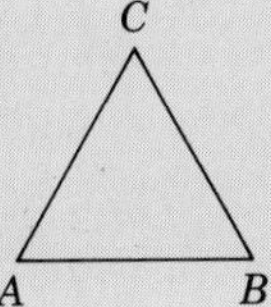

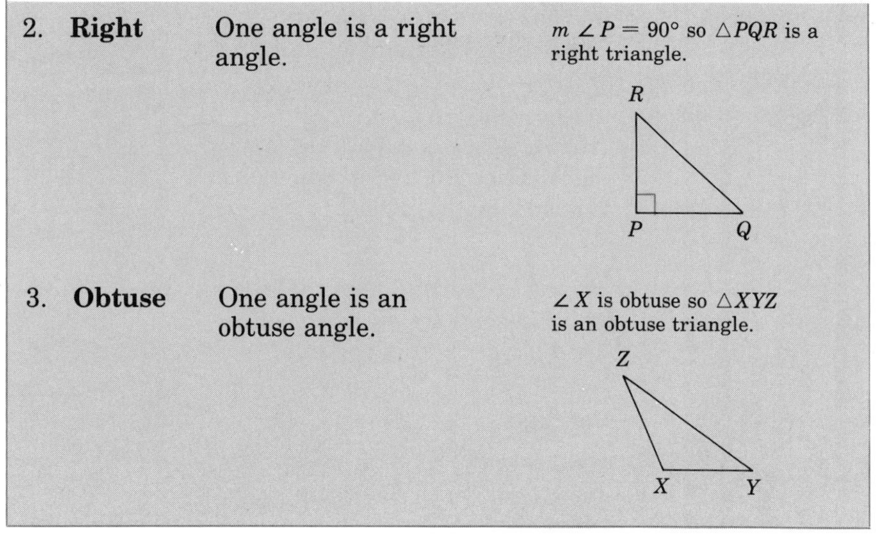

2. **Right**	One angle is a right angle.	$m \angle P = 90°$ so $\triangle PQR$ is a right triangle.
3. **Obtuse**	One angle is an obtuse angle.	$\angle X$ is obtuse so $\triangle XYZ$ is an obtuse triangle.

Every triangle is said to have six parts, namely three angles and three sides. Two sides of a triangle are said to **include** the angle at their common endpoint or vertex. The third side is said to be **opposite** this angle.

The sides in a right triangle have special names. The longest side, opposite the right angle, is called the **hypotenuse** and the other two sides are called **legs.** (See Figure IV.10.)

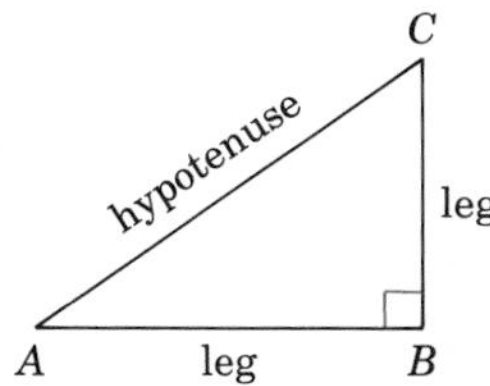

$\triangle ABC$ is a right triangle. $m\angle B = 90°$. $\overline{AC}$ is opposite $\angle B$.

Figure IV.10

Two important statements can be made about any triangle:

1. The sum of the measures of the angles is 180°.
2. The sum of the lengths of two sides must be greater than the length of the third side.

EXAMPLES

1. In the figure of ΔABC, $AB = AC$. What kind of triangle is ΔABC?

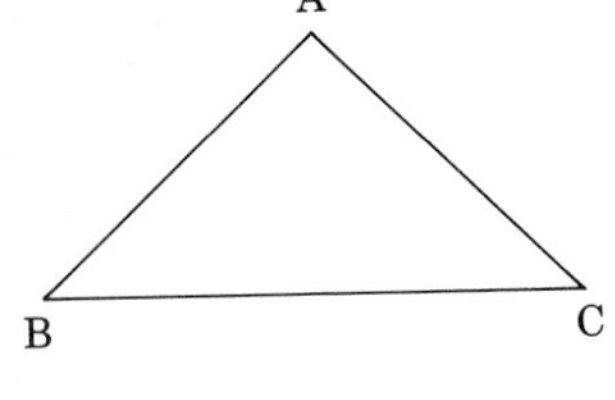

Solution

ΔABC is isosceles because two sides are equal. (You can measure the sides with a ruler.)

2. Suppose the lengths of the sides of ΔPQR are stated as shown in the figure. Is this possible?

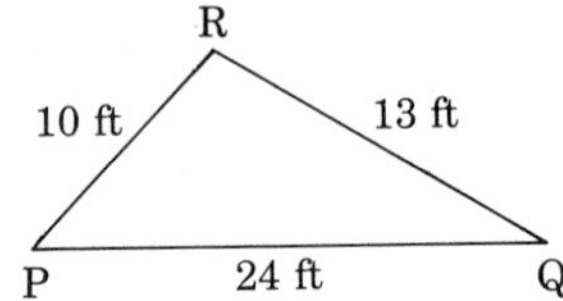

Solution

This is not possible because $PR + QR = 10 \text{ ft} + 13 \text{ ft} = 23 \text{ ft}$ and $PQ = 24$ ft, which is longer than the sum of the other two sides. In a triangle, the sum of the lengths of two sides must be greater than the length of the third side.

3. In the figure of ΔBOR, $m\angle B = 50°$ and $m\angle 0 = 70°$.
   (a) What is $m\angle R$?
   (b) What kind of triangle is ΔBOR?
   (c) Which side is opposite $\angle R$?
   (d) Which sides include $\angle R$?
   (e) Is ΔBOR a right triangle? Why?

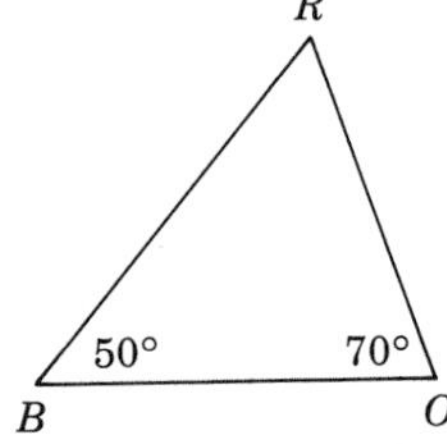

Solution

(a) The sum of the measures of the angles must be 180°. Since $50° + 70° = 120°$,

$$m\angle R = 180° - 120° = 60°.$$

(b) ΔBOR is an acute triangle since all the angles are acute. Also, ΔBOR is scalene because no two sides are equal.
(c) $\overline{BO}$ is opposite $\angle R$.
(d) $\overline{RB}$ and $\overline{RO}$ include $\angle R$.
(e) ΔBOR is not a right triangle because none of the angles is a right angle.

On an intuitive level, two triangles are said to be **similar** if they have the same "shape." They may or may not have the same "size." More formally (and more usefully), similar triangles have the following two important properties.

> **Two triangles are similar if:**
>
> 1. Their **corresponding angles are equal.**
> 2. Their **corresponding sides are proportional.** (See Figure IV.11.)

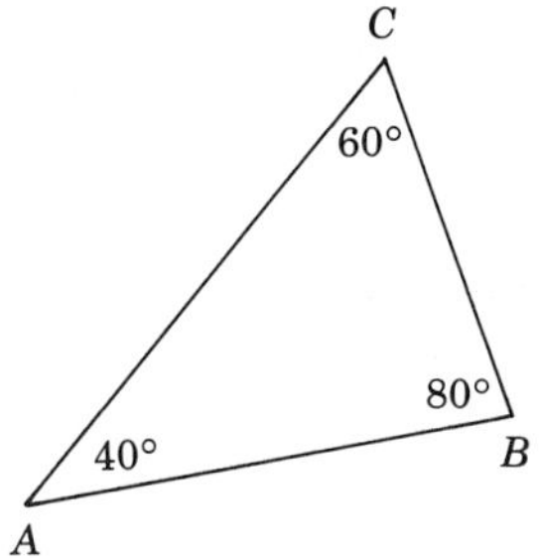

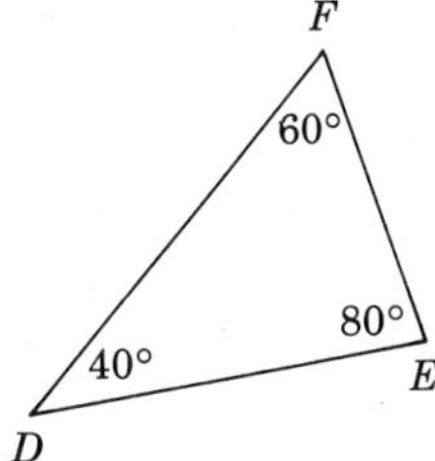

$\angle A$ corresponds to $\angle D$.
$\angle B$ corresponds to $\angle E$.
$\angle C$ corresponds to $\angle F$.

$\frac{AB}{DE} = \frac{BC}{EF} = \frac{AC}{DF}$
The corresponding sides are proportional.

We write $\triangle ABC \sim \triangle DEF$. ($\sim$ is read "is similar to.")

Figure IV.11

EXAMPLES

4. Consider the two triangles ΔABC and ΔAXY as shown in the figure with $\overline{CB}$ and $\overline{YX}$ both perpendicular to $\overline{AX}$. Are the triangles similar?

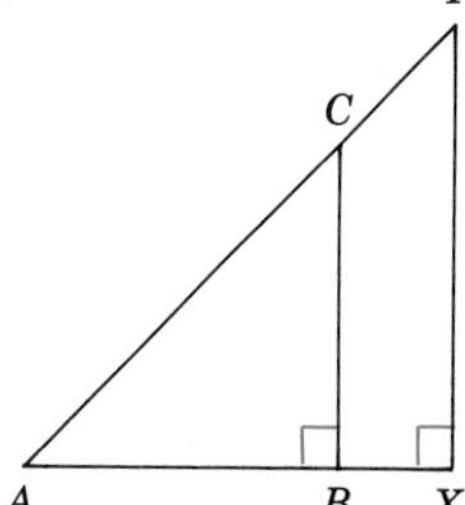

Solution

We can show that the corresponding angles are equal as follows:

$m\angle CAB = m\angle YAX$ because they are the same angle.
$m\angle CBA = m\angle YXA$ because both are right angles (90°).
$m\angle BCA = m\angle XYA$ because the sum of the measures of the angles in each triangle must be 180°.

Therefore, the corresponding angles are equal, and the triangles are similar.

5. Refer to the figure used in Example 4. If $AB = 4$ cm, $AX = 8$ cm, and $BC = 3$ cm, find XY.

Solution

Since $\overline{AB}$ and $\overline{AX}$ are corresponding sides (they are opposite equal angles) and $\overline{BC}$ and $\overline{XY}$ are corresponding sides (both are opposite equal angles), we have the proportion:

$$\frac{AB}{AX} = \frac{BC}{XY}$$

Thus,

$$\frac{4 \text{ cm}}{8 \text{ cm}} = \frac{3 \text{ cm}}{XY}$$

$$4 \cdot XY = 3 \cdot 8$$

$$\frac{\cancel{4} \cdot XY}{\cancel{4}} = \frac{24}{4}$$

$$XY = 6 \text{ cm}$$

Exercises IV.3

Name each of the following triangles with the indicated measures of angles and lengths of sides.

1. 4 cm, 6 cm, 8 cm

2.

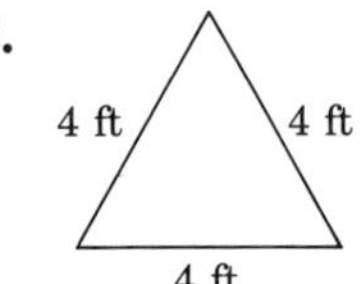

3.

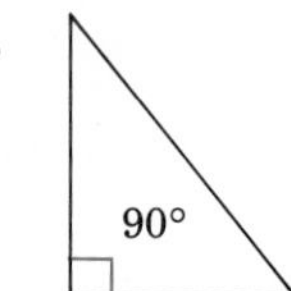

4. 110°

5.

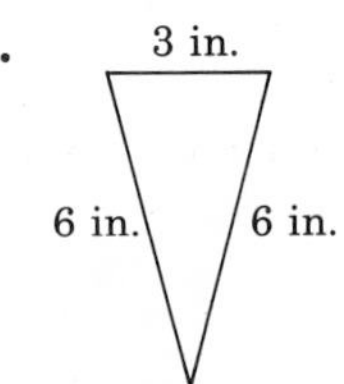

6.

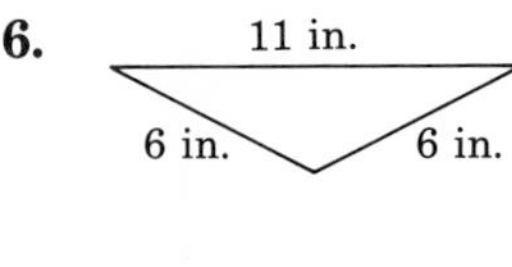

7.

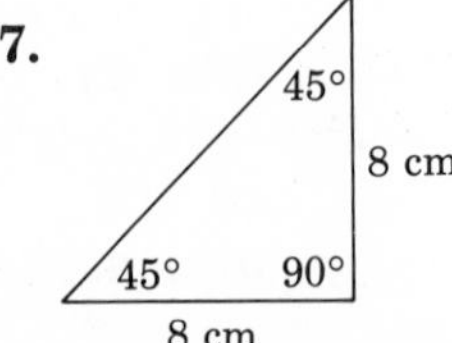

8.

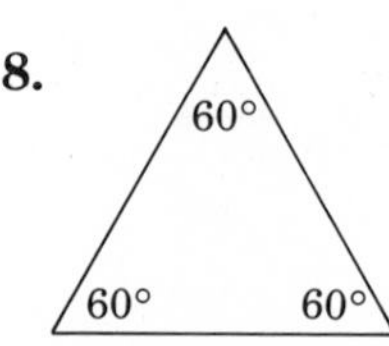

9.

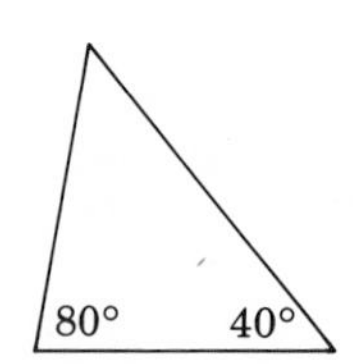

10. The Pythagorean Theorem states that, in a right triangle, the square of the hypotenuse is equal to the sum of the squares of the two legs. See the following figures. Are the triangles ΔABC and ΔDEF right triangles? If so, which sides are the hypotenuses? Are the triangles similar? If so, which angles are equal?

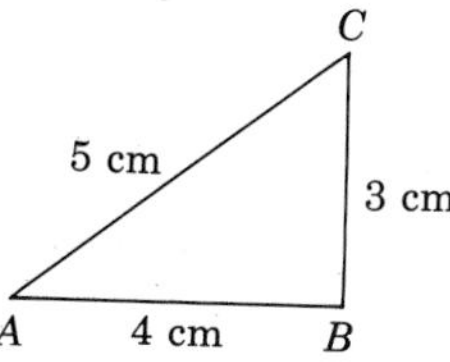

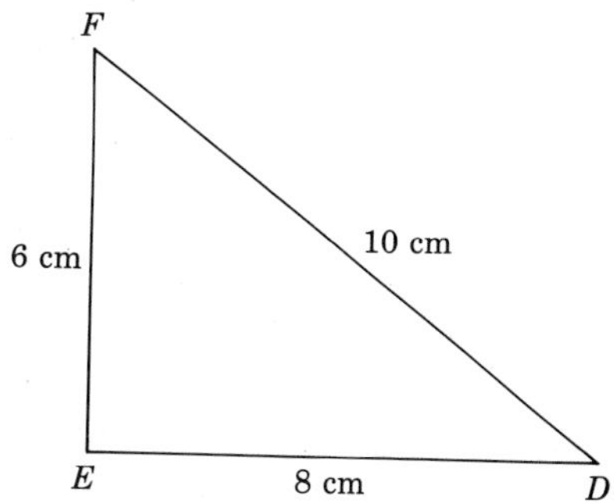

11. In the figure, $m\angle Q, = m\angle T$. Is $\Delta PRQ \sim \Delta PST$? Give reasons for your answer.

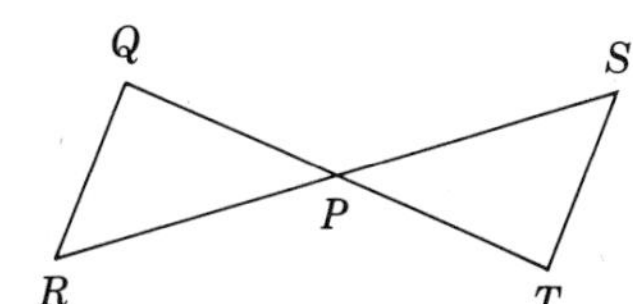

12. Suppose $\Delta XYZ \sim \Delta UVW$, and $m\angle Z = 30°$, and $m\angle W = 30°$. If both triangles are isosceles, what are the measures of the other four angles? (In an isosceles triangle, the angles opposite the equal sides must be equal.)

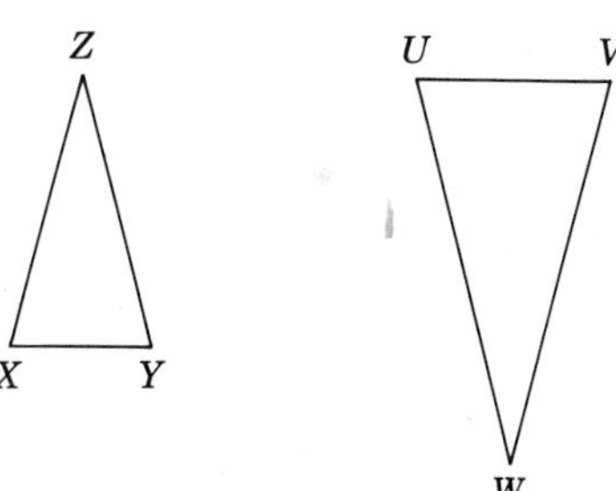

13. Can you form a ΔSTV if $ST = 12$ cm, $TV = 9$ cm, and $SV = 15$ cm? If so, what kind of triangle is ΔSTV?

14. Can you form a triangle with the lengths of the sides 4 ft, 6 ft, and 10 ft? Why or why not?

15. In the figure, the segments $\overline{AB}$ and $\overline{CD}$ are parallel and $\overline{ED}$ is perpendicular to both $\overline{AB}$ and $\overline{CD}$. Are the triangles ΔABE and ΔCDE similar? State your reasoning. What are the lengths of sides $\overline{CE}$ and $\overline{DE}$?

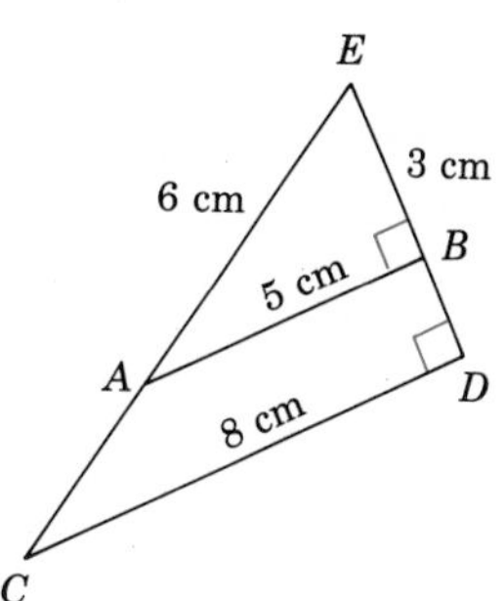

Answer Key

Numbers in brackets following each answer in the Chapter Tests and in Cumulative Tests refer to the section that covers the corresponding test question.

CHAPTER 1

Exercises 1.1, Page 4

1. 3(10) + 7(1)
thirty-seven

2. 8(10) + 4(1)
eighty-four

3. 9(10) + 8(1)
ninety-eight

5. 1(100) + 2(10) + 2(1)
one hundred twenty-two

6. 4(100) + 9(10) + 3(1)
four hundred ninety-three

7. 8(100) + 2(10) + 1(1)
eight hundred twenty-one

9. 1(1000) + 8(100) + 9(10) + 2(1)
one thousand, eight hundred ninety-two

10. 5(1000) + 4(100) + 9(10) + 6(1)
five thousand, four hundred ninety-six

11. 1(10,000) + 2(1000) + 5(100) + 1(10) + 7(1)
twelve thousand, five hundred seventeen

13. 2(100,000) + 4(10,000) + 3(1000) + 4(100) + 0(10) + 0(1)
two hundred forty-three thousand, four hundred

14. 8(100,000) + 9(10,000) + 1(1000) + 5(100) + 4(10) + 0(1)
eight hundred ninety-one thousand, five hundred forty

15. 4(10,000) + 3(1000) + 6(100) + 5(10) + 5(1)
forty-three thousand, six hundred fifty-five

17. 8(1,000,000) + 4(100,000) + 0(10,000) + 0(1000) + 8(100) + 1(10) + 0(1)
eight million, four hundred thousand, eight hundred ten

18. 5(1,000,000) + 6(100,000) + 6(10,000) + 3(1000) + 7(100) + 0(10) + 1(1)
five million, six hundred sixty-three thousand, seven hundred one

19. 1(10,000,000) + 6(1,000,000) + 3(100,000) + 0(10,000) + 2(1000) + 5(100) + 9(10) + 0(1)
sixteen million, three hundred two thousand, five hundred ninety

21. 8(10,000,000) + 3(1,000,000) + 0(100,000) + 0(10,000) + 0(1000) + 6(100) + 0(10) + 5(1)
eighty-three million, six hundred five

22. 1(100,000,000) + 5(10,000,000) + 2(1,000,000) + 4(100,000) + 0(10,000) + 3(1000) + 6(100) + 7(10) + 2(1)
one hundred fifty-two million, four hundred three thousand, six hundred seventy-two

23. 6(100,000,000) + 7(10,000,000) + 9(1,000,000) + 0(100,000) + 7(10,000) + 8(1000) + 1(100) + 0(10) + 0(1)
six hundred seventy-nine million, seventy-eight thousand, one hundred

25. 8(1,000,000,000) + 5(100,000,000) + 7(10,000,000) + 2(1,000,000) + 0(100,000) + 0(10,000) + 3(1000) + 4(100) + 2(10) + 5(1)
eight billion, five hundred seventy-two million, three thousand, four hundred twenty-five

26. 76
27. 132
29. 3842
30. 2005
31. 192,151
33. 21,400
34. 33,333
35. 5,045,000
37. 10,639,582
38. 281,300,501
39. 530,000,700
41. 90,090,090
42. 82,700,000
43. 175,000,002
45. 757

46. 1—hundred thousands
9—thousands
7—hundreds
5—tens

47. 7—millions
8—hundred thousands
4—ten thousands
6—tens
2—units

49. twenty-five thousand, one hundred twenty
50. ninety million; one hundred forty-nine million, seven hundred thirty thousand

Exercises 1.2, Page 8

1. 760
2. 30
3. 80
5. 300
6. 720
7. 990
9. 4200
10. 4500
11. 500
13. 600
14. 3800
15. 76,500
17. 7000
18. 6000
19. 8000
21. 13,000
22. 14,000
23. 62,000
25. 80,000
26. 130,000
27. 260,000
29. 120,000
30. 310,000
31. 180,000
33. 100
34. 0
35. 1000

Exercises 1.3, Page 12

1. 16
2. 15
3. 13
5. 21
6. 20
7. 19
9. 12
10. 17
11. 18
13. 12
14. 21
15. 17
17. 27
18. 18
19. 21
21. commutative
22. commutative
23. associative
25. associative
26. associative
27. identity
29. identity
30. associative
31. 162
33. 239
34. 835
35. 1298
37. 1236
38. 4168
39. 6869
41. 1,603,426
42. 1,463,930
43. 2,610,667
45. 2762 miles
46. $6,313,323
47. $18,463
49. 1518 students
50. 33,830 appliances

51.

+	5	8	7	9
3	8	11	10	12
6	11	14	13	15
5	10	13	12	14
2	7	10	9	11

Exercises 1.4, Page 18

1. 3 **2.** 13 **3.** 0 **5.** 9 **6.** 17 **7.** 9
9. 5 **10.** 6 **11.** 0 **13.** 13 **14.** 20 **15.** 20
17. 5 **18.** 45 **19.** 13 **21.** 94 **22.** 126 **23.** 218
25. 475 **26.** 376 **27.** 593 **29.** 188
30. 478 **31.** 1569 **33.** 1568 **34.** 1531
35. 0 **37.** 694 **38.** 5871 **39.** 2517
41. 2,806,644 **42.** 3,800,559 **43.** 1,006,958 **45.** 5,671,011
46. 222 **47.** 140 **49.** 32 years **50.** 44 points
51. $250,404 **53.** $934 **54.** $235,456 **55.** $39,100
57. $3700 **58.** $868

Exercises 1.5, Page 23

1. 72 **2.** 42 **3.** 48 **5.** 30 **6.** 45 **7.** 27
9. 24 **10.** 28 **11.** 56 **13.** 0 **14.** 0 **15.** 9
17. 0 **18.** 0 **19.** 6 **21.** 42 **22.** 60 **23.** 60
25. 30 **26.** 24 **27.** 40 **29.** 96 **30.** 105 **31.** 72
33. 0 **34.** 0 **35.** 0

45.

GIVEN NO.	ADD 5	TRIPLE	SUBTRACT 15
2	7	21	6
1	6	18	3
7	12	36	21
8	13	39	24
5	10	30	15
6	11	33	18

46.

•	5	8	7	9
3	15	24	21	27
6	30	48	42	54
5	25	40	35	45
2	10	16	14	18

47.

•	4	1	9	6
5	20	5	45	30
9	36	9	81	54
6	24	6	54	36
0	0	0	0	0

Exercises 1.6, Page 26

1. 250 **2.** 7600 **3.** 47,000 **5.** 720 **6.** 13
7. 3000 **9.** 400 **10.** 3600 **11.** 1200 **13.** 6300
14. 9000 **15.** 4000 **17.** 15,000 **18.** 3600 **19.** 5200
21. 16,000 **22.** 180,000 **23.** 10,000 **25.** 60,000
26. 25,000 **27.** 12,000 **29.** 240,000 **30.** 800,000
31. 48,000 **33.** 600,000 **34.** 16,000,000 **35.** 630,000
37. $5 \cdot 1000$ **38.** $3 \cdot 100$ **39.** $17 \cdot 10$ **41.** $63 \cdot 10{,}000$ **42.** $8 \cdot 10$

Exercises 1.7, Page 31

1. 224 **2.** 162 **3.** 432 **5.** 344 **6.** 432
7. 455 **9.** 252 **10.** 760 **11.** 2352 **13.** 330

14. 1189
15. 2412
17. 960
18. 2790
19. 7055
21. 544
22. 2548
23. 880
25. 375
26. 4371
27. 2064
29. 156
30. 2916
31. 5166
33. 2850
34. 7632
35. 29,601
37. 9800
38. 174,045
39. 125,178
41. 31,200
42. 66,960
43. 380,000
45. 496,400
46. 1,504,000
47. 217,300
49. 249,600
50. 897,000
51. 6821
53. $32,760; $93,600
54. $24,480; $42,600
55. 275 mi; 200 mi

Exercises 1.8, Page 37

1. 30
2. 10
3. 21
5. 12
6. 30
7. 5
9. 6 R4
10. 7 R2
11. 24
13. 10 R11
14. 14
15. 11 R7
17. 41
18. 45 R6
19. 47 R8
21. 5 R2
22. 2 R3
23. 6 R1
25. 6
26. 12
27. 9
29. 32 R2
30. 3 R10
31. 9
33. 8
34. 15 R5
35. 11 R8
37. 42 R3
38. 50
39. 20
41. 30
42. 400 R3
43. 300 R13
45. 301 R4
46. 2 R2
47. 3 R3
49. 61 R15
50. 54 R3
51. 2
53. 22 R74
54. 4 R192
55. 7 R358
57. 196 R370
58. 221 R308
59. 107 R215
61. yes; yes; 23
62. yes; yes; 62
63. 30
65. 22; yes
66. 203; yes
67. $64
69. $10,035
70. 102 chairs

Exercises 1.9, Page 40

1. 103
2. 54
3. 6
5. 485
6. 502
7. 3000
9. 586
10. 706
11. 82
13. 85
14. $42
15. $57
17. $932
18. 7 hr
19. $10,432

Review Questions: Chapter 1, Page 43

1. 4(100) + 9(10) + 5(1)
four hundred ninety-five
2. 1(1000) + 9(100) + 7(10) + 5(1)
one thousand, nine hundred seventy-five
3. 6(10,000) + 0(1000) + 3(100) + 0(10) + 8(1)
sixty thousand, three hundred eight
4. 4856
5. 15,032,197
6. 672,340,083
7. 630
8. 15,000
9. 700
10. 2600
11. commutative prop. of add.
12. associative prop. of mult.
13. associative prop. of add.
14. commutative prop. of mult.
15. $32 \div 8 = 4$; $2 \div 2 = 1$; division is not associative
16. 10,541
17. 1674
18. 2400
19. 0
20. 480,000
21. 508
22. 2384
23. 2102
24. 5952
25. 3,822,498
26. 14,388,000
27. 292 R2
28. 135 R81
29. 703 R7
30. 1059
31. 35
32. 9
33. $1485; $99
34. 83

35. If a is a whole number, then there is a unique whole number 0 with the property that $a + 0 = a$. If a is a whole number, then there is a unique whole number 1 with the property that $a \cdot 1 = a$.

36. 70

37.

GIVEN NO.	ADD 100	DOUBLE	SUBTRACT 200
3	103	206	6
20	120	240	40
15	115	230	30
8	108	216	16

38. 20 times

Test: Chapter 1, Page 46

1. 8(1000) + 9(100) + 5(10) + 2(1)
eight thousand, nine hundred fifty-two [1.1]
2. identity [1.5] **3.** $7 \cdot 9 = 9 \cdot 7$ [1.5] **4.** 1000 [1.2]
5. 140,000 [1.2] **6.** 12,009 [1.3] **7.** 1735 [1.3]
8. 13,781,661 [1.3] **9.** 488 [1.4] **10.** 1229 [1.4]
11. 5707 [1.4] **12.** 2584 [1.7] **13.** 220,405 [1.7]
14. 210,938 [1.7] **15.** 403 [1.8] **16.** 172 R388 [1.8]
17. 2005 [1.8] **18.** 74 [1.9] **19.** 54 [1.8]
20. \$306; \$51; \$224 [1.9]

CHAPTER 2

Exercises 2.1, Page 49

	Exponent	*Base*	*Power*		*Exponent*	*Base*	*Power*
1.	3	2	8	**2.**	5	2	32
3.	2	5	25	**5.**	0	7	1
6.	2	11	121	**7.**	4	1	1
9.	0	4	1	**10.**	6	3	729
11.	2	3	9	**13.**	0	5	1
14.	50	1	1	**15.**	1	62	62
17.	2	10	100	**18.**	3	10	1000
19.	2	4	16	**21.**	4	10	10,000
22.	3	5	125	**23.**	3	6	216
25.	0	19	1				

26. 2^2 **27.** 5^2 **29.** 3^3 **30.** 2^5
31. 11^2 **33.** 2^3 **34.** 3^2 **35.** 6^2
37. 9^2 or 3^4 **38.** 8^2 or 4^3 or 2^6 **39.** 10^2 **41.** 10^4
42. 6^3 **43.** 12^2 **45.** 3^5 **46.** 25^2 or 5^4
47. 15^2 **49.** 7^3 **50.** 10^5 **51.** 6^5
53. $2^2 \cdot 7^2$ **54.** $5^2 \cdot 9^3$ **55.** $2^2 \cdot 3^3$ **57.** $7^2 \cdot 13$
58. 11^3 **59.** $2 \cdot 3^2 \cdot 11^2$ **61.** 64 **62.** 9

63. 49 **65.** 225 **66.** 196 **67.** 324
69. 144 **70.** 400

Exercises 2.2, Page 53

1. 3 **2.** 15 **3.** 7 **5.** 22 **6.** 31
7. 3 **9.** 5 **10.** 10 **11.** 5 **13.** 5
14. 0 **15.** 3 **17.** 7 **18.** 0 **19.** 0
21. 6 **22.** 3 **23.** 3 **25.** 27 **26.** 0
27. 26 **29.** 0 **30.** 5 **31.** 69 **33.** 68
34. 80 **35.** 140 **37.** 5 **38.** 9 **39.** 0
41. 9 **42.** 93 **43.** 12 **45.** 24 **46.** 118
47. 70 **49.** 230 **50.** 34 **51.** 110 **53.** 110
54. 0 **55.** 2980

Exercises 2.3, Page 57

1. 2, 3, 4, 9 **2.** 3, 9 **3.** 3, 5 **5.** 2, 3, 5, 10
6. 3 **7.** 2, 4 **9.** 2, 3, 4 **10.** 3, 5
11. none **13.** 3, 9 **14.** 2, 3, 4, 5, 10 **15.** none
17. none **18.** 2 **19.** 3 **21.** 3, 5
22. 3 **23.** 2, 3, 4, 5, 9, 10 **25.** 2, 3 **26.** 2
27. none **29.** 2 **30.** 2, 3 **31.** 3, 5
33. 3, 9 **34.** 3, 5 **35.** 2, 3, 9 **37.** 2, 4, 5, 10
38. 2, 3, 4, 9 **39.** 3, 9 **41.** 2, 3, 4, 5, 10 **42.** none
43. 2, 4 **45.** 2, 3, 4, 5, 10 **46.** 2, 3, 5, 10 **47.** 3, 5
49. none **50.** 2 **51.** 2, 3, 9 **53.** 2, 3, 4
54. 5 **55.** 2, 3, 4, 9 **57.** 3, 9 **58.** 2
59. 2, 3, 4 **61.** yes; 18, 36, 54, 72, 90 **62.** no; 9, 18, 36, 45, 63

Exercises 2.4, Page 63

1. 5, 10, 15, 20, . . . **2.** 7, 14, 21, 28, . . . **3.** 11, 22, 33, 44, . . .
5. 12, 24, 36, 48, . . . **6.** 9, 18, 27, 36, . . . **7.** 20, 40, 60, 80, . . .
9. 16, 32, 48, 64, . . . **10.** 25, 50, 75, 100, . . .

11. (circled numbers shown in parentheses)

1	(2)	(3)	4	(5)	6	(7)	8	9	10
(11)	12	(13)	14	15	16	(17)	18	(19)	20
21	22	(23)	24	25	26	27	28	(29)	30
(31)	32	33	34	35	36	(37)	38	39	40
(41)	42	(43)	44	45	46	(47)	48	49	50
51	52	(53)	54	55	56	57	58	(59)	60
(61)	62	63	64	65	66	(67)	68	69	70
(71)	72	(73)	74	75	76	77	78	(79)	80
81	82	(83)	84	85	86	87	88	(89)	90
91	92	93	94	95	96	(97)	98	99	100

13. prime
14. composite; 4, 7
15. composite; 4, 8
17. prime
18. composite; 2, 8
19. composite; 7, 9
21. composite; 3, 17
22. prime
23. prime
25. prime
26. composite; 4, 13
27. composite; 3, 19
29. composite; 2, 43
30. prime
31. prime
33. 3, 4
34. 8, 2
35. 12, 1
37. 25, 2
38. 5, 4
39. 8, 3
41. 12, 3
42. 7, 1
43. 21, 3
45. 5, 5
46. 4, 4
47. 12, 5
49. 9, 3
50. 18, 4
51. 2

Exercises 2.5, Page 68

1. $2^3 \cdot 3$
2. $2^2 \cdot 7$
3. 3^3
5. $2^2 \cdot 3^2$
6. $2^2 \cdot 3 \cdot 5$
7. $2^3 \cdot 3^2$
9. 3^4
10. $3 \cdot 5 \cdot 7$
11. 5^3
13. $3 \cdot 5^2$
14. $2 \cdot 3 \cdot 5^2$
15. $2 \cdot 3 \cdot 5 \cdot 7$
17. $2 \cdot 5^3$
18. $3 \cdot 31$
19. $2^3 \cdot 3 \cdot 7$
21. $2 \cdot 3^2 \cdot 7$
22. $2^4 \cdot 3$
23. 17
25. $3 \cdot 17$
26. $2^4 \cdot 3^2$
27. 11^2
29. $3^2 \cdot 5^2$
30. $2^2 \cdot 13$
31. 2^5
33. $2^2 \cdot 3^3$
34. 103
35. 101
37. $2 \cdot 3 \cdot 13$
38. $2^2 \cdot 5^3$
39. $2^4 \cdot 5^4$
41. 1, 2, 3, 4, 6, 12
42. 1, 2, 3, 6, 9, 18
43. 1, 2, 4, 7, 14, 28
45. 1, 11, 121
46. 1, 3, 5, 9, 15, 45
47. 1, 3, 5, 7, 15, 21, 35, 105
49. 1, 97
50. 1, 2, 3, 4, 6, 8, 9, 12, 16, 18, 24, 36, 48, 72, 144

Exercises 2.6, Page 74

1. 24
2. 105
3. 36
5. 110
6. 252
7. 60
9. 200
10. 600
11. 196
13. 240
14. 112
15. 100
17. 700
18. 432
19. 252
21. 8
22. 210
23. 1560
25. 60
26. 120
27. 240
29. 2250
30. 918
31. 726
33. 2610
34. 324
35. 675
37. 120
38. 600
39. 1680
41. 120: $8 \cdot 15$, $10 \cdot 12$, $15 \cdot 8$
42. 30: $6 \cdot 5$, $15 \cdot 2$, $30 \cdot 1$
43. 120: $10 \cdot 12$, $15 \cdot 8$, $24 \cdot 5$
45. 270: $6 \cdot 45$, $18 \cdot 15$, $27 \cdot 10$, $45 \cdot 6$
46. 1140: $12 \cdot 95$, $95 \cdot 12$, $228 \cdot 5$
47. 4410: $45 \cdot 98$, $63 \cdot 70$, $98 \cdot 45$
49. 14,157: $99 \cdot 143$, $143 \cdot 99$, $363 \cdot 39$
50. 3375: $125 \cdot 27$, $135 \cdot 25$, $225 \cdot 19$
51. 70 min; 7 times, 5 times
53. 48 hr; 4 orbits and 3 orbits, respectively
54. once every 30 days; once every 30 days
55. 180 days; 18 trips, 15 trips, 12 trips, 10 trips

Review Questions: Chapter 2, Page 77

1. base, exponent, power
2. different factors
3. prime
4. composite
5. 16
6. 27
7. 169
8. 441
9. 15

10. 46
11. 35
12. 13
13. 2
14. 2
15. 3, 5, 9
16. 2, 3, 4, 9
17. none
18. 2, 3, 4, 5, 9, 10
19. 2, 4
20. 3
21. 3, 6, 9, 12, . . . ; yes, 3 is prime.
22. yes
23. 2, 3, 5, 7, 11, 13, 17, 19, 23, 29, 31, 37, 41, 43, 47, 53, 59
24. 6 and 4
25. 5 and 12
26. $2 \cdot 3 \cdot 5^2$
27. $5 \cdot 13$
28. $2^2 \cdot 3 \cdot 7$
29. $2^2 \cdot 23$
30. 168
31. 1800
32. 270
33. 1638; $18 \cdot 91$, $39 \cdot 42$, $63 \cdot 26$
34. 210 sec; 14 laps, 12 laps

Test: Chapter 2, Page 79

1. exponent—3; base—7; power—343 [2.1]
2. $7^2 = 49$; $11^2 = 121$; $13^2 = 169$; $17^2 = 289$; $19^2 = 361$ [2.1]
3. 53 [2.2]
4. 66 [2.2]
5. 14 [2.2]
6. 5 [2.2]
7. 2, 3, 4, and 9 [2.3]
8. none [2.3]
9. 2, 3, 4, 5, 9, and 10 [2.3]
10. 19, 38, 57, 76, 95 [2.4]
11. 6, 9, 15, 27, 39, 51 [2.4]
12. $2 \cdot 5 \cdot 7$ [2.5]
13. $2 \cdot 3^2 \cdot 13$ [2.5]
14. $2 \cdot 3 \cdot 5^3$ [2.5]
15. 1, 3, 5, 9, 15, 45 [2.5]
16. 120 [2.6]
17. 108 [2.6]
18. 75 [2.6]
19. 1008 [2.6]
20. 378 seconds; 7 laps and 6 laps [2.6]

CHAPTER 3

Exercises 3.1, Page 85

1. $\frac{1}{3}$
2. $\frac{1}{4}$
3. $\frac{1}{4}$
5. $\frac{1}{4}$
6. $\frac{1}{2}$

7.

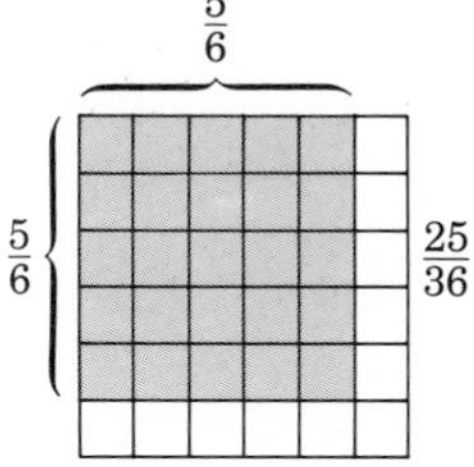

9.

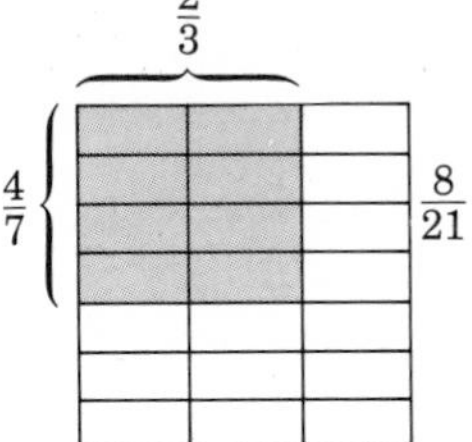

10.

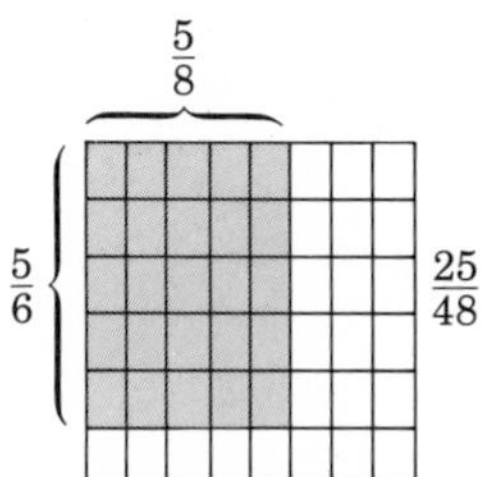

11. $\frac{6}{25}$
13. $\frac{1}{9}$

14. $\frac{1}{16}$ 15. $\frac{1}{4}$ 17. $\frac{3}{8}$ 18. $\frac{15}{32}$ 19. $\frac{9}{16}$ 21. $\frac{4}{81}$

22. $\frac{12}{35}$ 23. $\frac{5}{32}$ 25. 0 26. 0 27. $\frac{9}{25}$ 29. $\frac{9}{100}$

30. $\frac{12}{1}$ or 12 31. $\frac{10}{1}$ or 10 33. $\frac{14}{50}$ 34. $\frac{3}{50}$

35. $\frac{4}{21}$ 37. $\frac{99}{20}$ 38. $\frac{6}{385}$ 39. $\frac{48}{455}$

41. $\frac{1}{360}$ 42. $\frac{27}{100}$ 43. $\frac{840}{1} = 840$ 45. 0

46. 0 47. 0 49. $\frac{728}{45}$ 50. $\frac{9}{500}$

Exercises 3.2, Page 91

1. 9 2. 6 3. 8 5. 12 6. 10 7. 15
9. 10 10. 10 11. 15 13. 10 14. 2 15. 27
17. 40 18. 15 19. 10 21. 30 22. 9 23. 15
25. 70 26. 42 27. 50 29. 15 30. 32 31. 80
33. 60 34. 70 35. 66 37. 90 38. 900 39. 300

41. $\frac{1}{3}$ 42. $\frac{2}{3}$ 43. $\frac{3}{4}$ 45. $\frac{2}{5}$ 46. $\frac{4}{5}$ 47. $\frac{7}{18}$

49. $\frac{0}{25} = 0$ 50. $\frac{3}{4}$ 51. $\frac{2}{5}$ 53. $\frac{5}{6}$ 54. $\frac{1}{4}$ 55. $\frac{2}{3}$

57. $\frac{2}{3}$ 58. $\frac{3}{4}$ 59. $\frac{6}{25}$ 61. $\frac{2}{3}$ 62. $\frac{2}{3}$ 63. $\frac{12}{35}$

65. $\frac{2}{9}$ 66. $\frac{3}{7}$ 67. $\frac{25}{76}$ 69. $\frac{1}{2}$ 70. $\frac{4}{1} = 4$ 71. $\frac{6}{1} = 6$

73. $\frac{2}{17}$ 74. $\frac{3}{8}$ 75. $\frac{9}{20}$ 77. $\frac{10}{9}$ 78. $\frac{11}{15}$ 79. $\frac{5}{4}$

81. $\frac{5}{18}$ 82. $\frac{1}{6}$ 83. $\frac{1}{4}$ 85. $\frac{21}{4}$ 86. $\frac{189}{52}$ 87. $\frac{77}{4}$

89. $\frac{8}{5}$ 90. $\frac{3}{10}$ 91. $\frac{2}{7}$ 93. $\frac{2}{3}$ 94. $\frac{1}{80}$ 95. $\frac{21}{16}$

Exercises 3.3, Page 96

1. $\frac{9}{8}$ 2. $\frac{25}{24}$ 3. $\frac{5}{3}$ 5. $\frac{4}{7}$ 6. $\frac{7}{12}$ 7. $\frac{3}{4}$

9. $\frac{5}{9}$ 10. $\frac{21}{16}$ 11. $\frac{7}{12}$ 13. 5 14. 3 15. 7

17. 1 18. 1 19. $\frac{9}{16}$ 21. $\frac{3}{20}$ 22. $\frac{1}{10}$ 23. $\frac{1}{10}$

25. $\frac{2}{7}$ 26. $\frac{3}{11}$ 27. $\frac{1}{4}$ 29. $\frac{16}{5}$ 30. 5 31. $\frac{5}{9}$

33. $\frac{1}{5}$ 34. $\frac{2}{5}$ 35. $\frac{4}{9}$ 37. $\frac{65}{24}$ 38. $\frac{24}{39}$ 39. $\frac{176}{105}$

41. $\frac{22}{7}$ 42. 98 43. 125 45. 100 46. $\frac{50}{27}$ 47. $\frac{62}{43}$

49. 225 **50.** $\frac{5}{12}$

Exercises 3.4, Page 102

1. 16 **2.** 39 **3.** 54 **5.** 1000
6. $\frac{10}{10} = 1$ **7.** $\frac{5}{14}$ **9.** $\frac{3}{2}$ **10.** $\frac{3}{2}$
11. $\frac{10}{5} = 2$ **13.** $\frac{5}{3}$ **14.** $\frac{3}{5}$ **15.** $\frac{13}{18}$
17. $\frac{11}{16}$ **18.** $\frac{23}{20}$ **19.** $\frac{3}{5}$ **21.** $\frac{12}{12} = 1$
22. $\frac{11}{16}$ **23.** $\frac{17}{20}$ **25.** $\frac{17}{21}$ **26.** $\frac{3}{4}$
27. $\frac{9}{13}$ **29.** $\frac{23}{54}$ **30.** $\frac{151}{140}$ **31.** $\frac{23}{72}$
33. $\frac{9}{4}$ **34.** $\frac{3}{5}$ **35.** $\frac{1}{2}$ **37.** $\frac{5}{8}$
38. $\frac{49}{60}$ **39.** $\frac{5}{6}$ **41.** $\frac{343}{432}$ **42.** $\frac{31}{96}$
43. $\frac{13}{9}$ **45.** $\frac{1}{5}$ **46.** $\frac{7}{6}$ **47.** $\frac{317}{1000}$
49. $\frac{271}{10{,}000}$ **50.** $\frac{631}{100}$ **51.** $\frac{8191}{1000}$ **53.** $\frac{753}{1000}$
54. $\frac{63}{50}$ **55.** $\frac{89}{200}$ **57.** $\frac{613}{1000}$ **58.** $\frac{27{,}683}{10{,}000}$
59. $\frac{5134}{1000} = \frac{2567}{500}$ **61.** $\frac{5}{3}$ **62.** $\frac{7}{6}$ **63.** $\frac{119}{72}$
65. $\frac{19}{100}$ **66.** $\frac{531}{1000}$ **67.** $\frac{49}{36}$ **69.** $\frac{73}{96}$
70. $\frac{39}{100}$

Exercises 3.5, Page 107

1. $\frac{3}{7}$ **2.** $\frac{2}{7}$ **3.** $\frac{3}{5}$ **5.** $\frac{1}{2}$ **6.** $\frac{1}{4}$ **7.** $\frac{1}{3}$
9. $\frac{3}{5}$ **10.** $\frac{2}{3}$ **11.** $\frac{1}{2}$ **13.** $\frac{13}{30}$ **14.** $\frac{9}{15}$ **15.** $\frac{1}{12}$
17. $\frac{9}{32}$ **18.** $\frac{5}{16}$ **19.** $\frac{13}{20}$ **21.** $\frac{7}{54}$ **22.** $\frac{11}{18}$ **23.** $\frac{1}{40}$
25. $\frac{7}{60}$ **26.** $\frac{17}{4}$ **27.** $\frac{27}{8}$ **29.** $\frac{3}{16}$ **30.** $\frac{16}{3}$ **31.** $\frac{87}{100}$
33. $\frac{3}{50}$ **34.** $\frac{9}{1000}$ **35.** $\frac{1}{25}$ **37.** $\frac{3}{20}$ **38.** 0 **39.** 0
41. $\frac{1}{10}$ **42.** $\frac{1}{8}$ **43.** $\frac{1}{3}$ **45.** $\frac{17}{20}$ **46.** $\frac{5}{9}$ **47.** $\frac{5}{24}$
49. $\frac{1}{50}$ **50.** $\frac{29}{1000}$ **51.** $\frac{5}{16}$ **53.** $\frac{11}{12}$ **54.** $\frac{9}{25}$ **55.** $\frac{11}{10}$

Exercises 3.6, Page 112

1. $\frac{3}{4}$ by $\frac{1}{12}$ **2.** $\frac{7}{8}$ by $\frac{1}{24}$ **3.** $\frac{17}{20}$ by $\frac{1}{20}$
5. $\frac{13}{20}$ by $\frac{1}{40}$ **6.** $\frac{21}{25}$ by $\frac{11}{400}$ **7.** equal
9. $\frac{11}{48}$ by $\frac{1}{60}$ **10.** $\frac{37}{100}$ by $\frac{1}{20}$ **11.** $\frac{3}{5}, \frac{2}{3}, \frac{7}{10}$
13. $\frac{11}{12}, \frac{19}{20}, \frac{7}{6}$ **14.** $\frac{5}{42}, \frac{1}{3}, \frac{3}{7}$ **15.** $\frac{1}{4}, \frac{1}{3}, \frac{1}{2}$
17. $\frac{13}{18}, \frac{7}{9}, \frac{31}{36}$ **18.** $\frac{40}{36}, \frac{31}{24}, \frac{17}{12}$ **19.** $\frac{20}{10{,}000}, \frac{3}{1000}, \frac{1}{100}$
21. $\frac{2}{3}$ **22.** $\frac{1}{5}$ **23.** $\frac{13}{9}$ **25.** $\frac{187}{32}$ **26.** $\frac{43}{450}$
27. $\frac{1}{4}$ **29.** $\frac{8}{39}$ **30.** $\frac{14}{45}$ **31.** $\frac{15}{64}$ **33.** $\frac{29}{36}$
34. $\frac{31}{36}$ **35.** $\frac{27}{10}$

Review Questions: Chapter 3, Page 116

1. 0 **2.** undefined **3.** $\frac{3}{2}, \frac{2}{3}$ **4.** associative

5. $\frac{2}{5}$; $\frac{2}{3}$; $\frac{4}{15}$ **6.** $\frac{1}{30}$ **7.** $\frac{3}{49}$

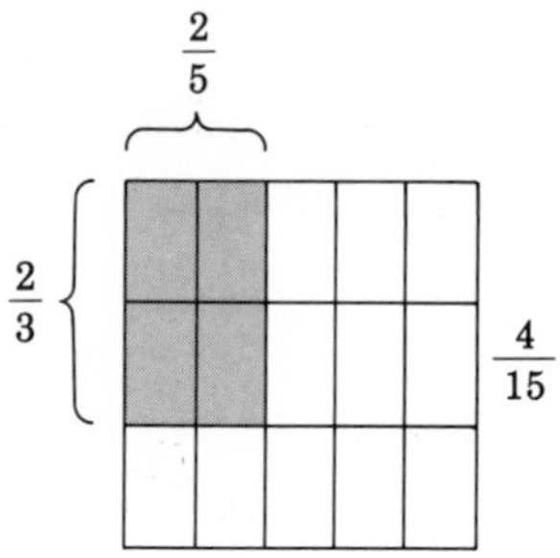

8. $\frac{1}{12}$ **9.** 2 **10.** 60 **11.** 65 **12.** $\frac{1}{2}$ **13.** $\frac{9}{8}$
14. 0 **15.** $\frac{5}{4}$ **16.** $\frac{5}{7}$ **17.** $\frac{2}{3}$ **18.** $\frac{1}{4}$ **19.** $\frac{49}{72}$
20. $\frac{7}{22}$ **21.** $\frac{25}{54}$ **22.** $\frac{7}{20}$ **23.** $\frac{1}{3}$ **24.** $\frac{7}{8}$ **25.** $\frac{1}{9}$
26. $\frac{5}{3}$ **27.** 1 **28.** $\frac{5}{4}$ **29.** $\frac{4}{5}$ **30.** $\frac{4}{5}$ by $\frac{2}{15}$ **31.** $\frac{11}{20}, \frac{5}{9}, \frac{7}{12}$
32. $\frac{25}{112}$ **33.** $\frac{32}{125}$ **34.** $\frac{19}{30}$ **35.** $\frac{7}{18}$

Test: Chapter 3, Page 118

1. reciprocals [3.3] **2.** $\frac{20}{56}$ [3.2] **3.** $\frac{3}{5}$ [3.2] **4.** $\frac{9}{11}$ [3.2] **5.** $\frac{29}{30}$ [3.4]

6. $\frac{10}{27}$ [3.1] **7.** $\frac{7}{36}$ [3.5] **8.** $\frac{2}{7}$ [3.3] **9.** $\frac{1}{12}$ [3.2] **10.** $\frac{4}{15}$ [3.5]
11. 25 [3.3] **12.** $\frac{3}{7}$ [3.4] **13.** $\frac{151}{120}$ [3.4] **14.** $\frac{32}{15}$ [3.3] **15.** $\frac{1}{2}$ [3.2]
16. 2 [3.5] **17.** $\frac{9}{20}$ [3.3] **18.** $\frac{19}{20}$ [3.6] **19.** $\frac{141}{70}$ [3.7] **20.** $\frac{7}{12}, \frac{3}{5}, \frac{5}{8}$ [3.6]

CHAPTER 4

Exercises 4.1, Page 124

1. $\frac{4}{3}$ **2.** $\frac{5}{2}$ **3.** $\frac{4}{3}$ **5.** $\frac{3}{2}$ **6.** $\frac{3}{2}$
7. $\frac{7}{5}$ **9.** $\frac{5}{4}$ **10.** $\frac{5}{4}$ **11.** $4\frac{1}{6}$ **13.** $1\frac{1}{3}$
14. $1\frac{1}{4}$ **15.** $1\frac{1}{2}$ **17.** $6\frac{1}{7}$ **18.** $2\frac{1}{8}$ **19.** $7\frac{1}{2}$
21. $3\frac{1}{9}$ **22.** $2\frac{1}{15}$ **23.** 3 **25.** $4\frac{1}{2}$ **26.** $2\frac{1}{17}$
27. $\frac{3}{1} = 3$ **29.** $1\frac{3}{4}$ **30.** $1\frac{17}{20}$ **31.** $\frac{37}{8}$ **33.** $\frac{76}{15}$
34. $\frac{8}{5}$ **35.** $\frac{46}{11}$ **37.** $\frac{7}{3}$ **38.** $\frac{34}{7}$ **39.** $\frac{32}{3}$
41. $\frac{34}{5}$ **42.** $\frac{71}{5}$ **43.** $\frac{50}{3}$ **45.** $\frac{101}{5}$ **46.** $\frac{47}{5}$
47. $\frac{92}{7}$ **49.** $\frac{17}{1} = 17$ **50.** $\frac{151}{50}$

Exercises 4.2, Page 127

1. $14\frac{5}{7}$ **2.** 10 **3.** 8 **5.** $12\frac{3}{4}$ **6.** $15\frac{9}{10}$
7. $8\frac{2}{3}$ **9.** $34\frac{7}{18}$ **10.** $10\frac{1}{2}$ **11.** $7\frac{2}{3}$ **13.** $42\frac{7}{20}$
14. $8\frac{10}{21}$ **15.** $10\frac{3}{4}$ **17.** $12\frac{7}{27}$ **18.** $14\frac{1}{16}$ **19.** $18\frac{23}{45}$
21. $28\frac{1}{2}$ **22.** $7\frac{1}{4}$ **23.** $9\frac{29}{40}$ **25.** $12\frac{47}{60}$ **26.** $4\frac{2}{3}$
27. $63\frac{11}{24}$ **29.** $53\frac{83}{192}$ **30.** $12\frac{5}{12}$ **31.** $18\frac{7}{8}$ **33.** $36\frac{23}{24}$
34. $9\frac{7}{20}$ **35.** $90\frac{1}{8}$ **37.** $35\frac{7}{10}$ km **38.** $89\frac{1}{8}$ ft **39.** $12\frac{1}{8}$ in.

Exercises 4.3, Page 131

1. $4\frac{1}{2}$ **2.** $5\frac{3}{4}$ **3.** $1\frac{5}{12}$ **5.** 4 **6.** 4
7. $3\frac{1}{2}$ **9.** $3\frac{1}{4}$ **10.** $5\frac{1}{8}$ **11.** $3\frac{5}{8}$ **13.** $4\frac{2}{3}$

14. $6\frac{1}{6}$ **15.** $6\frac{7}{12}$ **17.** $10\frac{4}{5}$ **18.** $3\frac{29}{32}$ **19.** $5\frac{7}{12}$

21. $1\frac{7}{16}$ **22.** $3\frac{9}{10}$ **23.** $\frac{1}{3}$ **25.** $\frac{5}{8}$ **26.** $57\frac{1}{6}$

27. $1\frac{11}{16}$ **29.** $1\frac{13}{16}$ **30.** 17 **31.** 5 **33.** $7\frac{3}{40}$

34. $16\frac{5}{24}$ **35.** $7\frac{4}{5}$ **37.** $\frac{3}{8}$ **38.** $\frac{1}{4}$ **39.** $\frac{1}{10}$

41. $\frac{1}{5}$ **42.** $\frac{3}{5}$ **43.** $\frac{5}{6}$ **45.** $1\frac{1}{4}$

46. $\frac{3}{5}$ hr or 36 min **47.** $1\frac{3}{20}$ hr or 1 hr 9 min **49.** $6\frac{3}{5}$ min or 6 min 36 sec

50. $4\frac{5}{8}$ dollars **51.** $3\frac{1}{4}$ **53.** $3\frac{1}{4}$ ft or 3 ft 3 in. **54.** $\frac{3}{4}$ hr or 45 min

Exercises 4.4, Page 137

1. $7\frac{7}{12}$ **2.** $1\frac{13}{35}$ **3.** $10\frac{1}{2}$ **5.** $22\frac{1}{2}$ **6.** 12

7. $31\frac{1}{6}$ **9.** 8 **10.** 12 **11.** 30 **13.** $21\frac{3}{4}$

14. 1 **15.** 1 **17.** $27\frac{1}{2}$ **18.** $10\frac{1}{5}$ **19.** $19\frac{1}{4}$

21. $\frac{1}{7}$ **22.** $\frac{1}{5}$ **23.** $\frac{1}{10}$ **25.** $\frac{36}{385}$ **26.** $39\frac{3}{8}$

27. 11 **29.** 242 **30.** $15\frac{99}{160}$ **31.** $11\frac{3}{8}$ **33.** 17

34. $1\frac{3}{10}$ **35.** $2\frac{1}{10}$ **37.** $1827\frac{1}{10}$ **38.** $1291\frac{5}{42}$ **39.** $3016\frac{26}{75}$

41. 40 **42.** 20 **43.** 20 **45.** 75 **46.** 100

47. $\frac{5}{16}$ **49.** $\frac{9}{14}$ **50.** $\frac{7}{10}$ **51.** 10 ft, 22 ft **53.** 177 mi

54. $1012\frac{7}{8}$ ft **55.** 90 pages; 18 hr

Exercises 4.5, Page 143

1. $\frac{1}{3}$ **2.** $\frac{9}{20}$ **3.** $\frac{5}{9}$ **5.** $\frac{10}{39}$ **6.** $\frac{7}{60}$

7. $\frac{29}{155}$ **9.** $\frac{28}{17} = 1\frac{11}{17}$ **10.** $\frac{62}{39} = 1\frac{23}{39}$ **11.** $\frac{200}{301}$

13. $\frac{7}{5} = 1\frac{2}{5}$ **14.** $\frac{37}{32} = 1\frac{5}{32}$ **15.** $\frac{41}{3} = 13\frac{2}{3}$ **17.** $\frac{63}{5} = 12\frac{3}{5}$

18. $\frac{9}{32}$ **19.** $\frac{20}{11} = 1\frac{9}{11}$ **21.** $\frac{1}{5}$ **22.** 0

23. $\frac{172}{9} = 19\frac{1}{9}$ **25.** $\frac{791}{120} = 6\frac{71}{120}$ **26.** $\frac{3}{2} = 1\frac{1}{2}$ **27.** $\frac{3}{5}$

29. $\frac{5}{2} = 2\frac{1}{2}$ **30.** $\frac{63}{19} = 3\frac{6}{19}$ **31.** $\frac{44}{25}$ **33.** $\frac{52}{75}$

34. $\frac{5}{2} = 2\frac{1}{2}$ **35.** $\frac{25}{21} = 1\frac{4}{21}$ **37.** 4 **38.** 3

39. $\frac{7}{45}$ Wrong **41.** $\frac{429}{185} = 2\frac{59}{185}$ Wrong **42.** $\frac{27}{40}$ **43.** $\frac{69}{35} = 1\frac{34}{35}$

45. $\frac{47}{40} = 1\frac{7}{40}$ **46.** $\frac{22}{45}$ **47.** $\frac{50}{27} = 1\frac{23}{27}$ **49.** 210 passengers

50. 180 dollars **51.** $\frac{543}{40} = 13\frac{23}{40}$ **53.** $\frac{42}{5} = 8\frac{2}{5}$ **54.** $\frac{163}{24} = 6\frac{19}{24}$

Review Questions: Chapter 4, Page 147

1. $6\frac{5}{8}$ **2.** 7 **3.** $3\frac{21}{50}$ **4.** $\frac{51}{10}$ **5.** $\frac{35}{12}$

6. $\frac{41}{3}$ **7.** $\frac{123}{4} = 30\frac{3}{4}$ **8.** $\frac{11}{15}$ **9.** 12 **10.** $\frac{2}{9}$

11. $\frac{49}{8} = 6\frac{1}{8}$ **12.** $\frac{56}{3} = 18\frac{2}{3}$ **13.** 104 **14.** $\frac{7}{10}$ **15.** $\frac{79}{12} = 6\frac{7}{12}$

16. $\frac{3}{2} = 1\frac{1}{2}$ **17.** $10\frac{2}{3}$ **18.** $120\frac{17}{18}$ **19.** $\frac{90}{7} = 12\frac{6}{7}$ **20.** $\frac{49}{15} = 3\frac{4}{15}$

21. $\frac{11}{70}$ **22.** $\frac{33}{4} = 8\frac{1}{4}$ **23.** $\frac{91}{120}$ **24.** $\frac{7}{4} = 1\frac{3}{4}$ **25.** 64

26. $\frac{5}{2} = 2\frac{1}{2}$ **27.** $\frac{39}{124}$ **28.** $\frac{11}{4} = 2\frac{3}{4}$ lb **29.** 625 dollars **30.** $\frac{61}{8} = 7\frac{5}{8}$

31. $7\frac{5}{8}$ yd **32.** 6000 dollars **33.** 6 ft

Test: Chapter 4, Page 149

1. $4\frac{6}{7}$ [4.1] **2.** $\frac{59}{8}$ [4.1] **3.** $43\frac{1}{14}$ [4.2]

4. $10\frac{7}{10}$ [4.3] **5.** $\frac{21}{2}$ or $10\frac{1}{2}$ [4.4] **6.** $\frac{5}{7}$ [4.5]

7. $7\frac{5}{42}$ [4.2] **8.** $1\frac{2}{11}$ [4.3] **9.** $\frac{101}{12} = 8\frac{5}{12}$ [4.2]

10. $\frac{37}{24} = 1\frac{13}{24}$ [4.3] **11.** $\frac{113}{8} = 14\frac{1}{8}$ [4.2] **12.** $\frac{17}{36}$ [4.3]

13. 9 [4.4] **14.** $\frac{16}{27}$ [4.5] **15.** $\frac{255}{16} = 15\frac{15}{16}$ [4.4]

16. $\frac{131}{10} = 13\frac{1}{10}$ [4.5] **17.** $\frac{25}{24} = 1\frac{1}{24}$ [4.5] **18.** $\frac{20}{7} = 2\frac{6}{7}$ [4.4]

19. $\frac{63}{20} = 3\frac{3}{20}$ [4.5] **20.** $\frac{63}{8} = 7\frac{7}{8}$ [4.5]

CUMULATIVE TEST: CHAPTERS 1–4, Page 151

1. two hundred three thousand, twenty-one [1]
2. associative; multiplication [1]
3. 1005 [1] **4.** 2973 [1] **5.** 2208 [1] **6.** 140,000 [1]

7. 221 dollars [1] **8.** 5372 [1] **9.** 714,000 [1]
10. distributive [1] **11.** 13 [1] **12.** 506 R12 [1]
13. 21 dollars [1] **14.** 32 [2] **15.** 267 [2]
16. 2, 3, 5, 9, and 10 [2] **17.** no [2] **18.** $2^3 \cdot 3^2 \cdot 5$ [2]
19. 210 [2] **20.** 900 [2] **21.** $\frac{66}{78}$ [3] **22.** $\frac{15}{32}$ [3]
23. $\frac{1}{4}$ [3] **24.** $\frac{17}{35}, \frac{3}{5}, \frac{5}{7}$ [3] **25.** $\frac{1}{16}$ [3]
26. $\frac{1}{9}$ [3] **27.** $\frac{11}{24}$ [3] **28.** $\frac{22}{15} = 1\frac{7}{15}$ [3]
29. $48\frac{1}{36}$ [4] **30.** $\frac{38}{35} = 1\frac{3}{35}$ [4] **31.** 9 [4]
32. $\frac{3}{2} = 1\frac{1}{2}$ [4] **33.** $\frac{45}{4} = 11\frac{1}{4}$ [4] **34.** 90 [4]
35. $\frac{17}{6} = 2\frac{5}{6}$ [4] **36.** $\frac{33}{32} = 1\frac{1}{32}$ [4] **37.** $15\frac{3}{4}$ gal [4]
38. 3 [4] **39.** $\frac{12}{5} = 2\frac{2}{5}$ [4] **40.** $\frac{29}{8} = 3\frac{5}{8}$ lb per week [4]

CHAPTER 5

Exercises 5.1, Page 156

1. 37.498 **2.** 18.76 **3.** 4.11 **5.** 87.003
6. 95.2 **7.** 62.7 **9.** 100.38 **10.** 250.623
11. $82\frac{56}{100}$ **13.** $10\frac{576}{1000}$ **14.** $100\frac{6}{10}$ **15.** $65\frac{3}{1000}$
17. 0.014 **18.** 0.17 **19.** 6.28 **21.** 72.392
22. 850.0036 **23.** 700.77 **25.** 600,500.402
26. five tenths **27.** ninety-three hundredths
29. thirty-two and fifty-eight hundredths
30. seventy-one and six hundredths
31. thirty-five and seventy-eight thousandths
33. eighteen and one hundred two thousandths
34. fifty and eight thousands
35. six hundred seven and six hundred seven thousandths
37. five hundred ninety-three and eighty-six hundredths
38. four thousand, seven hundred and six hundred seventeen thousandths
39. five thousand and five thousandths
41. nine hundred and four thousand, six hundred thirty-eight ten-thousandths
42. $356\frac{45}{100}$; three hundred fifty-six and $\frac{45}{100}$
43. $651\frac{50}{100}$; six hundred fifty-one and $\frac{50}{100}$

Exercises 5.2, Page 162

1. (a) 7 (b) 6 (c) 7; 8 (d) 34.8
2. (a) 3 (b) 2 (c) 3; 2 (d) 6.83
3. (a) 4 (b) 3 (c) 4; 3 (d) 1.0064; ten-thousandth

5. 89.0 6. 7.6 7. 18.0 9. 14.3 10. 0.0
11. 0.39 13. 5.72 14. 8.99 15. 7.00 17. 0.08
18. 6.00 19. 5.71 21. 0.067 22. 0.056 23. 0.634
25. 32.479 26. 9.430 27. 17.364 29. 0.002 30. 20.770
31. 479. 33. 18. 34. 20. 35. 382. 37. 440.
38. 701. 39. 6333. 41. 5160. 42. 6480. 43. 500.
45. 1000. 46. 380. 47. 5480. 49. 92,540. 50. 7010.
51. 7000. 53. 48,000. 54. 103,000. 55. 217,000. 57. 380,000.
58. 4,501,000. 59. 7,305,000. 61. 0.00058 62. 0.54 63. 470.
65. 500. 66. 6000. 67. 3.230 69. 80,000. 70. 78,420

Exercises 5.3, Page 167

1. 2.3 2. 11.5 3. 7.55 5. 72.31
6. 4.6926 7. 276.096 9. 44.6516 10. 481.25
11. 118.333 13. 7.148 14. 7.4914 15. 93.877
17. 103.429 18. 46.943 19. 137.150 21. 1.44
22. 8.93 23. 15.89 25. 64.947 26. 4.895
27. 4.7974 29. 2.9434 30. 34.186 31. \$95.50; \$4.50
33. 0.25 in. 34. \$94.85; \$5.15 35. 12.28 in.

Exercises 5.4, Page 172

1. 0.42 2. 0.24 3. 9.0 or 9 5. 21.6 6. 38.4
7. 0.42 9. 0.004 10. 0.009 11. 0.108 13. 0.0276
14. 0.0225 15. 0.0486 17. 0.0006 18. 0.0025 19. 0.375
21. 1.4 22. 3.975 23. 1.725 25. 5.063 26. 0.0648
27. 0.08 29. 0.0329 30. 0.1156 31. 0.000045 33. 0.00822
34. 0.00954 35. 0.535 37. 0.0009222 38. 0.0013845 39. 9.224655
41. 4.9077 42. 1.4094 43. 54.5034 45. 2.32986 46. 346
47. 2056 49. 693 50. 1610 51. 3820 53. 7.19
54. 18.6 55. 4178.2 57. 470 58. 50 59. 61.5
61. 37.125 ft 62. \$240.90 63. \$4209.50 65. \$617.98 66. 382.4 mi
67. 7.8; 8.0; no 69. \$418.75 70. \$3.90

Exercises 5.5, Page 178

1. 2.3 2. 0.6 3. 9.9 5. 0.1 6. 0.9
7. 2056 9. 20 10. 5 11. 5.2 13. 0.7
14. 1.1 15. 0.1 17. 283.6 18. 21.0 19. 56.4
21. 12.12 22. 5.70 23. 0.96 25. 0.22 26. 0.22
27. 5.04 29. 30.870 30. 4.246 31. 0.285 33. 0.079
34. 0.291 35. 0.007 37. 196.613 38. 0.784 39. 0.5036
41. 0.0186 42. 16.7 43. 13.81 45. 0.01699 46. 1670
47. 1381 49. 1,699,000 50. 70.4 51. 442.8 mi 53. 18.75 mi
54. 19.0 mph 55. \$1.25 57. \$239.65 58. \$23.85 59. \$295

Exercises 5.6, Page 183

1. $\frac{9}{10}$ **2.** $\frac{3}{10}$ **3.** $\frac{5}{10}$ **5.** $\frac{62}{100}$ **6.** $\frac{38}{100}$
7. $\frac{57}{100}$ **9.** $\frac{526}{1000}$ **10.** $\frac{625}{1000}$ **11.** $\frac{16}{1000}$ **13.** $\frac{51}{10}$
14. $\frac{72}{10}$ **15.** $\frac{815}{100}$ **17.** $\frac{1}{8}$ **18.** $\frac{9}{25}$ **19.** $\frac{9}{50}$
21. $\frac{9}{40}$ **22.** $\frac{91}{200}$ **23.** $\frac{17}{100}$ **25.** $\frac{16}{5} = 3\frac{1}{5}$ **26.** $\frac{5}{4} = 1\frac{1}{4}$
27. $\frac{25}{4} = 6\frac{1}{4}$ **29.** $0.\overline{6}$ **30.** 0.3125 **31.** $0.\overline{63}$ **33.** 0.6875
34. 0.5625 **35.** $0.\overline{428571}$ **37.** $0.1\overline{6}$ **38.** $0.2\overline{7}$ **39.** $0.\overline{5}$
41. 0.292 **42.** 0.485 **43.** 0.417 **45.** 0.031 **46.** 0.071
47. 1.231 **49.** 1.429 **50.** 13.333

Review Questions: Chapter 5, Page 186

1. 10 **2.** four tenths **3.** seven and eight hundredths
4. ninety-two and one hundred thirty-seven thousandths
5. eighteen and five thousand, five hundred twenty-six ten-thousandths
6. $81\frac{47}{100}$ **7.** $100\frac{3}{100}$ **8.** $9\frac{592}{1000}$ **9.** $200\frac{5}{10}$ **10.** 2.17
11. 84.075 **12.** 3003.003 **13.** 5900 **14.** 7.6 **15.** 0.039
16. 2.06988 **17.** 26.82 **18.** 64.151 **19.** 93.418 **20.** 17.79
21. 9.02 **22.** 3.9623 **23.** 104.272 **24.** 22.708 **25.** 0.72
26. 0.02 **27.** 0.0064 **28.** 235. **29.** 1.7632 **30.** 5964.1
31. 0.12 **32.** 0.1728 **33.** 171.55 **34.** 0.71 **35.** 880.53
36. 200. **37.** 2.961 **38.** 0.00567 **39.** 1.9435 **40.** 23.5
41. $\frac{7}{100}$ **42.** $\frac{81}{40} = 2\frac{1}{40}$ **43.** $\frac{3}{200}$ **44.** $0.\overline{3}$ **45.** 0.625
46. $2.\overline{4}$ **47.** 0.882 **48.** 0.980

Test: Chapter 5, Page 188

1. $\frac{9}{250}$ [5.6] **2.** 0.0925 [5.6] **3.** three hundred and three hundredths [5.1]
4. 5000.32 [5.1] **5.** 193.18 [5.2] **6.** 200 [5.2] **7.** 193.2 [5.2]
8. 99.385 [5.3] **9.** 15.92 [5.3] **10.** 41.77 [5.3] **11.** 0.02379 [5.3]
12. 29.527 [5.3] **13.** 13.934 [5.3] **14.** 0.014 [5.4] **15.** 7000 [5.5]
16. 5.57708 [5.4] **17.** 3.03 [5.5] **18.** 13,850 [5.4] **19.** 0.0021 [5.5]
20. 20.0 or 20 mpg [5.5]

CHAPTER 6

Exercises 6.1, Page 192

1. $\frac{\text{2 nickels}}{\text{3 nickels}}$ **2.** $\frac{\text{6 chairs}}{\text{5 students}}$ **3.** $\frac{\text{6 ft}}{\text{5 ft}}$ **5.** $\frac{\text{1 bookshelf}}{\text{6 feet}}$ **6.** $\frac{\text{10 cm}}{\text{10 cm}}$ or 1

7. $\frac{100 \text{ cm}}{100 \text{ cm}} = 1$ **9.** $\frac{21 \text{ dollars}}{4 \text{ stocks}}$ **10.** $\frac{19 \text{ mi}}{1 \text{ gal}}$ **11.** $\frac{1 \text{ min}}{24 \text{ min}}$ **13.** $\frac{4 \text{ quarters}}{4 \text{ quarters}}$ or 1

14. $\frac{5 \text{ quarters}}{12 \text{ quarters}}$ **15.** $\frac{1 \text{ oz}}{8 \text{ oz}}$ **17.** $\frac{1 \text{ in.}}{2 \text{ in.}}$ **18.** $\frac{20 \text{ mi}}{1 \text{ gal}}$ **19.** $\frac{1800 \text{ ft}}{1 \text{ sec}}$

21. 5.6¢/oz; 5.3¢/oz; 32 oz box **22.** 7.2¢/oz; 10.0¢/oz; 42 oz box
23. 8.7¢/oz; 11.8¢/oz; 8 oz container **25.** 43.7¢/oz; 64.5¢/oz; 8 oz jar
26. 37.5¢/oz; 46.2¢/oz; 12 oz jar
27. 10.6¢/fl oz; 8.3¢/fl oz; 11.5¢/fl oz; 12 fl oz can
29. 15.6¢/oz; 14.1¢/oz; 25.6 oz box **30.** 12.5¢/oz; 13.0¢/oz; 16 oz pkg.
31. 12.4¢/oz; 7.5¢/oz; 16 oz box
33. 2.1¢/sq ft; 2.2¢/sq ft; 2.4¢/sq ft; 200 sq ft box
34. 7.2¢/oz; 7.7¢/oz; 9.1¢/oz; 32 oz bottle **35.** 13.7¢/oz; 13.3¢/oz; 12 oz pkg.
37. 4.1¢/oz; 5.6¢/oz; 5.7¢/oz; 7.3¢/oz; 9 lb 3 oz box
38. 10.8¢/oz; 10.5¢/oz; 10.0¢/oz; 9.8¢/oz; 40 oz jar
39. 0.9¢/fl oz; 1.4¢/fl oz; 1.9¢/fl oz; 1 gal bottle
41. 10.7¢/oz; 12.2¢/oz; 13.8¢/oz; 14 oz box **42.** 8.2¢/oz; 9.2¢/oz; 5.0¢/oz; 24 oz jar
43. 7.7¢/oz; 5.3¢/oz; 32 oz jar **45.** 8.7¢/oz; 6.9¢/oz; 16 oz jar
46. 6.0¢/oz; 3.1¢/oz; 32 oz bottle **47.** 5.8¢/oz; 6.0¢/oz; 40 oz can
49. 3.0¢/oz; 3.9¢/oz; 30 oz container **50.** 2.6¢/oz; 3.4¢/oz; 31 oz can

Exercises 6.2, Page 197

1. means: 5, 20; extremes: 4, 25 **2.** means: 9, 12; extremes: 4, 27 **3.** means: 8, 3; extremes: $\frac{1}{2}$, 48

5. true **6.** false **7.** true **9.** false **10.** true **11.** true **13.** true
14. true **15.** true **17.** true **18.** true **19.** true **21.** false **22.** false
23. false **25.** true **26.** true **27.** false **29.** true **30.** true **31.** true
33. true **34.** true **35.** true **37.** true **38.** true **39.** true **41.** false
42. false **43.** true **45.** false

Exercises 6.3, Page 201

1. $x = 12$ **2.** $y = 2$ **3.** $z = 20$ **5.** $B = 40$ **6.** $B = 21$
7. $x = 50$ **9.** $A = \frac{21}{2}$ **10.** $x = 5$ **11.** $D = 100$ **13.** $x = 1$
14. $y = 28\frac{2}{9}$ **15.** $x = \frac{3}{5}$ **17.** $w = 24$ **18.** $w = 150$ **19.** $y = 6$
21. $x = \frac{3}{2}$ **22.** $R = 40$ **23.** $R = 60$ **25.** $A = 2$ **26.** $B = 80$
27. $B = 120$ **29.** $R = \frac{100}{3}$ **30.** $R = \frac{200}{3}$ **31.** $x = 22$ **33.** $x = 1$
34. $x = 60$ **35.** $x = 15$ **37.** $B = 7.8$ **38.** $B = 8.2$ **39.** $x = 1.56$
41. $R = 50$ **42.** $R = 20$ **43.** $B = 48$ **45.** $A = 27.3$ or $27\frac{3}{10}$

Exercises 6.4, Page 206

1. 45 mi **2.** $7.80 **3.** 11 gal
5. $180 **6.** $3.27 **7.** 20 yd

9. 437.5 g
10. \$120
11. \$12
13. \$1275
14. \$160,000
15. \$34
17. 42.86 ft or $42\frac{6}{7}$ ft
18. $7\frac{1}{2}$
19. 7.5 mph; 52.5 mph
21. 3360 mi
22. $11\frac{1}{4}$ hr
23. 9 hr
25. 22.3 gal
26. 259,200 revolutions
27. \$648
29. $1\frac{7}{8}$ in.
30. 30.48 cm
31. e.g., $\frac{10}{4}$
33. $1\frac{1}{2}$ cups
34. 18 biscuits
35. 4700 g
37. 2.7 kg
38. 0.004 mg
39. 500 mg
41. 4 drams
42. 3 drams
43. 1920 minims
45. $8\frac{1}{3}$ grains
46. 60 g
47. 75 minims
49. 0.5 ounce
50. 750 mL

Review Questions: Chapter 6, Page 210

1. $\frac{4 \text{ nickels}}{5 \text{ nickels}}$
2. $\frac{5 \text{ in.}}{9 \text{ in.}}$
3. $\frac{2.54 \text{ cm}}{1 \text{ in.}}$
4. $\frac{17 \text{ mi}}{1 \text{ gal}}$
5. $\frac{9 \text{ hr}}{24 \text{ hr}}$
6. $\frac{2 \text{ girls}}{3 \text{ boys}}$
7. 13.2¢/oz; 12.4¢/oz; "deli" cheese
8. 1.7¢/fl oz; 1.6¢/fl oz; 1 gal of milk
9. 4.7¢/oz; 4.8¢/oz; 1 lb 2 oz loaf
10. 21.5¢/oz; 18.7¢/oz; 8 oz bologna
11. means: 5, 16; extremes: 4, 20
12. means: 3, $\frac{1}{9}$; extremes: $\frac{1}{6}$, 2
13. means: 7, 0.75; extremes: 3, 1.75
14. true
15. true
16. false
17. $x = 5$
18. $y = 300$
19. $w = 5\frac{1}{6}$
20. $a = \frac{7}{15}$
21. 149.8 mi
22. 4.6855
23. 166.85 mi
24. \$700
25. 40,000 hairpins
26. 45 mph
27. $26\frac{2}{3}$ ft
28. $266\frac{2}{3}$ mi
29. 64 km per hr
30. 459 girls

Test: Chapter 6, Page 211

1. 7, 4 [6.1]
2. 7, 72 [6.2]
3. true [6.2]
4. $\frac{2 \text{ days}}{5 \text{ days}}$ [6.1]
5. $\frac{1 \text{ sec}}{150 \text{ sec}}$ [6.1]
6. $\frac{2 \text{ nickels}}{5 \text{ nickels}}$ [6.1]
7. 6.8¢/oz; 6.5¢/oz; 1 quart is the better buy [6.1]
8. $x = 27$ [6.3]
9. $y = \frac{50}{3} = 16\frac{2}{3}$ [6.3]
10. $z = 75$ [6.3]
11. $x = \frac{1}{96}$ [6.3]
12. $x = 1.122$ [6.3]
13. $y = 19.5$ [6.3]
14. $x = \frac{18}{7} = 2\frac{4}{7}$ [6.3]
15. $x = \frac{1}{5}$ [6.3]
16. 1.2 mi [6.4]
17. 24.236 mi [6.4]
18. \$108 [6.4]
19. 1440 revolutions [6.4]
20. 371 Democrats [6.4]

CHAPTER 7

Exercises 7.1, Page 215

1. 60% **2.** 20% **3.** 65% **5.** 3.5% **6.** 0.5%
7. 110% **9.** 99% **10.** 40% **11.** 30% **13.** 40%
14. 50% **15.** 7% **17.** 90% **18.** 15% **19.** 25%
21. 45% **22.** 65% **23.** 75% **25.** 53% **26.** 68%
27. 77% **29.** 125% **30.** 110% **31.** 150% **33.** 200%
34. 250% **35.** 236% **37.** 16.3% **38.** 27.2% **39.** 13.4%
41. 20.25% **42.** 93.5% **43.** 0.5% **45.** 0.25% **46.** $3\frac{1}{2}\%$
47. $10\frac{1}{4}\%$ **49.** $24\frac{1}{2}\%$ **50.** $17\frac{3}{4}\%$
51. **(a)** 18% **(b)** 17%; Investment (a) is better.
53. **(a)** 10% **(b)** 10%; Neither investment is better.
54. **(a)** 10% **(b)** 10%; Neither investment is better.
55. **(a)** 15% **(b)** 12%; Investment (a) is better.

Exercises 7.2, Page 219

1. 2% **2.** 9% **3.** 10% **5.** 70% **6.** 90%
7. 36% **9.** 83% **10.** 75% **11.** 25% **13.** 40%
14. 65% **15.** 2.5% **17.** 4.6% **18.** 5.5% **19.** 0.3%
21. 110% **22.** 130% **23.** 125% **25.** 200% **26.** 108%
27. 105% **29.** 230% **30.** 215% **31.** 0.02 **33.** 0.1
34. 0.18 **35.** 0.15 **37.** 0.25 **38.** 0.3 **39.** 0.35
41. 0.101 **42.** 0.115 **43.** 0.132 **45.** 0.0525 **46.** 0.065
47. 0.1375 **49.** 0.2025 **50.** 0.185 **51.** 0.0025 **53.** 0.0017
54. 0.005 **55.** 1.25 **57.** 1.3 **58.** 1.2 **59.** 2.22
61. 0.06 **62.** 0.15 **63.** 0.085 **65.** 0.45 **66.** 20%
67. 12% **69.** 2% **70.** 0.5% or $\frac{1}{2}\%$

Exercises 7.3, Page 226

1. 3% **2.** 16% **3.** 7% **5.** 50% **6.** 75%
7. 25% **9.** 55% **10.** 70% **11.** 30% **13.** 20%
14. 40% **15.** 80% **17.** 26% **18.** 4% **19.** 48%
21. 12.5% **22.** 62.5% **23.** 87.5% **25.** $55\frac{5}{9}\%$ **26.** $28\frac{5}{7}\%$
27. $42\frac{6}{7}\%$ **29.** $63\frac{7}{11}\%$ **30.** $45\frac{5}{11}\%$ **31.** $107\frac{2}{14}\%$ **33.** 105%
34. 125% **35.** 175% **37.** 137.5% **38.** 250% **39.** 210%
41. $\frac{1}{10}$ **42.** $\frac{1}{20}$ **43.** $\frac{3}{20}$ **45.** $\frac{1}{4}$ **46.** $\frac{3}{10}$
47. $\frac{1}{2}$ **49.** $\frac{3}{8}$ **50.** $\frac{1}{6}$ **51.** $\frac{1}{3}$ **53.** $\frac{33}{100}$

54. $\frac{1}{200}$ **55.** $\frac{1}{400}$ **57.** 1 **58.** $1\frac{1}{4}$ **59.** $1\frac{1}{5}$
61. $\frac{3}{1000}$ **62.** $\frac{1}{40}$ **63.** $\frac{5}{8}$ **65.** $\frac{3}{400}$

Exercises 7.4, Page 232

1. 7 **2.** 3.1 **3.** 9 **5.** 9 **6.** 18 **7.** 36
9. 150 **10.** 85 **11.** 700 **13.** 75 **14.** 84 **15.** 42
17. 150% **18.** 40% **19.** 20% **21.** 50% **22.** 20% **23.** $33\frac{1}{3}\%$
25. 12.5 **26.** 23.56 **27.** 110 **29.** 130% **30.** 150% **31.** 72
33. 520 **34.** 38 **35.** 200% **37.** 16.32 **38.** 26.88 **39.** 58.5
41. 43.92 **42.** 11.4 **43.** 72 **45.** 80 **46.** 16 **47.** 40
49. 10 **50.** 25 **51.** 37.5 **53.** 70 **54.** 163.2 **55.** 76.3

Exercises 7.5, Page 238

1. \$1200 **2.** \$346.50 **3.** \$4.23 **5.** \$5700 **6.** \$195; \$6305
7. \$710 **9.** \$12,000 **10.** 0.03 **11.** \$4.30
13. 1st salesman—\$405; 2nd salesman—\$525
14. 20 problems **15.** 153 free throws **17.** \$750; \$600; \$636 **18.** \$1.82; \$32.02
19. \$400; \$80; $33\frac{1}{3}\%$ on cost; 25% on selling price
21. rent—32%; food—35%; taxes—12% **22.** 88%; 250 pages; 30 pages **23.** \$10.53; 60%
25. \$1875; \$2025 **26.** 10%; $11\frac{1}{9}\%$; 20 lb are 10% of 200 lb but $11\frac{1}{9}\%$ of 180 lb
27. \$656.40 **29.** \$22,018 **30.** the \$26 book; \$1.26

Review Questions: Chapter 7, Page 242

1. hundredths **2.** 85% **3.** 18% **4.** 37% **5.** $16\frac{1}{2}\%$
6. 15.2% **7.** 115% **8.** 6% **9.** 30% **10.** 67%
11. 2.7% **12.** 300% **13.** 120% **14.** 0.35 **15.** 0.04
16. 0.0025 **17.** 0.0025 **18.** 0.071 **19.** 1.32 **20.** 60%
21. 15% **22.** 16% **23.** 37.5% **24.** $41\frac{2}{3}\%$ **25.** $126\frac{2}{3}\%$
26. $\frac{7}{50}$ **27.** $\frac{2}{5}$ **28.** $\frac{33}{50}$ **29.** $\frac{1}{8}$ **30.** 4
31. $\frac{67}{200}$ **32.** 15.6 **33.** 2.55 **34.** $233\frac{1}{3}$ **35.** $42\frac{6}{7}$
36. 20% **37.** $33\frac{1}{3}\%$ **38.** 25% **39.** 1.095 **40.** 50
41. $254\frac{6}{11}$ **42.** 0.975 **43.** 200% **44.** \$11.93
45. 50% on cost; $33\frac{1}{3}\%$ on selling price
46. 8 problems **47.** \$1950 **48.** 0.012

49. $5\frac{5}{9}$%—movie; $38\frac{8}{9}$%—clothes; $22\frac{2}{9}$%—anniv.
50. $10,000; $9150

Test: Chapter 7, Page 244

1. 4.5% [7.1]
2. 3.6% [7.2]
3. 500% [7.2]
4. 45% [7.3]
5. 137.5% = $137\frac{1}{2}$% [7.3]
6. 0.08 [7.2]
7. 1.23 [7.2]
8. 0.1225 [7.2]
9. $\frac{7}{2} = 3\frac{1}{2}$ [7.3]
10. $\frac{39}{400}$ [7.3]
11. $\frac{31}{500}$ [7.3]
12. 13.2 [7.4]
13. 5 [7.4]
14. 72 [7.4]
15. 540 [7.4]
16. 97 [7.4]
17. 35% [7.4]
18. $755.25 [7.5]
19. 50% on cost; $33\frac{1}{3}$% on selling price [7.5]
20. $3570 [7.5]

CHAPTER 8

Exercises 8.1, Page 248

1. $50
2. $240
3. $75
5. $43.33
6. $112.50
7. $2500
9. $\frac{1}{2}$ yr or 6 month
10. $1\frac{2}{3}$ yr or 20 month
11. $48
13. $13.50
14. $275
15. $11.25
17. $48
18. $40.25; $2340.25
19. $730
21. $463.50; $36.50
22. $20; $26.67
23. $37,500
25. 10%
26. (a) $16 (b) $100 (c) 30 days (d) $8\frac{1}{2}$% = 8.5%
27. (a) $7.50 (b) 60 days (c) 18% (d) $100
29. 19.2%; $562.50
30. 180 days; $77.07
31. $712,500
33. $\frac{1}{2}$ yr or 6 month
34. $337,500
35. (a) $17.50 (b) 60 days or 2 month (c) $11\frac{1}{2}$% (d) $1133.33

Exercises 8.2, Page 257

1. $13,791.70; $14,631.61
2. $9459.52
3. $306.83; $5306.83; $6.83 more
5. $370.80 yearly; $376.54 semiannually; $379.52 quarterly
6. $19,201.32
7. $800; $1441.96
9. $19,965; no ($17,395.43 @ 5% semiannually); no ($2569.57 more for 10% annually); no ($20,173.40 @ 5% semiannually for 6 years is $208.40 more than 10% annually for 3 years)
10. $74.18 more compounded monthly
11. $1251.39; $900 interest

13. Double in 12 years;

YEAR	PRINCIPAL	INTEREST	TOTAL
1	$10,000.00	$ 600.00	$10,600.00
2	10,600.00	636.00	11,236.00
3	11,236.00	674.16	11,910.16
4	11,910.16	714.61	12,624.77
5	12,624.77	754.49	13,382.26
6	13,382.26	802.93	14,185.19
7	14,185.19	851.11	15,036.30
8	15,036.30	902.20	15,938.50
9	15,938.50	956.29	16,894.79
10	16,894.79	1,003.68	17,908.47
11	17,908.47	1,074.51	18,982.98
12	18,982.98	1,138.99	20,121.96

14. Almost double in 6 years;

YEAR	PRINCIPAL	INTEREST	TOTAL
1	$10,000.00	$1200.00	$11,200.00
2	11,200.00	1344.00	12,544.00
3	12,544.00	1505.28	14,049.28
4	14,049.28	1685.92	15,735.20
5	15,735.20	1888.23	17,623.42
6	17,623.42	2114.81	19,738.23

15. **(a)** $466.56 **(b)** $832.00 **(c)** $970.45 **(d)** Each principal was larger in part **(c)**. **(e)** $2934.39 **(f)** $634.76

Exercises 8.3, Page 260

For each of the following problems, you are given a copy of a checkbook register, the corresponding bank statement, and a reconciliation sheet. You are to find the true balance of the account on both the checkbook register and the reconciliation sheet as shown in Examples 1 and 2. Follow the directions on the reconciliation sheet.

1.

YOUR CHECKBOOK REGISTER

Check No.	Date	Transaction Description	Payment (−)	(✓)	Deposit (+)	Balance
			Balance brought forward			0
	7-15	Deposit ( )		✓	700.00	+700.00 700.00
1	7-15	Quiet Town Apt. (rent/deposit)	520.00	✓		-520.00 180.00
2	7-15	Pa Bell Telephone (phone installation)	32.16	✓		-32.16 147.84
3	7-15	XYZ Power Co. (gas/elect. hookup)	46.49	✓		-46.49 101.35
4	7-16	Foodway Stores (groceries)	51.90	✓		-51.90 49.45
	7-20	Deposit ( )		✓	350.00	+350.00 399.45
5	7-23	Comfy Furniture (sofa, chair)	300.50			-300.50 98.95
	8-1	Deposit ( )			350.00	+350.00 448.95
		Interest ( )		✓	2.50	+2.50 451.45
		Service ( )	2.00	✓		-2.00 449.45
					True Balance	449.45

BANK STATEMENT
Checking Account Activity

Transaction Description	Amount	(✓)	Running Balance	Date
Beginning Balance	700.00	✓	700.00	7-15
Check #1	520.00	✓	180.00	7-16
Check #4	51.90	✓	128.10	7-17
Check #2	32.16	✓	95.94	7-18
Check #3	46.49	✓	49.45	7-18
Deposit	350.00	✓	399.45	7-20
Interest	2.50	✓	401.95	7-31
Service Charge	2.00	✓	399.95	7-31
Ending Balance			399.95	

RECONCILIATION SHEET

A. First, mark ✓ beside each check and deposit listed in both your checkbook register and on the bank statement.

B. Second, in your checkbook register, add any interest paid and subtract any service charge listed on the bank statement.

C. Third, find the total of all outstanding checks.

Outstanding Checks

No.	Amount
5	300.50
Total	300.50

Statement Balance	399.95
Add deposits not credited	+350.00
Total	749.95
Subtract total amount of checks outstanding	-300.50
True Balance	449.45

2.

YOUR CHECKBOOK REGISTER

Check No.	Date	Transaction Description	Payment (−)	(✓)	Deposit (+)	Balance
			Balance brought forward			1610.39
1234	12-7	Pearl City (pearl ring)	524.00	√		-524.00 1086.39
1235	12-7	Comp-U-Tate (home computer)	801.60	√		-801.60 284.79
1236	12-8	Sporty Hutz (skis)	206.25	√		-206.25 78.54
1237	12-8	Guild Card Shop (christmas cards)	25.50	√		-25.50 53.04
	12-10	Deposit ( )		√	1000.00	+1000.00 1053.04
1238	12-14	Toys-R-We (stuffed panda)	80.41	√		-80.41 972.63
1239	12-24	Meat Markette (turkey)	18.39	√		-18.39 954.24
1240	12-24	Poodle Shoppe (pedigreed puppy)	300.00	√		-300.00 654.24
1241	12-31	Homey Sav. & Loan (mortgage payment)	600.00			-600.00 54.24
		Service Charge ( )	-0-			— 54.24
			True Balance			54.24

BANK STATEMENT
Checking Account Activity

Transaction Description	Amount	(✓)	Running Balance	Date
Beginning Balance		√	1610.39	12-01
Check #1234	524.00	√	1086.39	12-08
Check #1236	206.25	√	880.14	12-09
Check #1237	25.50	√	854.64	12-09
Deposit	1000.00	√	1854.64	12-10
Check #1235	801.60	√	1053.04	12-11
Check #1238	80.41	√	972.63	12-15
Check #1239	18.39	√	954.24	12-27
Check #1240	300.00	√	654.24	12-28
Ending Balance			654.24	

RECONCILIATION SHEET

A. First, mark ✓ beside each check and deposit listed in both your checkbook register and on the bank statement.

B. Second, in your checkbook register, add any interest paid and subtract any service charge listed on the bank statement.

C. Third, find the total of all outstanding checks.

Outstanding Checks

No.	Amount
1241	600.00
Total	

Statement Balance	654.24
Add deposits not credited	+ —
Total	654.24
Subtract total amount of checks outstanding	−600.00
True Balance	54.24

3.

YOUR CHECKBOOK REGISTER

Check No.	Date	Transaction Description	Payment (−)	(✓)	Deposit (+)	Balance
			Balance brought forward			756.14
271	6-15	Parts, Parts, Parts (spark plugs)	12.72	✓		−12.72 743.42
272	6-24	Firerock Tire Co. (2 tires)	121.40	✓		−121.40 622.02
273	6-30	Gus' Gas Station (tune-up)	75.68			−75.68 546.34
	7-1	Deposit ( )			250.00	+250.00 796.34
274	7-1	Prudent Ins. Co. (car insurance)	300.00			−300.00 496.34
		Service Charge ( )	1.00	✓		−1.00 495.34
		( )				
		( )				
		( )				
		( )				
					True Balance	495.34

BANK STATEMENT
Checking Account Activity

Transaction Description	Amount	(✓)	Running Balance	Date
Beginning Balance		✓	756.14	6-01
Check #271	12.72	✓	743.42	6-16
Check #272	121.40	✓	622.02	6-26
Service Charge	1.00	✓	621.02	6-30
Ending Balance			621.02	

RECONCILIATION SHEET

A. First, mark ✓ beside each check and deposit listed in both your checkbook register and on the bank statement.

B. Second, in your checkbook register, add any interest paid and subtract any service charge listed on the bank statement.

C. Third, find the total of all outstanding checks.

Outstanding Checks

No.	Amount
273	75.68
274	300.00
Total	375.68

Statement Balance	621.02
Add deposits not credited	+250.00
Total	871.02
Subtract total amount of checks outstanding	−375.68
True Balance	495.34

5.

YOUR CHECKBOOK REGISTER

Check No.	Date	Transaction Description	Payment (−)	(✓)	Deposit (+)	Balance
			Balance brought forward			967.22
772	4-13	C.P. Hay (accountant)	85.00	✓		− 85.00 882.22
	4-14	Deposit ( )		✓	1200.00	+1200.00 2082.22
773	4-14	E.Z. Pharmacy (aspirin)	4.71	✓		− 4.71 2077.51
774	4-15	I.R.S. (income tax)	2000.00			−2000.00 77.51
775	4-30	Heavy Finance Co. (loan payment)	52.50			− 52.50 25.01
	5-1	Deposit ( )			600.00	+600.00 625.01
		Interest ( )		✓	2.82	+ 2.82 627.83
		Service Charge ( )	4.00	✓		4.00 623.83
		( )				
		( )				
					True Balance	623.83

BANK STATEMENT
Checking Account Activity

Transaction Description	Amount	(✓)	Running Balance	Date
Beginning Balance		✓	967.22	4-01
Deposit	1200.00	✓	2167.22	4-14
Check #772	85.00	✓	2082.22	4-15
Check #773	4.71	✓	2077.51	4-15
Interest	2.82	✓	2080.33	4-30
Service Charge	4.00	✓	2076.33	4-30
Ending Balance			2076.33	

RECONCILIATION SHEET

A. First, mark ✓ beside each check and deposit listed in both your checkbook register and on the bank statement.

B. Second, in your checkbook register, add any interest paid and subtract any service charge listed on the bank statement.

C. Third, find the total of all outstanding checks.

Outstanding Checks

No.	Amount
774	2000.00
775	52.50
Total	2052.50

Statement Balance	2076.33
Add deposits not credited	+ 600.00
Total	2676.33
Subtract total amount of checks outstanding	−2052.50
True Balance	623.83

6.

YOUR CHECKBOOK REGISTER

Check No.	Date	Transaction Description	Payment (−)	(✓)	Deposit (+)	Balance
			Balance brought forward			1403.49
86	9-1	Now Stationers (school supplies)	17.12	√		−17.12 1386.37
87	9-2	Young-At-Heart (clothes)	192.50	√		−192.50 1193.87
88	9-4	H.S.U. Bookstore (books)	56.28	√		−56.28 1137.59
89	9-7	Regents Office (tuition)	380.00			−380.00 757.59
90	9-7	Off-Campus Apts. (rent)	240.00	√		−240.00 517.59
91	9-27	State Telephone Co. (phone bill)	24.62			−24.62 492.97
92	9-30	Up-N-Up Foods (groceries)	47.80			−47.80 445.17
		Service Charge ( )	4.00	√		−4.00 441.17
		( )				
		( )				
					True Balance	441.17

BANK STATEMENT
Checking Account Activity

Transaction Description	Amount	(✓)	Running Balance	Date
Beginning Balance		✓	1403.49	9-01
Check #86	17.12	√	1386.37	9-02
Check #87	192.50	√	1193.87	9-05
Check #88	56.28	√	1137.59	9-05
Check #90	240.00	√	897.59	9-10
Service Charge	4.00	√	893.59	9-30
Ending Balance			893.59	

RECONCILIATION SHEET

A. First, mark ✓ beside each check and deposit listed in both your checkbook register and on the bank statement.

B. Second, in your checkbook register, add any interest paid and subtract any service charge listed on the bank statement.

C. Third, find the total of all outstanding checks.

Outstanding Checks

No.	Amount
89	380.00
91	24.62
92	47.80
Total	452.42

Statement Balance	893.59
Add deposits not credited	+ —
Total	893.59
Subtract total amount of checks outstanding	− 452.42
True Balance	441.17

7.

YOUR CHECKBOOK REGISTER

Check No.	Date	Transaction Description	Payment (−)	(✓)	Deposit (+)	Balance
			Balance brought forward			602.82
14	6-20	Aisle Bridal (flowers)	102.40	✓		−102.40 / 500.42
	6-22	Deposit ( )		✓	1000.00	+1000.00 / 1500.42
15	6-24	Tuxedo Junction (tux)	55.65	✓		−55.65 / 1444.77
16	6-28	D. Lohengrin (organist)	25.00			−25.00 / 1419.77
17	6-28	D-Lux Limo (limo rental)	75.00			−75.00 / 1344.77
18	6-30	K.K. Katering (food caterer)	700.00			−700.00 / 644.77
19	7-1	Luv-Lee Stationers (thank-you cards)	15.20			−15.20 / 629.57
		Service Charge ( )	1.00	✓		−1.00 / 628.57
		( )				
		( )				
					True Balance	628.57

BANK STATEMENT
Checking Account Activity

Transaction Description	Amount	(✓)	Running Balance	Date
Beginning Balance		✓	602.82	6-01
Deposit	1000.00	✓	1602.82	6-22
Check #14	102.40	✓	1500.42	6-22
Check #15	55.65	✓	1444.77	6-26
Service Charge	1.00	✓	1443.77	6-30
Ending Balance			1443.77	

RECONCILIATION SHEET

A. First, mark ✓ beside each check and deposit listed in both your checkbook register and on the bank statement.

B. Second, in your checkbook register, add any interest paid and subtract any service charge listed on the bank statement.

C. Third, find the total of all outstanding checks.

Outstanding Checks

No.	Amount
16	25.00
17	75.00
18	700.00
19	15.20
Total	815.20

Statement Balance	1443.77
Add deposits not credited	+ —
Total	1443.77
Subtract total amount of checks outstanding	− 815.20
True Balance	628.57

9.

YOUR CHECKBOOK REGISTER

Check No.	Date	Transaction Description	Payment (−)	(✓)	Deposit (+)	Balance
			Balance brought forward			147.02
203	2-3	Food Stoppe (groceries)	26.90	✓		-26.90 120.12
204	2-8	Ekhon Oil (gasoline bill)	71.45	✓		-71.45 48.67
205	2-14	Rose's Roses (flowers)	25.00	✓		-25.00 23.67
206	2-14	I.M.R.U. (alumni dues)	20.00			-20.00 3.67
	2-15	Deposit ( )		✓	600.00	+600.00 603.67
207	2-26	SRO (theater tickets)	52.50			- 52.50 551.17
208	2-28	MPG Mtg. (house payment)	500.00			-500.00 51.17
		Service Charge ( )	3.00	✓		- 3.00 48.17
		( )				
		( )				
			True Balance			48.17

BANK STATEMENT
Checking Account Activity

Transaction Description	Amount	(✓)	Running Balance	Date
Beginning Balance		✓	147.02	2-01
Check #203	26.90	✓	120.12	2-04
Check #204	71.45	✓	48.67	2-14
Deposit	600.00	✓	648.67	2-15
Check #205	25.00	✓	623.67	2-15
Service Charge	3.00	✓	620.67	2-28
Ending Balance			620.67	

RECONCILIATION SHEET

A. First, mark ✓ beside each check and deposit listed in both your checkbook register and on the bank statement.

B. Second, in your checkbook register, add any interest paid and subtract any service charge listed on the bank statement.

C. Third, find the total of all outstanding checks.

Outstanding Checks

No.	Amount
206	20.00
207	52.50
208	500.00
Total	572.50

Statement Balance	620.67
Add deposits not credited	+ —
Total	620.67
Subtract total amount of checks outstanding	− 572.50
True Balance	48.17

10.

YOUR CHECKBOOK REGISTER

Check No.	Date	Transaction Description	Payment (−)	(✓)	Deposit (+)	Balance
			Balance brought forward			4071.82
996	10-1	Red-E Credit (loan payment)	200.75	✓		−200.75 3871.07
997	10-10	United Ways (charity donation)	25.00			−25.00 3846.07
998	10-21	MacIntosh Farms (barrel of apples)	42.20	✓		−42.20 3803.87
999	10-26	Fun Haus (costume rental)	35.00	✓		−35.00 3768.87
1000	10-28	Yum Yum Shoppe (Halloween candy)	12.14			−12.14 3756.73
1001	10-29	B-Sharp, Inc. (piano tuners)	20.00			−20.00 3736.73
1002	10-30	Food-2-Gooo! (party platter)	78.50			−78.50 3658.23
1003	10-31	Cash ( )	300.00	✓		−300.00 3358.23
1004	10-31	Principal S&L (house payment)	1250.60			−1250.60 2107.63
		Interest ( )		✓	16.29	16.29 2123.92
				True Balance		2123.92

BANK STATEMENT
Checking Account Activity

Transaction Description	Amount	(✓)	Running Balance	Date
Beginning Balance		✓	4071.82	10-01
Check #996	200.75	✓	3871.07	10-05
Check #998	42.20	✓	3828.87	10-23
Check #999	35.00	✓	3793.87	10-30
Check #1003	300.00	✓	3493.87	10-31
Interest	16.29	✓	3510.16	10-31
Ending Balance			3510.16	

RECONCILIATION SHEET

A. First, mark ✓ beside each check and deposit listed in both your checkbook register and on the bank statement.

B. Second, in your checkbook register, add any interest paid and subtract any service charge listed on the bank statement.

C. Third, find the total of all outstanding checks.

Outstanding Checks

No.	Amount
997	25.00
1000	12.14
1001	20.00
1002	78.50
1004	1250.60
Total	1386.24

Statement Balance	3510.16
Add deposits not credited	+ ———
Total	3510.16
Subtract total amount of checks outstanding	−1386.24
True Balance	2123.92

Exercises 8.4, Page 272

1. Soc. Sci.; Chem. & Phys. and Humanities; 3300, 21.2%
2. fac. sal.—$5,625,000, admin. sal.—$2,500,000, nonteach. sal.—$1,625,000, mainten.—$1,250,000, stud. prog.—$625,000, sav.—$500,000, supplies—$375,000; $3,125,000; 22%; $2,750,000
3. 1977; 18 in. 1979 and 1980; 16 in. average
5. news—300 min, commerc.—180 min, soaps—180 min, sitcoms—156 min, drama—144 min, movies—120 min, children's prog.—120 min
6. wheat—$1500 loss; steel—$1000 gain; steel—$2000 gain; wheat—$1500 loss
7. City C: 20 runners; 40%; City C: 12 runners; 24%
9. highest—July; lowest—March; average—16.58
10. The scales represent two different types of quantities; Sue; Bob and Sue; 54.5% studying; 75% studying; yes, longer work hours seem to produce lower GPAs and vice versa.
11. highest—August; $3.5 million; $0.3 million or $300,000; $5.3 million; Dec. 1980—29.41%; Dec. 1981—31.94%

Exercises 8.5, Page 280

1. (a) 79 (b) 84 (c) 85 (d) 38
2. (a) 45 (b) 48 (c) 52 (d) 33
3. (a) 6.2 (b) 6 (c) 6 (d) 6
5. (a) $400 (b) $375 (c) $325 (d) $225
6. (a) $81\frac{5}{6}$ (b) $83\frac{1}{2}$ (c) 82 (d) 12
7. (a) 81 (b) 79 (c) 88 (d) 26
9. (a) 22 (b) 19 (c) 18 (d) 21
10. (a) 32 (b) 30 (c) 30 (d) 40
11. (a) 18.5 in. (b) 14.9 in. (c) none (d) 22.0 in.

Review Questions: Chapter 8, Page 284

1. $97.50
2. $807
3. $7.50
4. $2000
5. Principal—$2000; Rate—$8\frac{1}{2}$%; Time—18 mo; Interest—$12
6. annually—$803.40; semiannually—$815.82; quarterly—$822.25

7.

YEAR	PRINCIPAL	INTEREST	TOTAL
1	$10,000.00	$ 800.00	$10,800.00
2	10,800.00	864.00	11,664.00
3	11,664.00	933.12	12,597.12
4	12,597.12	1,007.77	13,604.89
5	13,604.89	1,088.39	14,693.28
6	14,693.28	1,175.46	15,868.74
7	15,868.74	1,269.50	17,138.24
8	17,138.24	1,371.06	18,509.30
9	18,509.30	1,480.74	19,990.04

8. housing—$10,500; food—$7000; taxes—$3500; transportation—$3500; clothing—$3500; entertainment—$2800; education—$2450; savings—$1750
9. $4000
10. $\frac{3}{5}$
11. $18,000
12. 87
13. 88
14. 88.5
15. 26

Test: Chapter 8, Page 286

1. $2825 [8.1]
2. $6400 [8.1]
3. 9% [8.1]
4. 16 yr [8.1]
5. $42.65 [8.2]
6. 65¢ [8.1]
7. $\frac{13}{25}$ [8.4]
8. $750,000 [8.4]
9. $468,000 [8.4]
10. $$15\frac{1}{2}$ million [8.4]
11. 10% [8.4]
12. 100% [8.4]
13. (a) 2 hr (b) 2 hr (c) 2 hr (d) 4 hr [8.5]
14. (a) 10.14 cm (b) 8.89 cm (c) 7.62 cm (d) 17.78 cm [8.5]

CUMULATIVE TEST II: 1–8, Page 288

1. 200,016 [1]
2. 300.004 [5]
3. 16.00 [5]
4. $0.\overline{42}$ [5]
5. 0.525 [5]
6. 180% [7]
7. 0.015 [7]
8. $\frac{4}{3} = 1\frac{1}{3}$ [3]
9. $\frac{41}{11} = 3\frac{8}{11}$ [4]
10. $\frac{8}{85}$ [3]
11. $\frac{152}{15} = 10\frac{2}{15}$ [4]
12. $46\frac{5}{12}$ [4]
13. 12 [4]
14. $\frac{9}{5} = 1\frac{4}{5}$ [4]
15. 5,600,000 [1]
16. 398.988 [5]

17. 75.744 [5] **18.** 0.01161 [5] **19.** 80.6 [5] **20.** 7 [2]
21. 2, 3, and 4 [2] **22.** $2^2 \cdot 3^2 \cdot 11$ [2] **23.** 210 [2] **24.** zero [1]
25. 7160 [6] **26.** 50 [7] **27.** 18.5 [7] **28.** 250 [7]
29. $x = \frac{3}{8}$ [6] **30.** 196 [6] **31.** 325 [1] **32.** 79 [1]
33. $200 [8] **34.** $4536 [8] **35.** 9% [8] **36.** 18 yr [8]
37. $5306.04 [8] **38.** 45 [8] **39.** 36 [8] **40.** 64 [8]

CHAPTER 9

Exercises 9.1, Page 296

1. milli, centi, deci, deka, hecto, kilo
2. 100 cm **3.** 500 cm **5.** 600 cm **6.** 2000 mm **7.** 300 mm
9. 1400 mm **10.** 16 mm **11.** 18 mm **13.** 350 mm **14.** 40 dm
15. 160 dm **17.** 210 cm **18.** 3000 m **19.** 5000 m **21.** 6400 m
22. 1.1 cm **23.** 2.6 cm **25.** 4.8 cm **26.** 0.06 dm **27.** 0.12 dm
29. 0.03 m **30.** 0.145 m **31.** 0.256 m **33.** 0.32 m **34.** 1.5 m
35. 1.7 m **37.** 2.4 km **38.** 0.5 km **39.** 0.4 km **41.** 462 cm
42. 0.063 m **43.** 0.052 m **45.** 6410 mm **46.** 0.3 cm **47.** 0.5 cm
49. 0.057 m **50.** 20 km **51.** 35 km **53.** 2300 m **54.** 0.0005 km
55. **(a)** 2 **(b)** 3 **(c)** 4 **(d)** 1 **(e)** 6 **(f)** 5 **57.** 104 mm **58.** 53.2 m
59. 31.4 cm **61.** 19.468 cm **62.** 110 m **63.** 13.5 cm **65.** 16 000 m

Exercises 9.2, Page 305

1. 300 mm^2 **2.** 560 mm^2 **3.** 870 mm^2 **5.** 6 cm^2
6. 0.28 cm^2 **7.** 14 cm^2 **9.** 400 cm^2 = 40 000 mm^2
10. 730 cm^2 = 73 000 mm^2 **11.** 5700 cm^2 = 570 000 mm^2
13. 1700 dm^2 = 170 000 cm^2 = 17 000 000 mm^2
14. 290 dm^2 = 29 000 cm^2 = 2 900 000 mm^2
15. 3 dm^2 = 300 cm^2 = 30 000 mm^2 **17.** 1.42 cm^2
18. 58 cm^2 **19.** 2 m^2 **21.** 780 m^2 **22.** 30 000 m^2 **23.** 4 m^2
25. 869 a = 86 900 m^2 **26.** 781 a = 78 100 m^2 **27.** 16 a = 1600 m^2
29. 0.01 ha **30.** 0.15 ha **31.** 50 000 a = 500 ha **33.** 3000 a = 30 ha
34. **(a)** 3 **(b)** 2 **(c)** 4 **(d)** 1 **(e)** 6 **(f)** 5
35. 875 cm^2 **37.** 20 mm^2 **38.** 78.5 m^2 **39.** 7.065 cm^2 **41.** 38.5 mm^2
42. 5 cm^2 or 500 mm^2 **43.** 6 cm^2 **45.** 57.12 mm^2 **46.** 21.195 m^2
47. 32.28 dm^2 **49.** 7536 m^2 **50.** 15.8 km^2 **51.** 3.14 m^2
53. 1.2 m^2 = 12 000 cm^2 **54.** 2500 cm^2 = 0.25 m^2 **55.** 1750 cm^2 = 0.175 m^2

Exercises 9.3, Page 313

1. 1000 mm^3
1000 cm^3
1000 dm^3
1 000 000 000 m^3

2. 10 cm
100 mm
100 cm^2
10 000 mm^2
1000 cm^3
1 000 000 mm^3

3. 10 dm
100 cm
100 dm^2
10 000 cm^2
1000 dm^3
1 000 000 cm^3

5. 73 000 dm^3 **6.** 900 dm^3 **7.** 0.4 cm^3 **9.** 0.063 m^3 **10.** 8 700 000 cm^3 **11.** 0.005 cm^3 **13.** 0.000 019 dm^3 **14.** 5000 mm^3 **15.** 2000 cm^3 **17.** 5300 mL **18.** 30 mL **19.** 0.03 L **21.** 48 000 L **22.** 72 kL **23.** 0.32 hL **25.** 0.29 kL **26.** 0.569 L **27.** 7 200 000 mL **29.** 9500 L **30.** 0.72 hL **31.** 70 dm^3 **33.** 381.51 cm^3 **34.** 904.32 dm^3 **35.** 12.56 dm^3 **37.** 224 cm^3 **38.** 9106 dm^3 **39.** 282.6 cm^3

Exercises 9.4, Page 318

1. 7000 mg **2.** 2000 g **3.** 0.0345 g **5.** 4 t **6.** 5.6 kg **7.** 73 000 000 mg **9.** 540 mg **10.** 700 mg **11.** 5000 kg **13.** 2000 kg **14.** 0.896 g **15.** 896 000 mg **17.** 75 kg **18.** 3 g **19.** 7 000 000 g **21.** 0.00034 kg **22.** 780 mg **23.** 0.016 g **25.** 3940 mg **26.** 0.0923 kg **27.** 5600 kg **29.** 3.547 t **30.** 2.963 t

Review Questions: Chapter 9, Page 318

1. 1500 cm **2.** 0.35 dm **3.** 3700 mm^2 **4.** 0.17 cm^2 **5.** 300 a **6.** 30 000 m^2 **7.** 5000 cm^3 **8.** 36 000 mL **9.** 13 000 cm^3 **10.** 68 000 mm^3 **11.** 5000 g **12.** 3400 mg **13.** 6710 kg **14.** 0.019 g **15.** 8000 g **16.** 4.29 kg **17.** 12 cm; 3 cm^2 **18.** 212 mm; 2280 mm^2 **19.** 25.42 m; 44.13 m^2 **20.** 33 mm; 63 mm^2 **21.** 294 L **22.** 0.02931 L (rounded off) **23.** 1256 cm^2 **24.** 87 920 mm^3 **25.** 0.02 m^2

Test: Chapter 9, Page 321

1. 20 cm by 180 cm [9.1] **2.** 10 kg by 9990 g [9.4] **3.** volume is the same [9.3] **4.** 0.37 m [9.1] **5.** 2300 cm [9.1] **6.** 2000 cm^3 [9.3] **7.** 1.2 kg [9.4] **8.** 5600 kg [9.4] **9.** 7500 m^2 [9.2] **10.** 11 m [9.1] **11.** 4000 mm^3 [9.3] **12.** 9.6 cm^2 [9.2] **13.** 0.0835 g [9.4] **14.** 251.2 cm—perimeter; 5024 cm^2—area [9.1, 9.2] **15.** 32 m—perimeter; 40 m^2—area [9.1, 9.2] **16.** 0.33912 L [9.3] **17.** 1 000 000 mm^3 [9.3] **18.** 105 cm^2 [9.2] **19.** 0.11304 m^3 [9.3] **20.** 113.04 dm^3 [9.3]

CHAPTER 10

Exercises 10.1, Page 324

1. 60 in. **2.** 36 in. **3.** 18 in. **5.** 4 ft **6.** 10 ft **7.** $2\frac{1}{2}$ ft **9.** 9 ft **10.** 12 ft **11.** 7 ft **13.** 15,840 ft **14.** 21,120 ft **15.** 2 yd **17.** 2 mi **18.** 3 mi **19.** 32 fl oz

21. 6 pt
22. 3 pt
23. 20 qt
25. 3 gal
26. 5 gal
27. $3\frac{3}{4}$ gal
29. 80 oz
30. 48 oz
31. 38 oz
33. 5000 lb
34. 6000 lb
35. 300 min
37. 90 min
38. 210 min
39. $\frac{1}{2}$ hr
41. 72 hr
42. 96 hr
43. 1800 sec
45. 300 sec
46. 180 sec
47. 4 min
49. 2 days
50. 3 days

Exercises 10.2, Page 328

1. 4 ft 8 in.
2. 5 ft 6 in.
3. 7 lb 4 oz
5. 6 min 20 sec
6. 15 min 30 sec
7. 3 days 6 hr
9. 9 gal 1 qt
10. 3 gal 2 qt
11. 5 pt 4 fl oz
13. 9 ft 7 in.
14. 12 ft 1 in.
15. 18 lb 2 oz
17. 18 min 10 sec
18. 19 min 45 sec
19. 4 hr 11 min 15 sec
21. 6 days 2 hr 35 min
22. 4 days 2 hr
23. 13 gal 3 qt
25. 7 gal 3 qt 4 fl oz
26. 7 gal 2 qt 7 fl oz
27. 12 yd 2 ft 6 in.
29. 2 yd 2 ft 2 in.
30. 1 yd 7 in.
31. 3 gal 2 qt 14 fl oz
33. 2 hr 45 min
34. 1 hr 40 min
35. 4 min 50 sec
37. 4 lb 8 oz
38. 9 lb 12 oz
39. 3 ft 8 in.

Exercises 10.3, Page 336

1. 77°F
2. 176°F
3. 50°F
5. 10°C
6. $37\frac{7}{9}$°C
7. 0°C
9. 157.48 cm
10. 190.5 cm
11. 2.745 m
13. 96.6 km
14. 161 km
15. 644 cm
17. 124 mi
18. 40.3 mi
19. 21.7 mi
21. 197 in.
22. 39.4 in.
23. 19.35 cm^2
25. 55.8 m^2
26. 27.9 m^2
27. 83.6 m^2
29. 405 ha
30. 101.25 ha
31. 741 acres
33. 53.82 ft^2
34. 11.96 yd^2
35. 4.65 $in.^2$
37. 9.46 L
38. 18.92 L
39. 94.625 L
41. 10.6 qt
42. 26.5 qt
43. 11.088 gal
45. 3277.4 cm^3
46. 353.15 ft^3
47. 4.54 kg
49. 1102.5 lb
50. 3.5 oz

Exercises 10.4, Page 342

1. $\frac{3}{10}$
2. $116\frac{2}{3}$
3. 16
5. $4\frac{1}{2}$
6. 2
7. 270
9. 20
10. 2
11. $1\frac{1}{2}$
13. 5
14. $1\frac{4}{5}$
15. 6
17. 3
18. $6\frac{2}{3}$
19. 150
21. 15 mL Ampicillin
22. 1 mL milk of magnesia
23. 2 mL castor oil
25. 4 mL Robitussin®
26. 5 mL castor oil
27. $\frac{2}{3}$ mL Polaramine®
29. $7\frac{1}{2}$ mL castor oil

Review Questions: Chapter 10, Page 344

1. 33
2. $\frac{39}{2} = 19\frac{1}{2}$
3. $\frac{37}{6} = 6\frac{1}{6}$
4. $\frac{2}{3}$
5. 76
6. 2600
7. $\frac{27}{8} = 3\frac{3}{8}$
8. $0.4 = \frac{2}{5}$
9. 192
10. 58
11. $2.8 = 2\frac{4}{5}$
12. $3.5 = 3\frac{1}{2}$
13. $25.6 = 25\frac{3}{5}$
14. $\frac{19}{2} = 9\frac{1}{2}$
15. $\frac{13}{16}$
16. $\frac{11}{8} = 1\frac{3}{8}$
17. 2 yd 7 in.
18. 4 yd 1 ft 6 in.
19. 4 lb 6 oz
20. 3 t 1120 lb
21. 6 days 4 hr
22. 1 hr 3 min 13 sec
23. 1 qt 1 pt 8 fl oz
24. 5 gal 1 pt
25. 5 lb 11 oz
26. 18 hr 43 min 16 sec
27. 1 yd 2 ft 5 in.
28. 4 gal 2 qt 1 pt
29. 2 lb 7 oz
30. 2 gal 1 qt
31. 49 min
32. 2 yd 2 ft 8 in.
33. $\frac{200}{9} = 22\frac{2}{9}$
34. $37.4 = 37\frac{2}{5}$
35. 12.7
36. 68.58
37. 10.0076
38. 7.88
39. 1.3725
40. 0.68625
41. 22.96
42. 3.488
43. 322
44. 20.5275
45. 54.56
46. 0.248
47. 2.43
48. 20.9
49. 22.6044
50. 1.482
51. 18.16
52. 510.3
53. 15.89
54. 0.1764
55. 80
56. 11
57. $\frac{7}{2} = 3\frac{1}{2}$
58. $\frac{15}{2} = 7\frac{1}{2}$
59. 15 mL
60. $2\frac{1}{2} = 2.5$ mL
61. 13 mL
62. $7\frac{1}{2} = 7.5$ mL

Test: Chapter 10, Page 347

1. 92 [10.1]
2. $\frac{11}{9} = 1\frac{2}{9}$ [10.1]
3. 1680 [10.1]
4. 4.27 [10.3]
5. 19.7 [10.3]
6. 567 [10.3]
7. 17.199 [10.3]
8. 5.203 [10.3]
9. 6.6 [10.3]
10. 6.723 [10.3]
11. 124 [10.3]
12. 0.7 [10.4]
13. 40 [10.4]
14. 20°C [10.3]
15. 2 gal 2 qt 2 fl oz [10.2]
16. 7 days 12 hr 48 min [10.2]
17. 1 yd 1 ft 1 in. [10.2]
18. 2 qt 7 fl oz [10.2]
19. 1 lb 8 oz [10.2]
20. 75 mL [10.4]

CHAPTER 11

Exercises 11.1, Page 353

1. -14
2. -12
3. $+10$ or 10
5. $-1\frac{1}{3}$
6. $-2\frac{1}{4}$
7. $+5.3$ or 5.3
9. -30
10. -40
11. 0 1 2
13. -3 -2 -1 0 1
14. -3 -2 -1 0
15. 1 2 3 4 5 6 7 8 9 10

17. 0 1 2 3 4 5   **18.** 0 1 2 3 4

19. $-2\frac{1}{2}$ -1.3 $-\frac{1}{8}$ 0 0.75   **21.** >   **22.** <

23. <   **25.** >   **26.** >   **27.** <   **29.** =   **30.** >
31. true   **33.** false   **34.** true   **35.** true   **37.** false   **38.** true
39. true   **41.** false   **42.** false   **43.** $x = 6$ or $x = -6$

45. $x = 4.3$ or $x = -4.3$ **46.** $x = \frac{3}{2}$ or $x = -\frac{3}{2}$ **47.** no values for x

49. $x = \frac{2}{3}$ or $x = -\frac{2}{3}$ **50.** $x = \frac{1}{4}$ or $x = -\frac{1}{4}$

Exercises 11.2, Page 358

1. 2   **2.** 1   **3.** 10   **5.** 19   **6.** -10
7. -9 **9.** 0   **10.** 25   **11.** -4 **13.** 2
14. -3 **15.** -16 **17.** 17   **18.** -4 **19.** 0
21. 4   **22.** -6 **23.** 14   **25.** -1 **26.** 4
27. -15 **29.** 41   **30.** 6   **31.** 13   **33.** 10
34. 8   **35.** -12 **37.** -10 **38.** 0   **39.** 0
41. -235 **42.** -165 **43.** 120   **45.** -121 **46.** $-5\frac{1}{2}$
47. $-3\frac{4}{5}$ **49.** $2\frac{1}{10}$ **50.** $-5\frac{1}{4}$ **51.** -18.81 **53.** 1.6
54. -6.1 **55.** -39.28 **57.** 5.9   **58.** -8 **59.** -1.1

Exercises 11.3, Page 362

1. 3   **2.** 13   **3.** 11   **5.** -7 **6.** -13
7. -9 **9.** 4   **10.** 10   **11.** -10 **13.** 1
14. 3   **15.** 18   **17.** 17   **18.** 23   **19.** -31
21. -5 **22.** -4 **23.** 0   **25.** -5 **26.** -2.1
27. -12.51 **29.** 0   **30.** $7\frac{1}{4}$ **31.** 30   **33.** 4
34. 5   **35.** -9 **37.** 23.9   **38.** -11.9 **39.** -28.78
41. 8   **42.** 12   **43.** 6   **45.** 10   **46.** 17
47. -4 **49.** 6   **50.** 6   **51.** -9 **53.** -1
54. -15 **55.** -8 **57.** -18 **58.** -10 **59.** 3
61. 1   **62.** 2   **63.** 0   **65.** 15   **66.** 4
67. 4   **69.** -3 **70.** ~~28.3~~ 22.3   **71.** -28.6 **73.** ~~$14\frac{1}{4}$~~ $21\frac{1}{4}$
74. $\frac{1}{2}$ **75.** 0   **77.** 5   **78.** 0   **79.** -5

Exercises 11.4, Page 367

1. -15 **2.** -24 **3.** 24   **5.** -20 **6.** -24
7. 28   **9.** -50 **10.** -33 **11.** -21 **13.** -48

14. -36 **15.** 63 **17.** 0 **18.** 0 **19.** 90
21. 24 **22.** 30 **23.** 60 **25.** -42 **26.** -64
27. -45 **29.** -60 **30.** -70 **31.** 0 **33.** -1
34. -27 **35.** 16 **37.** -3 **38.** -9 **39.** -2
41. 4 **42.** 10 **43.** 5 **45.** -5 **46.** -5
47. -3 **49.** 2 **50.** 3 **51.** 4 **53.** -3
54. -3 **55.** -8 **57.** 2 **58.** 4 **59.** 1
61. undefined **62.** undefined **63.** 0 **65.** undefined **66.** 0
67. -19.6 **69.** 0.0901 **70.** 1.898 **71.** -5.5 **73.** $\frac{14}{27}$
74. $-\frac{5}{4} = -1\frac{1}{4}$ **75.** $-\frac{5}{4} = -1\frac{1}{4}$

Exercises 11.5, Page 370

1. -23 **2.** -17 **3.** 33 **5.** -8 **6.** -1
7. -12 **9.** -37 **10.** -23 **11.** -48 **13.** -100
14. -48 **15.** -81 **17.** -13.44 **18.** -1.92 **19.** -7
21. 16 **22.** -9 **23.** 104 **25.** -0.27 **26.** -0.23
27. 12.6 **29.** $-\frac{8}{21}$ **30.** $-\frac{4}{5}$ **31.** positive **33.** positive
34. positive **35.** negative **37.** 0 **38.** 0 **39.** undefined

Review Questions: Chapter 11, Page 373

1. $+4$ or 4 **2.** $-\frac{3}{8}$ **3.** 0 **4.** 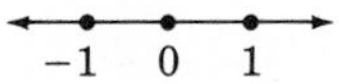

5. 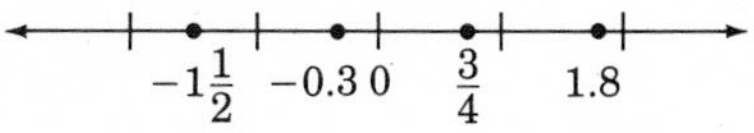

6. 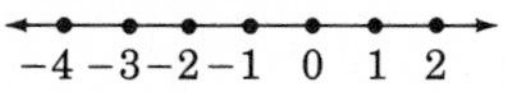

7. $<$ **8.** $>$ **9.** $=$ **10.** true
11. true **12.** false **13.** $x = 0.4$ or $x = -0.4$ **14.** no values
15. $x = \frac{9}{10}$ or $x = -\frac{9}{10}$ **16.** 4 **17.** -5 **18.** -20
19. 0 **20.** $-4\frac{3}{10}$ **21.** -3.3 **22.** -124
23. 5.62 **24.** 5 **25.** 16 **26.** -21
27. 0 **28.** 26 **29.** 6 **30.** -50
31. -74 **32.** -10.29 **33.** $\frac{35}{24} = 1\frac{11}{24}$ **34.** 0
35. -72 **36.** 64 **37.** $\frac{1}{2}$ **38.** 0
39. -0.024 **40.** -125 **41.** -17 **42.** $\frac{1}{3}$
43. 0 **44.** undefined **45.** 16 **46.** -290
47. -6 **48.** 8 **49.** $\frac{7}{4} = 1\frac{3}{4}$ **50.** $-\frac{5}{4} = -1\frac{1}{4}$
51. undefined **52.** 0

Test: Chapter 11, Page 375

1. -0.34 [11.1]

2. 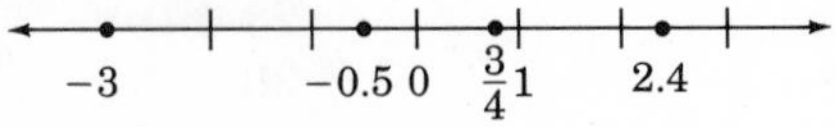

[11.1]

3. $>$ [11.1]
4. $<$ [11.1]
5. $x = 17$ or $x = -17$ [11.1]
6. -4.6 [11.2]
7. -9 [11.2]
8. $-9\frac{1}{4}$ [11.3]
9. 16 [11.3]
10. 0.028 [11.4]
11. -90 [11.4]
12. $\frac{1}{4}$ [11.4]
13. 4.41 [11.4]
14. 0.8 [11.4]
15. undefined [11.4]
16. -5 [11.4]
17. $-\frac{13}{6} = -2\frac{1}{6}$ [11.4]
18. $-8\frac{9}{14}$ [11.5]
19. -32 [11.5]
20. 0, negative [11.5]

CHAPTER 12

Exercises 12.1, Page 380

1. $8x$
2. x
3. $6x$
5. $-7a$
6. $-7y$
7. $-12y$
9. $-9x$
10. $-3x$
11. $-8x$
13. 0
14. $-p$
15. $-c$
17. $-9a$
18. $-2c$
19. 0
21. $5x - 7$
22. $-x + 2$
23. $-x + 5$
25. $-11a - 2$
26. $-3x - 2$
27. $3x + 1$
29. $4x + 7$
30. $4y - 1$
31. $5y^2$
33. $2x^2 + 3x + 1$
34. $5x^2 - 2x + 3$
35. $-y$
37. $3a^2 + 2ab$
38. $2y^2 + 5xy$
39. $5x + 4y$
41. -5
42. 0
43. 0
45. 9
46. -15
47. 22
49. -11
50. -22
51. -3
53. 12
54. 4
55. -8
57. 26
58. -9
59. -12

Exercises 12.2, Page 385

1. $x = 6$
2. $x = 7$
3. $y = 22$
5. $y = -5$
6. $x = -3$
7. $x = -2$
9. $y = 2$
10. $x = 7$
11. $x = 6$
13. $y = -4$
14. $x = -6$
15. $x = -6$
17. $y = 5$
18. $y = 3$
19. $x = 13$
21. $x = 1$
22. $x = 4$
23. $y = 2$
25. $x = -3$
26. $y = 1$
27. $y = 2$
29. $x = -6$
30. $x = -1$
31. $y = -3$
33. $x = 9$
34. $x = -7$
35. $y = -3$
37. $x = -1$
38. $x = 3$
39. $x = -3$
41. $x = -5$
42. $y = 4$
43. $y = 5$
45. $x = 0$
46. $x = 0$
47. $x = 3$
49. $x = 3$
50. $x = 8$
51. $x = -7$
53. $x = 5$
54. $y = 3$
55. $y = 4$
57. $x = 5$
58. $x = 4$
59. $x = 7$
61. $x = -12$
62. $y = 75$
63. $x = -\frac{5}{2}$
65. $x = \frac{220}{3}$

Exercises 12.3, Page 390

[NOTE: There may be correct alternative answers for 1–30.]

1. $x + 5$	**2.** $x + 6$	**3.** $x + 10$	**5.** $x - 8$
6. $x - 4$	**7.** $x - 14$	**9.** $x + 11$	**10.** $6 - x$
11. $2x$	**13.** $\frac{x}{-7}$	**14.** $\frac{-18}{x}$	**15.** $4x - 3$
17. $4(x - 3)$	**18.** $5(x - 7)$	**19.** $-2(x + 5)$	**21.** $13 + 2x$
22. $x + 3x$	**23.** $6(x + 1)$	**25.** $-3 + 3x$	**26.** $8x + 5$
27. $9(x - 3)$	**29.** $7x - 3$	**30.** $16 + 2x$	**31.** 19
33. 54	**34.** -32	**35.** 15	**37.** -4
38. 1	**39.** 4	**41.** 15	**42.** -1
43. 5	**45.** $-15, -14, -13$	**46.** 7, 9, 11	**47.** 25, 27, 29
49. 14, 16, 18, 20	**50.** 5	**51.** 2	**53.** -3
54. 3	**55.** $-3, -2$	**57.** 7 cm	**58.** 14 cm
59. 125 yd			

Exercises 12.4, Page 395

1. $d = 125$ mi	**2.** $A = 400\ \text{ft}^2$	**3.** $A = 28.26\ \text{cm}^2$
5. $y = 4$	**6.** $P = \$3000$	**7.** $w = 10$ cm
9. $\alpha = 90°$	**10.** $h = 5$ mm	**11.** $m = \frac{f}{a}$
13. $h = \frac{L}{2\pi r}$	**14.** $P = \frac{I}{RT}$	**15.** $x = \frac{y - b}{m}$
17. $a = P - b - c$	**18.** $\beta = 180 - \alpha - \gamma$	**19.** $l = \frac{P - 2w}{2}$ or $l = \frac{P}{2} - w$
21. $w = \frac{V}{lh}$	**22.** $h = \frac{V}{\pi r^2}$	**23.** $y = -2x + 4$
25. $y = x - 7$	**26.** $y = 3x - 5$	**27.** $y = \frac{-3x + 6}{2}$ or $y = -\frac{3}{2}x + 3$
29. $x = \frac{2y - 3}{6}$ or $x = \frac{1}{3}y - \frac{1}{2}$		**30.** $y = \frac{5x + 1}{3}$ or $y = \frac{5}{3}x + \frac{1}{3}$

Review Questions: Chapter 12, Page 398

1. $15x$	**2.** $12y$	**3.** $3x$	**4.** $-6x$
5. $14w$	**6.** 0	**7.** $-2y$	**8.** $-34p$
9. $-8a + 6$	**10.** $-2x + 14$	**11.** $14y - 7$	**12.** $-10x - 10$
13. $2x^2$	**14.** $-3y^2$	**15.** $5x^2 - 3x + 1$	**16.** $4y^2 + 2y - 1$
17. $2x^2 - 3$	**18.** $5y^2 + 4$	**19.** $9x + y$	**20.** $7x - 6y - 6$
21. 27	**22.** 4	**23.** 25	**24.** 21
25. -31	**26.** $x = 4$	**27.** $y = 16$	**28.** $x = -4$
29. $y = 3$	**30.** $x = -7$	**31.** $y = -2$	**32.** $x = 4$
33. $y = -5$	**34.** $x = -6$	**35.** $y = 5$	**36.** $x = 3$
37. $x = 4$	**38.** $y = -5$	**39.** $x = 0$	**40.** $x = 9$
41. $y = -3$	**42.** $x = 0$	**43.** $y = 0$	**44.** $x = 1$
45. $y = 2$	**46.** $x = -12$	**47.** $y = 3$	**48.** $x = 12$

49. $y = \frac{15}{2}$ **50.** $x + 8$ **51.** $5x - 3$ **52.** $\frac{x}{9}$

53. $-3(x + 2)$ **54.** $10 + 4x$ **55.** $18 - 2x$ **56.** $10x - 15 = -35; -2$

57. $2x + 14 = x - 13; -27$ **58.** $x + (x + 1) + (x + 2) = 78; 25, 26, 27$

59. $\frac{1}{3}x = (x + 2) + (x + 4) - 41; 21, 23, 25$ **60.** $7x = 5(x + 2); 5$

61. $[x + (-8)] = 4x - 6; -\frac{2}{3}$ **62.** $680 = 2(200) + 2w; 140\ m$

63. $\beta = 100°$ **64.** $r = 25$ mph **65.** $d = \frac{C}{d}$ **66.** $y = \frac{3x - 2}{4}$ or $y = \frac{3}{4}x - \frac{1}{2}$

Test: Chapter 12, Page 400

1. $-3x - 7$ [12.1] **2.** $4x^2 - 4x - 4$ [12.1] **3.** -1 [12.1]

4. 1 [12.1] **5.** $x = 40$ [12.2] **6.** $y = 0$ [12.2]

7. $x = 6$ [12.2] **8.** $y = 0$ [12.2] **9.** $y = -12$ [12.2]

10. $x = -10$ [12.2] **11.** $x = -3$ [12.2] **12.** $y = 20$ [12.2]

13. $\frac{x}{4} - 3$ [12.3] **14.** $1 + 2(x + 3)$ [12.2] **15.** $\frac{3}{4}x - 5$ [12.3]

16. 82, 84, 86 [12.3] **17.** -16 [12.3] **18.** $6\frac{3}{4}$ cm [12.3]

19. $T = 2$ yr [12.4] **20.** $y = \frac{4x + 2}{3}$ or $y = \frac{4}{3}x + \frac{2}{3}$ [12.4]

CHAPTER 13

Exercises 13.1, Page 404

1. perfect square **2.** perfect square

3. perfect square **5.** perfect square

6. perfect square **7.** not a perfect square

9. not a perfect square **10.** not a perfect square

11. $1.732 \cdot 1.732 = 2.999824$ and $1.733 \cdot 1.733 = 3.003289$

13. -10 **14.** $\sqrt{3}$ **15.** $-10, \frac{3}{4}, 7.2, \sqrt{3}$ **17.** $2\sqrt{7}$ **18.** $2\sqrt{6}$

19. $4\sqrt{2}$ **21.** $12\sqrt{2}$ **22.** $11\sqrt{3}$ **23.** $11\sqrt{2}$ **25.** $10\sqrt{3}$

26. $8\sqrt{2}$ **27.** $5\sqrt{5}$ **29.** $7\sqrt{2}$ **30.** $11\sqrt{5}$ **31.** $5\sqrt{6}$

33. 14 **34.** $20\sqrt{2}$ **35.** $4\sqrt{5}$ **37.** $2\sqrt{10}$ **38.** 16

39. 19 **41.** $6\sqrt{2}$ **42.** $13\sqrt{2}$ **43.** $\sqrt{3}$ **45.** $-2\sqrt{10}$

46. $-3\sqrt{11}$ **47.** $3\sqrt{6}$ **49.** $6\sqrt{2} + 6\sqrt{3}$ **50.** $4\sqrt{10}$ **51.** $6\sqrt{b} - 3\sqrt{a}$

53. $4\sqrt{2x}$ **54.** $7\sqrt{3x}$ **55.** $4\sqrt{5a}$

Exercises 13.2, Page 410

1. true **2.** true **3.** false **5.** false

6. true **7.** false **8.** false **9.** false

10. true **11.** $-1 < x < 2$, open interval

13. $x < 0$, open interval **14.** $0 \le x \le 3$, closed interval

15. $-1 \le x$, half-open interval **17.** -2 0 open interval

18. closed interval

19.  closed interval

21.  half-open interval

22.  closed interval

23.  open interval

25.  half-open interval

26. 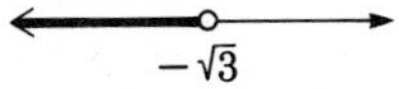 open interval

27. 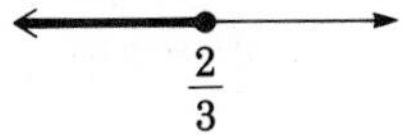 half-open interval

29. $x > 4$

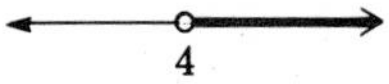

30. $y \geq 2$

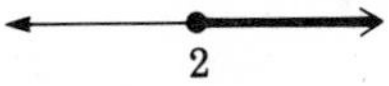

31. $y \leq -3$

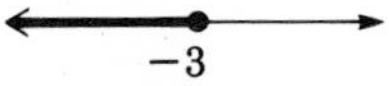

33. $y > -\frac{6}{5}$

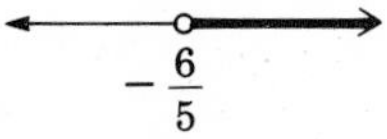

34. $y > \frac{2}{3}$

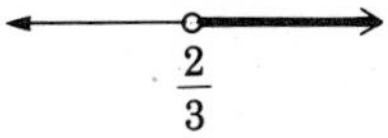

35. $x < \frac{11}{7}$

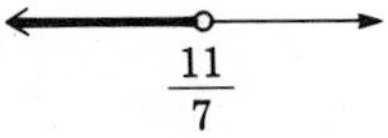

37. $-5 > x$

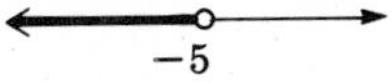

38. $x \geq 7$

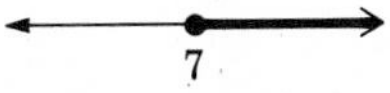

39. $x \leq 5$

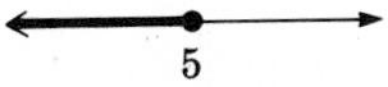

41. $x > -6$

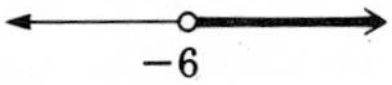

42. $5 < x$

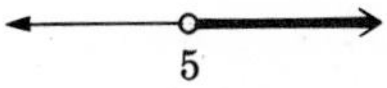

43. $x \leq -1$

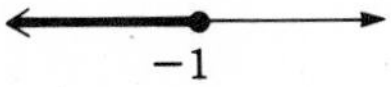

45. $x > -1$

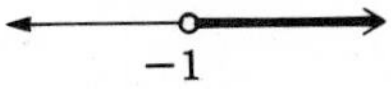

46. $0 \geq x$

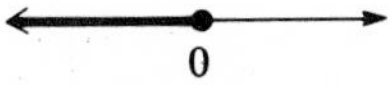

47. $x < -5$

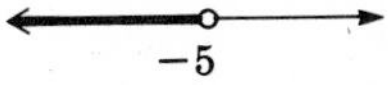

49. $1 \leq x \leq 4$

50. $-1 \leq y \leq -\frac{1}{2}$

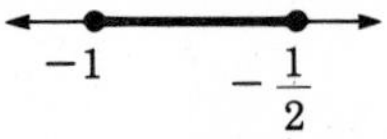

51. $\frac{1}{3} \leq x \leq 2$

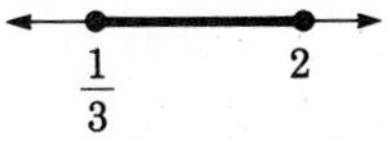

53. $-3 \leq y \leq 1$

54. $-5 \geq x > -6$

55. $3 < x \leq 5$

Exercises 13.3, Page 415

1. $\{(-5, 1), (-3, 2), (-1, 1), (1, 2), (2, 1)\}$
2. $\{(-3, 1), (-1, 4), (0, 5), (1, 4), (3, -3)\}$
3. $\{(-3, 2), (-2, -1), (-2, 1), (0, 0), (2, 1), (3, 0)\}$
5. $\{(-3, -4), (-3, 3), (-1, -1), (-1, 1), (1, 0)\}$
6. $\{(-1, -4), (-1, -3), (-1, 0), (-1, 2), (-1, 5)\}$
7. $\{(-1, -5), (0, -4), (1, -3), (2, -2), (3, -1), (4, 0)\}$
9. $\{(-1, 4), (0, 3), (1, 2), (2, 1), (3, 0), (4, -1), (5, -2)\}$
10. $\{(-3, 0), (-2, -1), (-1, -2), (0, -3), (0, 0), (1, -2), (2, -1), (3, 0)\}$

11.

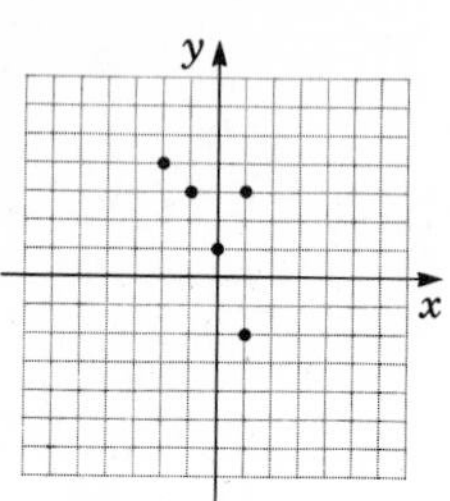

13.

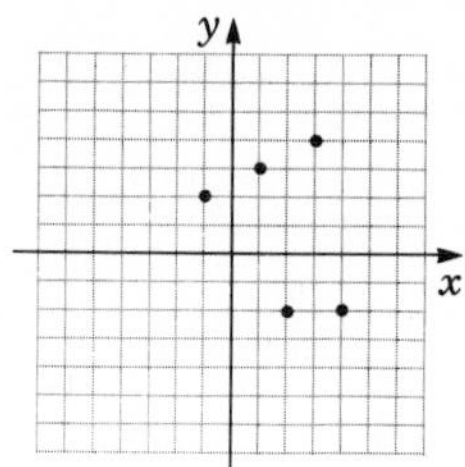

14.

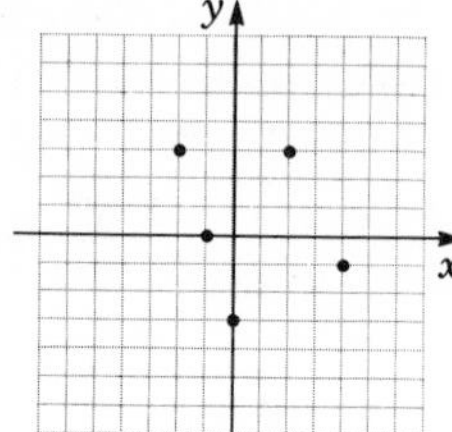

15.

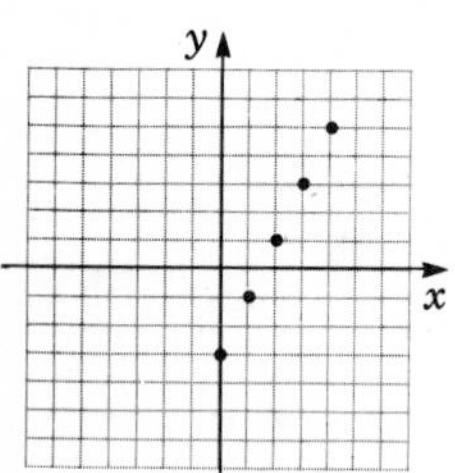

17.

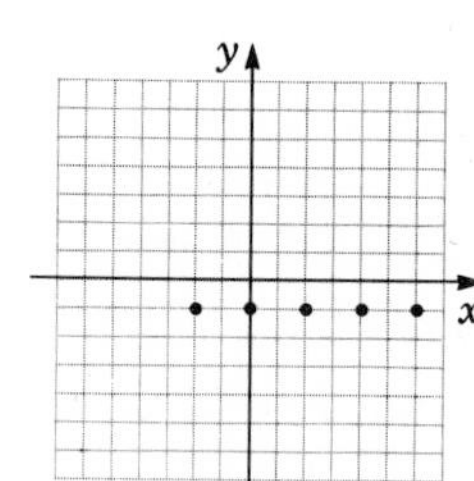

18.

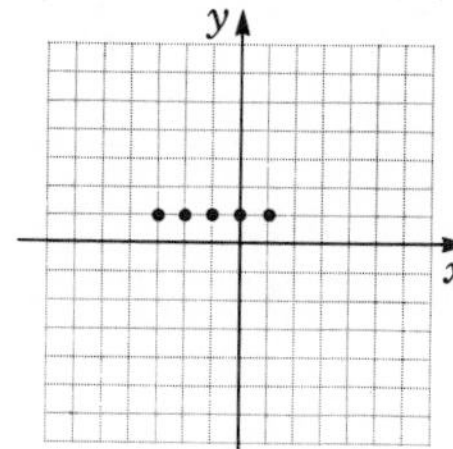

19.

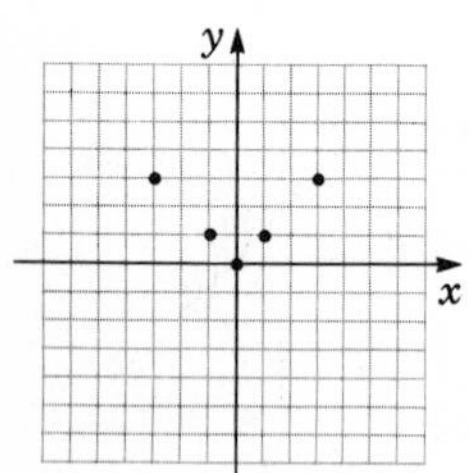

21.

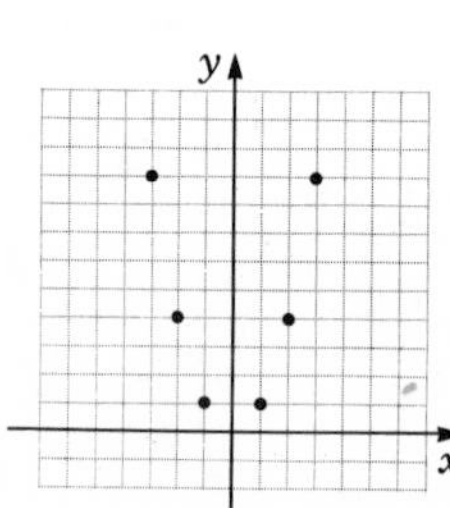

22.

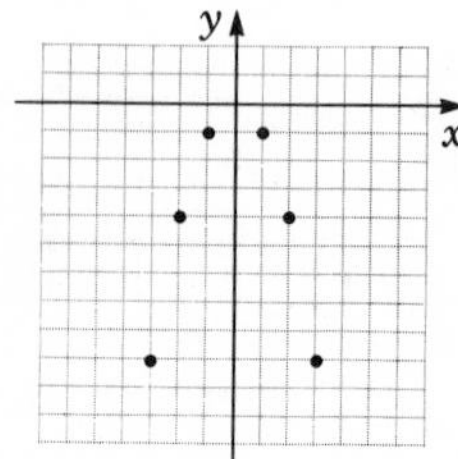

23.

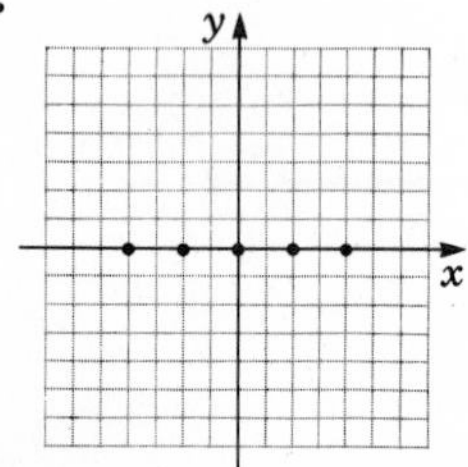

25.

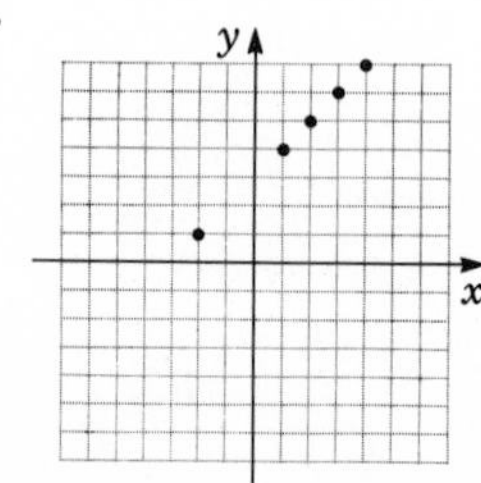

26.

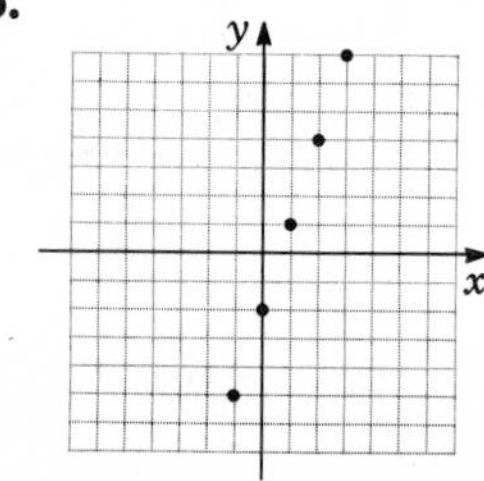

27.

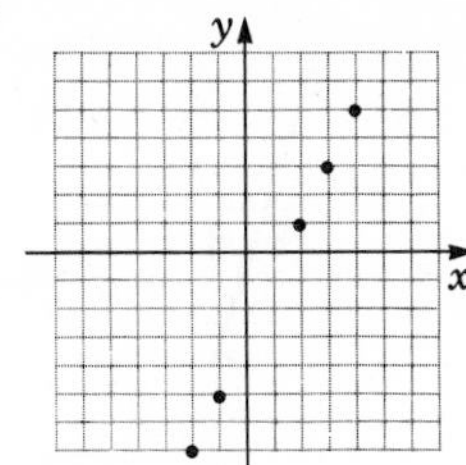

29.

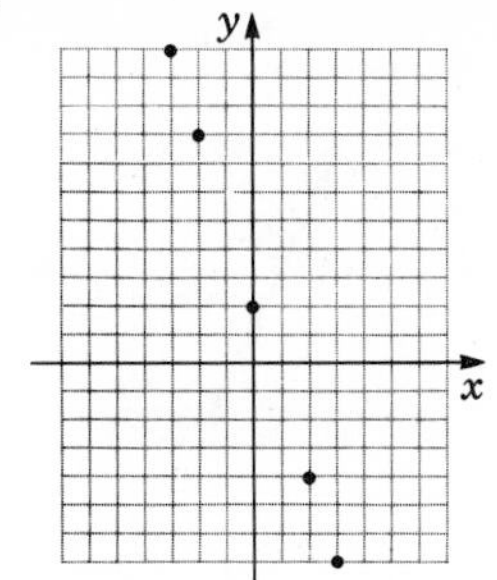

30.

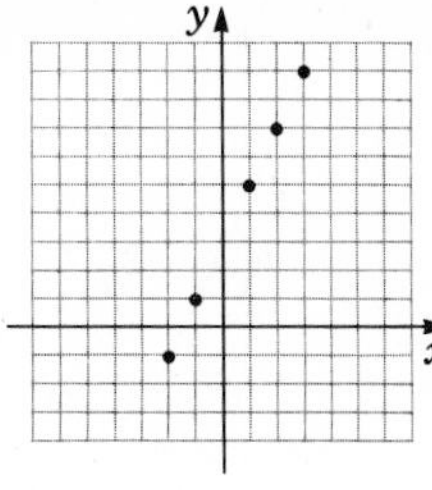

Exercises 13.4, Page 420

1.

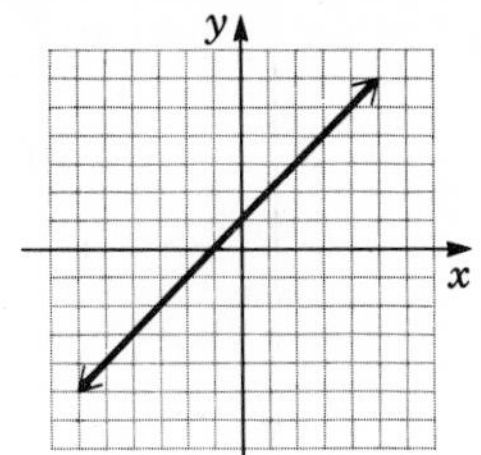

2.

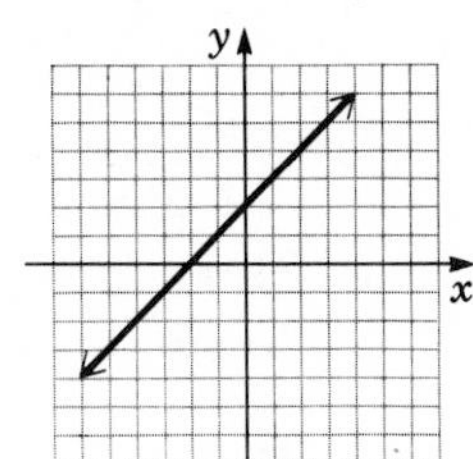

3.

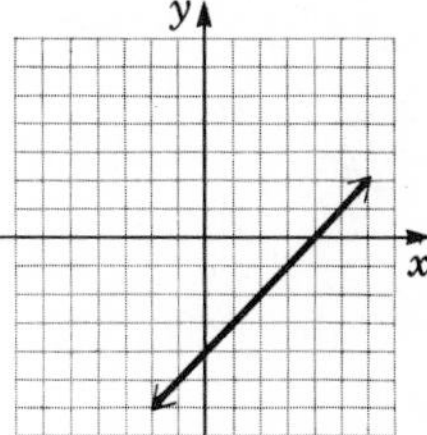

5.

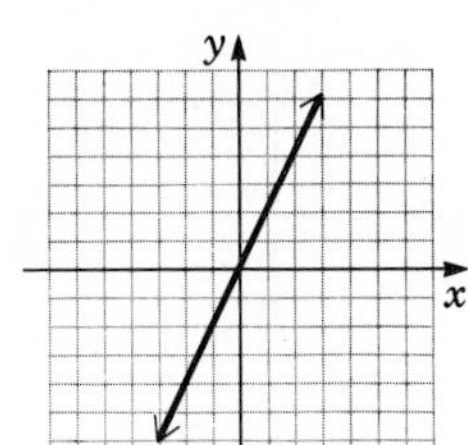

6.

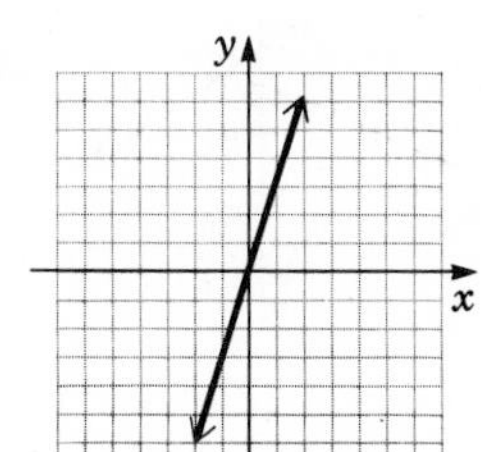

7.

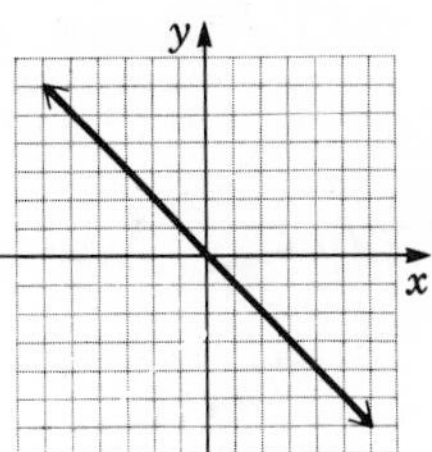

9.

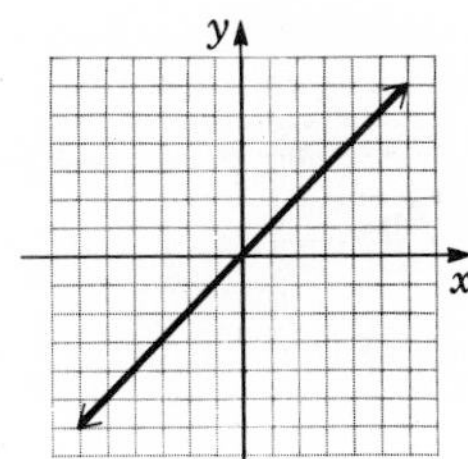

10.

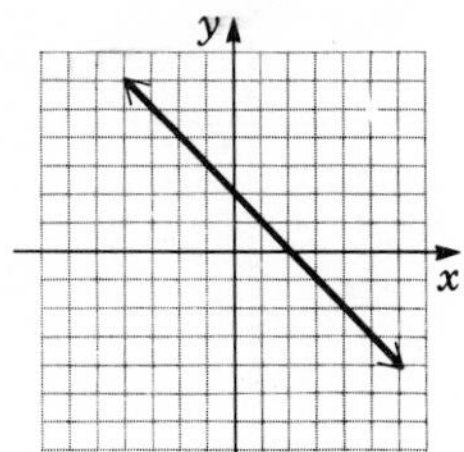

11.

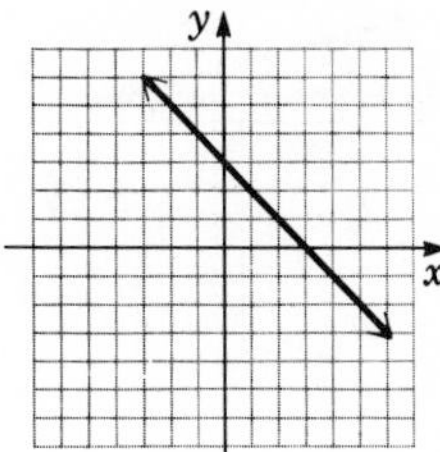

13.

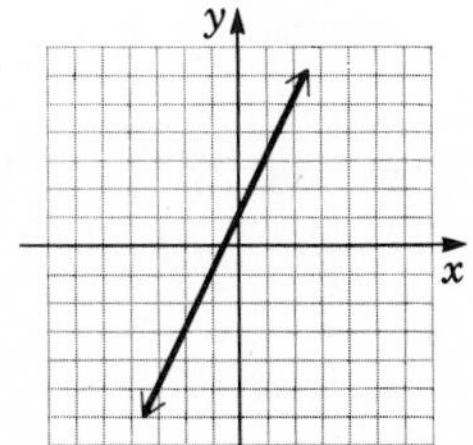

14.

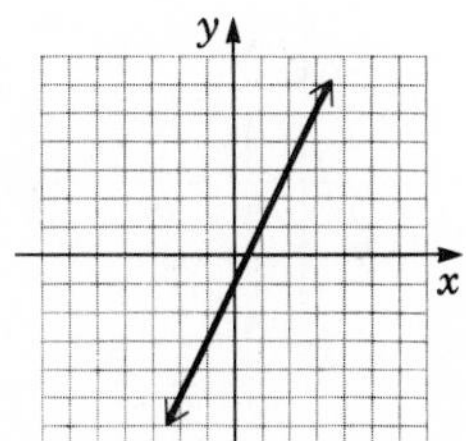

15.

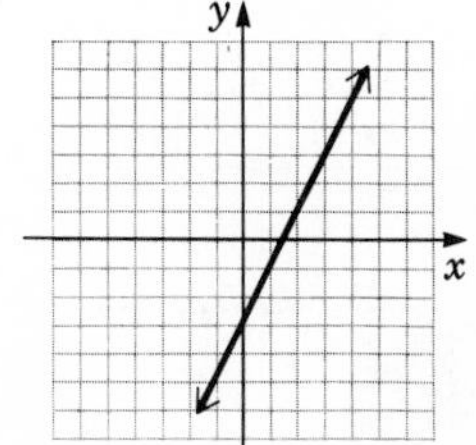

17.

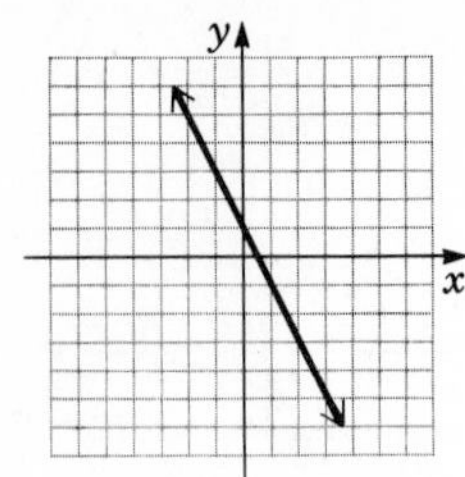

18.

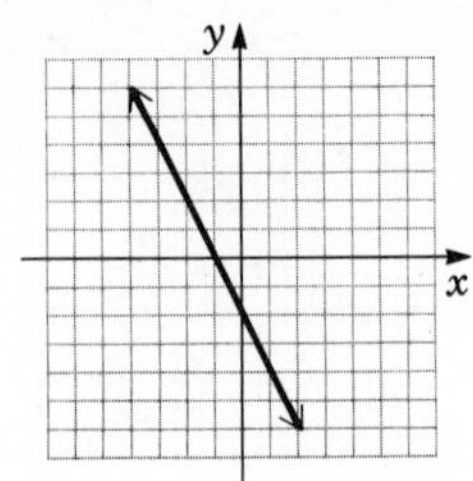

19.

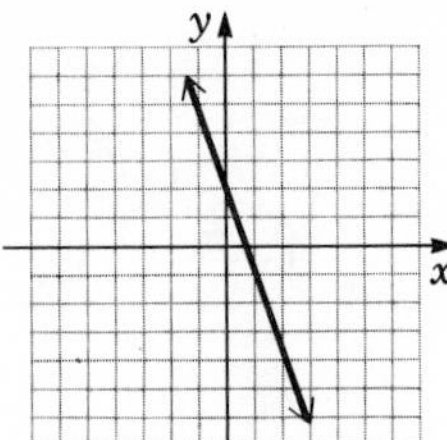

21.

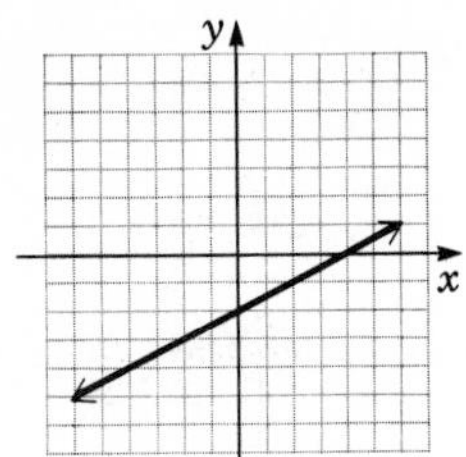

22.

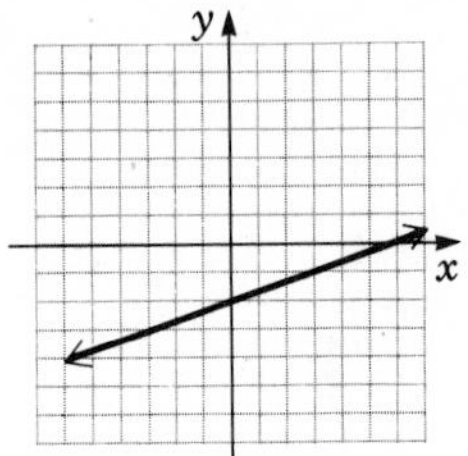

23.

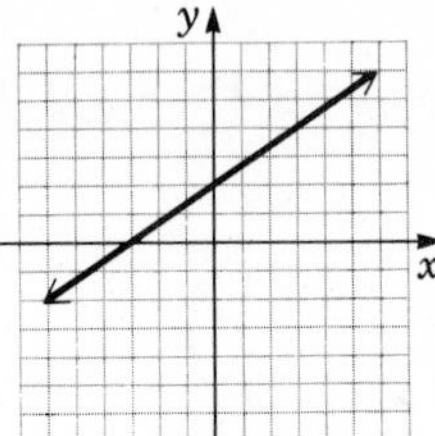

25.

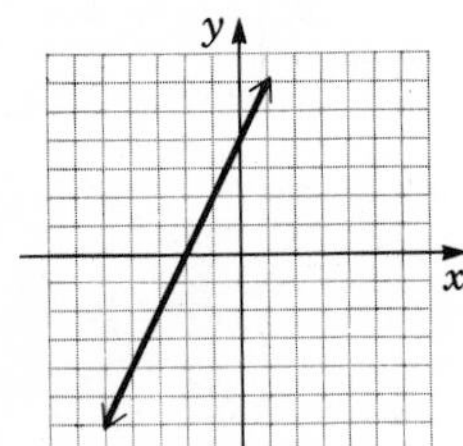

26.

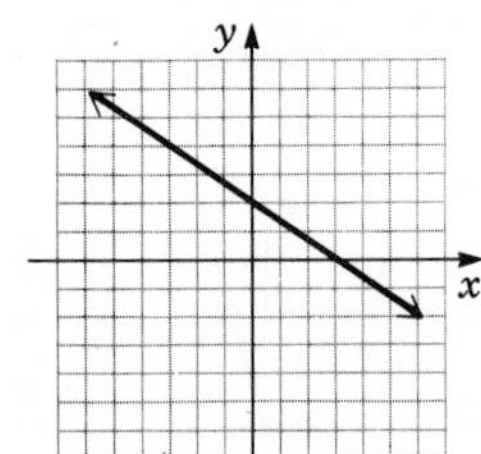

27.

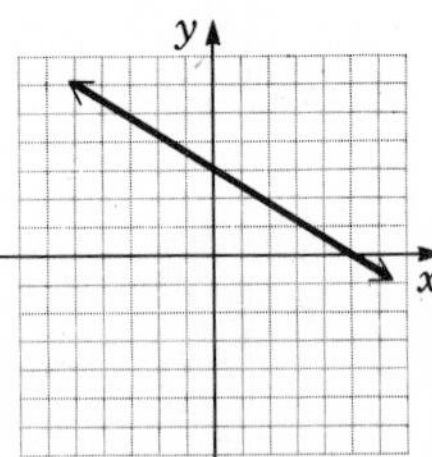

29.

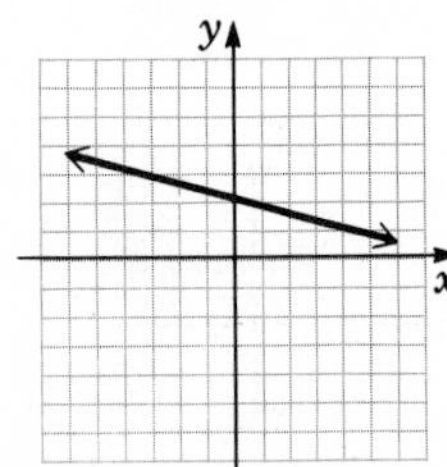

30.

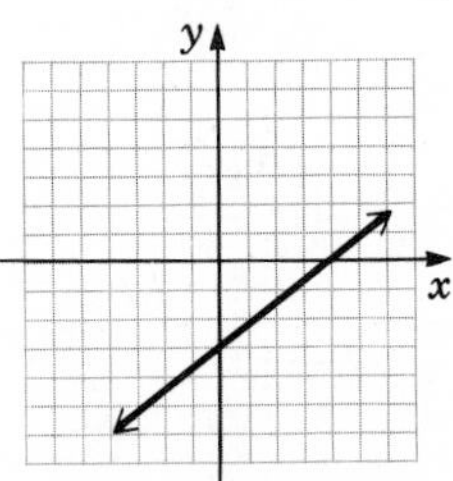

31.

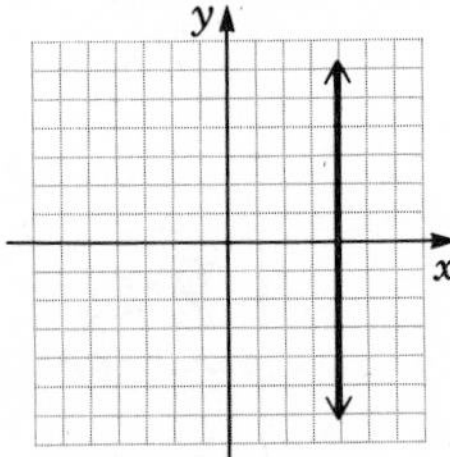

33.

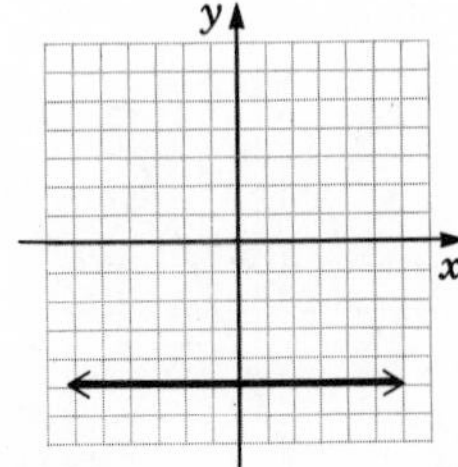

34.

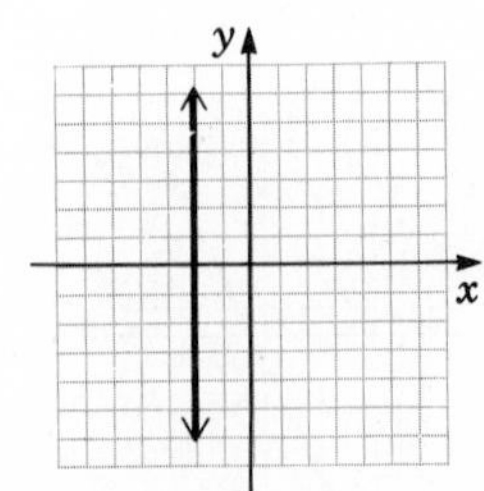

35.

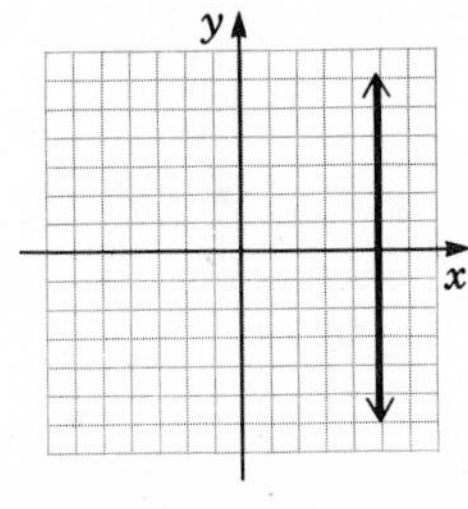

37.

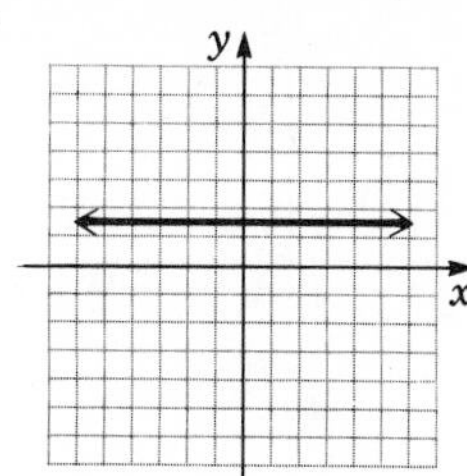

38.

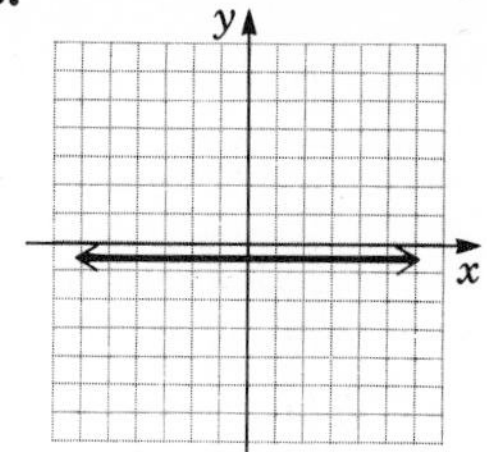

39.

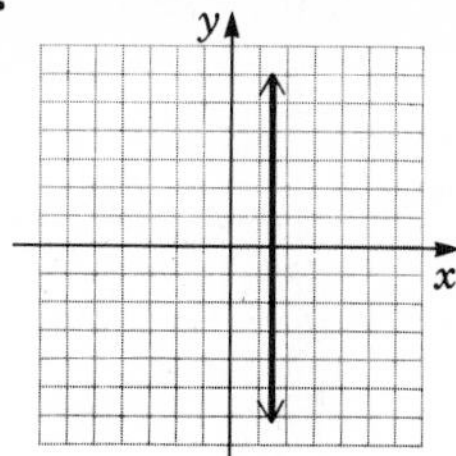

Exercises 13.5, Page 424

1. Vertex: (0, 0)

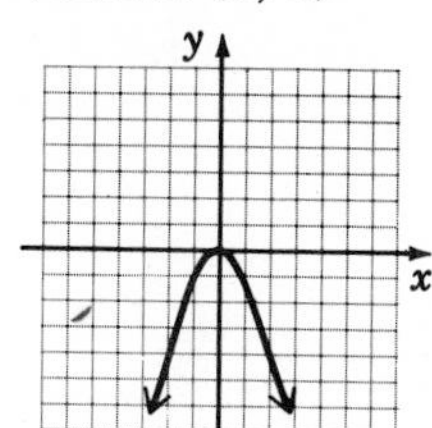

2. Vertex: (0, 0)

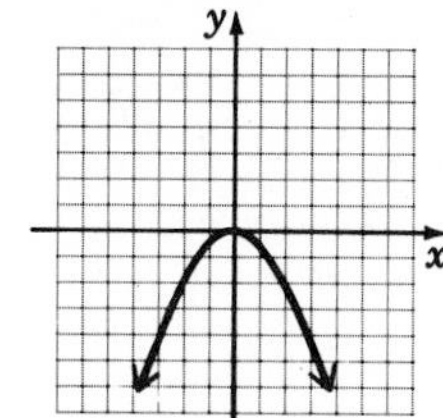

3. Vertex: (0, 0)

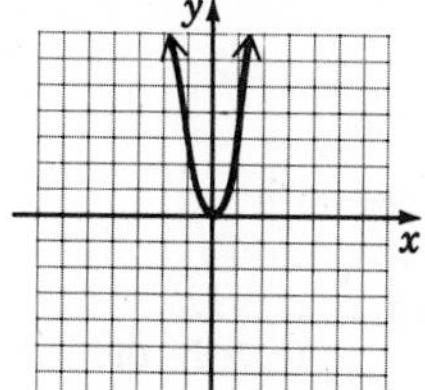

5. Vertex: (0, 0)

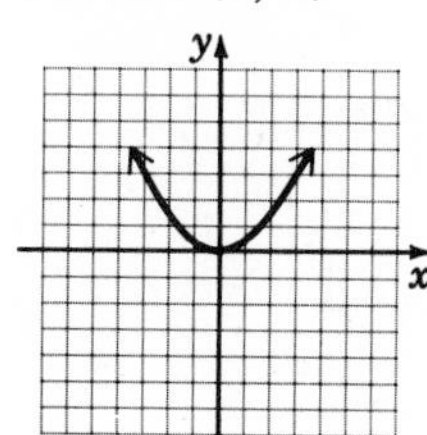

6. Vertex: (0, 0)

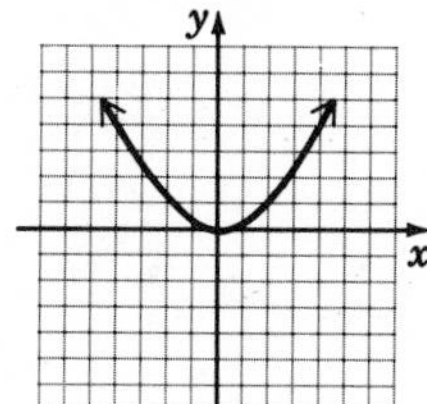

7. Vertex: (0, 1)

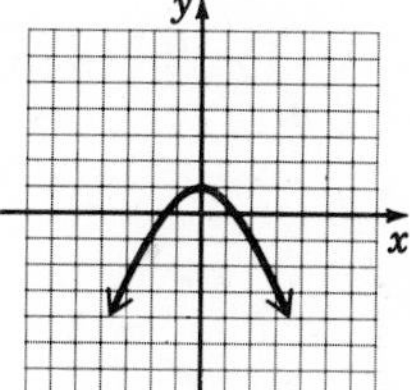

9. Vertex: (0, −2)

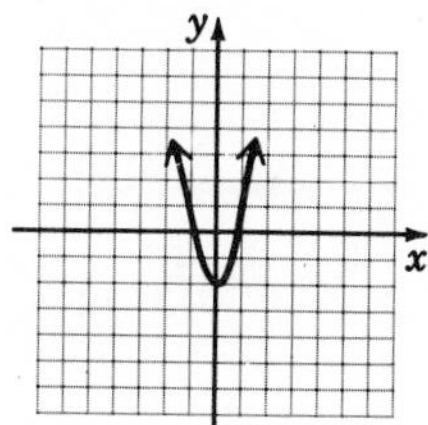

10. Vertex: (0, 2)

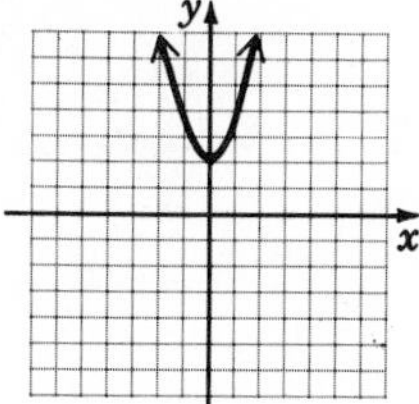

11. Vertex: (0, 2)

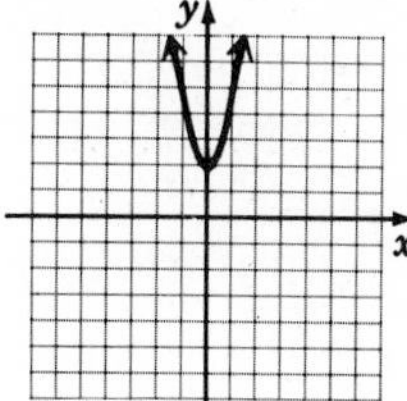

13. Vertex: (0, 1)

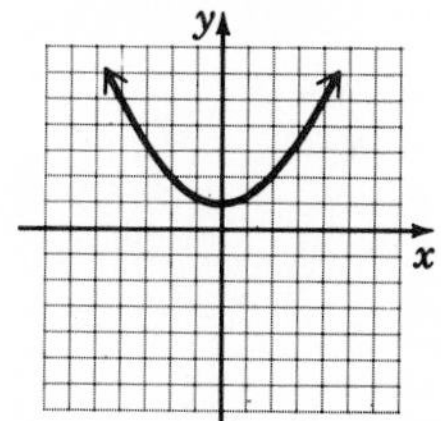

14. Vertex: (0, 3)

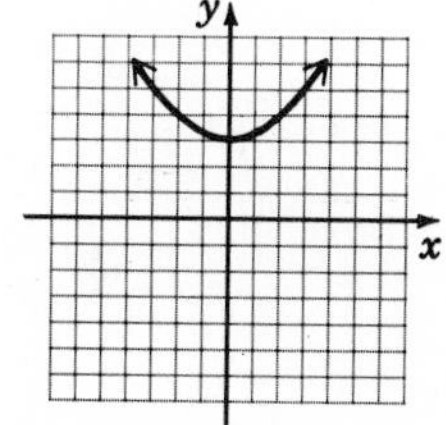

15. Vertex: (0, 1)

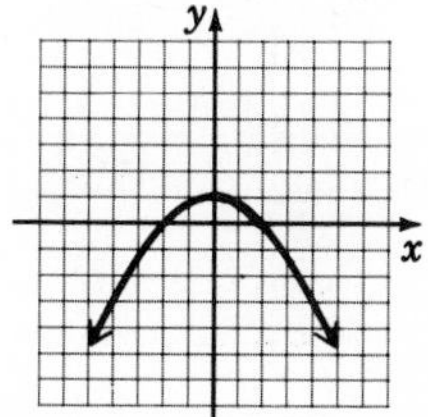

17. Vertex: (0, −6)

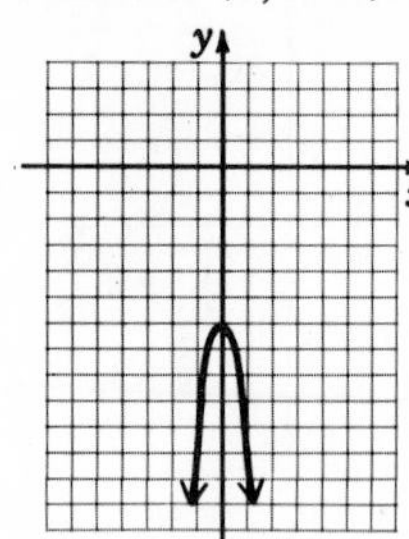

18. Vertex: (0, −5)

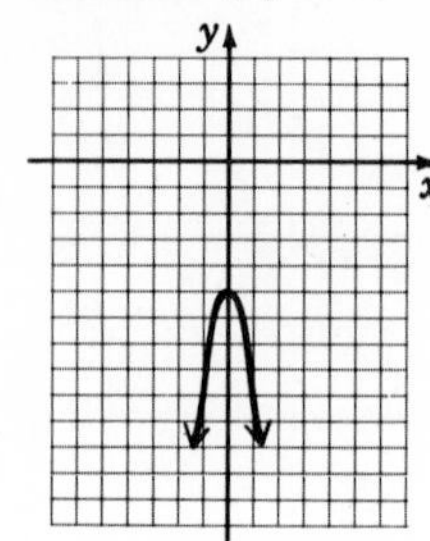

19. Vertex: (0, −6)

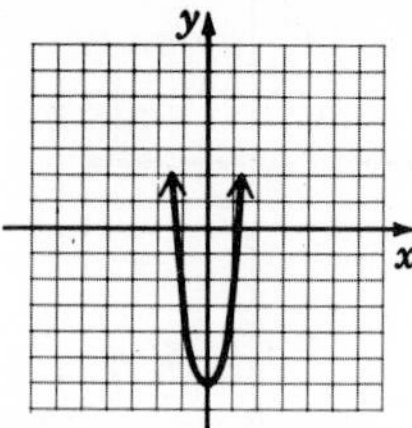

Exercises 13.6, Page 431

1.

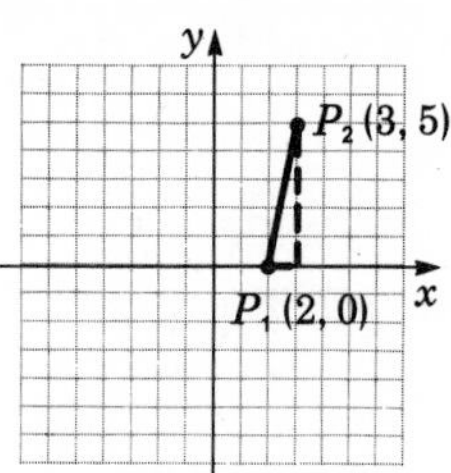

$d = \sqrt{(3-2)^2 + (5-0)^2}$
$= \sqrt{26}$

2.

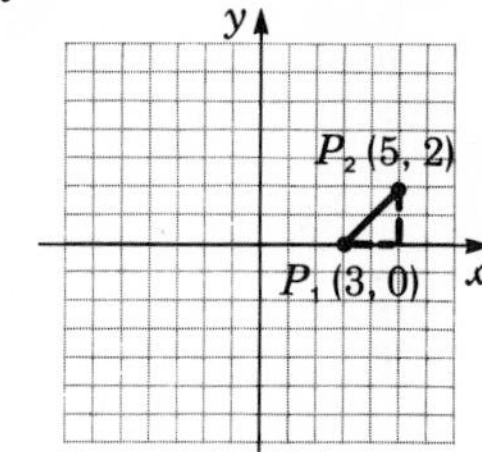

$d = \sqrt{(5-3)^2 + (2-0)^2}$
$= \sqrt{8} = 2\sqrt{2}$

3.

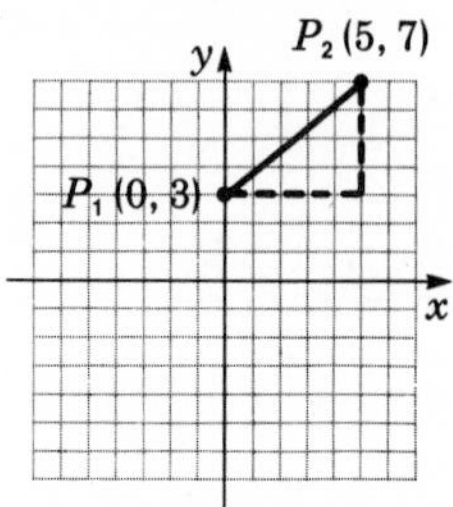

$d = \sqrt{(5-0)^2 + (7-3)^2}$
$= \sqrt{41}$

5.

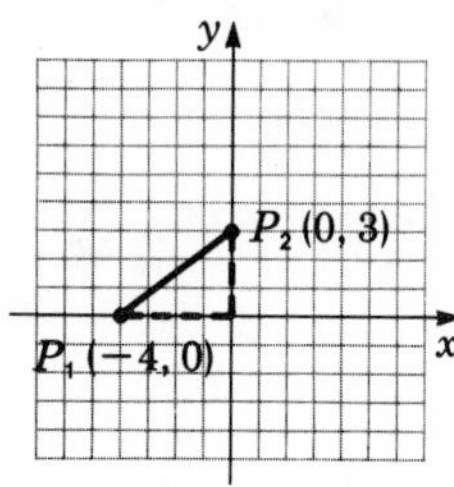

$d = \sqrt{(0-(-4))^2 + (3-0)^4}$
$= 5$

6.

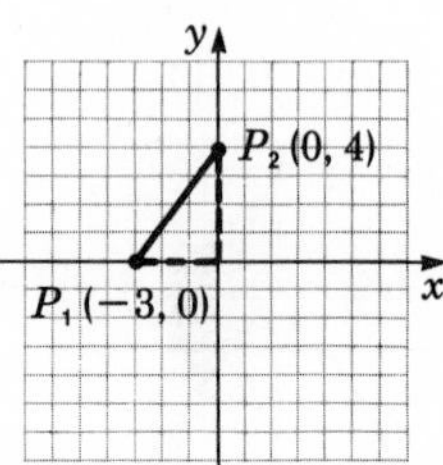

$d = \sqrt{(0-(-3))^2 + (4-0)^2}$
$= 5$

7.

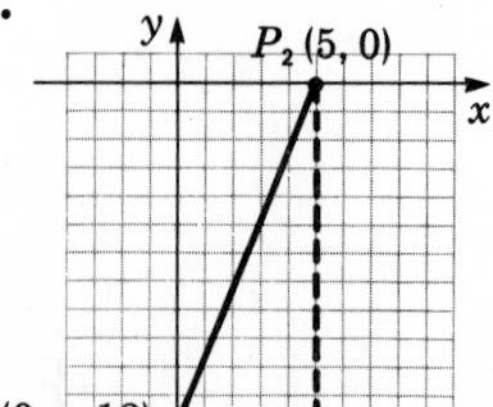

$d = \sqrt{(5-0)^2 + (0-(-12))^2}$
$= 13$

9.

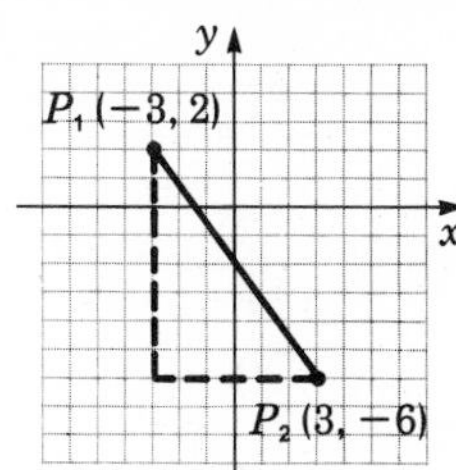

$$d = \sqrt{(3 - (-3))^2 + (-6 - 2)^2}$$
$$= 10$$

10.

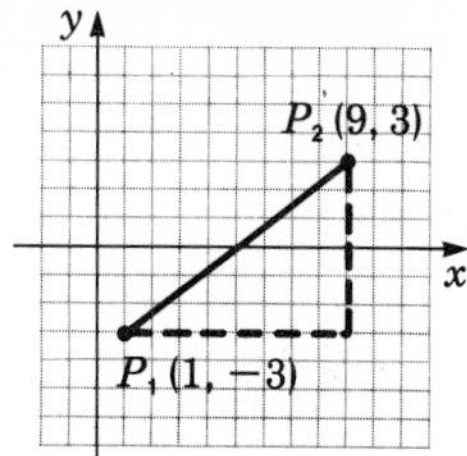

$$d = \sqrt{(9 - 1)^2 + (3 - (-3))^2}$$
$$= 10$$

11.

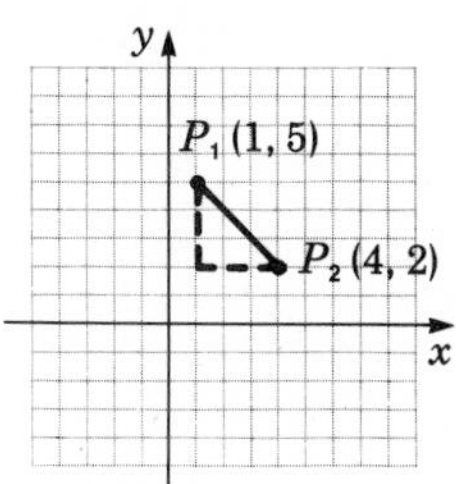

$$d = \sqrt{(4 - 1)^2 + (2 - 5)^2}$$
$$= \sqrt{18} = 3\sqrt{2}$$

13.

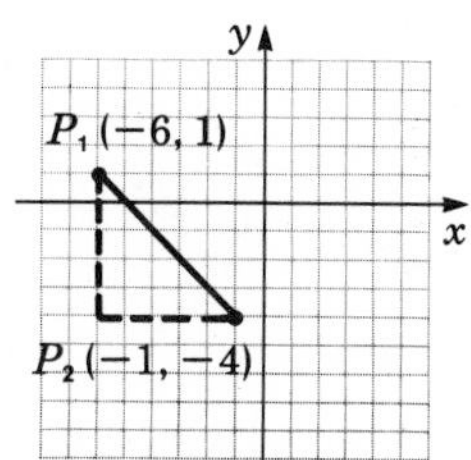

$$d = \sqrt{(-1 - (-6))^2 + (-4 - 1)^2}$$
$$= \sqrt{50} = 5\sqrt{2}$$

14.

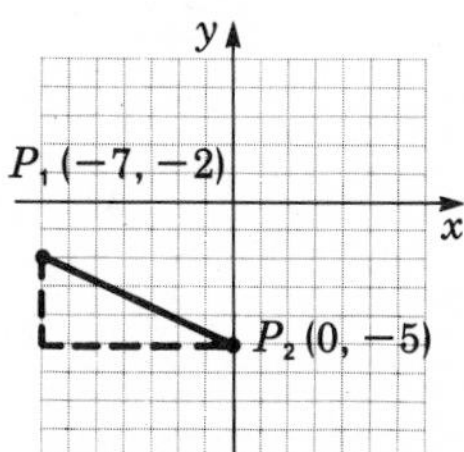

$$d = \sqrt{(0 - (-7))^2 + (-5 - (-2))^2}$$
$$= \sqrt{58}$$

15.

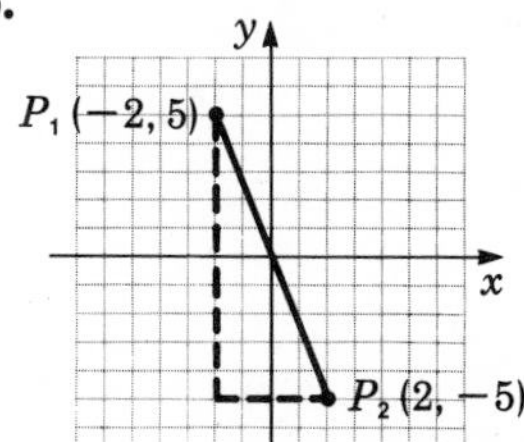

$$d = \sqrt{(2 - (-2))^2 + (-5 - 5)^2}$$
$$= \sqrt{116} = 2\sqrt{29}$$

17.

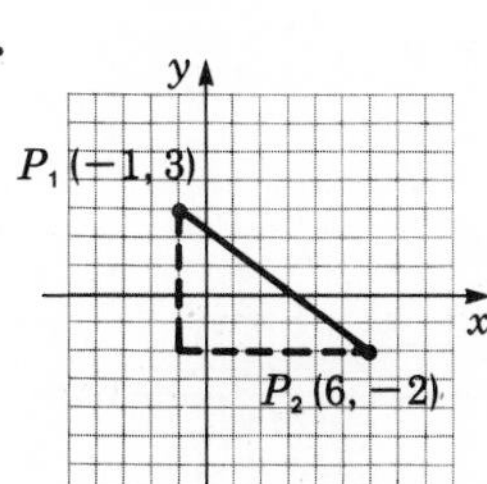

$$d = \sqrt{(6 - (-1))^2 + (-2 - 3)^2}$$
$$= \sqrt{74}$$

18.

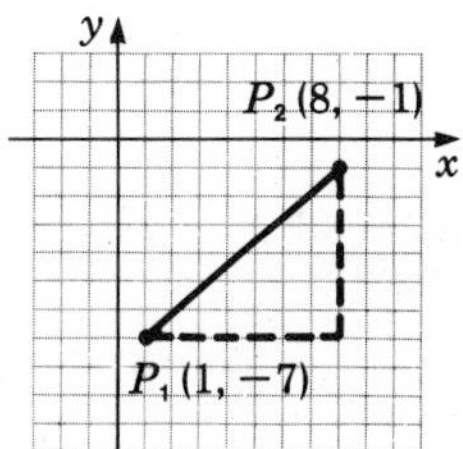

$$d = \sqrt{(8 - 1)^2 + (-1 - (-7))^2}$$
$$= \sqrt{85}$$

19.

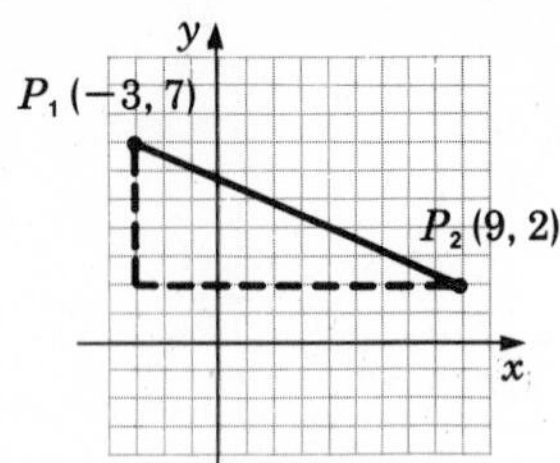

$d = \sqrt{(9 - (-3))^2 + (2 - 7)^2}$
$= 13$

21.

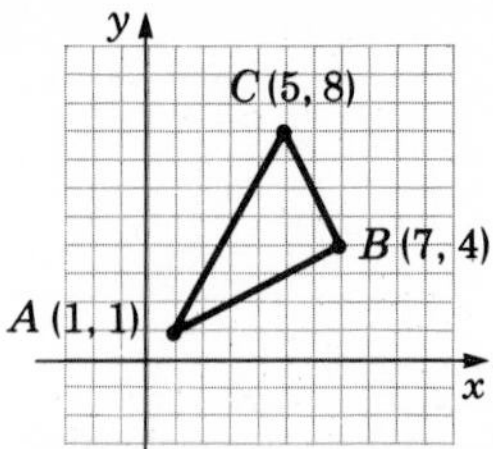

$AB = \sqrt{45}$, $AC = \sqrt{65}$, $BC = \sqrt{20}$
The triangle ABC is a right triangle since $(AB)^2 + (BC)^2 = (AC)^2$ or $45 + 20 = 65$.

23. $M = (3, 0)$

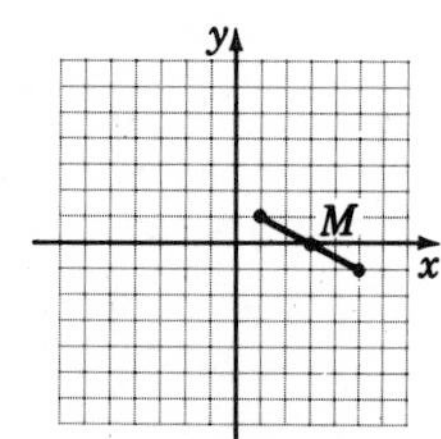

25. $M = (1, 1)$

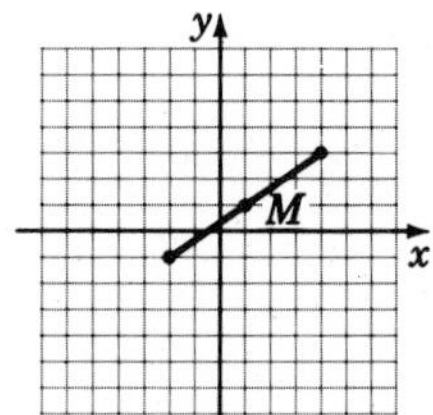

26. $M = \left(\frac{1}{2}, -\frac{3}{2}\right)$

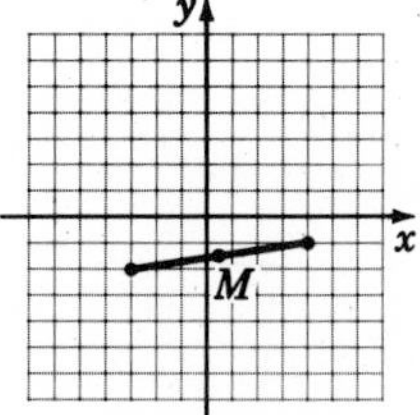

27. $M = (-1, 3)$

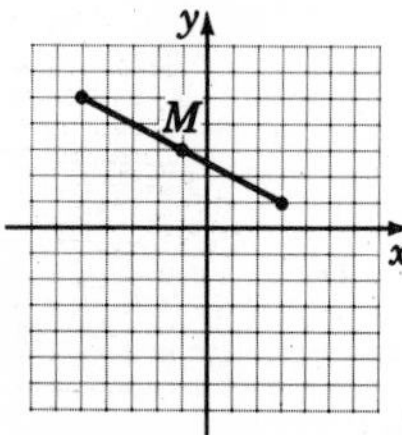

29. $M = \left(\frac{9}{2}, 3\right)$

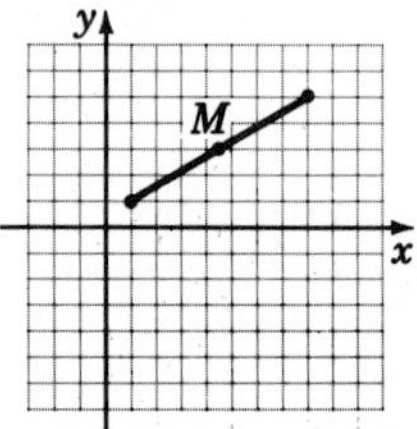

30. $M = \left(\frac{1}{16}, -\frac{1}{2}\right]$

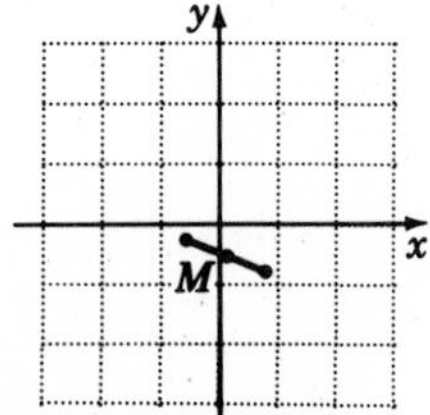

31. $M = \left(-\frac{7}{32}, 0\right)$

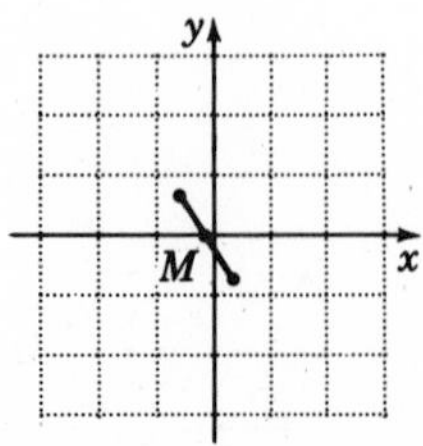

33. $M = (2.1, 1.6)$

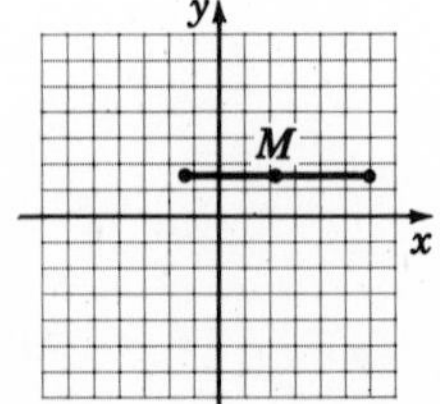

34. $M = (2.3, 1.9)$

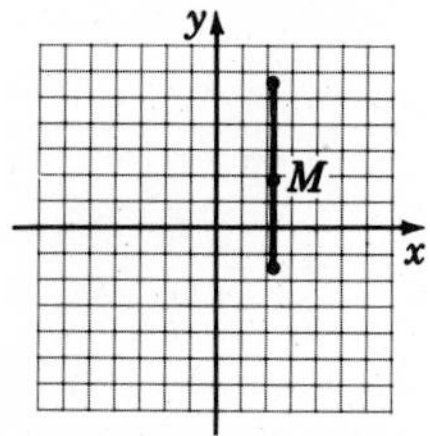

35. $M = (2.8, 1.1)$

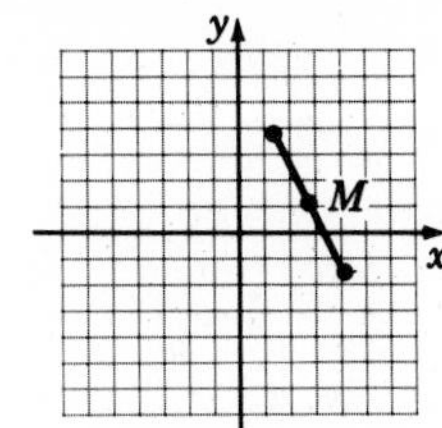

Review Questions: Chapter 13, Page 434

1. $(2.645)^2 = 6.996025$
$(2.646)^2 = 7.001316$

2. 0 is an integer.

3. $-\frac{1}{2}$, 0, 6.13 are rational numbers.

4. $-\sqrt{5}$, $-\frac{1}{2}$, 0, 6.13 are real numbers.

5. 13
6. 15
7. $8\sqrt{2}$
8. $3\sqrt{7}$
9. $3\sqrt{6}$
10. $11\sqrt{2}$
11. $10\sqrt{3}$
12. $15\sqrt{2}$
13. $5\sqrt{3}$
14. $2\sqrt{2} - 3\sqrt{5}$
15. $5\sqrt{3} + 3\sqrt{5}$
16. $-9\sqrt{3a}$

17. $-\frac{1}{2}$ $\frac{3}{4}$ open interval

18. 0 5 half-open interval

19. $\sqrt{2}$ half-open interval

20. −3 3.1 closed interval

21. $\frac{1}{3}$ open interval

22. 14 15 half-open interval

23. $-2 < x \leq 3$, half-open interval

24. $x < \frac{1}{2}$, open interval

25. $x < -4$ −4

26. $y \leq 7$ 7

27. $y \geq \frac{8}{3}$ $\frac{8}{3}$

28. $-3 > x$ −3

29. $-2 < x$ −2

30. $2 \leq x \leq 3$ 2 3

31. $\{(-4, 1), (-3, 0), (-2, -1), (0, 0), (1, 1), (2, 0)\}$

32. $\{(-1, 1), (0, 1), (1, 1), (2, 1)\}$
33. $\{(-1, 0), (0, 2), (1, 4), (2, 2), (3, 0)\}$

34.

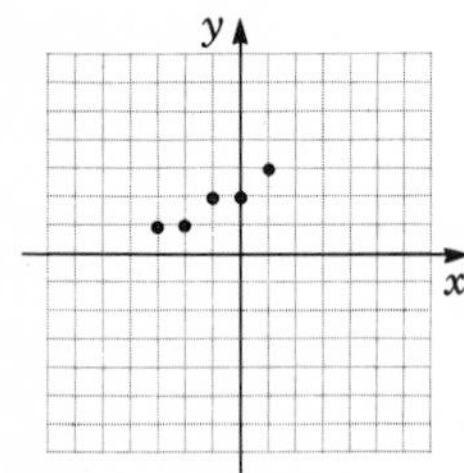

35.

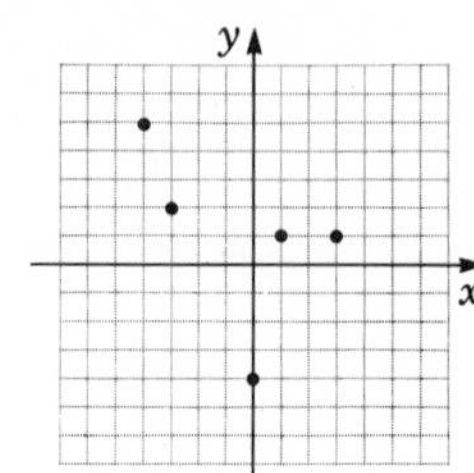

36.

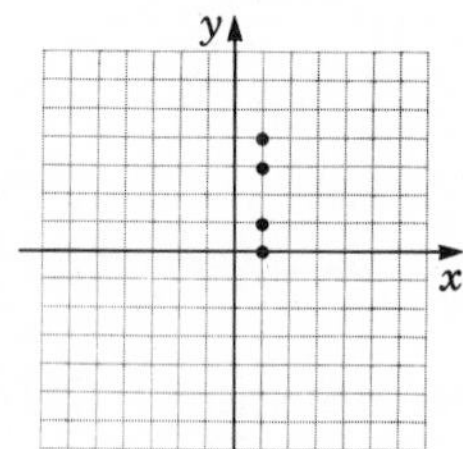

37.

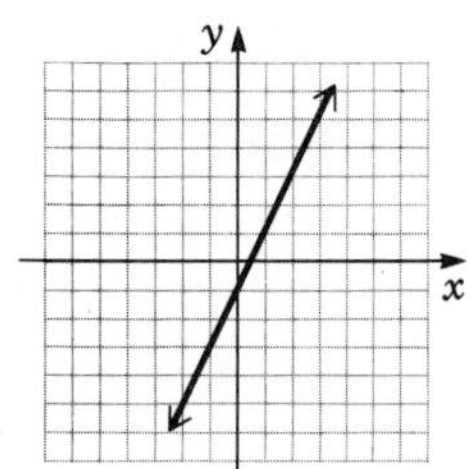

38.

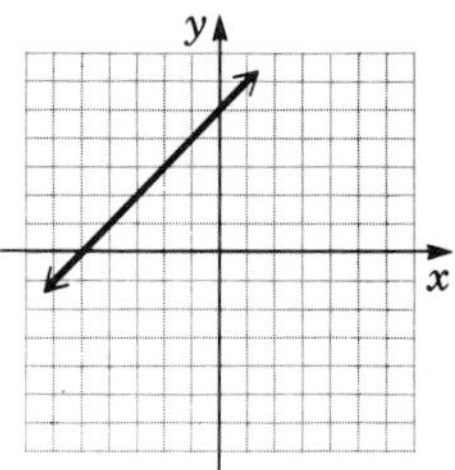

39.

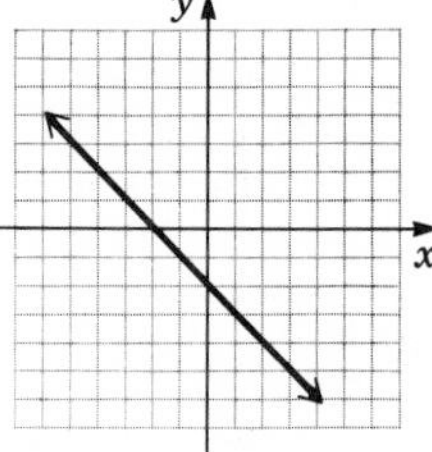

40.

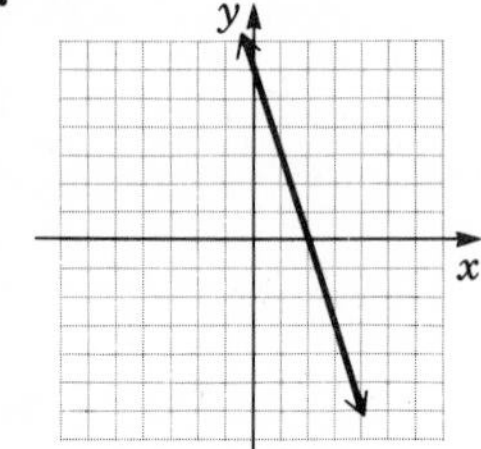

41.

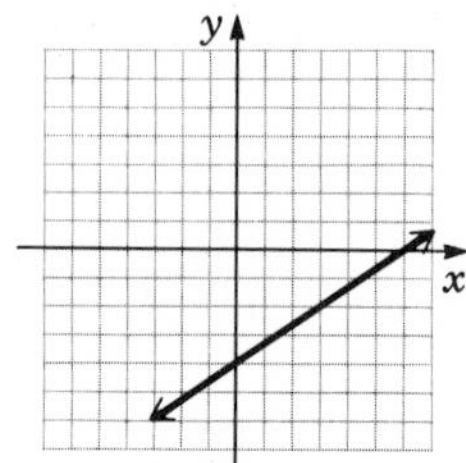

42.

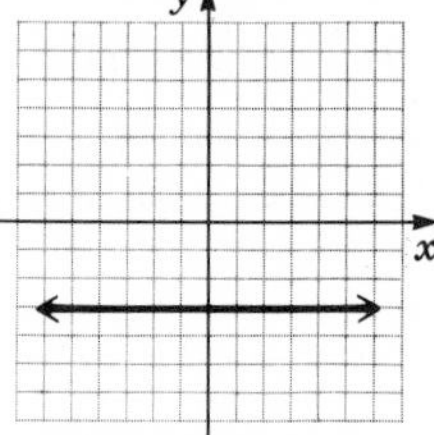

43. Vertex: $(0, 4)$

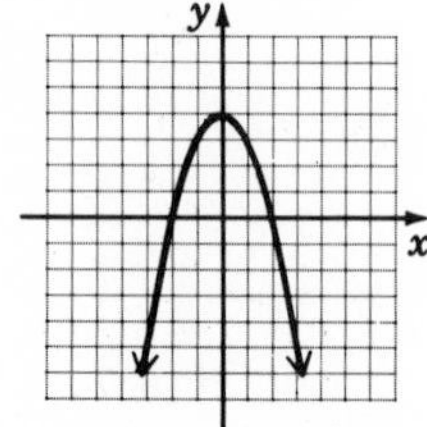

44. Vertex: $(0, -3)$

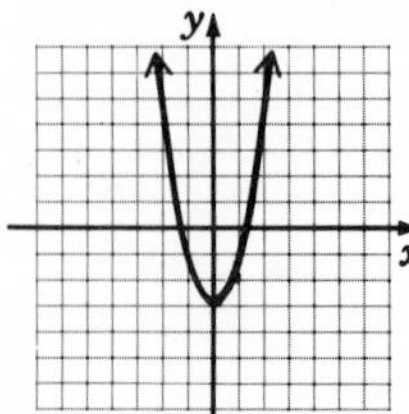

45. $d = \sqrt{(4-0)^2 + (6-3)^2} = 5$

46. $d = \sqrt{(5-(-2))^2 + (-1-1)^2} = \sqrt{53}$

47.

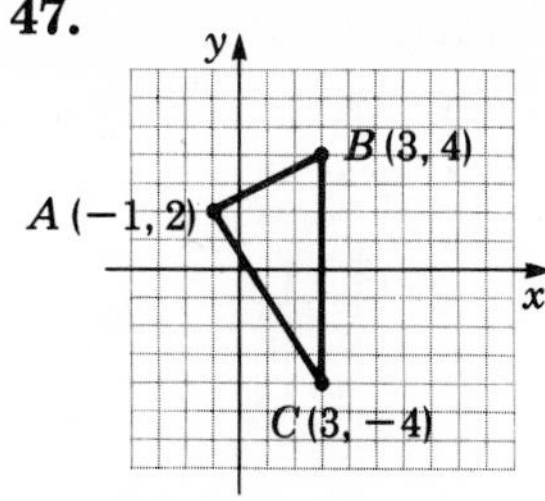

$AB = \sqrt{20}$, $AC = \sqrt{52}$, $BC = 8$
$(AB)^2 + (AC)^2 \neq (BC)^2$ or $20 + 52 \neq 64$.
So, the triangle is not a right triangle.

48. $M = \left(3, \frac{1}{2}\right)$

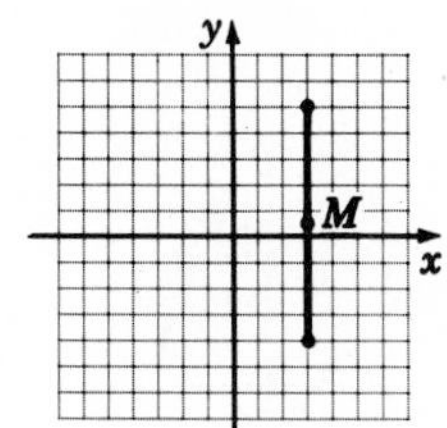

49. $M = (0, 0)$

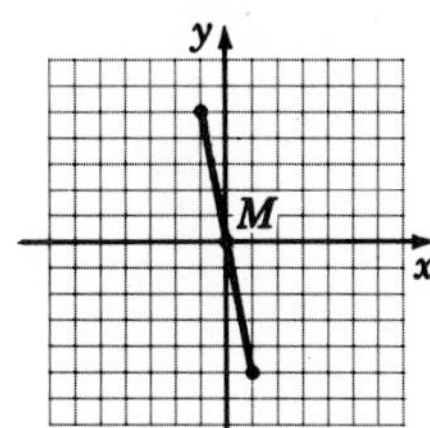

Test: Chapter 13, Page 436

1. $-5, -\sqrt{9}, -1.2, 0, 1\frac{3}{4}$ [13.1]

2. $-5, -\sqrt{9}, 0$ [13.1]

3. $5\sqrt{5} + \sqrt{2}$ [13.1]

4. $2\sqrt{3} - 3\sqrt{2}$ [13.1]

5. $5\sqrt{2}$ [13.1]

6. $x \geq 4$; half-open interval [13.2]

7. 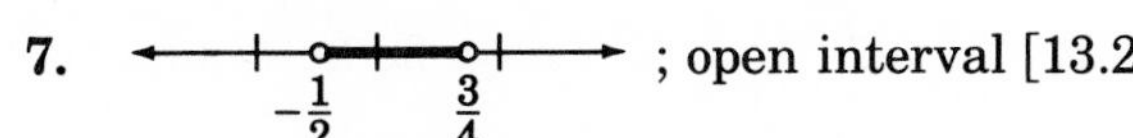

; open interval [13.2]

8. $x \geq -1$; [13.2]

9. $-1 < x < 1$; [13.2]

10. $\{(-4, 2), (-3, 0), (-1, -1), (0, 3), (2, 1), (4, -1)\}$ [13.3]

11. [13.3]

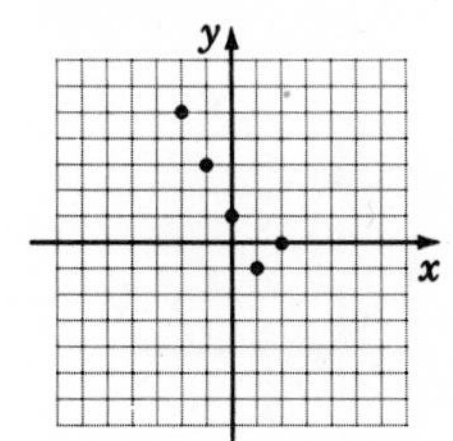

12. [13.4]

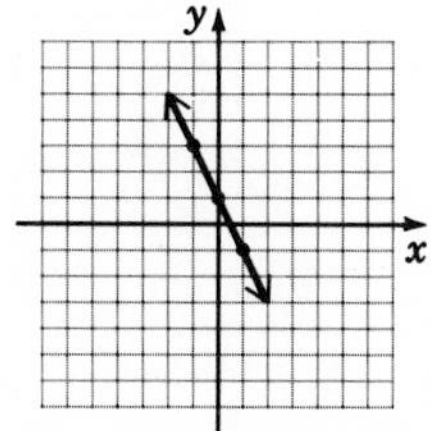

13. [13.4]

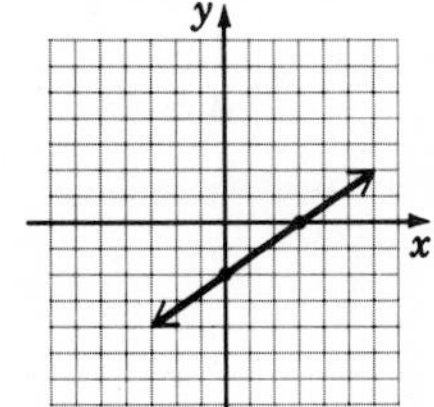

14. [13.4]

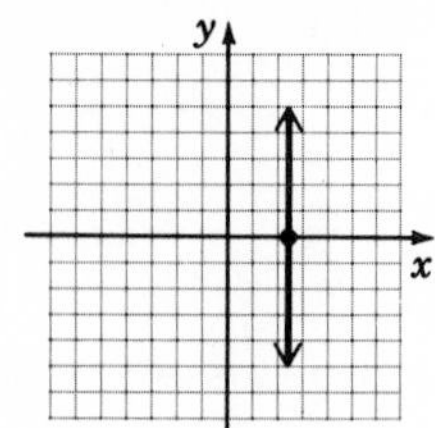

15. Vertex: (0, 2) [13.5]

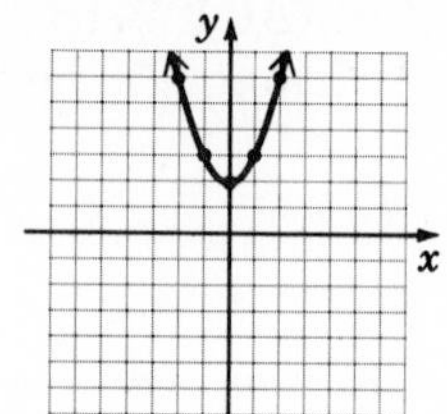

16. Vertex: (0, 5) [13.5]

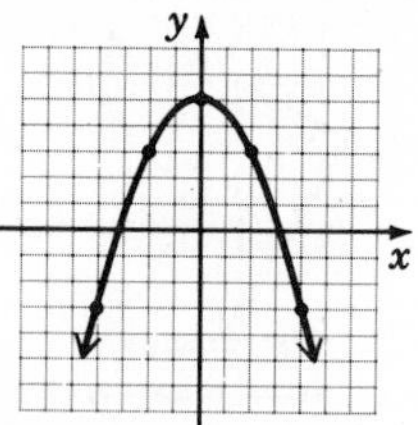

17. $d = \sqrt{32} = 4\sqrt{2}$ [13.6]

18. $d = \sqrt{25} = 5$ [13.6]

19. $AB = 5$, $AC = 61$, $BC = 6$, and

$$\begin{aligned}(AB)^2 + (BC)^2 &= (5)^2 + (6)^2\\ &= 25 + 36\\ &= 61\\ &= (AC)^2\end{aligned}$$

So, the triangle is a right triangle. [13.6]

20. $M = (-1, 1)$ [13.6]

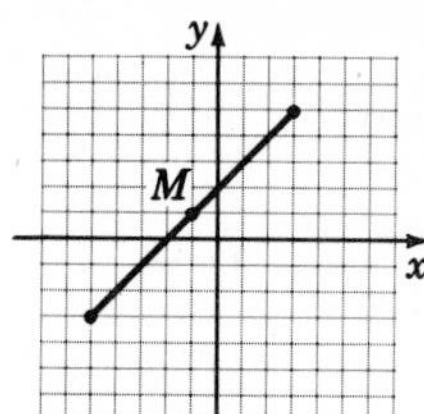

CUMULATIVE III: CHAPTERS 1–13, Page 438

1. $1\frac{1}{6}$ [3]
2. $30\frac{13}{40}$ [4]
3. 23.69 [5]
4. 6 yd 2 in. [10]
5. 9.068 [11]
6. $\frac{7}{36}$ [3]
7. $1\frac{1}{2}$ [4]
8. 74.875 [5]
9. 2 days 23 hr 45 min [10]
10. 4.47 [11]
11. 72,000,000 [1]
12. $\frac{1}{7}$ [3]
13. 16 [4]
14. 1.9684 [5]
15. -14.4 [11]
16. $\frac{5}{6}$ [3]
17. $7\frac{1}{2}$ [4]
18. 0.608 [5]
19. undefined [11]
20. $2^3 \cdot 3^2 \cdot 11$ [2]
21. 2, 3, 4, and 9 [2]
22. false [6]
23. 6000 [1]
24. 723.0 or 723 [5]
25. 7.88 [8]
26. 22 [2]
27. 0 [12]
28. $3\sqrt{5} + 20\sqrt{3}$ [13]
29. $d = 6\sqrt{2}$ [13]
30. $<$ [11]
31. $(2x + 6) - 5$ [12]
32. $x = -12$ [12]
33. $x = 18$ [12]
34. $-\frac{1}{2} < x \le \frac{1}{4}$ [13]

$-\frac{1}{2}$ $\frac{1}{4}$

35. $x = \dfrac{y - b}{m}$ [12]
36. [13]

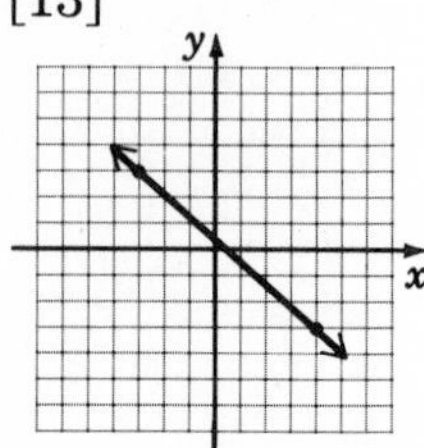

37. [13]

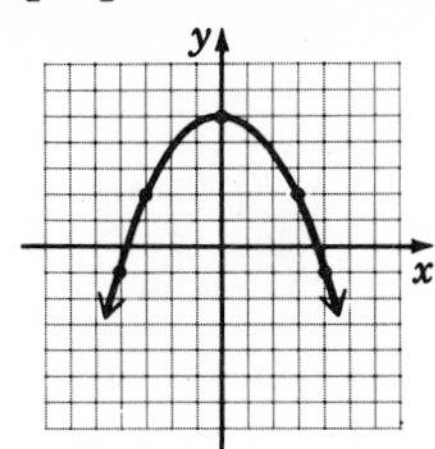

38. 40 [7]

39. 210% [7]

40. 567.5 [10]
41. 1.778 [10]
42. 76.1° F [10]
43. 64 L [9]
44. 84 cm² [9]
45. 125.6 cm [9]
46. 0.785 m² [9]
47. 225 mi [6]
48. 2 mi per day [1]
49. $1674 [8]
50. −15, −14 [12]

APPENDIX I

Exercises I.1, Page 444

1. 254 **2.** 163,041 **3.** 21,327 **5.** 140 **6.** 256
7. 53 **9.** 2362 **10.** 97 **11.** 744 **13.** 978

14. (64)

(a) ∩∩∩∩∩∩IIII

(b) ·:·:

(c) Γ$^{\Delta}$ΔIIII
(d) LXIV

15. (532)

(a) ९९९९९∩∩∩II

(b) ≐

(c) Γ^{H}ΔΔΔII
(d) DXXXII

17. (846)

(a) ९९९९९९९९∩∩∩∩IIIIII

(b) ∸

(c) Γ^{H}HHHΔΔΔΔΓI
(d) DCCCXLVI

Exercises I.2, Page 448

1. 4983 **2.** 32 **3.** 1802 **5.** 47,900 **6.** 222 **7.** 46 **9.** 999

10. (a) VVVVVVV <<<<<VV

(b) $\overline{\text{YOB}}$
(c) 四
百
七
十
二

11. (a) VVVVVVVVV <<<<< VVVVVV

(b) $\overline{\phi\text{CF}}$
(c) 五
百
九
十
六

13. (a) ∨ ∨ ∨∨∨∨∨
(b) $\overline{/\Gamma X\Xi E}$
(c) 三
千
六
百
六
十
五

14. (a) ∨∨ ∨ <<<∨∨∨
(b) $\overline{/Z\Sigma C\Gamma}$
(c) 七
千
二
百
九
十
三

15. (a) ∨∨∨ <<<<<∨∨
(b) $\overset{A}{/M}\Omega NB$
(c) 十
千
八
百
五
十
二

APPENDIX II

Exercises II.1, Page 451

1. $35 = 3(10^1) + 5(10^0)$

2. $761 = 7(10^2) + 6(10^1) + 1(10^0)$

3. $8469 = 8(10^3) + 4(10^2) + 6(10^1) + 9(10^0)$

5. $62{,}322 = 6(10^4) + 2(10^3) + 3(10^2) + 2(10^1) + 2(10^0)$

6. $11_{(2)} = 1(2^1) + 1(2^0) = 1(2) + 1(1) = 2 + 1 = 3$

7. $101_{(2)} = 1(2^2) + 0(2^1) + 1(2^0) = 1(4) + 0(2) + 1(1) = 4 + 0 + 1 = 5$

9. $1011_{(2)} = 1(2^3) + 0(2^2) + 1(2^1) + 1(2^0) = 1(8) + 0(4) + 1(2) + 1(1)$
$= 8 + 0 + 2 + 1 = 11$

10. $1101_{(2)} = 1(2^3) + 1(2^2) + 0(2^1) + 1(2^0) \times 1(8) + 1(4) + 0(2) + 1(1)$
$= 8 + 4 + 0 + 1 = 13$

11. $110{,}111_{(2)} = 1(2^5) + 1(2^4) + 0(2^3) + 1(2^2) + 1(2^1) + 1(2^0)$
$= 1(32) + 1(16) + 0(8) + 1(4) + 1(2) + 1(1)$
$= 32 + 16 + 0 + 4 + 2 + 1 = 55$

13. $101{,}011_{(2)} = 1(2^5) + 0(2^4) + 1(2^3) + 0(2^2) + 1(2^1) + 1(2^0)$
$= 1(32) + 0(16) + 1(8) + 0(4) + 1(2) + 1(1)$
$= 32 + 0 + 8 + 0 + 2 + 1 = 43$

14. $11{,}010_{(2)} = 1(2^4) + 1(2^3) + 0(2^2) + 1(2^1) + 0(2^0)$
$= 1(16) + 1(8) + 0(4) + 1(2) + 0(1) = 16 + 8 + 0 + 2 + 0$
$= 26$

15. $1000_{(2)} = 1(2^3) + 0(2^2) + 0(2^1) + 0(2^0) = 1(8) + 0(4) + 0(2) + 0(1)$
$= 8 + 0 + 0 + 0 = 8$

17. $11{,}101_{(2)} = 1(2^4) + 1(2^3) + 1(2^2) + 0(2^1) + 1(2^0)$
$= 1(16) + 1(8) + 1(4) + 0(2) + 1(1) = 16 + 8 + 4 + 0 + 1$
$= 29$

18. $10{,}110_{(2)} = 1(2^4) + 0(2^3) + 1(2^2) + 1(2^1) + 0(2^0)$
$= 1(16) + 0(8) + 1(4) + 1(2) + 0(1) = 16 + 0 + 4 + 2 + 0$
$= 22$

19. $111{,}111_{(2)} = 1(2^5) + 1(2^4) + 1(2^3) + 1(2^2) + 1(2^1) + 1(2^0)$
$= 1(32) + 1(16) + 1(8) + 1(4) + 1(2) + 1(1)$
$= 32 + 16 + 8 + 4 + 2 + 1 = 63$

21. 111

Exercises II.2, Page 453

1. $24_{(5)} = 2(5^1) + 4(5^0) = 2(5) + 4(1) = 10 + 4 = 14$
2. $13_{(5)} = 1(5^1) + 3(5^0) = 1(5) + 3(1) = 5 + 3 = 8$
3. $10_{(5)} = 1(5^1) + 0(5^0) = 1(5) + 0(1) = 5 + 0 = 5$
5. $104_{(5)} = 1(5^2) + 0(5^1) + 4(5^0) = 1(25) + 0(5) + 4(1) = 25 + 0 + 4 = 29$
6. $312_{(5)} = 3(5^2) + 1(5^1) + 2(5^0) = 3(25) + 1(5) + 2(1) = 75 + 5 + 2 = 82$
7. $32_{(5)} = 3(5^1) + 2(5^0) = 3(5) + 2(1) = 15 + 2 = 17$
9. $423_{(5)} = 4(5^2) + 2(5^1) + 3(5^0) = 4(25) + 2(5) + 3(1) = 100 + 10 + 3 = 113$
10. $444_{(5)} = 4(5^2) + 4(5^1) + 4(5^0) = 4(25) + 4(5) + 4(1) = 100 + 20 + 4 = 124$
11. $1034_{(5)} = 1(5^3) + 0(5^2) + 3(5^1) + 4(5^0) = 1(125) + 0(25) + 3(5) + 4(1)$
$= 125 + 0 + 15 + 4 = 144$
13. $244_{(5)} = 2(5^2) + 4(5^1) + 4(5^0) = 2(25) + 4(5) + 4(1) = 50 + 20 + 4 = 74$
14. $3204_{(5)} = 3(5^3) + 2(5^2) + 0(5^1) + 4(5^0) = 3(125) + 2(25) + 0(5) + 4(1)$
$= 375 + 50 + 0 + 4 = 429$
15. $13{,}042_{(5)} = 1(5^4) + 3(5^3) + 0(5^2) + 4(5^1) + 2(5^0)$
$= 1(625) + 3(125) + 0(25) + 4(5) + 2(1)$
$= 625 + 375 + 0 + 20 + 2 = 1022$
17. yes; 13 **18.** yes; 28 **19.** {0, 1, 2, 3, 4, 5, 6, 7}

Exercises II.3, Page 456

1. $1000_{(2)} = 8$ **2.** $311_{(5)} = 81$ **3.** $11{,}000_{(2)} = 24$
5. $432_{(5)} = 117$ **6.** $1000_{(2)} = 8$ **7.** $1011_{(2)} = 11$
9. $433_{(5)} = 118$ **10.** $1302_{(5)} = 202$ **11.** $10{,}100_{(2)} = 20$
13. $10{,}100_{(2)} = 20$ **14.** $1222_{(5)} = 187$ **15.** $2241_{(5)} = 321$
17. $110{,}111_{(2)} = 55$ **18.** $23{,}240_{(5)} = 1695$ **19.** $3002_{(5)} = 377$
21. $30{,}141_{(5)} = 1921$ **22.** $1{,}101{,}001_{(2)} = 105$ **23.** $110{,}001_{(2)} = 49$
25. $10{,}001{,}010_{(2)} = 138$

APPENDIX III

Exercises III, Page 459

1. 4 **2.** 4 **3.** 17 **5.** 10 **6.** 6 **7.** 3 **9.** 6 **10.** 11 **11.** 2
13. 8 **14.** 11 **15.** 4 **17.** 12 **18.** 11 **19.** 25 **21.** 1 **22.** 20 **23.** 1
26. relatively prime **27.** relatively prime **29.** relatively prime
30. not rel. prime; GCD is 7 **31.** relatively prime **33.** relatively prime
34. not rel. prime; GCD is 11 **35.** relatively prime

APPENDIX IV

Exercises IV.1, Page 466

1. $\overline{AB}, \overline{BC}, \overline{CD}, \overline{BD}, \overline{DE}, \overline{EC}, \overline{CA}, \overline{EA}$
2. **(a)** $m\angle 2 = 75°$ **(b)** $m\angle 2 = 87°$ **(c)** $m\angle 2 = 45°$ **(d)** $m\angle 2 = 15°$
3. **(a)** $m\angle 4 = 135°$ **(b)** $m\angle 4 = 90°$ **(c)** $m\angle 4 = 70°$ **(d)** $m\angle 4 = 45°$

5. (a) two rays (b) two rays (c) no
6. (a) obtuse (b) acute (c) right (d) $\angle BOC$ and $\angle AOD$
(e) $\angle BOC$ and $\angle BOD$; $\angle AOC$ and $\angle AOD$
7. 25° 9. 40° 10. 95° 11. 55° 13. 110°
14. (a) $\angle AOF$ and $\angle AOB$; $\angle AOF$ and $\angle FOE$; $\angle DOE$ and $\angle DOA$; $\angle COE$ and $\angle COA$; $\angle BOE$ and $\angle BOA$
(b) $\angle AOB$ and $\angle BOC$; $\angle COD$ and $\angle DOE$; $\angle FOE$ and $\angle BOC$
15. $m\angle A = 55°$; $m\angle B = 45°$; $m\angle C = 80°$
17. $m\angle L = 115°$; $m\angle M = 90°$; $m\angle N = 105°$; $m\angle O = 110°$; $m\angle P = 120°$
18. $m\angle D = 100°$; $m\angle E = 45°$; $m\angle F = 35°$

Exercises IV.2, Page 471

1. (a) Two lines are parallel if they are in the same plane and do not intersect.
(b) Two lines are perpendicular if they intersect to form a right angle.
(c) Two lines intersect if there is one point on both lines.

2. (a) (b)

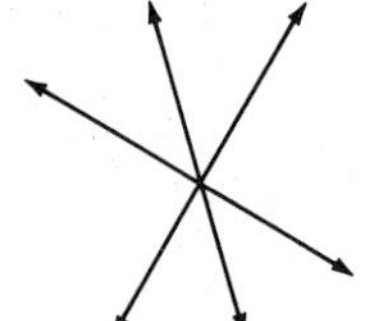

(c)

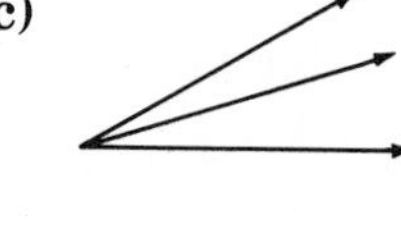

3. (a) (b)

5. Three points determine a plane.

6. (a) $\angle CPB$ and $\angle BPD$; $\angle BPC$ and $\angle APC$; $\angle BPD$ and $\angle DPA$.
(b) There are no vertical angles in the figure.
7. (a) one plane (b) one line (c) one line
8. $m\angle 6 = 55°$ because $\angle 8$ and $\angle 6$ are vertical angles, and vertical angles are equal.
9. $m\angle 7 = 125°$ because $\angle 7$ and $\angle 8$ are supplementary and the sum of the measures of supplementary angles is 180°.
10. Yes, $m\angle 2 = m\angle 6$. Since $m\angle 1 = 125°$ and $\angle 1$ and $\angle 2$ are supplementary, $m\angle 2 = 55°$. Since $m\angle 8 = 55°$ and $\angle 6$ and $\angle 8$ are vertical angles, $m\angle 6 = 55°$.
12. (a) $m\angle 2 = 70°$, $m\angle 3 = 90°$, $m\angle 4 = 20°$, $m\angle 5 = 70°$
(b) $\angle 3$ is supplementary to $\angle 6$.
(c) $\angle 2$ and $\angle 5$ are complementary to $\angle 1$.
13. $\angle 1$ and $\angle 4$; $\angle 3$ and $\angle 5$; $\angle 4$ and $\angle 6$, $\angle 1$ and $\angle 6$
14. (a) $\angle QPR$ and $\angle RPB$ are acute.
(b) $\angle APR$ is obtuse.
(c) $\angle APQ$ and $\angle BPQ$ are right angles.
(d) none
(e) $\angle BPR$ and $\angle QPR$ are complementary.
(f) $\angle APQ$ and $\angle BPQ$ are supplementary; $\angle APR$ and $\angle BPR$ are supplementary.
(g) $\angle APQ$ and $\angle BPQ$ are adjacent; $\angle QPR$ and $\angle BPR$ are adjacent; $\angle APR$ and $\angle BPR$ are adjacent.

Exercises IV.3, Page 478

1. scalene (and obtuse)
2. equilateral (and acute)
3. scalene (and right)
5. isosceles (and acute)
6. isosceles (and obtuse)
7. isosceles (and right)
9. scalene (and acute)
10. Yes, both triangles are right triangles because $4^2 + 3^2 = 5^2$ and $8^2 + 6^2 = 10^2$.
The hypotenuses are $\overline{AC}$ and $\overline{DF}$.
Yes, the triangles are similar because

$$\frac{3}{6} = \frac{4}{8} = \frac{5}{10}$$

The corresponding angles are: $\angle B$ and $\angle E$, $\angle A$ and $\angle D$, $\angle C$ and $\angle F$.
11. Yes, the triangles are similar.
$\angle RPQ = \angle TPS$ because vertical angles are equal.
$\angle R = \angle S$ because the sum of the measures of the angles in both triangles is 180°. Since two of the angles are the same in both triangles, the third angles must be the same.
13. Yes, because the sum of any two sides is longer than the third side. The triangles would be scalene.
14. No, because $4 + 6 = 10$. The sum of two sides must be greater than the third side in a triangle.
15. $\angle E = \angle E$ in both triangles.
$\angle EBA = \angle EDC$ because both angles are right angles. $\angle EAB = \angle ECD$ because the sum of the angles in each triangle is 180°. $\frac{6}{CE} = \frac{5}{8}$ and $\frac{3}{DE} = \frac{5}{8}$. So, $CE = \frac{48}{5} = 9.6$ cm and $DE = \frac{24}{5} = 4.8$ cm.

Index

2 3 4 5 6 7 8 9 0

POWERS, ROOTS, AND PRIME FACTORIZATIONS

NO.	SQUARE	SQUARE ROOT	CUBE	CUBE ROOT	PRIME FACTORIZATION
1	1	1.0000	1	1.0000	—
2	4	1.4142	8	1.2599	prime
3	9	1.7321	27	1.4423	prime
4	16	2.0000	64	1.5874	2 · 2
5	25	2.2361	125	1.7100	prime
6	36	2.4495	216	1.8171	2 · 3
7	49	2.6458	343	1.9129	prime
8	64	2.8284	512	2.0000	2 · 2 · 2
9	81	3.0000	729	2.0801	3 · 3
10	100	3.1623	1000	2.1544	2 · 5
11	121	3.3166	1331	2.2240	prime
12	144	3.4641	1728	2.2894	2 · 2 · 3
13	169	3.6056	2197	2.3513	prime
14	196	3.7417	2744	2.4101	2 · 7
15	225	3.8730	3375	2.4662	3 · 5
16	256	4.0000	4096	2.5198	2 · 2 · 2 · 2
17	289	4.1231	4913	2.5713	prime
18	324	4.2426	5832	2.6207	2 · 3 · 3
19	361	4.3589	6859	2.6684	prime
20	400	4.4721	8000	2.7144	2 · 2 · 5
21	441	4.5826	9261	2.7589	3 · 7
22	484	4.6904	10,648	2.8020	2 · 11
23	529	4.7958	12,167	2.8439	prime
24	576	4.8990	13,824	2.8845	2 · 2 · 2 · 3
25	625	5.0000	15,625	2.9240	5 · 5
26	676	5.0990	17,576	2.9625	2 · 13
27	729	5.1962	19,683	3.0000	3 · 3 · 3
28	784	5.2915	21,952	3.0366	2 · 2 · 7
29	841	5.3852	24,389	3.0723	prime
30	900	5.4772	27,000	3.1072	2 · 3 · 5
31	961	5.5678	29,791	3.1414	prime
32	1024	5.6569	32,768	3.1748	2 · 2 · 2 · 2 · 2
33	1089	5.7446	35,937	3.2075	3 · 11
34	1156	5.8310	39,304	3.2396	2 · 17
35	1225	5.9161	42,875	3.2711	5 · 7
36	1296	6.0000	46,656	3.3019	2 · 2 · 3 · 3
37	1369	6.0828	50,653	3.3322	prime
38	1444	6.1644	54,872	3.3620	2 · 19
39	1521	6.2450	59,319	3.3912	3 · 13
40	1600	6.3246	64,000	3.4200	2 · 2 · 2 · 5
41	1681	6.4031	68,921	3.4482	prime
42	1764	6.4807	74,088	3.4760	2 · 3 · 7
43	1849	6.5574	79,507	3.5034	prime
44	1936	6.6333	85,184	3.5303	2 · 2 · 11
45	2025	6.7082	91,125	3.5569	3 · 3 · 5
46	2116	6.7823	97,336	3.5830	2 · 23
47	2209	6.8557	103,823	3.6088	prime
48	2304	6.9282	110,592	3.6342	2 · 2 · 2 · 2 · 3
49	2401	7.0000	117,649	3.6593	7 · 7
50	2500	7.0711	125,000	3.6840	2 · 5 · 5